Vegetation and Climate

Siegmar-W. Breckle • M. Daud Rafiqpoor

Vegetation and Climate

 Springer

Siegmar-W. Breckle
Department of Ecology
Bielefeld, Germany

M. Daud Rafiqpoor
University of Bonn
Bonn, Germany

ISBN 978-3-662-64038-8   ISBN 978-3-662-64036-4   (eBook)
https://doi.org/10.1007/978-3-662-64036-4

This book is a translation of the original German edition „Vegetation und Klima" by Breckle, Siegmar-W., published by Springer-Verlag GmbH, DE in 2019. The translation was done with the help of artificial intelligence (machine translation by the service DeepL.com). A subsequent human revision was done primarily in terms of content, so that the book will read stylistically differently from a conventional translation. Springer Nature works continuously to further the development of tools for the production of books and on the related technologies to support the authors.

This Springer imprint is published by the registered company Springer-Verlag GmbH, DE part of Springer Nature.
The registered company address is: Heidelberger Platz 3, 14197 Berlin, Germany

Photo 1 Cultivation of sunflowers (*Helianthus annuus*) as a monoculture on fields in the Eastern part of North Rhine-Westphalia (photo: Breckle)

Accompanying Word

Prof. em. Dr. Michael Succow. Professor of Geobotany and Landscape Ecology; winner of the "Right Livelihood Award" ("Alternative Nobel Prize"); Honorary Prize of the German Federal Foundation for the Environment; Chairman of the Michael Succow Foundation for the Protection of Nature. https://de.wikipedia.org/wiki/Michael_Succow

The vegetation cover—the plant dress of our earth—is currently undergoing a strong change worldwide. No longer by natural processes, but by us humans, by our civilization! The emerging dramatic changes in the climate, the loss of the natural fertility of our soils, the humus, the loss of the fullness of life, the biodiversity force us to rethink our handling of nature—which will also be our basis of life in the future—and to act in a new way. This requires knowledge combined with responsibility.

Knowledge about the diversity and function, i.e. the natural balance of "our" earth, its ecosystems and their interaction. But also knowledge about how nature—driven by the principles of evolution—is able to perfect itself further and further, to optimize. Not to maximize and thus often enough to fail, as our daily actions in dealing with nature show again and again.

The dilemma of our time can be summed up in three sentences: If we leave nature unchanged, we cannot exist. If we destroy it, we perish. The narrow, ever-narrowing path between change and destruction can only be successfully trodden by a society whose economic activities are integrated into the natural balance and whose ethics make it feel part of nature.

Knowledge of nature, contact with nature, experience of nature, love of nature and the resulting responsibility for nature form the basis for the necessary reform, a rethinking of the way we deal with our biosphere, this so thin living skin of our earth. Today, more than ever, our own future viability is at stake, that is, the future viability of human civilization! We can be sure that the nature project will continue, but possibly without us!

Why am I writing these thoughts in the preface for this book in particular: Because I am very happy and grateful about its appearance—now even more in the updated edition. This textbook, once conceived and implemented by Heinrich Walter, then continued by Siegmar Breckle, accompanied my life. It fascinated me as a biology student at the University of Greifswald and was later indispensable for understanding the large ecosystems in various parts of the world. Today, this synopsis continues to be an important source of knowledge for the work of my foundation in the protection and sustainable use of ecosystems on a global scale. This book, so condensed in content, manages to show ecosystem knowledge, the "interplay" of the individual bio- and geo-components, starting with the individual organism, the auto-ecology and then continuing at the ecosystem level, the syn-ecology. It helps to understand the so wonderfully ecologically built house earth, to understand its vulnerability and to lead to sustainable action. Admirable is the bringing together and linking of an enormously widening knowledge, especially in recent times, about the function and functionality of the ecosystems that support us. The experience of the regenerative power of many ecosystems gives us hope, if we give them time and space.

In the meantime, this university textbook or handbook reaches not only many students all over the world but also land users, environmentalists, politicians, . . . I am particularly pleased that the new edition, completely revised by Siegmar-W. Breckle and the renowned Afghan vegetation ecologist M. Daud Rafiqpoor, has now also been published in Dari. Let us hope that this "standard work" will be translated into many more languages, because all over the world we need more than ever textbooks at our colleges and universities that promote knowledge of nature and ecosystemic thinking, that enlighten us about the "miracle of nature" in all its complexity. Only in this way will it be possible to deal with nature "entrusted to us" in a more responsible, future-oriented and sustainable way in the future and to better understand and use the self-healing powers of nature.

Prof. em. Dr. Michael Succow
Michael Succow Foundation for the Protection of Nature
Greifswald, July 2021
http://www.succow-stiftung.de

Accompanying Word

Prof. Dr. Wilhelm Barthlott, emeritus Professor of Botany at the University of Bonn, long-time Director of the Botanical Gardens and founding Director of the Nees Institute for Biodiversity of Plants, member of several scientific academies, honoured with the German Environmental Prize and other awards https://en.wikipedia.org/wiki/Wilhelm_Barthlott

With the beginning of the twenty-first century we have finally arrived in the age of the Anthropocene: global changes caused by many billion humans are dramatically and progressively changing our **environment** (e.g. climate change, extinction of species) and through social and other media, influence our **"environment"** and thoughts. Many of the classic statements of Enlightenment philosophy, once intended for an *empty world* with less than one billion people, may no longer be valid for today's *full world* of 8 billion inhabitants. Even the century-old rule that *"growth is progress"* applies only to a very limited extent today. In an interconnected hitherto Eurocentric but cosmopolitan world, Europe (including later the USA) with its ideologies and values is, for the first time between the late Middle Ages and the late twentieth century, is no longer the measure of all things for the rest of the hyper-globalized earth.

The clearly measurable current climate change is of existential importance to us. A few seemingly neglectable Celsius degrees of average temperature are dramatically changing our environment and our livelihoods. Few still dismiss the reference to planetary boundaries as "alarmism". Often, this lacks the scientific knowledge that is well researched in the disciplines of geography and biology in their comprehensive sense.

Sea levels rise with the consequence of future mass migration and probably wars; vegetation and agriculture change. We, humans, are totally dependent on plants: They provide directly (cereals, vegetables, fruits) or indirectly (meat production is based on herbivores) all our food, but also provide us with materials (wood, coal, even fossil oil, cotton, wool, medicines) and even the air we breathe: Oxygen is produced exclusively by plants.

We share with far more than 10 million different living organisms (scientifically known to date are only 1.8 million species) our planet, our common house, and they are the basis of our ecosystems and thus our life. Due to the exponentially increasing population and the high consumption of natural resources, these systems are threatened worldwide today. "Global change" is the term, incipient "Climate change" the omen. The prognoses are bad: Sea level rise alone will probably cause many millions of climate refugees from the coastal areas of the earth by the end of this century. We are also at the beginning of an extinction catastrophe of earth-historical dimensions with regard to biodiversity.

We can still shape our future. Data is provided by the natural sciences—but the decisions are determined by politics, economy and media, culture, religion, and often emotions. Education and training are the basis.

With the textbook "Vegetation and Climate", a modern, easy-to-read overview—without being obtrusive about the incipient changes—of the fundamental relationships between the plant world, ecology and climate is now available from highly renowned and experienced authors.

Vegetation, soil and climate are the most important components of ecological systems. The book presents a compact synthesis of our current knowledge of the ecology of the Earth and is thus the basis for understanding the big picture in a global perspective. Education and training are the keys to our future: How we shape it depends on them. Now, with the completely revised and up-to-date new edition of the textbook classic "Vegetation and Climate Zones by H. Walter & S.-W. Breckle (1999)", a comprehensive modern textbook for students is available. "Vegetation and Climate" will provide many young people worldwide with a basis for understanding the complex ecological relationships for shaping our future.

Prof. Wilhelm Barthlott, University of Bonn
Bonn, July 2021

Preface

This richly illustrated textbook has a long history. The first paperback edition by H. Walter entitled "Vegetationszonen und Klima" (Vegetation Zones and Climate) was published by Ulmer-Verlag in 1970. It represented a short summary of the large two-volume work Vegetation of the Earth (Walter: Volume I, second ed., 1964; Volume II, 1968). In the following editions, the ecological principles were increasingly elaborated. In parallel, the "Vegetation of the Earth" underwent a comprehensive and extensive revision with emphasis on ecological aspects and with a very consistent structure. It was then published in four volumes as "Ecology of the Earth" (Walter & Breckle Volume I, 1983, Volume II, 1984; Volume III, 1986; Volume IV, 1991). In the meantime, the "Ecology of the Earth" has also been partly published in second or third editions. The sixth edition of the pocketbook "Vegetation Zones and Climate" was completed by H. Walter in 1989 shortly before his death. The seventh edition of "Vegetation and Climate Zones" was revised and published by S.-W. Breckle in 1999.

Subsequently, the publishing landscape has changed significantly. Many new textbooks have appeared. In the meantime, the Internet now dominates the textbook market in many cases. Nevertheless, the fragmented knowledge and the mixture with "fake news" still cannot replace a good textbook.

The complete re-editing of "Vegetation and Climate" is related to the translation of the book into Dari as a basis for a modern ecology textbook for Afghanistan. This gave rise to the idea of producing a translation of the seventh edition of "Vegetation and Climate Zones" into the local language (Dari) but adapted to the ecological and vegetation conditions of the country. However, it soon became clear that this would only be possible if the German edition were first completely revised and supplemented with suitable colour illustrations; only then would a good translation be feasible. Thus, this richly illustrated new edition is at the same time the basis for a new, modern ecology textbook in Afghanistan.

Almost everywhere in the country there is still a lack of the simplest textbooks; poorly legible, copied pages are often the only thing available. Or they are foreign textbooks that have been translated literally and hardly address the local conditions. In Afghanistan's universities, teaching is still frontal. Students depend on their notes during lectures; there are no good and modern standard textbooks for students and for teachers to prepare for classes and exams. This is why the translation of this pocketbook was so important. Also, it enables us to deal with some terms that are wrong or not clearly clarified in Dari in biology. In the previous few Dari-language books, there are no adequate translations for scientific terms; they have been coined here in the Dari translation itself and all collected in a glossary.

The intensive occupation with the natural environment of Afghanistan had made possible the publication of a photo overview volume with an extensive introduction to the flora and vegetation of this country (Breckle & Rafiqpoor 2010) and an inventory of the flora as a checklist (Breckle et al. 2013); it resulted in about 5000 species and an endemism of 25%. Both books are bilingual, English and Dari. Both books were funded by the German Academic Exchange Service (Deutschen Akademischen Austauschdienst, DAAD) from the German Federal Foreign Office's funds for peace-making activities in Afghanistan. They were sent to schools, universities and institutions in the country for free distribution.

This ecology textbook "Vegetation and Climate" has been published separately as a German edition and a Dari edition. The Deutsche Gesellschaft für Internationale Zusammenarbeit GmbH (GIZ) has gratefully assumed the costs for printing and transport of the Dari edition for free distribution in Afghanistan. For the country, which has been ravaged by decades of war and civil strife, this is another building block for better education, which has seen some progress everywhere in recent years despite the current setbacks. Perhaps these scientific books are also a good example for other countries.

In contrast to today's mostly technically oriented and analytical biological research down to the smallest details of gene function, ecology strives for synthesis, the representation of the large interrelationships. The results of analytical research are the many, also important, building blocks that, figuratively speaking, have to be put together to form a building or mosaic picture. That is why the ecologist needs a comprehensive **interdisciplinary knowledge**. No matter how good one's knowledge in a special field may be, it is not enough. Therefore, many experiences and scientific findings of own research journeys to all floral regions and climatic zones of the earth and essential results from the scientific literature are collected in this book in order to obtain possibilities of comparison on a global scale. This principle taught by Heinrich Walter, which he summarized in the guiding principle *"the laboratory of the ecologist is God's nature and his working field the whole world"*, was also decisive for this edition. With numerous additional colour pictures, many examples of "nature" are vividly presented and the natural vegetation is highlighted as the basis in the various climatic zones. The activities of man can only be touched upon marginally, even if the anthropogenic changes and destruction of natural areas have assumed a threatening scale today.

The new edition of the textbook "Vegetation and Climate" is particularly characterized by a consistent structure. After the introductory chapters, which provide the basic knowledge of scientific ecology, the large areas of the earth, the zonobiomes, are dealt with. The introductory chapters lay the foundation for understanding the geobotanical and ecological treatment of the Earth's natural large spaces. The zonobiomes, the large areas determined by climate, are presented comparatively; numerous colour graphics and photographs provide a clear picture of the zonobiomes. The special features are highlighted and certain focal points are discussed in more detail. In the last part, some conclusions are drawn with regard to the activities of man, which are often controlled only by actionism without being sustainable. The consequences leave little hope. This can best be countered by good education and, in addition to the all-important digitization that is now ubiquitous, mechanization must be geared to human survival and well-being and to sustainable land use. The basics for ecological understanding can be provided by this textbook.

Many colleagues, especially Wilhelm Barthlott, Bonn; Hans Breckle, Karlsbad; Eberhard Fischer, Koblenz; Helmut Freitag, Göttingen; Reinhard Fritsch, Gatersleben; Jürgen Homeier, Göttingen; Frank Joisten, Stettiner Hof; Michael Keusgen, Marburg; Ernst Kluge, Frankfurt/M; Georg Miehe, Marburg; Stefan Porembski, Rostock; Christian Opp, Marburg; Khaled Rafiqpoor, Roermond; Michael Richter, Erlangen; Michael Succow, Greifswald; Kim Vanselow, Erlangen; Karsten Wesche, Görlitz have supported us and also provided additional photographic material, for which we sincerely thank them all.

The German edition is published by Springer Verlag. Special thanks go to the publisher for the very good cooperation, the great obligingness and the help.

We would like to thank Uta Breckle and Sadeka Rafiqpoor especially for their help, but also for their inexhaustible patience.

Bielefeld and Bonn, July 2019

Amendment to the English Edition

Springer Nature, as one of the leading knowledge publishers, is convinced that this book will play a major role in the future. This statement was based on the great demand for the e-version of the book in particular. As part of a novel attempt to translate selected German-language books into English with the help of artificial intelligence, "Vegetation and Climate" was selected in order to make the book accessible to a broad international readership. Aware of the shortcomings compared to a native speaker, the swift translation of the book was to be done with the help of **AI** software. The automatic translation of the text was astonishingly good, only little had to be changed. Linguistic perfection is certainly not quite optimal yet, but it enables the rapid provision of scientific literature in other languages. The transcription of the graphics was somewhat more difficult, a lot of "manual work" was still required. The ambiguity of some terms had to be critically questioned. We are convinced that being able to provide students and interested parties with a well-founded textbook will continue to be of great importance in the future—despite the wealth of information on the internet. The preparation of the English edition has enabled us to eliminate some minor errors from the German edition and to improve one or the other graphic, as well as to add more instructive photos.

Mr. Simon Shah-Rohlfs from the Springer Nature team has made valuable suggestions for the realization of the project and Apurva Sarvade and Janani Mourougane and their teams in India have managed the technical processing. We are once again indebted to Springer-publisher and all those involved who have contributed to the success of this interesting project and hope that, especially in times of the ever-increasing importance of scientific knowledge, this illustrated textbook will be well received and used worldwide.

Bielefeld and Bonn, July 2022

Siegmar-W. Breckle
M. Daud Rafiqpoor

Photo 2 Golden temple and sophisticated traditional garden and landscape architecture in the warm temperate climate of zonobiome V in Kyoto, Japan (photo Breckle)

Contents

Physical Units and Conversion Factors

Basic Units

Length	Metre	m
Mass	Kilogram	kg
Time	Second	s
Temperature	Kelvin	K
Light intensity	Candela	Cd
Amount of substance	Mol	Mol

Other Units

Force	Newton	N

$$1 \ N = 1 \ kg \times m \times s^{-2} = 0.102 \ kp$$

Pressure	Pascal	Pa

$$1 \ Pa = 10^5 \ Pa = 0.9869 \ at = 750 \ Torr = 750 \ mm \ Hg$$

Energy	Joule	J

$$1 \ J = 1 \ N \cdot m = 10^7 erg$$
Heat units
$$1 \ kcal = 4.187 \ kJ = 1.163 \ Wh$$
$$1 \ J = 0.102 \ kp \cdot m = 2.29 \cdot 10^{-4} kcal = 2.78 \cdot 10^{-7} kWh$$

Power	Watt	W

$$1 \ W = 1 \ J \cdot s^{-1} = 1 \ N \cdot m \cdot s^{-1}$$
$$= 0.102 \ kp \cdot m \cdot s^{-1} = 0.236 \ cal \cdot s^{-1}$$
$$= 0.86 \ kcal \cdot h^{-1}$$

Radiation, illuminance	Lux	lx

$$1 \ lx = 1 \ 1m \times m^{-1} = ca. \ 10^{-2} W \cdot m^{-2}$$

Luminous flux	Lumen	lm

Luminous intensity		$cd\text{-}m^{-2}$

$$1 \text{ lx (red light)} = \text{ca. } 4\text{T}0^{-3}\text{W m}^{-2}$$
$$1 \text{ lx (blue light = white light)} = \text{ca.} 10^{-2}\text{W m}^{-2}$$
$$1 \text{ W} \times \text{m}^{-2}(\text{PhAR}) = 3 - 5 \text{ Einstein} \times \text{m}^{-2} \times \text{s}^{-1}$$
$$1 \text{ Einstein} = 1 \text{ mol Photons} = 75 \text{ kcal (blue)} = 3 \times 10^5\text{J}$$

Further Conversions

$$1 \text{ g TG} \times \text{m}^{-2} = 10^{-2}\text{t} \times \text{ha}^{-1}$$
$$1 \text{ g organic mass} = 0.45 \text{ g C} = 1.5 \text{ g CO}_2$$

Transformation Energy for Changes in the State of Water

Solid $\leftrightarrow$ liquid (melting; freezing): $0.3337 \text{ MJ-kg}^{-1}$ (79.5 cal-g^{-1}).

Liquid $\leftrightarrow$ gaseous (evaporation, vaporization; condensation): 2.26 MJ-kg^{-1} (539 cal-g^{-1}).

Gaseous $\leftrightarrow$ solid (sublimation): 2.86 MJ kg^{-1} (684 cal-g^{-1}).

Internationally Defined Prefixes for Units and the Associated Factors (English Designations)

10^1	Ten	Deka	As
10^2	Hundred	Hecto	h
10^3	Thousand	Kilo	k
10^6	Million	Mega	M
10^9	*Billion*	Giga	G
10^{12}	Trillion	Tera	T
10^{15}	*Quadrillion*	Peta	P
10^{18}	Quintillion	Exa	E
10^{-1}	Dezi	d	(tenths)
10^{-2}	Zenti	c	(hundredths)
10^{-3}	Milli	m	(thousandths)
10^{-6}	Micro	µ	(millionths)
10^{-9}	Nano	n	
10^{-12}	Piko	p	
10^{-15}	Femto	f	
10^{-18}	Atto	a	

Abbreviations and Symbols

a	Year
A	A horizon for soils (with predominantly organic material)
a s l	Above sea level
B	B-horizon in soils (transitional horizon between organic overburden and weathered bedrock)
BHD	Breast Height Diameter of tree stems, trunks in centimetres
C	C horizon for soils (subsoil: weathered bedrock in the soil profile)
°C	degree Celsius
cal	Calorie
CAM	Diurnal acid metabolism in photosynthesis *(Crassulaceae Acid Metabolism)*
CEC	*Cation Exchange Capacity*
d	Day (24 h)
DI	Diversity Index
Dm, dw	Dry matter, dry weight
E	Einstein (light quantum quantity)
E	East
E_a	Current actual evaporation
E_p	Potential evaporation
ET	Evapotranspiration (total evaporation)
FK	Field capacity
FW	Fresh weight
g	Gram
G	G horizon in soils (waterlogged, low-oxygen gley horizon)
GPP	Gross primary production
h	Hour (60 min)
ha	Hectare (10^4 m^2)
J	Joule
K	Kelvin
kg	Kilogram
kW	Kilowatt
l	Litres
LAI	*Leaf area index*
LR	Available light, photic ratio
lx	Lux
m	Metre
M	Mass (substance production)
mg	Milligram
min	Minute (60 s)
ml	Millilitre
mm	Millimetre

mol	Mol
μm	Micrometre
N	Newton
N	North
OB	Orobiome
P	Precipitation
Pa	Pascal (1 Pa = 10^{-5} bar)
PB	Pedobiome
pF	Water potential
pH	Negative decadic logarithm of hydrogen ion concentration (H^+, acid strength)
Ph	Photosynthesis
PhAR	Photosynthetically active radiation
ppb	*Parts per billion* (10^{-9})
ppm	*Parts per million* (10^{-6})
π*	potential osmotic pressure
R	*Respiration*
RF	Relative humidity
RQ	Respiration quotient (carbohydrates = 1, fats = 0.7)
s	Second
S	South
sZB	Subzonobiome
t	Time
t	Ton (10^3 kg)
T	Transpiration
Torr	mm Hg, obsolete pressure measurement (= 10^5 Pa).
UV	Ultraviolet (short-wave light)
W	West
WC	Water content
WSD	Water saturation deficit
ZB	Zonobiome
ZE	Zonoecotone

Part I

General Part

Preliminary Remarks

Contents

© Springer-Verlag GmbH Germany, part of Springer Nature 2022
S.-W. Breckle, M. D. Rafiqpoor, *Vegetation and Climate*, https://doi.org/10.1007/978-3-662-64036-4_1

Photo 3 Nomad tents of a Kochi family in the surroundings of Mazare Sharif, N-Afghanistan (photo: F. Joisten)

1.1 Scientific Ecology

Ecology is a biological science and thus just like life (according to our current knowledge) in our solar system, it is limited to the earth. Life as a whole is connected with open cycles and energy flow—i.e. a build-up of substances with binding of solar energy as well as a decomposition with the release of the bound energy mostly in the form of heat.

The smallest independent unit of life is the **cell**, whose compartments, structure and function are the subjects of molecular biology, biochemistry and physiology. Ultrastructural research using the latest techniques plays a major role here today, as does the recording and manipulation of genetic material.

Unicellular organisms are primarily the object of study in microbiology. The next higher living unit is the organism with its multicellular tissues and organs. We distinguish plant, fungal and animal organisms, which are morphologically, anatomically and functionally very diverse. Phytology (botany) deals with the former, zoology with the latter. The green plants are autotrophic and constructive, the colourless as well as the animal organisms heterotrophic and transforming or decomposing. Heterotrophic are also the fungi, which today is considered a separate group of organisms and with which the mycology deals.

The ecological factors act at different levels of complexity, naturally also at the molecular level (Table 1.1). They cause certain effects and interactions. At the level of individuals, adaptation takes place via modifications, mutations and selection. This is, among other things, the topic of **autecology**. At the ecosystem level, these adaptations and constantly changing population structures mean ever-changing dynamics, for example for material cycles and energy flows (flux). Populations are recorded by **demecology**, **synecology** examines biotic communities and their composition (static view), **ecosystem biology investigates the** dynamics in biotic communities and thus

also the properties that determine energy flows (flux) and material cycles.

The highest living units are the communities of plant and animal organisms, each composed of populations that, together with the abiotic environmental factors (climate and soil), form ecosystems characterized by a constant circulation of substances and energy. The study of these ecosystems from the smallest to the global—the **biosphere**—is the task of ecology in the broadest sense.

This book provides a brief, accessible introduction to this global ecology.

Heinrich Walter, the founder of this textbook (Breckle 2002), expressed the connections between humans and the biosphere as follows:

> The biosphere comprises the natural world into which man has been placed and which, thanks to his mental capacities (good sense and reasoning)-, he is able to regard objectively—thus raising himself above it. On the one hand, he is a child of this external, apparent world, and dependent upon nature, but on the other hand through the world within himself, has access to the divine.
>
> Only an awareness of both sides of his nature enables man develop into a wise and harmonious being with the hope of divine fulfilment upon death. It is not the sole calling of man to use nature to his own ends. He also bears the responsibility for maintaining the earth's ecological balance, of tending and preserving it, to the best of his ability.

In order to fulfil this task and not to engage in overexploitation, which ultimately calls his own existence into question, man must recognize the ecological laws of nature and take them into account, even if there are still and again and again people who believe that they can abolish nature and rely entirely on technology, or vice versa, people who follow dogmatic or fundamentalist currents completely uncritically and in a frightening manner.

We will deal primarily with natural ecological systems and conditions, as it would go beyond the scope of this abridged version to also deal in detail with all the secondary ecosystems created by humans and the various stages of

Table 1.1 The different levels of complexity and examples of impacts

Complexity level	Examples of reactions and possible effects (e.g., salt exposure)
Interactions and effects in biomes, in the biosphere (large scale ecosystems)	Salt and other material cycles, material balances, energy fluxes, sedimentation, accumulation in erosion basins, geomorphological long-term processes
Interactions and effects in ecosystems	Salt and mineral cycles, mass balance, accumulations, mass balances, energy yield, species composition (frequency and dominance)
Effects on populations	Reproduction, age distribution, competitive ability, selection
Interactions with intact whole plants, individuals	Mineral metabolism, vitality, water balance, adjustments of growth, developmental stages, hormonal balance
Interactions with tissues	Formative effects, defect formation, osmotic stress, ion effects
Interactions with cells	Formative effects, altered differentiation, premature senescence.
Effects on cell organelles	Respiration, photosynthesis, biosynthesis of secondary plant compounds
Effects on bio-membranes	Permeability, potential changes
Bioeffects on macromolecules	Gene regulation, enzyme activities, DNA changes

degradation, especially since the ecological laws of nature are best discernible in natural ecosystems, which are therefore in a dynamic equilibrium. Natural ecosystems are the models and the reference point for sustainability. They have developed and optimized over millions of years of evolution.

1.2 Importance of Systematics and Taxonomy for Biology

The destruction of tropical ecosystems not only increases degraded areas and renders them completely infertile through erosion, but much more serious is the loss of biodiversity (Boerboom and Wiersum 1983; Mutke and Barthlott 2008; Barthlott et al. 2014; Barthlott and Rafiqpoor 2016). This destruction results in a disproportionately large loss of terrestrial plant and animal species and correspondingly coordinated communities. Species loss due to primaeval forest die-offs is occurring many times more rapidly than, say, dinosaur extinctions or changes during glacial periods. Worldwide ecological field research is still urgently needed (Breckle 2004).

Currently, about 1.8 million animal and plant species have been described, i.e. scientifically documented. However, as one must assume today, this is only a fraction of the species occurring on the globe. The diversity of certain areas and the comparison of different recording methods can be estimated by extrapolation, resulting in figures of five to ten million species. Other approaches, such as fractal geometry, yield species numbers of up to 30 million. The real numbers are very uncertain to estimate, but each new expedition material from the tropics always yields a wealth of new species. Scientific processing of the material often lags years behind. The number of specialists for many animal groups is so small that they cannot keep up with the processing of the material, and most of it remains unprocessed. The systematic affiliation, the taxonomic-nomenclatural correct naming, or even more the phylogenetic relationships are in many animal groups only very roughly or not yet known. In the case of higher plants, the state of knowledge is much better, due to the smaller number of species. But already in the case of algae and even more so in the case of fungi, so many unknown new species are still to be expected that it would be urgently necessary not only to considerably accelerate teaching, i.e. teaching and research, in systematics at universities and research centres but at least to reintroduce it at all. Actually, almost all biologists work with organisms—but some biochemists, physiologists, geneticists often seem to no longer know which organisms they actually work with.

Helmut Gams (1960): *"All the findings of the various sub-disciplines of biology, i.e. as far as possible all the characteristics, should ultimately be used to achieve a constant improvement in the natural system of organisms"* (oral comm.).

Systematics is the biological science of the future; it organizes diversity, biodiversity, the conservation of which is now recognized as a fundamental problem (Breckle 1999).

Box 1.1 The Tasks of Systematics and Taxonomy
Systematics and taxonomy are essential foundations for understanding between the biological disciplines. Systematics brings order into diversity. On the one hand, it must conservatively serve the understanding by a fixed framework, on the other hand, it must progressively by a flexible framework express the advances in knowledge of phylogenetics also in nomenclature. Without well-founded methods in systematics and taxonomy, not only ecology but also the whole of biology hangs in the air and has no point of reference.

Box 1.2 The Documentation
In addition to the task of presenting scientific facts, processes and structures to the public in exhibitions, museums also have the important task of scientific documentation.

1.3 Importance of Scientific Documentation (e.g. in Museums)

In the systematic-taxonomic treatment of species diversity, documentation is of decisive importance. Type material, on the basis of which the species diagnoses are described, must be stored safely for the future in museums or in the large herbaria as the essential documentation centres.

With the help of the new possibilities of information exchange, catalogues and taxonomic overviews, identification keys, range maps can be deposited on the Internet and thus made accessible to all users online. But even for this, there is a lack of sufficient numbers of capable young biologists, and certainly not of the political insight to set the right priorities for the future.

There are still many amateur scientists who spend their free time studying a particular group of organisms, not to mention the many ornithologists. This has been shown in the floristic mapping of Central Europe. Many of these private collectors have special knowledge and possess valuable small collections. The museums must be enabled to acquire such material as a donation or as an estate, or even by purchase. Today this often fails due to a lack of insight and financial, human or spatial resources and valuable, perhaps irretrievable material ends up in the trash.

1.4 Importance of Excursions for Young Scientists

Students can only find their way around organismic biology if they are offered opportunities to get to know organisms in their environment in the field. In recent years, for example, universities have reduced their mandatory program of beginner field trips from a meagre five afternoons to an irresponsible three. Some no longer require field trips at all. Grants for field trips lasting several days have been drastically cut.

Obviously, there are more and more biologists who have never had the good fortune to participate in a good several days Field Trip and realize that this is the most intensive way of learning, of grasping not only biological but general scientific relationships. Not just **seeing** something presented to you in front of the "goggle-box", the television, but **looking** and synthetically **grasping and recognizing** connections, for example, about the geological, the geomorphological situation, the natural resources as the basis for the possibilities of agriculture and forestry in the area under consideration, the flora and fauna and their interdependence, the spatio-temporal dynamics of producers, consumers and degradation processes, phenology, the historical basis of landscape formation, the possibilities of sustainable conservation, all this can be explained to students standing on a hill. But whether faculties (or ministries) today still want or are even capable of doing this? Biology without a due share of field biology is an amputated biology. On field trips, the participants are in the middle of the action. Only then can they also meet possible dangers, only then are appropriate precautions without fearful hysteria (for example, against ticks, or spiders, or snakes) a natural prevention, and only then they also learn to move in nature in accordance with nature. Today this can be supplemented very sensibly to a moderate extent and well supported by critical use of online archives and information.

Especially also for other disciplines, excursions are regarded of crucial importance today. Surprisingly and fortunately, some student councils have grasped this more quickly than some teaching staff who have been reformed several times and teach so-called modern and modulated subjects.

Box 1.3 The Importance of Field Trips
Excursions are the most intensive form of learning. Through analytical grasping and synthetic linking of contexts, one learns to look and understand correctly, using all the senses, to synthesize and understand the links and connections between the organisms and their environment from which they develop, the concepts, vision and understanding of the whole regarded ecosystems.

References

Barthlott, W. & Rafiqpoor, M.D. 2016: Biodiversität im Wandel – Globale Muster der Artenvielfalt. In: Lozán, J.L., Breckle, S.-W., Müller, R. & Rachor, E. (Hrsg.): Warnsignal Klima: Die Biodiversität: 44-50. In Kooperation mit GEO-Verlag. Wissenschaftliche Auswertungen. www.warnsignal-klima.de

Barthlott, W., Erdelen, W. & Rafiqpoor, M.D. 2014: Biodiversity and Technical Innovations: Biomimicry from the Macro- to the Nanoscale. In: Lanzerath, D. & M. Friehle (eds.): Concept and Value in Biodiversity. Routledge Studies in Biodiversity Politics and Management, 2014: 300-315. ISBN 978-1-415-66057-0

Boerboom, J.H. A., & Wiersum, K.F. 1983: Human impact on tropical moist forest. In: Holzner, W., Werger, M.J.A., & Ikusima, I. (eds.): Man's impact on vegetation. Junk, The Hague: 83–106

Breckle, S.-W. 1999: Wie wichtig ist Systematik für Biologen und Ökologen? Cour. Forsch.-Inst. Senckenberg **215**: 49–54

Breckle, S.-W. 2002: Salinity, halophytes and salt affected natural ecosystems. In: Läuchli, A. & Lüttge, U. (eds.): Salinity: Environment – Plant – Molecules. Kluwer Acad. Publ. Dordrecht: 35-77

Breckle, S.-W. 2004: Flora, Vegetation und Ökologie der alpin-nivalen Stufe des Hindukusch (Afghanistan). In: Breckle, S.–W., Schweizer B, Fangmeier, A. (eds.): Proceed. 2nd Symposium AFW Schimper–Foundation: Results of worldwide ecological studies. Stuttgart–Hohenheim: 97–117

Mutke, J. & Barthlott, W. (2008): Biodiversität und ihre Veränderungen im Rahmen des Globalen Umweltwandels. In: Lanzerath D., Mutke, J., Barthlott, W., Baumgärtner, S., Becker, C. & Spranger, T.M. (Hrsg.): Biodiversität. [Ethik in den Biowissenschaften – Sachstandsberichte des DRZE, 5]. Freiburg i.B.: 25-74

Part A: Ecological Basics (Autecology)

2

Contents

© Springer-Verlag GmbH Germany, part of Springer Nature 2022
S.-W. Breckle, M. D. Rafiqpoor, *Vegetation and Climate*, https://doi.org/10.1007/978-3-662-64036-4_2

Photo 4 Part of the Great Wall of China north of Beijing in a dust storm with loess (Photo: Breckle)

2.1 Ecological Factors

The concept of dividing the landscape into zonobiomes requires clear ecological criteria that take into account the interrelationships of ecological factors. The ecological factors determine for organisms their environment, in which they realise one of their basic functions of existence, namely reproduction. The persistence of organisms in an ecosystem is determined by competition and adaptation. In a floristically uniform area, the structure of the vegetation in the system "**climate-soil-plant**" is determined by environmental factors and this mainly indirectly by influencing the competitive power of the species occurring in the area. In this interplay, the individual ecological factors often complement each other very differently in their selection effect. Figure 2.1 illustrates the mutual relationships of the ecological factors.

The soil type and the vegetation type are shaped by the climate, but for the vegetation the flora (and secondarily the fauna) and for the soil the bedrock and the vegetation (as well as the edaphon) are of great importance. There are such close interrelationships between soil and vegetation that one may almost call it a unit. Both the soil and the vegetation exert a certain influence on the climate, but directly only in the area of the air layer near the ground; that is, they influence the microclimate. All factors acting on plants, or on an organism in general, constitutes its environment, the physico-chemical factors (without competition) being called its **site**, while the place where it grows is called the growing **site**, **biotope** or **ecotope**. The factors determining the growth and development of the plant can be divided into five groups of primary factors:

1. **Radiation**: Light intensity and day length—the light factor
2. **Warmth**: Temperature relations—the temperature factor
3. **Water**: Hydrological conditions—the water factor
4. **Chemical factors**: Nutrients or toxins, poisons
5. **Mechanical factors**: Wind, fire, animal browsing and trampling.

It makes no difference to the plants whether, for example, the favourable warmth conditions are due to the large-scale climate or to the growing location (habitat) on a sheltered south-facing slope (in the Northern hemisphere). Similarly, it makes no difference to the plant whether the necessary soil moisture is due to a favourable distribution of precipitation or the low evaporation on a north-facing slope or, finally, to the soil structure and proximity to groundwater; the main thing is that the plant does not suffer from a lack of water.

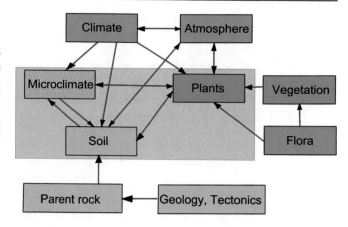

Fig. 2.1 Scheme of interactions between various environmental domains (areas) and plants

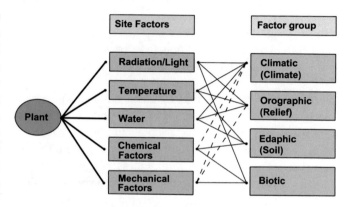

Fig. 2.2 Scheme of the various ecological factors and their effects on the plant

The five groups of factors condition in their mutual effect the expression of complex site factors (secondary factors, complex factors), namely climatic, orographic, edaphic (soil) and biotic, as this is shown in the scheme (Fig. 2.2) with some essential site parameters.

2.1.1 Radiation, Light

All life on earth is kept going by the flow (flux) of energy radiated by the sun and supplied to the biosphere (apart from the special case of deep-sea black smokers with chemosynthesis). Photosynthesis of the plant binds radiation energy in the form of latent chemical energy. It benefits all links in the food chain for the operation of life processes. Radiation is generally the primary source of energy for building organic matter. By regulating the heat and water balance of the earth, it creates the energetic prerequisite for the fulfilment of the life requirements of organisms. However, radiation is not only a source of energy for the plant, but can also be a stress factor. Radiation effects are triggered by the absorption of light quanta and each radiation-dependent process is mediated by very specific photoreceptors. They exhibit a typical absorption spectrum. Important factors are the time, duration, exposure to solar radiation and the spectral composition of the illumination.

Radiation and Plant

Radiation hits the above-ground parts of the plant from all sides: Direct and scattered sunlight (sky radiation), diffuse radiation when the sky is overcast, and finally radiation reflected from the ground. Many plants arrange their assimilation surfaces in such a way that as few leaves as possible are constantly exposed to direct radiation. Most leaves are in partial shade, where they receive scattered light (Larcher 2001). Upright leaves (e.g. of many monocots and globose pads), leaves in profile position (e.g. of *Iris* and *Lactuca species*), drooping leaves (e.g. of *Eucalyptus*) *as* well as assimilation organs with curved surfaces (round leaves, scale leaves, needle leaves, assimilation shoot axes) are tangent to the sun's rays at acute angles. This protects the leaves from strong light damage and overheating, but they receive more light in the morning and evening. In the crowns of individual trees and shrubs, a brightness gradient develops from the crown edge to the interior of the crown. Depending on the species-specific ability to develop pronounced shade leaves, a distinction is made between **light crowns** (pine, larch, birch, umbrella acacia) and **shade crowns** (many conifers, beech, broad-leaved evergreens). In light crowns, the innermost leaves receive on average 10–20% of the outdoor brightness, while in shade crowns leaves are still active with a light consumption of 1–3% (Larcher 2001).

Absorption of Radiation Through the Leaves

Of the radiation falling on a leaf, part is remitted (i.e. diffusely reflected → **remission**), part is absorbed (**absorption**), and the rest is transmitted (**transmission**). The light retained by the leaf (**remission**) consists of the light reflected from the surface and the scattered radiation from inside the leaf (Fig. 2.3).

The reflectivity depends on the surface condition of the leaves. Felt hairs increase the reflection considerably. In the visible range, leaves remit on average only 6–10% of the radiation. Highly glossy leaves can reflect visible light up to 12–15%. Due to this scattered light, the interior of such "glossy light forests"is somewhat brightened. **Green** light is

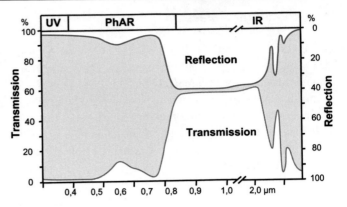

Fig. 2.3 Relative reflection, transmission and absorption of a poplar leaf (*Populus deltoides*) as a function of the wavelength of the incident radiation. After Gates 1965. Optical parameters of leaves of different plant species are given in Gausman & Allen (1973) (modified after Larcher 2001)

remitted more (10–20%), **orange** and **red** the least (3–10%). **Ultraviolet** radiation is reflected little by leaves (no more than 3%); in the **infrared** range, however, leaves reflect 70% of the incident radiation (Larcher 2001).

The radiation penetrating the leaf is largely absorbed (**absorption**). When passing through the leaf, the radiation is attenuated in such a way that the radiation gain of cell layers lying behind each other drops exponentially. Depending on the leaf structure and the equipment of the mesophyll cells with chloroplasts, leaves usually absorb 60–80% of the photosynthetically active radiation (PhAR). The leaves of some herbaceous species that occur in the deep shade of tropical rainforests contain lenticular cells (*ocelli* → from Latin *ocellus* = little eyes) in their upper epidermis that focus the weak light onto the chloroplasts arranged in a contracted manner in the mesophyll. The absorption in the **visible** range is mainly due to the chloroplast pigments. Most of the **ultraviolet** radiation is retained by cuticular and corked outer layers of the epidermis and by phenolic compounds in the cell sap of the outermost cell layers, so that at most 2–5%, but usually less than 1%, of the UV radiation can enter the deeper leaf layers. The epidermis and hairs are effective UV filters for the assimilating parenchyma: e.g., shield hairs on *Elaeagnus leaves* absorb 40% of UV-B radiation (Larcher 2001). Lichens deposit coloured compounds ("lichen substances") in the upper bark layers that are both UV and light filters. **Infrared** is absorbed little by leaves in the range up to 2000 nm, but almost completely (97%) in the range of long-wave temperature radiation above 7000 nm. Accordingly, the plant behaves like a black body in relation to thermal radiation (Larcher 2001).

The radiation transmittance (**transmission**) of leaves depends on the structure and thickness of the leaf. Soft-leaved leaves allow 10–20% of the sun's rays to pass through, very thin leaves up to 40%, thick and coarse leaves are almost impermeable to radiation (<3%). The best transmission is in the green, but especially in the near infrared. Light filtered through foliage is therefore particularly rich in wavelengths around 500 nm and >800 nm. Under a leaf canopy there is a red-green shadow, in the forest darkness only dark red and infrared shadows. Through canopy sheaths and through the bark of thin twigs, up to 0.5–2% of the incident light, mainly long-wavelength light, penetrates both into the interior of the forest and into the interior of the plants, i.e. in the leaves and meristems, where the phytochrome system is active. Apical meristems in buds receive more dark red radiation (700–840 nm) than light red (600–690 nm), and the light red/dark red ratio changes with bud formation and seasonally. These signals are perceived by the phytochrome system, whereupon gene activation triggers the corresponding changes in developmental behaviour and differentiation (Larcher 2001).

Plants adapt to the local radiation climate and to the prevailing quantity and quality of the radiation available at their growing site by modulative (short-term), modificative and evolutionary means. **Modulative adaptations** occur rapidly and are reversible; after returning to the initial situation, the initial behaviour is soon restored. Examples of photomodulations are: Nastic movements such as the closing stomate cell movements; leaf movements that cause a favourable exposure of the leaf blade to light incidence; the diurnal and weather-related opening and closing of flowers. Modulative radiation adaptations that directly affect photosynthesis proceed via changes in the chloroplasts (Larcher 2001).

Modificatively, plants adapt to average radiation conditions over weeks or months during adolescence. Phenotypic differentiations of organs and tissues are usually not traceable. If light conditions later change, then new shoots sprout and the originally attached, now unadapted leaves age and are shed. Plants that develop in bright light form a vigorous axial system. Their leaves have multiple staggered mesophyll, chloroplast-rich cells, and a dense vein network. As a result of structural adaptation and more active metabolic processes, high-light-adapted plants produce greater dry matter growth, higher dry matter energy content, and better fertility (flowering frequency, flower set, fruit yield). Low-light-adapted plants develop longer internodes and thin leaves with a large surface area. This enables them to cope on sites with low energy supply (Larcher 2001).

Evolutive adaptations to the available radiation are hereditary and determine the site preference of different plant species and photo-ecotypes. The classification of plants into dim-light plants, shade plants (heliophytes) and strong-light plants (which grow in places without shade, e.g. in high mountains, in deserts and on seashores) reflects ecological differentiation by selection and adaptability. The plant's response norm is hereditarily determined. Thus, although sun plants are shade-adaptable, they are not to the same extent as genetically programmed shade plants; the same applies in the opposite direction (Larcher 2001).

Modulative, modificative and evolutionary adaptations overlap and thus give plants the opportunity to use the available radiation as much as possible through finely graduated adaptation. Due to the diversity of growth forms, light-ecological niches in the multi-storey crown horizons of the tree layer of dense forests are exploited by lianas and epiphytes. In addition, secondary effects of radiation (e.g. heat and influences on the water balance) play a role in all adaptations to site brightness. Sun plants are therefore always adapted to higher temperatures, dry air and a temporary stress of the water status (Larcher 2001).

2.1.2 Temperature, Frost, Heat

The radiation supply is also an important factor for the temperature conditions. Ecologically, thermal conditions are of great importance for vegetation at a **site**, as life only takes place within certain temperature ranges. Temperature extremes are tolerated differently by different organisms. The heat resistance limit of most plant species is between 50 and 60 °C (Table 2.1). To a certain extent, plants can protect themselves from heat stress by radiation reflection, by transpiration cooling or also physiologically (heat shock proteins).

But much more significant is the cold (**frost**). The cold resistance limit is not as sharp as the heat resistance limit. Besides the plants **sensitive to chilling** or **cold** (mostly of tropical origin), there are the **freeze-sensitive** plants (which avoid ice formation in the tissues, for example by increasing the concentration of cell sap) and the freeze-tolerant plants, which instead of a large central vacuole often form many small vacuoles in which membrane damage by ice crystals is kept small.

With regard to temperature, we distinguish among animal organisms, on the one hand, the cold-blooded or **poikilothermic** species (such as amphibians), whose body temperature depends on the external temperature and changes in the same

Table 2.1 Temperature resistance of leaves of shoot plants of different climatic regions. Limit temperature at 50% damage (TL50 in °C) after 2-h or longer exposure to cold and half-hour heat treatment (from Larcher 2001)

Plant group	Cold damage in the hardened state (°C)	Heat damage during the growing season (°C)
Tropics		
Trees	+5 to −2	45–55
Forest undergrowth	+5 to −3	45–48
High mountain plants	−5 to −15(−20)	at 45
Subtropics		
Woody evergreens	−8 to −12	50–60
Seasonal green woody plants	(−10 to −15)[a]	
Subtropical palms	−5 to −14	55–60
Succulents	−5 to −10(−15)	58–67
C_4 grasses	−1 to −5(−8)	60–64
Winter annual desert herbs	−6 to −10	50–55
Temperate zone		
Evergreen woody plants of winter mild coastal areas	−7 to −15(−20)	46–50(55)
Relict species of tertiary tree flora	−8 to −20(−15 to −30)[a]	
Dwarf shrubs of Atlantic heaths	−20 to −25	45–50
Deciduous trees and shrubs with wide distribution	(−25 to −35)[a]	at 50
Herbaceous plants of sunny locations	−10 to −20(−30)	47–52
Herbaceous plants of shady locations	−10 to −20(−30)	40–45
Tumbleweeds	(−30 to N2[b])[a]	60–65
Halophytes	−10 to −20	
Succulents	−10 to −25	(42)55–62
Water plants	−5 to −12	38–44
Homoiohydric ferns	−10 to −40	46–48
Winter cold areas		
Evergreen conifers	−40 to −90	44–50
Boreal deciduous trees	(until N_2)[a]	42–45
Arctic-alpine dwarf shrubs	−30 to −70	48–54
Herbaceous plants of the high mountains and the Arctic	(−30 to N_2)[a]	44–54

[a]Vegetative buds
[b]Temperature of liquid nitrogen (−196 °C)

sense with it; on the other hand, the warm-blooded or **homoiothermic** species (such as humans), which have their own body temperature, largely independent of the external temperature and fairly constant. In these organisms it makes no sense to measure the external temperature in order to relate it directly to the course of vital functions.

All plants are poikilothermic organisms, even if occasionally, as in the case of the *Arum* family (Araceae), the inflorescence can generate their own heat (Barthlott et al. 2009). The temperature of the surrounding air therefore gives an indication of the governing temperature conditions in the plasma. Certain smaller deviations due to purely physical reasons occur, especially in the case of strong radiation. In ecophysiological studies, they must be taken into account; after all, chloroplasts or mitochondria, for example, can often have over 10 K excess temperature in the leaf during the sunny day compared to the ambient air. In ecological overviews, one will usually have to be content with stating the air temperature.

In most poikilothermic animals, development is very dependent on temperature (Fig. 2.4), but usually modified by the water factor, for example humidity. The development time can often be specified very precisely by a corresponding mathematical function (Fig. 2.4), for example by a hyperbolic function. The example in Fig. 2.4 gives not only the duration of embryonic development but also the mortality of the eggs as a function of air temperature and relative humidity. From Fig. 2.5, it can also be seen that a certain temperature range at relatively high humidity represents the optimum range. Accordingly, one can easily imagine how different, depending on external conditions, the reproduction rates and thus the influence of some insect species in certain biotopes can be from year to year, even without other biotic interactions.

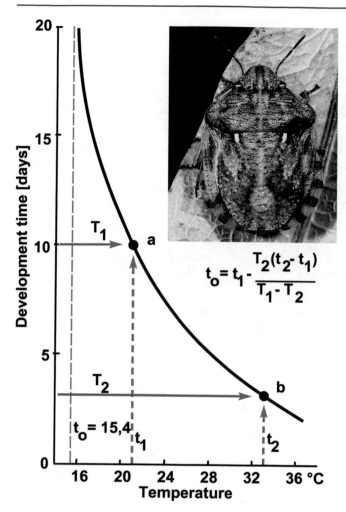

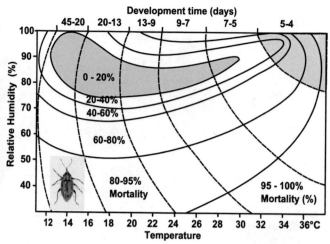

Fig. 2.4 Temperature dependence of the duration of embryonic development in the bug *Eurygaster maura* (Pentatomidae) (modified after Tischler 1984)

Fig. 2.5 Dependence of the duration of embryonic development and mortality of eggs of the alfalfa weevil (*Hypera postica*, Curculionidae) on temperature and relative humidity (modified after Tischler 1984)

Freezing is closely coupled with the behaviour of the tissue or cell water in the cell. Freezing of the vacuole with formation of ice crystals usually means a strong rupture of the membranes and thus considerable cell damage. In addition, the supply of water is blocked, so that prolonged exposure to frost often causes the plants to dry out (frost desiccation) rather than real freezing damage.

The different zonobiomes are characterized on the one hand by water availability, on the other hand due to the temperature factor. It is not so much the mean values of temperature that are important, but rather the extremes. And it matters whether frosts in an area occur regularly with the change of seasons or whether they occur episodically. One frost in 20 years in the coffee-growing regions of Brazil damaging the *Coffea*-shrubs causes the world market price of coffee to rise.

Plants prepare themselves for the annually recurring winter cold. The main frost categories are shown for the whole Earth in Fig. 2.6. Since water freezes by definition at 0 °C and increases in volume in the process, this has very special significance for living organisms. The zero degree limit, i.e. the occurrence of frost, therefore has a decisive influence on the various biomes (Table 2.1).

The following applies to the individual zonobiomes: Zonobiome I to III are frost-free (except in the higher altitudes of the mountains). In zonobiome IV and V light (episodic, partly periodic) frosts may occur occasionally. Zonobiome VI already regularly exhibits a typical, albeit short and not very severe, winter with frost. In zonobiome VII, on the other hand, with a continental climate, winters are very pronounced and sometimes severe (cold semi-deserts and deserts). In zonobiome VIII in the taiga, the winter can already be several or many months long and very severe; ZB IX of the tundra is characterized by winter; it is by far the longest season in the annual cycle. The occurrence of frost determines the occurrence of different resistant types of plants. In the equatorial zone with a minimum of not less than +5 °C, cold-sensitive plants predominate. In Zone D (Fig. 2.6), on the other hand, only completely freeze-resistant plants can survive, while in Zones C and B there are also limited freeze-tolerant plants and trees that are at least protected by freeze depression and good super-cooling.

Only about 30% of the earth's land surface is frost-free, while 42% is regularly subject to severe frost with a mean annual minimum below −20 °C.

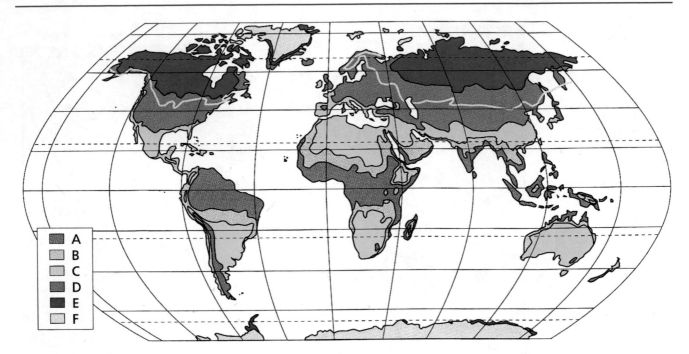

Fig. 2.6 The occurrence of frost on Earth. (**C**) Frost-free except high mountains; from (**D–F**) increasing frost frequency and intensity at higher latitudes. (**A**) Frost-free areas; (**B**) Frost-free, but up to +5 °C annual minimum possible; (**C**) Episodic frosts down to −10 °C; (**D**) Winter-cold areas with mean annual minimum between −10 and −40 °C; white line = −30 °C annual minimum isotherm; (**E**): Long winters, mean annual minimum fall below −40 °C; (**F**) Polar ice and permafrost areas (from Larcher 2001)

2.1.3   Water

For the structure of the biosphere, of all site or environmental factors, the temperature and water conditions are of primary importance. Light is nowhere in the minimum, because the long polar night meets the plants in the winter dormancy. The light factor therefore plays no decisive role in the large-scale structure of the vegetation of the earth.

Global Water Supply

The heat or temperature decreases fairly steadily from the tropics to the poles. Important here, as briefly discussed, is the frost line between the tropical and extratropical regions. The water factor has an even greater differentiating effect. Precipitation is distributed very unevenly over the Earth (Fig. 2.7).

The amount of mean annual precipitation varies between over 10,000 mm (Fig. 2.8, left: Cherrapunji, India) and practically zero (Fig. 2.8, right: Iquique, Chile) in the extreme deserts.

Figure 2.9 shows the major vegetation zones for which, in addition to the distributions of precipitation, the temperature conditions are also of particular importance, as expressed in the more or less zonal arrangement parallel to the latitudes (Fig. 2.50).

But not only on a large scale, but also on a small scale, temperature and water have a strongly differentiating effect on the plant cover due to the changing humidity of the biotopes. In general, water plays a very special ecological role in the life of plants, a much greater one than in the case of animals, because plants are location-bound. The water balance can be described quantitatively both at the level of the cell and the plant as a whole, as well as at the level of the ecosystem.

Water Balance Types and Drought Resistance

Depending on the water supply at the site, a distinction is made between hygrophytes, mesophytes and xerophytes. The hygrophytes as colonizers of evenly moist or wet sites (as well as some shade-loving herbs in the forest) have hardly any water shortage. The mesophytes are already better adapted to certain dry periods. Most species of the temperate latitudes belong to them. The xerophytes have developed many adaptations to the more or less severe and prolonged water shortage at their location. For the mechanisms of drought resistance, Levitt (1972) has characterized different possibilities: Most plants avoid drought by spatial or temporal avoidance; for true drought tolerance, special adaptations are needed, as we will see in some examples of xerophytes (see Sect. 2.1.3.5).

Soil Water

The availability of water for plants does not depend solely on the water content of the soil. The grain size distribution and thus the pore volume and the size of the capillary spaces in the soil also have a major influence. The maximum amount of water that a soil can absorb is equal to the pore volume, but then there is no soil air and therefore no oxygen left in the

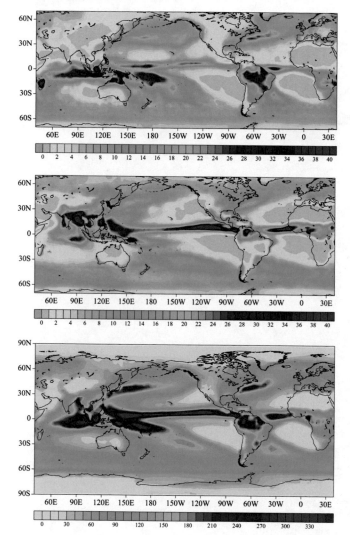

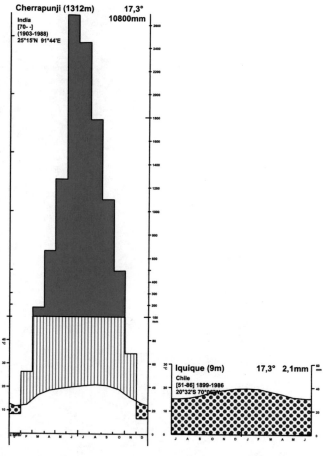

Fig. 2.8 Climate diagrams Cherrapunji in India and Iquique in Chile: One of the wettest and one of the driest climate stations (note that monthly precipitation data above 100 mm are compressed by a factor of 10)

Fig. 2.7 Global seasonal distribution of precipitation (in cm/month) comparing January (top), July (middle), and year (bottom) (source: Shepherd 2004; http://is.gd/h916at)

soil. Due to gravity, however, part of the water seeps into the depths. The field capacity (FC) is strongly dependent on the grain size distribution, as Fig. 2.10 demonstrates.

Beyond the FC, there is a fraction of soil water that is very tightly bound to the soil particles by adsorption forces (by electrostatic, as well as absorption and cohesion forces) in the very small pore spaces. These fractions are not accessible to plant roots. If a soil contains only these tightly bound fractions of water, it is called a "permanent wilting point" (PWP) (Fig. 2.10). In particularly fine-grained clay soils, the total water content may be more than 20%, yet none of it is available to the plant root because of the fine-grained nature of the soil. Accordingly, the water tension (water potential, cm water column, as log pF) of a large water content is particularly high. The PWP is on average slightly above pF = 4 (=10^4 cm water column), but is variable.

However, the water availability limit is not the same for all plants. Xerophytes and halophytes, which can develop very

high suction forces through their roots, are quite capable of still absorbing some water, which means that the permanent wilting point is different for different types of plants.

Thus, with regard to the water factor, the conditions are similarly complicated for plants as they are for animals with regard to temperature.

Water State of the Cell

First of all, one has to distinguish between plants able to change their water content (poikilohydric plants) and those maintaining a rather constant water balance (homoiohydric plants).

Plasma, the cell content, is physiologically active only when it is highly hydrated, i.e. imbibed and hydrated. If cells dry out, then the plasma enters a latent life state (i.e., it exhibits no measurable signs of life) or it dies. The thermodynamics of imbibition (swelling bodies) teaches us that the swelling state depends on the relative activity of the water (**a**), where **a = p/p$_0$**, i.e. equivalent to the relative vapour pressure.

By definition, pure water has a hydrature of 100% (i.e. it is available to the plant without restriction). The hydrature corresponds to the humidity (also given in %). A certain water vapour pressure is established above salt solutions,

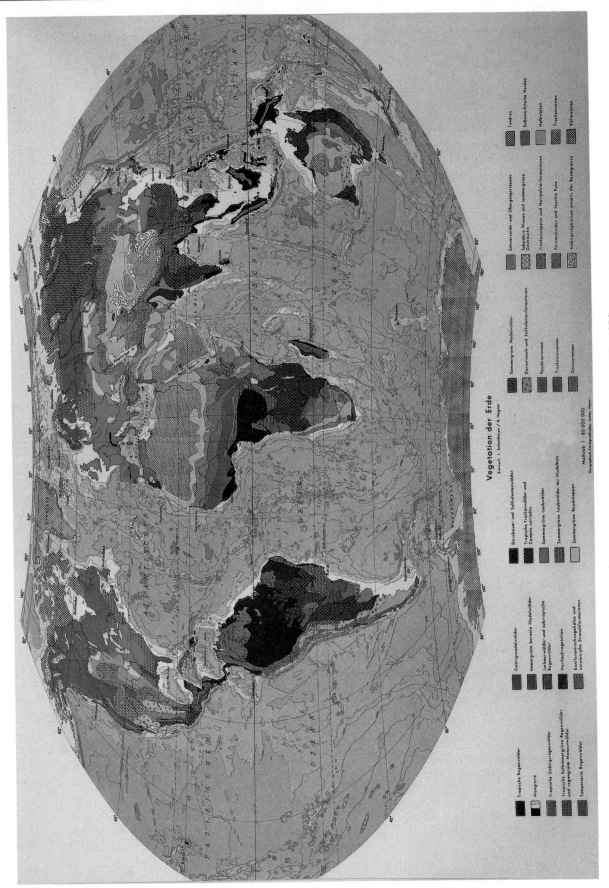

Fig. 2.9 The vegetation zones of the Earth (without edaphic or anthropogenic modifications) (from Schmithüsen Atlas 1976)

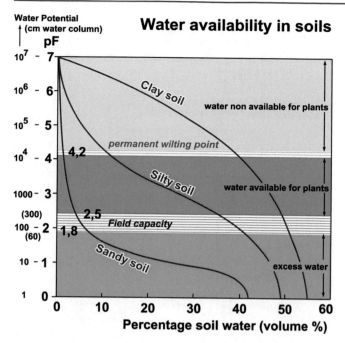

Fig. 2.10 Relationship between water potential and water content (pF curves) for three different soil types (sandy soil, silt (loess) soil and clay soil), with logarithmic ordinate (after Scheffer & Schachtschabel 1992). In sandy soils 2%, in silty soils approx. 10% and in clay soils 35–40% of the soil water is not available to plants

which is lower than that above pure water, and the hydrature is correspondingly lower.

Since the vital functions depend to a great extent on the swelling state of the protoplasm, it is important to know its hydrature (or activity of water). In the case of poikilohydric plants, in so far as these plants occur outside water, the hydrature depends entirely on the moisture content of the surrounding air. The lower plants (bacteria, algae, fungi and lichens) belong to them. If they are in contact with water or if the surrounding air is saturated with water vapour, the protoplasm of these species is almost maximally swollen and active. In dry air, on the other hand, severe de-swelling occurs, and the plasma passes into the latent state without dying. The cells of these organisms have no or only very small vacuoles, and the volume changes of the cell contents are therefore small during desiccation and the plasma structure is not damaged. The lower limit of hydration (humidity) at which growth can still be detected is very high in most bacteria, usually from 98 to 94%; in the unicellular algae and moulds it varies widely, and in only a few organisms does it fall as low as 70%, a value corresponding to the absolute minimum of hydration for life.

The productivity of poikilohydric organisms is low, their share in the vegetation mass on land is small today. They have therefore received little attention up to now, although they are often much more widespread on the soil surface, namely also in the deserts, than is assumed. Before the conquest of the country by higher plants they might have been already widespread on periodically moistened surfaces,

as today on periodically flooded clay surfaces in the deserts (Takyre). These are uninhabitable by higher plants because they offer no root space. However, fossil remains of lower plants are preserved only exceptionally, they are found relatively rarely in the older geological rock formations.

The homoiohydric terrestrial plants play a much greater role. They include all cormophytes, which originally evolved from green algae. Their cells are characterized by a large central vacuole. As a result, the plasma is directly adjacent to the cell sap in the vacuole, and the hydrature of the plasma is largely in equilibrium with that of the cell sap, thus not directly dependent on water conditions outside the cells. The cell sap of the vacuoles constitutes in the higher plants, as has been mentioned, an "inner aqueous medium," the vacuome, and the cell wall of cellulose an "outer aqueous medium," which in the course of phylogenetic development enabled them to pass from life in water to life on land, and to adapt themselves more and more to arid conditions. As long as terrestrial plants succeed in keeping the concentration of cell sap in the vacuole low, the plasma remains imbibed, that is, it has a high hydrature, regardless of the humidity of the surrounding air. This is more likely to be the case the more secure the supply of water from the moist soil through the root and transport system. In the mosses these facilities are only imperfectly developed, and they are therefore generally confined to very moist sites. In the ferns, too, the transport system is still not very efficient. They therefore avoid dry sites, all the more so because the development of the gametophytes is still strongly dependent on moisture. As far as mosses and some ferns (*Ceterach, Notholaena, Cheilanthes* and others as well as *Selaginella species*) have penetrated into desert areas, they had to change secondarily to the poikilohydric way of life, i.e. they tolerate desiccation during drought without dying ("resurrection plants"). They regained this desiccation ability, which is otherwise lacking in plants with strongly vacuolated cells, by a cell reduction with reduction of the vacuoles, which solidify even at low water losses, thus preventing deformation and damage of the plasma during desiccation.

Xerophytes

The most perfect adaptation of the water balance to terrestrial life has been achieved by the angiosperms. They have penetrated to extreme deserts. The measurement of their cell sap concentration shows that they are nevertheless able to maintain a low cell sap concentration and thus a high hydrature of the plasma without slowing down too much the gas exchange necessary for photosynthesis. An increase in cell sap concentration and thus dehydration of the plasma and increased osmotic adjustment by appropriate substances (compatible solutes) is generally not a useful adjustment for desert plants, but the sign of a disturbed water balance and a threat to their existence. For the knowledge of the water activity in the plasma, i.e. its hydrature and swelling state, the measurement of the external factors (precipitation, humidity, soil water etc.) is just as insufficient as the measurement of the external temperature in warm-blooded animals.

The determination of the cell sap concentration (and thus of the potential osmotic potential), which is directly related to the relative vapour pressure (= hydrature), provides information as to whether or not the plant is affected by the change in external conditions, in particular by a period of drought, with regard to the imbibition state of the plasma. The measurement of the water potential, on the other hand, is necessary when dealing with the flow through the plant from the roots to the transpiring organs. This is illustrated by characterizing the individual resistances in the plant in the hydraulic flow model diagram (Fig. 2.11).

Some of these flow resistances are constant, others are more or less variable. Especially the stomatal resistance is to be emphasized, because it allows a regulation of the water losses within wide limits. Corresponding to Ohm's law, the water flow (current) also depends on the resistances and the potential (voltage). The total "voltage" corresponds to the difference in suction force between the ground and the atmosphere. This difference in water potential is almost always very large, even in temperate climates. The suction force (water potential) says nothing about the hydrature state of the plasma, on which the course of all life phenomena depends. Both are closely related, as described by the osmotic characteristics.

While one must provide the usual information on external factors for site characterization, one must additionally refer to cell sap concentration and its change for characterizing the hydrature of the protoplasm, especially when discussing arid areas where the water factor plays a dominant role. Therefore, one must look more closely at the adaptations to drought and point to the osmotic conditions.

Studied species are regarded as stable units in the experiment, but they are very changeable during longer observation. Each plant constantly adapts morphologically to the respective environmental conditions. This is necessary for survival. These phenomena are associated with growth and only become noticeable after weeks or months. Ecologically they are particularly significant and in arid regions very noticeable if one examines a plant after a rainy season, i.e. during the drought period until the beginning of the next rainy season.

Adaptations to water deficiency must take into account the different osmotic state variables of the plant parts:

The suction stress (**S**) = − water potential (**Φ**), the potential osmotic pressure (**π***) = − osmotic potential (**Φ$_s$**) and the turgor pressure (**P**). The equations apply:

$$\mathbf{S = \pi^* \text{-} P} \text{ or } \mathbf{\Phi = \Phi_s + P}$$

The state variables are measured in the pressure dimension (today in MPa). **S** and **Φ** as well as **π*** and **Φ$_s$** are numerically

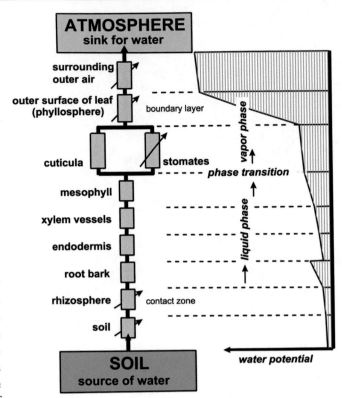

Fig. 2.11 The flow of water through a plant from the soil to the atmosphere can be compared with a scheme borrowed from electrical resistance analogue. The current (**I**) is driven by the voltage (**U**), in this case the water potential difference between the soil and the atmosphere, and limited by the sum of the resistances (**R**) in the plant, some of which are constant and some are variable (stomatal resistance as a control option for the water flux). Ohm's law (**U = R·I**) is applicable (according to Hillel 1980)

always the same and differ only by the sign (**Φ** and **Φ$_s$** are always negative).

It is important to be clear about the importance of the different quantities for the water balance of plants: If one is dealing only with the more physical process of water flow through the plant from the soil to the atmosphere, then one has to measure S or **Φ** respectively. If, on the other hand, one is dealing with the biological processes of the adaptations associated with growth, as is the case here, then **π*** or **Φ$_s$** is the decisive quantity, because it is directly related to the hydrature of the plasma, i.e. its imbibition state, as already mentioned above, and the growth processes of the plants are controlled by the latter.

The adaptations of the plants, which can only be detected after a longer period of time, can be regarded as feedback control loops, which are necessary for the maintenance of a certain equilibrium under changed conditions, in our case a balanced water balance. This is the controlled variable. The disturbance variable is increasing dryness during drought, the set point is a balanced water balance (i.e. water intake = water output). The living plasma acts as a sensor, because when the water balance is disturbed, an increase in **π*** (decrease in **Φ$_s$**)

occurs as a result of an increase in cell sap concentration, and this entails a decrease in the hydrature of the plasma, including the hydrature of the plasma of the meristematic cells at the shoot and root apex, which must be regarded as a manipulated variable. Their change, as part of a signaling chain, results in the newly formed organs being morphologically better adapted: The internodes become shorter, the leaves smaller and more xeromorphic, which causes reduced transpiration and allows the water balance to be balanced (already shown by Walter & Kreeb in 1970).

A first example from the Sonoran Desert will explain what has been said: The composite half-shrub *Encelia farinosa*, about 50 cm high, has large soft hygromorphic leaves during the rainy season, which are greenish and weakly hairy; their π^* is 2.2 to 2.3 MPa. In the drought season, water supply becomes more difficult, with π^* increasing to 2.8 MPa; this also causes a slight decrease in hydrature of the protoplasm of the meristem cells. The new leaves then formed by the meristem are smaller, more mesomorphic as well as more hairy, replacing the hygromorphic ones. If the drought continues, π^* increases to 3.2 MPa and the next leaves are

even smaller, thicker and densely white haired, allowing further transpiration reduction. In extremely long droughts, all leaves are shed as soon as 4.0 MPa is reached. Only the terminal buds remain with small leaf systems that do not develop further. The plant's water output is then so low that even with minimal water uptake from the soil, the plants are in a state of water balance.

As soon as the next rainy season starts, the potential osmotic pressure (π^*) drops again to the initial value of little more than 2.0 MPa, the hydrature of the meristem cells increases and the newly formed leaves become large and hygromorphic; as a result of intensive photosynthesis, strong growth sets in with heavy transpiration but still maintained water balance. This cycle repeats itself again and again. Similarly, this is true for many small shrubs and deep-rooted desert plants. In Fig. 2.12, the succulent leaves of *Zygophyllum dumosum* in the Negev Desert still have both leaflets and are green and turgescent (left). These spring leaves shrivel and turn brown in summer, and the succulent leaf midrib remains green longer (right). Eventually it also wilts and falls off. Then only the woody branches of the small

Fig. 2.12 Differently shaped leaves of *Zygophyllum dumosum*. (**a**) hygromorphic leaves; (**b**) xeromorphic leaves during desiccation as a result of water deficiency. The still existing partly mesomorphic leaves bear a grey pubescence (Photos: Breckle)

shrub remain, but they consume very little water and thus survive the long summer drought.

It is interesting that roots react differently to a decrease in hydrature of the meristem cells than shoots. Roots become thinner, but longer and do form less side roots. Inhibition of growth only sets in after more severely decreased hydrature while growth of shoots is immediately inhibited.

In general, it must be stated that higher π^*, i.e. lower osmotic potential, promotes the transition from vegetative growth to generative. This is also the case with ephemerals; for dwarf plants with higher cell sap concentration always flower first. This confirms the experience of gardeners that when the water supply is more difficult, the plants blossom more, whereas when the water supply is plentiful, they grow mainly vegetatively.

2.1.4 Chemical Factors and the Soil

Nutrients and Trace Elements, Mineral Supply

The nutrients and trace elements (micronutrients) and thus the mineral supply are another ecological factor that can control the occurrence of plants. In addition to the main elements C, H and O, several other chemical elements play a role as bio-elements through their participation in the structure of organisms. The bio-elements of plants are listed in Table 2.2.

In animals, iodine must be added as an essential element. Essential elements are generally characterized by the fact that they: (1) Are necessary for very specific functions in the metabolism of plants, and (2) Cannot be replaced by any

Table 2.2 Bio-elements in plants (macro- and micro-nutrients, trace elements and essentiality. In addition to C, O and H, the following elements are significant in plants

Item	Uptake as …	Enrichment in …	Relocatability	Symptoms of deficiency
N	NO_3^- NH_4^+	Young shoots, leaves, seeds	Good, in organic form (amino acids)	Stunted growth, premature yellowing (light green)
P	HPO_4^{2-} $H_2PO_4^-$	Reproduct. organs	Good, in organic form (amino acids)	Delayed flowering, peak drought, bronze discolouration
S	SO_4^{2-}	Leaves, seeds	Good, in organic form (amino acids), hardly as sulphate	Similar N, intercostal chlorotic leaves
K	K^+	Meristematic tissues	Very good	Leaf edge wilt, root rot
Mg	Mg^{2+}	Leaves	Pretty good	Scanty growth, intercostal chlorotic, older leaves
Ca	Ca^{2+}	Leaves, bark	Very bad	Disturbed division growth, peak drought, leaf deformations
Fe	Fe^{2+} FeIII-Chelate	Leaves	Bad	Chlorosis of young leaves, hardly any bud formation
Mn	Mn^{2+} Mn-Chelate	Leaves	More or less bad	Growth retardation, chlorosis, necrosis of young leaves
Zn	Zn^{2+} Zn-Chelate	Roots, shoots	Pretty Bad	Dwarfism, white-green old leaves, fructification disturbances
Cu	Cu^{2+} Cu-Chelate	Lignified axles	Bad	Peak drought, wilt, spotted chlorosis of young leaves
Mo	MoO_4^{2-}	Roots, leaves	Badly	Growth disturbances, shoot deformations, leaf edge browning
B	HBO_3^{2-} $H_2BO_3^-$	Leaves, shoot tips, vegetation tips	Bad	Meristem, phloem necrosis, cork disease
Ni	Ni^{2+} Ni-Chelate	Leaves of grass	Bad	Fructification disorders in grasses
Cl	Cl-	Leaves	Good	Partly wilting, thickening of roots
Na	Na+	Leaves, xylem parenchyma	Good	Growth disturbances (for C4 plants)
Se	SeO_4^{2-}	?	Bad	?
Al	$H_xAlO_y^{(+,-)}$	Wood, bark (leaves)	Very low	? (Ferns)
Si	$H_xSiO_y^{(+,-)}$	Wood, bark, needle leaves	Very low, almost zero	Leaf curvatures in grasses, palms, conifers, *Equisetum*
Co	Co^{2+} Co-Chelate	Leaves?	Bad	?? (legume nodules)
V	VO_3^-	?	?	??
F	F–	Leaves	Pretty good	??

other element. (3) A deficiency of the respective element produces a very specific deficiency symptom, which can only be eliminated by adding this element and no other.

However, the range of variation between oversupply and malnutrition is very different for each element and also different for each plant species (Adriano 1986). Accordingly, the demands of plants on the nutrients in the soil are species-specific; some plants indicate the availability of the nutrient elements in the soil by their presence: e.g. nitrogen indicator plants, such as *Urtica*, *Rubus*, etc.

The availability of nutrients and trace elements varies greatly depending on the soil type (Marschner 1986). Certain quantities are absorbed by the plants, which in the case of agricultural crops are withdrawn by the harvest and thus usually far exceed the natural replenishment through weathering of the soil minerals and parent rock. This is the reason that fertilizers have to be applied in agriculture. In general, the soil factor (i.e. the edaphic basis of mineral supply to plants) is an important prerequisite for the flourishing of plants and thus for the normal development and formation of ecosystems and thus shapes the character of the same. The provision of essential nutrients to plants exerts a major influence on thriving, mainly due to the fact that, via water availability to plants at the site, nutrient supply can vary widely.

The necessary, i.e. essential minerals for plants (and animals) ultimately come from the parent rock, from which the individual minerals are released through weathering. The nutrients become available through an increase in surface

Fig. 2.14 Dust particles on water lily leaves transported and deposited by atmospheric circulation from the Sahara to Germany and contracted by raindrops (Photo: Breckle)

area of soil minerals and through chemical restructuring. In the course of the processes of soil formation (in interaction with the plants) the material cycles in the ecosystem are fed. Ongoing losses through discharge into the groundwater (Fig. 2.13) or dust drift etc. must be supplemented by replenishment, essentially through weathering. Only then can the ecosystem remain sustainable, i.e. sustain itself over a long period of time. The balance of substances in the soil-plant system is balanced in the long term by inputs and weathering and by outputs (Fig. 2.13).

Dust input is also known. Thus, a not insignificant part of the nutrients of the Amazon rainforest is also likely to originate from long-distance transport of fine dust (for example mainly from the Sahara).

Seen as a whole, a very considerable transport of fine material is constantly taking place on the globe. In special weather conditions, Sahara dust can also reach Central Europe (Fig. 2.14). The rock particles, minerals, etc. that are released and crushed during weathering sediment for a time or are transported further until they finally end up in the world ocean or, for example, build up large river deltas. The sediment load (Fig. 2.15) from the different areas of the earth depends on the one hand on the relief energy and the differences in altitude, and on the other hand on the structure of the material; for example, the easily eroded loess from China is transported in large quantities (via the Yellow River into the Yellow Sea). The dust storms from these loess particles often cause considerable damage to the health of the population in Beijing in connection with increasing fine dust pollution due to increasing traffic. However, the fine dust discharge can sometimes be detected thousands of kilometres away (e.g. in Hawaii).

The material crushed by weathering is transported either by wind or by flowing water. The flow or wind velocity and the grain size of the particles to be transported are of great importance. With increasing particle size, the sedimentation

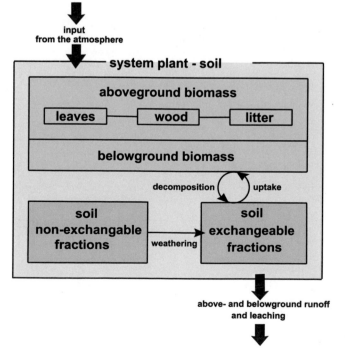

Fig. 2.13 The plant-soil system with the close interconnection of the compartments

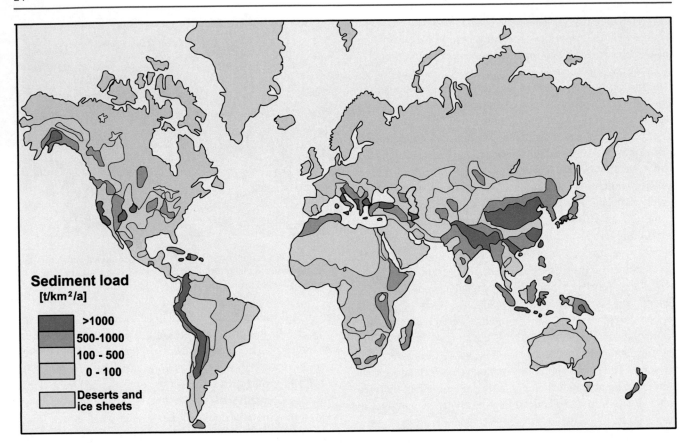

Fig. 2.15 Global overview of sediment loads from medium-sized drainage basins (modified from White et al. 1992)

process becomes more and more predominant and can only be overcome by very high flow velocities (Fig. 2.16).

The wind causes mainly grain sizes around 0.1 mm to move along the ground surface and sediment. These have a particularly low critical shear stress velocity (Fig. 2.17), so that saltation (jumping, hopping of grains) is facilitated, leading to the formation of the large sand dunes. In general,

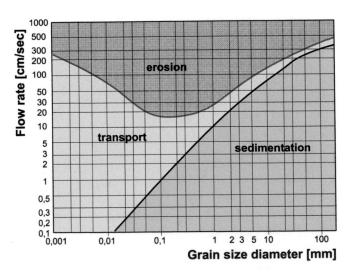

Fig. 2.16 Particle transport by flowing water and its dependence on grain size (in mm) and water flow velocity (in cm s^{-1}) (modified after Kuntze et al. 1994)

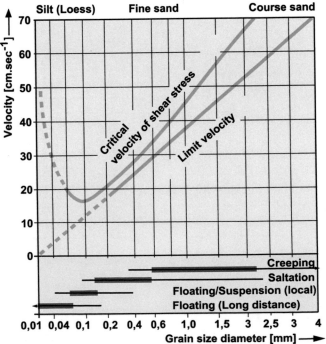

Fig. 2.17 Particle transport by wind and its dependence on grain size (in mm) and soil particle velocity (in cm s^{-1}). Grain sizes less than 0.01 mm (loess) remain suspended longer and can be transported far by long-distance transport. Fine sand with grain sizes between 0.1 and 0.5 mm is transported along the ground surface mainly by saltation (dunes, ripple marks) (Modified after White et al. 1992)

Table 2.3 Removal of a sandy loam soil layer by erosion. The time in years required to remove a 10 cm layer in the southeastern USA at a slope inclination of 10° is given

Vegetation cover, land use	Time (in years)
Natural, intact deciduous forest vegetation	320,000
Dense lawn	46,000
Arable farming with crop rotation	60
Cotton cultivation	25
Corn cultivation	20
Unvegetated bare ground	10

the processes of erosion and accumulation play a significant role in changing the site characteristics of ecosystems over the long term. Also, the greatly increased removal of soil material from cultivated areas leads to rapid changes and, in some circumstances, to degradation that does not allow further cultivations.

Table 2.3 shows the erosion rates for a humid area in the USA. According to this, it can be seen that a closed vegetation cover self-evidently is the best soil protection. If one calculates the amount of valuable soil washed away, this example results in 1500 t lost in 10 years for an area of only 1 ha. The soil losses in Germany are somewhat lower, since precipitation is usually less intense and these high values were determined for sloping areas.

The minerals pattern and the nutrients available in the system through long-term soil formation determine the productivity of ecosystems quite significantly. The parent rocks, however, hardly play a role in ecosystems of higher age with mature, deep soils, but on younger sites one can very well distinguish the different vegetation units (and thus ecosystems) on limestone, on crystalline, on gypsum, etc.

The availability of nutrients depends, on the one hand, on the inner surface of the soil minerals (and humus) and, on the other hand, of course, on the existing occupancy of the ion exchange sites located on the large inner surfaces. This ion exchange is partly in equilibrium with the pH value; in acidic soils, an increasing proportion of the ion-exchange sites is occupied by protons (Fig. 2.18). As a result, the proportion of exchangeable mineral nutrients (Ca, Mg, K) decreases.

In more acidic soils, there is the additional disadvantage that the trivalent Al^{3+} blocks further sites and thus, for example at pH 3, there are hardly any nutrient cations left in such a soil. In cool, humid climates, soil formation tends towards such acidic, nutrient-depleted soils (Taiga, ZB VIII).

In hot, humid climates, weathering of the parent rock and the formation and conversion of clay minerals proceeds much more rapidly (Fig. 2.19). Whereas in temperate climates clay usually occurs as three-layered clay minerals in the soil (and thus provide a relatively large cation exchange capacity), tropical soils are often characterized by the two-layered clay mineral kaolinite, which has only 5–10% in ion exchange capacity compared to three-layered clay minerals. This extreme cation poverty and depletion is one of the most important reasons for the "ecological disadvantage of the tropics"(Weischet 1980).

In dry areas, the accumulations on the soil surface or in the soil at certain soil depths lead to deposits which can be very solid and which can appear as lime or gypsum crusts or also as laterites (iron, aluminium oxides) etc. In humid areas, leaching processes gradually lead to a depletion and acidification (podsolation). Both processes decisively shape the formation of the individual ecosystems in the corresponding

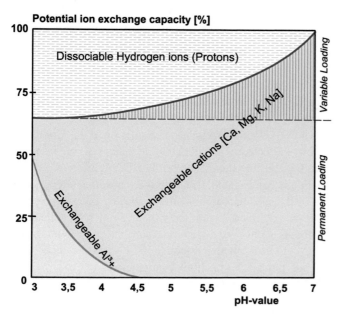

Fig. 2.18 The percentages of different cations in the potential exchange capacity (related to pH 7 = 100%) as a function of pH, in a soil with 20–30% clay, predominantly three-layer minerals, and 2–3% humus content (Modified after Scheffer & Schachtschabel 1992)

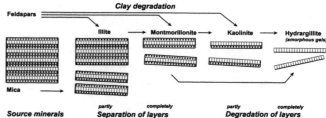

Fig. 2.19 The formation and decay of clay minerals. Illite and montmorillonite are three-layer clay minerals, kaolinite is a two-layer clay mineral (modified after Lerch 1991)

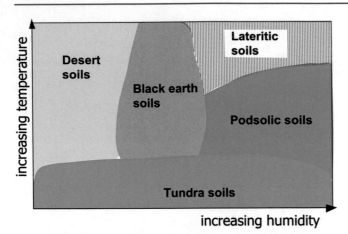

Fig. 2.20 The main groups of soils in the ecogram of moisture and temperature

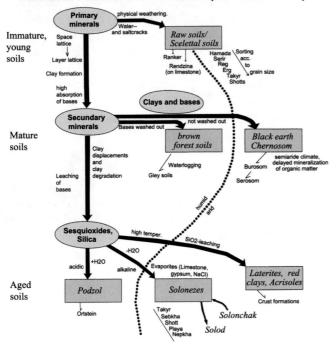

Fig. 2.21 Schematic of the genesis of soils on silicate rocks depending on various influencing factors

zonobiomes. The main types of soils are schematically assigned to certain climatic factors in the ecogram (Fig. 2.20).

However, soil genesis is a very long-lasting process. Some important processes are explained schematically in Fig. 2.21. The distinction between climatypical ecosystems and those that are more strongly characterized by pedological processes is only possible in a blurred way because certain pedological processes are themselves zonobiome-specific. Some pedobiomes are therefore actually zonobiome-specific biomes (for example in zonobiome II, where crust formation, laterites, etc. occur).

Soil formation takes place from the parent rock in interaction with the developing vegetation. The soil formation processes, influenced by the climate, lead to certain soil types or groups of soils. It can be seen that the historical aspect is also significant for soils.

Natural ecosystems very gradually create their specific soil type, which is then in harmony with the long-term climatic conditions and zonal vegetation at the site.

Salt: Halophytes and Salt Soils, Halobiomes

A very important group in many deserts are the salt plants or halophytes. They are bound to the occurrence of saline soils. Many halophytes are succulent, but they should not be grouped with the true succulents. Their succulence is the result of strong saline, or chloride, storage; for this reason their cell sap concentration is often very high and may exceed 5 MPa. In addition to the effect of salt (NaCl), the effects of other ions must always be considered on saline sites, for example, hydrogen carbonate (alkali soils, sodic soils), sulphate, borate. So it is not only the more specific case of exposure to salt (NaCl) in dryland soils that leads to vegetation differentiation.

The halophytes or salt plants colonize the saline soils on the sea coasts and in the deserts. The saline soils may have been conquered by plants evolutionarily relatively late. On these soils, land plants had to solve not only the water problem, but also the physiological effect of the salts.

It is appropriate to start from the plants themselves when defining halophytes: True halophytes are plants that accumulate larger amounts of salts in their organs and are not harmed by them, but are even promoted by them if the concentrations are not extremely high; the corresponding salts are mostly NaCl, sometimes also Na_2SO_4 or organic Na salts.

The concentration of cell sap in the vacuoles cannot be lower than that of the soil solution, which is usually very high in saline soils. If osmotically active substances were additionally formed in the cell sap, such as sugar, for example, the hydrature of the plasma would have to drop very sharply, which would be unfavourable. The solution of the problem is therefore effected in another way: So many salts are taken up from the soil into the cells that the concentration of the soil solution is equilibrated. Through these absorbed electrolytes (Na^+, Cl^-) no dehydration of the plasma takes place, but rather an additional hydration, which causes a succulence of the organs. Conversely, additional substances are synthesized in the cytoplasm, which establish the osmotic balance there, but are plasma compatible ('compatible solutes'). These substances can originate from quite different substance classes. They are often typical for certain plant families or genera, i.e. taxon-specific (Popp 1995).

In larger concentrations, salts are toxic. The halophytes must therefore be salt-resistant, but this is only possible to a

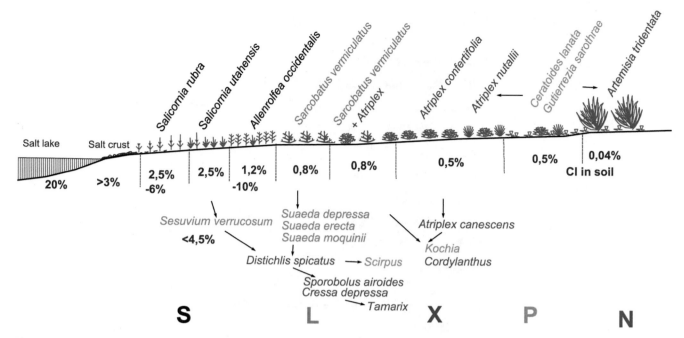

Fig. 2.22 Vegetation profile at the Great Salt Lake (Utah, USA) with indication of chloride contents in the soil (TG) in the individual vegetation belts (modified Kearney et al. 1914, Breckle 1976; abbrev. See Fig. 2.25)

certain extent, so that very heavily salinated soils remain vegetation-free (Fig. 2.22, for example on the shores of salt lakes).

Due to the different behaviour of plants towards high salt loads from the soil, different types of adaptation can be distinguished.

The **Non-halophytes (Halophobes)** (Fig. 2.22 **N**)—the majority of plants—die due to lack of osmotic adaptation when exposed to salts from water deficiency. Salts are toxic to salt-sensitive species. Therefore, these cannot grow on saline soils.

The **Facultative halophytes (Pseudo-halophytes)** (Fig. 2.22 **P**) are to a certain extent capable of osmotically adapting their uptake system in the root by salt uptake, but of fixing the salt in the root area and thus keeping the shoot relatively low in salt. Such salt-tolerant plants can withstand a salt concentration that is not too high, but develop better on non-saline soils.

For all halophytes applies, what is also pointed out for the mangroves, that the roots act like an ultrafilter, i.e. they take up practically only almost pure water from the salty soil solution and supply it to the leaves through the conductive pathways. In the vessels of the halophytes, high cohesive tensions were detected.

In **Euhalophytes**, too, the root system acts like an ultrafilter that allows only a few salts to pass through into the transporting system. However, these salts gradually accumulate in the shoot system and cause their halosucculence by formative differentiation processes: leaf succulence (Fig. 2.22 **L**) (for example *Suaeda*) or/and shoot succulence

(Fig. 2.22 **S**) (for example *Salicornia*). Euhalophytes are stimulated in growth by some salt enrichment. On ordinary soils containing only traces of NaCl, they snatch it, so that even then their salinity is relatively high. This stimulation comes from the chloride ion, which has a swelling effect on protein bodies. The consequence of this is a hypertrophy of the cells by a strong absorption of water, that is, a succulence of the organs. The higher the chloride content of the cell sap, the more pronounced is the succulence. Only the chloride ion has this effect, but not the sulphate ion, which has a deswelling effect on proteins. There are halophytes which, in addition to chlorides, also store larger quantities of sulphates in the cell sap; these halophytes are not or only weakly succulent. We must therefore distinguish between **Chloride halophytes** and **Sulphate halophytes.** They can grow side by side on one and the same soil. Salt uptake is usually species-specific (Breckle 1976). In studies of the halophyte problem, it is therefore not sufficient to examine the soils for their salt content; for the plant, only the salts with which the plasma comes into contact are of importance. One must always know the concentration and composition of the salts in the cell sap. How different the composition of the cell sap of halophytes and non-halophytes is, is shown in Fig. 2.23.

For the Euhalophytes, too, there is an upper limit to the salt concentration in the cell sap, which varies from species to species. If this becomes too high, the plants will deteriorate, which in the case of the Chenopodiaceae is usually indicated by a red colouration (N-containing dyes: Betalains Fig. 2.24), until they finally die. There is another group of halophytes in

Fig. 2.23 Content of inorganic ions in the cell sap of green organs of various Halophytes and Non-halophytes from Northern and Central Afghanistan (after Breckle, 1986). Species 1–18 are Chenopodiaceae. Species 1–6 are chloride halophytes, leaf or stem succulent; anion contents (Cl^- + SO_4^{2-}) exceed cation contents (Na^+ + K^+). Species 7–12 are Alkali halophytes with significantly lower inorganic anion contents, here larger amounts of organic anions are detectable in the cell sap, also clearly leaf or stem succulent. Species 13–18 are Pseudo-halophytes occurring on lower salt sites and are less succulent. Potassium predominates over sodium. Species 19–24 are non-Chenopodiaceae; 19 and 20 are Sulphate halophytes, withstand some salt. The others are Non-halophytes with low salt tolerance. 23 + 24 show in comparison the analysis values of maize leaves of a salt-loaded and an unloaded field; at 23 already with clearly yellowed leaves and salt damage

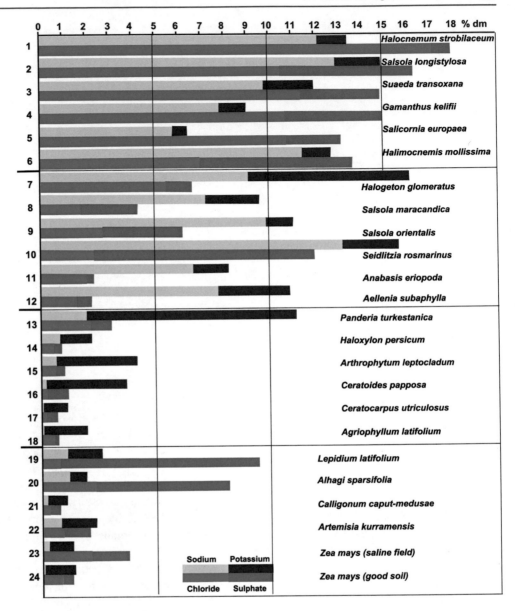

Fig. 2.24 Colour change due to betalain incorporation in *Salicornia europaea at the* beginning (**a**, photo: Breckle) with increasing drought and with complete salt accumulation beyond the concentration limit on the recently dried lake bottom of the Aral Sea (**b**, photo: Wucherer)

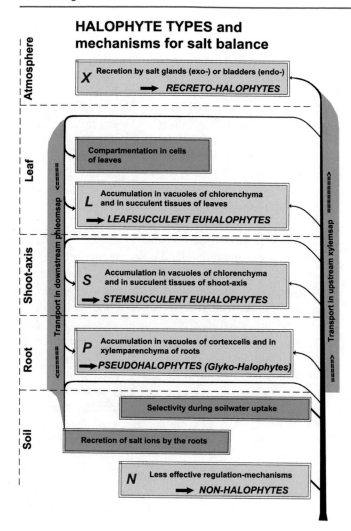

Fig. 2.25 Schematic classification of the different halophyte types based on the regulation of internal salinity (after Breckle 1976)

whose cell sap Na$^+$ is present in a significantly higher equivalent concentration than Cl$^-$ and SO$_4^{2-}$ combined. Thus, Na ions must be equilibrated by anions of organic acids. After these plants die, decomposition breaks down the organic acids to carbonates. The sodium enters the soil as Na$_2$CO$_3$ (soda ash), making it alkaline. We refer to these halophytes as **Alkali halophytes**.

Among the halophytes there are also species provided with salt glands, mostly not succulent. These **Recretohalophytes** (Fig. 2.25 **X**) are species that continuously excrete the salt they ingest, such as *Limonium, Reaumuria, Frankenia, Glaux, Spartina*, and other halophilous grasses. Salt glands are also present in an important tree, the Tamarisk (*Tamarix*), which is represented by many species in arid regions. When branches of this tree are shaken, salt dust falls from them. Since *Tamarix* excretes mainly NaCl,

sulphates predominate in the cell sap and the leaf organs are not succulent.

Salt excretion is also possible by storage in isolated bladder hairs (*Atriplex* etc.), which form a coating or can also be shed. Salt excretion is also possible by shedding, for example, old leaves rich in salt. The latter is also known in Facultative halophytes, such as *Juncus,* where the leaves turn yellow early, or in rosette plants (*Limonium* etc.), where new leaves are continuously formed. In addition to this more auto-ecological characterization of halophytes (Fig. 2.25), a distributional ecological characterization of the different halophyte types is also used: Obligate halophytes—Facultative halophytes—site-indifferent halophytes—Non-halophytes, which of course largely coincides with the auto-ecological typing.

Along a salt gradient in the terrain, for example around a salt lake, the halophytes usually occur in a certain zonation. On the very inside, Stem-succulent euhalophytes predominate, followed by Leaf-succulent ones on the outside, then there is often a zone with a particularly large number of Recretohalophytes, followed further out by the Pseudohalophytes (Fig. 2.25 P) and then on the outside (without salt pollution) the Non-halophytes. Such a halo-catena is best developed in areas where floristically there are many different halophyte species, as in Central Asia (Breckle 1986, 2002a).

For many halophytes of the arid regions, as already mentioned, the problem is not water, because they grow on wet salt soils of salt pans (Hygrohalophytes), but the salt balance. But there are also those that occur on dry saline soils and often suffer from water deficiency, notwithstanding strong salt storage (Xerohalophytes); these include *Atriplex, Haloxylon, Zygophyllum species*, and others in which one can often observe a precise reduction in transpiring surface area during drought, tailored to water availability; for example, *Zygophyllum dumosum* sheds the leaflets first, then the petioles (Fig. 2.12), others the young terminal shoots or even the green bark of leafless previous year's shoots.

In all arid regions there is a constant risk of **soil salinization** (Waisel 1972) Although the input of rainwater means only a small supply of salt (on average, rainwater contains 0.001% NaCl), a considerable amount accumulates in the long term if there is no corresponding discharge, as is more or less the case in all arid areas (by definition: potential evaporation exceeds precipitation). Arid areas are accordingly characterized geomorphologically (Fig. 2.46). They have endorheic basins; runoff generally does not reach the world ocean, but reaches only local basins that represent the erosion base. There, the salt of precipitation water and salt released by leaching of the surrounding rocks is enriched (salt pans or salt lakes, for example Dead Sea, Aral Sea, Great Salt Lake in Utah, Lake Chad, Dashte-Nawor, Hamune-Puzak in Afghanistan, etc.).

In all arid regions, irrigation (even with irrigation water that contains, for example, only 0.02% NaCl [= 200 ppm] and is thus of the best quality) leads to slow salinization (Breckle 2009, 2021), unless care is taken to ensure that the enriched salt is washed out of the fields again and again, just as the Nile in Egypt with its annual floods—before the Aswan Dam was built!—has provided for desalination in the Nile valley for thousands of years.

In arid areas, the vegetation mosaic is strongly influenced by soil salinity. The different biomes there are characterized by their salt load. It is not uncommon to find pronounced gradients of increasing salinity (and decreasing soil grain size) towards the basin landscapes. An example from the Great Salt Lake area is given in Fig. 2.22. Further examples of this are brought in the discussion of arid zonobiomes III and VII.

The long periods of drought in arid areas cause rivers to flow only periodically or even episodically. Since the potential evaporation is higher than the annual precipitation, in arid areas there are sinks without drainage in which all the water evaporates that reaches them through the tributaries. The salts dissolved in the water, as already stated, accumulate more and more in the course of time. A saturated solution may form and the salt crystallize out. Salt lakes or salt basins are characteristics of arid climates. Ultimately, the world ocean is also a terminal lake into which all soluble matter has been transported over billions of years. Most of the soluble salts consist of NaCl, because the hydrocarbonates precipitate early after loss of CO_2 as $CaCO_3$, the sulphates somewhat later as gypsum ($= CaSO_4$). The potassium salts crystallize, if at all, at the latest; thus a typical sequence of these evaporites is formed as a sedimentary sequence.

Sodium ions are released by weathering from silicates, whereas chloride ions are present in seawater in quantities of almost 20 g/litre (sulphate only 2.7 g), but chlorine-containing minerals are rare. Thus, little chloride ions can be released by weathering of minerals. Nevertheless, NaCl can always be detected in river water. It is also likely to have been enriched by HCl-containing exhalations from volcanoes over the long history of the Earth.

The NaCl of the saline soils of arid areas can be of various origins:

1. It is sea salt trapped in rocks that were deposited as marine sediments (evaporites). During the weathering of these rocks, the salt is dissolved by rainwater and transported into the drainless depressions. Deserts with marine sedimentary rocks (Jurassic, Cretaceous, Tertiary), for example the northern Sahara and the Egyptian Desert, are therefore highly brackish, whereas arid areas with igneous rocks or terrestrial sandstones have much less saline soils.

2. The arid areas that were lake or sea basins in the recent geological past that slowly dried up, for example the areas around the Great Salt Lake (Utah; Lake Bonneville as a glacial lake), around the Caspian and Aral lakes (Central Asia), around the Tuz Gölü (Central Anatolia), Dead Sea in the Middle East (Lake Lisan as a glacial lake), Lago Enriquillo (Hispaniola), Dasht-e Nawor (Afghanistan) and others, are also depleted.

3. When there is a strong surf on arid sea coasts, seawater is finely atomized, the salt water droplets dry out and the salt dust is blown as aerosol many kilometres inland. It is either deposited as such or added to the soil by rain or fog. This process also takes place in humid areas, but in these the deposited salt is constantly washed out and returned to the sea by rivers (cyclic salt). In arid areas without runoff, on the other hand, the salt accumulates. The brackish conditions of the Outer Namib and the arid parts of W-Australia can be traced back to this cause. If salt flats have developed in the depressions, the wind can blow salt dust from them further. But even far from the coasts, rain (with 10–20 ppm NaCl) steadily brings traces of salt with it.

4. Brackage can also occur when spring water loaded with salt comes to the surface, for example in the northern Caspian lowlands. In this case, it is salt from sea basins that dried up in earlier geological times (Permian, Muschelkalk), which forms deposits at greater depths. In arid areas this salt accumulates, in humid areas (salt springs for example in Bad Salzuflen, Salzdetfurth, Salzgitter, Salzburg, all in Middle Europe, with the term "Salz" = salt) it is in turn rapidly discharged to the sea.

In the deserts, after each rainfall, a shift of salt takes place from the higher parts of the relief to the lower ones, so that the depressions dry up. If the sedimentary rocks are very saline and precipitation is very low, as for example around Cairo-Heluan or in central Iran, the soil of the plateau sites may also contain salt. In the rainless central Sahara, no salt displacement takes place, thus salt accumulation in depressions is completely absent.

For the plants, it is not the salt content of the soil—calculated on the dry weight—that is important, but the salt concentration of the soil solution in the rooted soil region. In weakly saline soils, which are dry at the same time, the

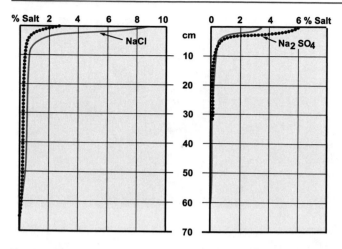

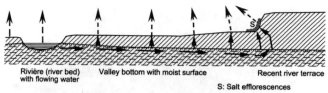

Fig. 2.28 Salt accumulation in the Swakop Valley (Namib Desert). The arrows indicate direction and strength of water flow in the soil; the dashed arrows indicate evaporation. Salt concentration increases towards the edge of the valley; salt blooms out at S at the base of the terrace where water flow ceases (modified after Walter 1990)

Fig. 2.26 Salinity at different soil depths in an irrigated bed (left) with groundwater rise and an unirrigated bed in the Swakop Valley (Namibia). NaCl = drawn out line, Na_2SO_4 = dashed line. The salts accumulate only at the surface (modified after Walter 1990)

concentration is often higher than in strongly brackish but wet soils.

Salt displacement is also brought about by evaporation from the soil surface if the groundwater is less than 1 m below the surface, so that it can rise capillary to the soil surface; a salt crust forms at the surface (Fig. 2.26), even if the groundwater contains only very small amounts of salt (Fig. 2.27). The salt always precipitates where the capillary water flow finds its end; these are the highest points of the micro-relief (Fig. 2.28).

Where the rule of *"no irrigation without drainage"* is not followed, crops collapse due to salinization in a few decades, as many "short-lived" development projects show and have shown. A particular example is the Helmand project near Kandahar, SW Afghanistan.

The presence of a salt crust in times of drought does not necessarily impede the growth of plants if they are rooted in the non-brackish groundwater. In the Pampa de Tamarugal in the Atacama Desert, *Prosopis* trees grow in holes in a half-meter-thick salt crust only because their roots reach groundwater streams with fresh water.

Any field irrigated in arid regions without some drainage constitutes a drainless basin and must in time dry up even if the water used for irrigation contains only very small quantities of salt. In this way vast cultivated areas in Mesopotamia and the Indus region have become salt deserts. This has not yet been the case with the undrained cotton fields of the Gezira in Sudan, because the water of the Blue Nile used for irrigation is particularly low in salt. Small amounts of salt are removed from the field each time the crop is harvested.

Salinization has become one of the most significant constraints on global crop production. More than 20% of the agricultural land in the world can no longer be used productively due to salinization. There have been tremendous efforts to breed more salt-resistant crops during the last decades (Flowers & Yeo 1995, Munns 2005, Breckle 2002b, 2009, 2021, Ibrahimova et al. 2021). Still the results are disappointing; The main reason for this is that a plant's salt tolerance is not determined by a few genes, but rather is a comprehensive physiological-biochemical response of the entire plant with far-reaching regulations and formative adaptations. As a rule, this means significantly reduced growth (Cheeseman 2015), but greater competitiveness on salty soils. The potential of alternative splicing mechanisms and targeting gene-editing technologies in understanding salt stress responses and developing salt-tolerant crops (Wani et al., 2020) is apparently overestimated, and often classic eco-physiological knowledge from long-term research on halophytes often seems to be completely forgotten. Nevertheless, halophyte research in the arid zonobiomes and the remediation of their saline soils are of great importance.

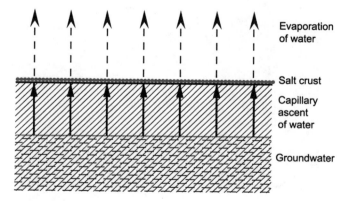

Fig. 2.27 Formation of a salt crust by capillary rise (arrows drawn out) of groundwater (dashed horizontally) and evaporation of water (dashed arrows); salt accumulation at the soil surface (modified after Walter 1990)

2.1.5 Mechanical Factors

Wind, Trampling

Wind and storm, frost formation, driving snow and avalanches, sand drifts and soil movements on slopes, all

Fig. 2.29 The mechanical action of wind on tree crowns in the coastal mountains near Carácas (Venezuela) (**a**, photo: Breckle) and in *Pinus pinaster* forests on the SW coast of the island of Sardinia (**b**, photo: Rafiqpoor)

these are mechanical influences on organisms. But one must also include footfall, trampling and browsing by livestock. A factor that also occurs naturally worldwide is fire, which ultimately destroys ecosystems mechanically and thus affects many organisms.

Many plants in the mountains are adapted to wind and snow breakage; they have very elastic branches and lean on the slope (Fig. 2.29).

In the case of sand fill in dune areas, there too are good adaptations of dune plants that grow upwards with the sand fill, but also withstand free blowing of the upper root zone well, as they develop very far-reaching roots. *Calligonum* is one such genus with numerous species in the Asian deserts (Fig. 2.30).

Regular trampling paths in lawns or meadows can be recognized by the fact that tread-resistant plants, often with rosettes lying against the ground, appear preferentially (Fig. 2.31).

Herbivory, i.e. the grazing of plants, is an essential process in ecosystems, which will be discussed in the context of material cycles. But every grazing means a mechanical damage of the plants. This can be seen very well in trees standing on pastures or trees at the edge of forests. The lower 1–2 m are kept practically free by grazing (Fig. 2.32), higher branches are not reached, apart from some climbing goats.

Fire

One must discuss the fire factor somewhat in more detail. There are a number of plant species in all drier climates that are adapted to fire. This is particularly evident in Australia, but also in all Mediterranean countries. In Australia, we speak of Pyrophytes (fire plants) when these species are virtually dependent on fire to continue to thrive. The fruits, for example, only release their seeds after a fire has passed

Fig. 2.30 In the desert areas of Central Asia, the roots of *Calligonum* are exposed from the dune sand by the wind. In this state, the water supply of the plant is provided by the main root and the fine root system not yet exposed by the sand (photo: Breckle)

Fig. 2.31 *Plantago major* resists mechanical pressure from trampling damage on trampling paths now worldwide (photo: U. Breckle)

Fig. 2.32 The lower parts of the trees at the forest edge have been stripped bare by cattle browsing up to a certain height level (**a**; photo: Breckle). Goats even climb up the trees and eat the leaves at all elevation levels (**b**: *Argania spinosa* in Morocco; photo: Breckle). Otherwise, the trees remain intact above a certain height that is not reached by other animals

Fig. 2.33 *Banksia* trees represent a good example of Pyrophytes. Their seeds in the cobs in the middle picture are only released (**c**) and become germinable by the action of fire (photos: **a** Breckle; **b** and **c**: Rafiqpoor)

over them (Fig. 2.33) and only then are they capable of germination.

This is due to the fact that a fire burns the hard woody components of the fruit or seed coat and thus mechanically loosens them. The next time it rains, the water can penetrate better and the swelling pressure then opens the fruit wall or seed coat. This can be observed in several species of the genus *Eucalyptus*, but also in many Proteaceae. In the Mediterranean region, the cork oak *(Quercus suber)* is particularly fire-resistant. Its thick bark protects the cambium (Fig. 2.34). After a fire, new branches sprout from it. But many other plants also sprout new shoots from underground storage organs, well fertilized by the ash.

Fig. 2.34 In the Mediterranean regions of Eurasia, cork is extracted from the bark of *Quercus suber* (**a**) (Here in Algarve, Portugal, photo: Rafiqpoor). The black soil and burn marks on the branches of the trees testify to fire exposure during the summer 2017 fire in the mountainous area of Monchique in the Algarve region, southern Portugal. It is the relatively thick cork layer that protects the cambium of the tree against fire exposure (**b**, photo: http://bit.do/bJDAo)

2.2   The Climate

2.2.1   General Questions

In everyday life we generally speak of rain, showers, hail, fog and thaw, high- and low-pressure areas or of the greenhouse effect and global warming. What is **weather** and what is **climate**? If you look out of the window in Kabul, or in Tehran, or in Cairo, or in Antofagasta, you may find that the sun shines from a bright blue sky for most of the year. During winter and transitional seasons, one might more or less rarely also observe overcast skies, snow or rain showers, thunderstorms with or without hail, or dust storms. Usually, these short-term observations are then talked about bad weather, but not climate. On the radio and television there are also daily only weather reports, no climate reports. These brief remarks indicate that in determining whether **weather** or **climate**, we are talking about a temporal dimension. So **weather change** (e.g. change from rain showers to sunshine) and **climate change** are two different dimensions in climatology. The first happens quickly, occasionally even several times a day (so-called "April weather" in Europe), the second

slowly and can only be detected over years or better over decades. Weather is therefore something that is happening. It can be interpreted, analysed and put into data. Data collection is always done at climate stations, which in some countries has a dense network, in others (like Afghanistan) a less dense one. The collected climate measurement series usually contain data on radiation, cloud cover, temperature, precipitation, relative humidity, air pressure, wind, etc., i.e. data on the ecological factors that are also important for organisms. From these long series of measurements, it is possible to calculate what the climate was like in recent years. And the further back the first measurements in a country go and the more data are available, the more precisely the climate there can be reconstructed.

So, statistically, climate is a term of something longer lasting. To understand this, one must take into account the spatial and temporal scales and meaningfully move from the small (or short-term) to the large (i.e. long-term).

Weather is the **short-term** in this system. Weather is the instantaneous physical state of the atmosphere produced by the meteorological elements and their interaction at a particular time in a particular place (or area). **Long period weather (German: Witterung)** is a **medium term** in this system. It describes the general, average or also predominant character of the weather events of a certain period of several days or weeks, rarely also months. A distinction is made between characteristic weather types or weather patterns, each of which is determined by prevailing weather conditions.

Climate is the **long term** in this system. It is defined as the summary of weather phenomena characterising the mean state of the atmosphere at a given place or in a more or less large area, with all its periodic variations in the course of the year. It is represented by the overall statistical properties (means, extremes, frequencies, durations, etc.) over a sufficiently long period of time. Generally, a period of 30 years is taken as the reference period. In general, however, it must be said that the longer the measurement period, the more reliable the statement about the climatic character of an area! For the compilation of ecological climate diagrams, the entire available measurement series of the meteorological stations are used if possible.

Climate is therefore the result of the interaction of several **climatic elements**, of which radiation is the most primary and fundamental.

2.2.2   The Radiation Budget and Astronomical Basics

The **driving force** for all climate-determining processes is solar radiation. The interaction of **solar** (position of the earth as a planet in the solar system), **meteorological** (physical and chemical processes in the atmosphere) and **geographical** (land/water distribution, relief, ocean currents, etc.) conditions basically creates the climate.

The radiation incident on the earth from the sun is, as a primary energy source, also the prerequisite for almost all **life processes** on earth, if one disregards for once the "black smokers" of the deep-sea trenches and their living creatures. The solar radiation arriving at the upper boundary of the atmosphere on a cm^2 perpendicularly facing the sun is 8.4 J (joules) per second; this corresponds to an average of 1367 $W/m^2/s$. This amount of radiation, known as the **solar constant,** varies by about $\pm 3.4\%$ over the course of the year, depending on the position of the Earth relative to the Sun. At the time of solar proximity (**perihelion:** currently 3 January: second-fifth depending on leap year) it increases to 1420 $W/m^2/s$ and at solar remoteness (**aphelion:** 5 July: third-sixth depending on leap year) it is 1325 $W/m^2/s$. The slightly elliptical orbit of the Earth around the Sun with a point a bit farer from the Sun and a point a bit closer to the Sun, but above all the inclination of the Earth's axis to the Earth's orbit around the Sun (Fig. 2.35) also explain the sequence of the seasons and their occurrence in the different latitudes and the differences between the northern and southern hemispheres. On average, only about half of the radiation at the upper atmospheric boundary reaches the Earth's surface.

In Fig. 2.36, the radiation balance for the vertically directed energy system **Earth-atmosphere-space** is shown

Fig. 2.35 Scheme for explaining the seasons based on the astronomical conditions of the Earth's orbit and the inclination of the Earth's axis relative to the orbital plane (ecliptic) (solstice line: summer or winter point; equinoctial line: equinoxes, spring or autumn point) (modified from Schönwiese 1994)

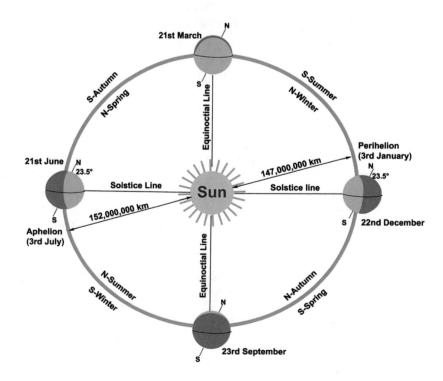

Fig. 2.36 Scheme of the radiation balance of the Earth-atmosphere-space system (modified after Lauer 1999)

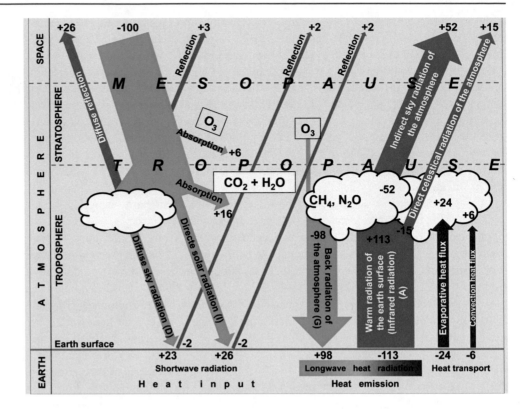

schematically with three balance areas: **Earth's surface**, the **troposphere** to the tropopause, and the **stratosphere** to the mesopause. On average for the entire system, the **Earth-space** radiation balance is balanced. The radiation balance (**S**) is the difference between the absorbed global radiation and the effective radiation and can be expressed in:

$$S = (I + D)\,(1\text{-}\alpha) - (A\text{-}G) \pm L \pm V$$

S = Radiation balance
I = Direct solar radiation
D = Diffuse sky radiation
G = Back radiation of the atmosphere
A = Warm radiation of the earth surface (infrared radiation)
V = Evaporative heat flux (latent heat) in the exchange between earth and the lower atmosphere
L = Convection heat flux (sensible heat) in the exchange between earth and the lower atmosphere
α = Planetary albedo: Unused energy loss as diffuse and direct reflection (26 + 3 + 2 + 2 = 33%: Fig. 2.36)

The individual energy flows in this system are shown as a percentage of the incoming solar radiation at the top of the atmosphere (100%). The sum of the individual components results in gains (input) and losses (emission) of the Earth-atmosphere-space system. Basically, the solar radiation arriving at the Earth minus the reflected portion is equal to the thermal radiation radiated from the Earth. If more heat is radiated from the Earth as a result of the anthropogenic greenhouse effect, there is excess heat in the atmosphere and consequently global warming.

The diagram (Fig. 2.37) shows the spatial distribution of extra-terrestrial solar radiation on Earth as a function of latitude. It is noticeable that especially the equatorial regions between 10°N and 10°S receive significantly less solar radiation during the course of the year with $30\text{--}35 \times 10^6\,\mathrm{J\,m^{-2}\,d^{-1}}$ compared to the marginal tropics, because here, due to the almost constant cloud cover and the resulting high reflection and absorption in the atmosphere, the areas of maximum radiation are shifted towards the marginal tropics around the 30th degree of latitude in the area of the subtropical dry regions. On the other hand, with a radiation amount of $>45 \times 10^6\,\mathrm{J\,m^{-2}\,d^{-1}}$, the radiation in the polar regions is surprisingly high at the time of the short summers (polar day = 24 h). This has consequences for the generative cycle (fructification and maturation processes) of tundra and taiga vegetation under long-day radiation climatic conditions.

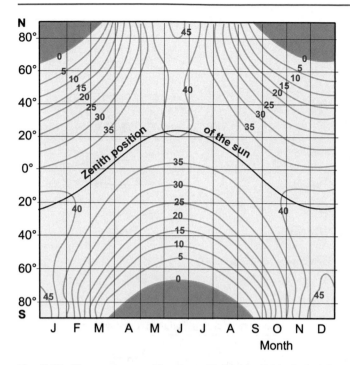

Fig. 2.37 The extra-terrestrial solar radiation on Earth during the course of the year as a function of latitude (numerical values in 10^6 J m^{-2} d^{-1}) (modified after Schönwiese 1994)

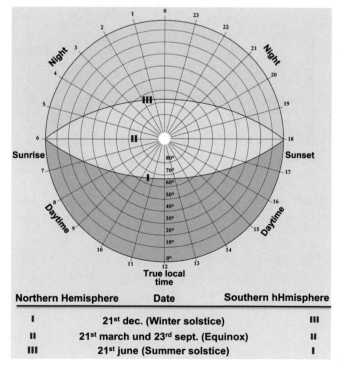

Fig. 2.38 Diagram of the theoretical daily sunshine duration (from Junghans 1969, modified after Lauer & Frankenberg 1986)

Despite the radiation-reducing effect of cloud cover, the sum of summer and winter insolation is highest at the equator because of the almost year-round high position of the sun.

The solar zenith in summer is located in the area of the respective tropics (approximately 23½°, solstices: summer and winter point: Fig. 2.38 **I & III**). Conversely, the length of day in summer reaches the maximum value of 24 h (polar day) from the Arctic Circle (ca. 66½°) to the poles, due to the inclination of the Earth's axis against the Earth's orbit. These and the astronomical facts explained below are indispensable for the understanding of the seasons as well as the ecological basics of geobotany and biogeography.

The index numbers (**I, II, III**) Fig. 2.38 refer to the reference dates of the equinoxes and the winter and summer solstices. Strictly speaking, only at the vernal and autumnal equinoxes, when the sun is above the equator, is the length of the day 12 h long (equinoxes: Fig. 2.38 **II**). Seen from Earth, the **day/night change** appears quite different depending on latitude because of astronomical conditions (Earth's orbit around the Sun with 365% days and tilt of Earth's axis with 24-h rotation). The variations in day length for the two hemispheres result from the difference between the day length at the summer solstice and the day length at the winter solstice. The times of sunrise and sunset can be read very easily from the equator to the poles for selected dates in Fig. 2.38. The number of astronomically possible hours of sunshine corresponds to the length of the day. The apparent path of the sun in the sky (Fig. 2.39) is always almost exactly 12 h long at the equator; at mid-latitudes, for example Frankfurt/Main, the variation in day length between summer and winter is already considerable at about 11 h, and at the pole it is night for half a year (polar night) and day for half a year (polar day).

The atmospheric processes and the respective latitude-dependent radiation angles ultimately determine what radiation remains at the Earth's surface (Fig. 2.40).

In general, the diurnal variation of insolation leads to constant changes in air temperature as a result of the different components of insolation and radiation in the radiation balance (see above). The relative energy turnover reaches its maximum energy gain around midday, the energy losses are particularly large immediately after sunset (Fig. 2.41).

The proportion of global radiation is greatest in the Earth's arid regions of marginal tropics, especially in the deserts (Sahara) (Fig. 2.42), while the radiation balance is lowest, since the balance values are extremely low due to the high effective radiation and very low water content in the air. Maximum values of the radiation balance occur in the tropics (red), especially over the oceans (dark red).

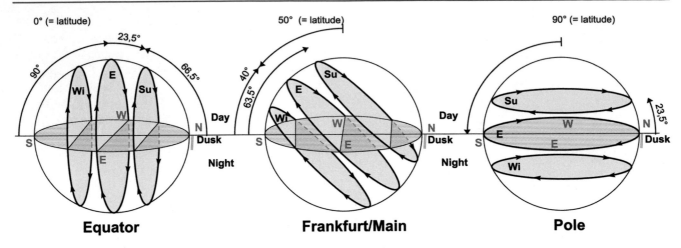

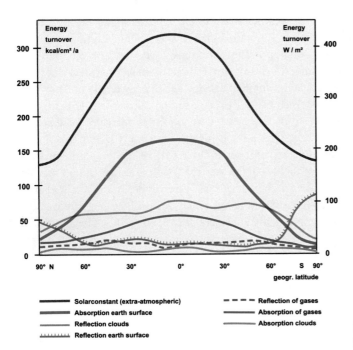

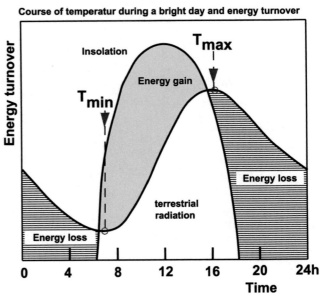

Fig. 2.39 Apparent path of the sun along the sky at different latitudes (modified after Schönwiese 1994)

Fig. 2.40 The turnover of solar irradiation energy in the atmosphere and at the Earth's surface and the solar constant as a function of latitude (modified after Schönwiese 1994)

Fig. 2.41 Scheme of the diurnal variation of solar radiation and terrestrial radiation. The sine curve also corresponds to the temperature curve on clear days (modified after Schönwiese 1994)

From Fig. 2.43 it can be seen that equatorial rainforests use >1.6% of global radiation for photosynthetic carbon acquisition, while these values are close to zero in the Sahara, where the maximum global radiation occurs.

The values of radiation use by the plant cover coincide with a high photosynthetic output of vegetation formations in the low latitudes (Lauer & Rafiqpoor 2002). Accordingly, phytomass (Fig. 2.16) and net production (Fig. 2.15) are particularly high in the tropics.

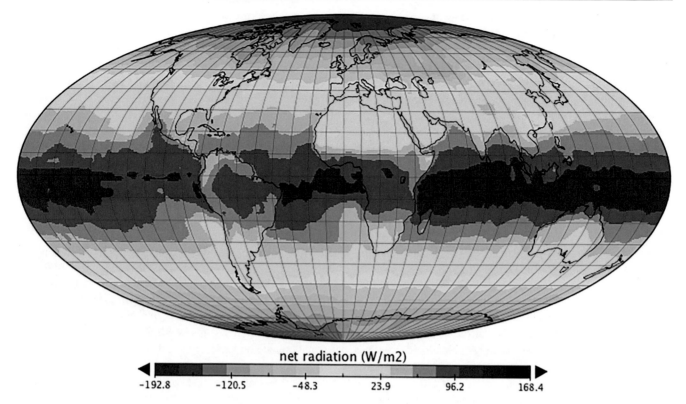

net radiation (W/m2)

−192.8   −120.5   −48.3   23.9   96.2   168.4

Fig. 2.42 Net radiation at the Earth's surface as the difference between incoming radiation and radiation reflected from the Earth for the month of March (zenith position of the Sun) for the period 1985–1989 prepared by David BICE, Professors of Geosciences, College of Earth and Mineral Science, The Pennsylvania State University as part of the NASA ERBE experiment (source: http://bit.do/bUMme)

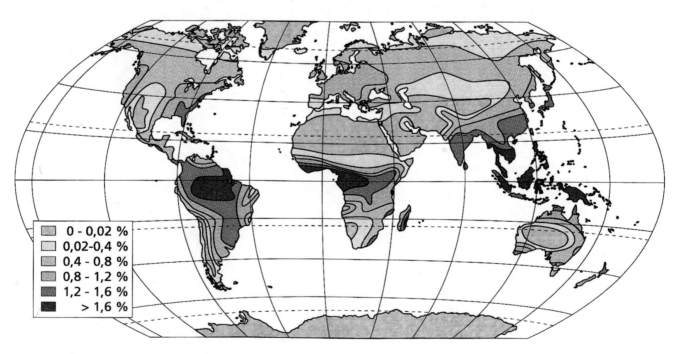

0 - 0,02 %
0,02-0,4 %
0,4 - 0,8 %
0,8 - 1,2 %
1,2 - 1,6 %
> 1,6 %

Fig. 2.43 Spatial distribution of percent radiation use by plant cover (from Larcher 2001)

2.2.3 The Heat Balance

The short-wave radiation emitted by the sun is converted into long-wave thermal radiation (sensible heat) at the earth's surface. Temperature is therefore an important climatological parameter as a measure of the thermal state of the air, which is essentially determined by the heat emission of the earth's surface. It only reflects the instantaneous heat state. The air temperature measured with a thermometer at a shaded and well ventilated weather hut is the so-called **sensible (sensitive)** heat (L) as an expression of the molecular kinetic energy. It provides the heating and is transported further by turbulent air movement (convection). The **latent** heat (V), or also the evaporation-heat, characterizes the potential-energy, that is consumed during the evaporation-process, i.e. is taken from the sensible heat, is released again through condensation-processes and is transferred into the sensible heat again. These two thermal components, converted by vertical mass transfer, are two important links in the Earth's heat balance. They consume most of the energy from the radiation balance. Since 72% of the Earth's surface is covered by sea, the transport of latent heat of evaporation plays a major role in the global water balance (see below) and participates with 80–85% in the Earth's heat budget (Lauer 1999).

For the thermal differentiation of the zonobiomes of the earth, the ecophysiologically important length of the thermal vegetation period as an expression of the heat balance of an area is of considerable importance. A calendrical month is considered to be thermally favourable if the natural and "cultivated" flora dominating in an area—from the point of view of the heat balance—achieves a clear material gain or fructifies during this month. This happens in the different zonobiomes plant stand typical at different heat levels. Walter (1960) defines the duration of the thermal growing season in the extratropics as the frost-free period of the year on a monthly basis, which is limited by reaching or falling

below certain temperature thresholds. The growing season, when woody plants begin to produce matter, begins after they green up, and for many species, when the daily mean reaches the 10 °C mark. When the temperature falls below a specific level in autumn, foliage discolouration sets in and with it the completion of assimilation activity. The optimal temperature range between the beginning and end of the growing season is usually no wider than 10–25 °C for most plants (Larcher 1980). Table 2.4 gives the threshold values for different formations of natural vegetation in the outer tropics and Table 2.5 for cultivated plants.

From such data series, Lauer & Rafiqpoor (2002) determined the length of thermal growing season at global level. The values for the length of the thermal vegetation period vary between 0 and 12 months. One can divide these values into classes according to landscape ecology: 0 = *hecistotherm*, 1–2 = *oligotherm*, 3–4 = *microtherm*, 5–6 = *mesotherm*, 7–9 = *macrotherm*, 10–12 = *megatherm*. The spatial representation of the lines of equal length of the thermal vegetation period (Isothermomena) gives a differentiated picture of the thermal climate on Earth (Fig. 2.44).

Table 2.4 Temperature thresholds for the growth and photosynthetic production of different formations of the natural vegetation and cultivated plants of the subtropics and mid-latitudes (after Lauer & Rafiqpoor 2002)

Vegetation formation	Thermal threshold (°C)
Boreal coniferous forest	5
Coniferous moist forest	5
Temperate deciduous forest	7
Mixed deciduous forest	10
Steppes (subtropics)	10
Patagonian Steppe	7
Semi-deserts (subtropics)	10
Deserts (subtropics)	10
High mountain formations	11
Subtropical moist forest	12
Montane coniferous forest	10
Deciduous broad leafed forest	10
Sclerophyllous woods	12
Steppes (mid-latitudes)	11
Pampa	11
Semi-deserts (mid-latitudes)	11
Deserts (mid-latitudes)	11
Tundra	5
Subpolar frost-debris zone	3

Table 2.5 Threshold values of the minimum and optimum temperature requirements of important crops in the subtropics (modified from Lauer & Rafiqpoor 2002)

Crops	Minimum temperature range (°C)	Optimum temperature range (°C)
Coconut Palm	24	26–27
Yams	20	25–30
Sugar cane	18–20 °C, at 15 °C growth stops	25–28
Cassava	20	>27
Cocoa	>20	>27
Coffee	18	>22
Pineapple	>18	>20
Tea	18	28
Millet	12–15	32–37
Sweet Potato	10	26–30
Tobacco	15–20	25–30
Rice	12–18	30–32
Olives	12–15	18–22
Sesame	12–15	25–27
Pumpkin	10–15	37–40
Cotton	18	30
Peanut	15	30
Corn	12–15	30–35
Potato	8–10	16–24
Winter wheat	4–6	15–30
Spring wheat	6–8	20–30
Rye	4–6	15–25
Barley	4–6	15–25
Oats	4–6	20–30
Sugar and beta beets	4–5	20–30
Meadow grasses	3–4	c. 25

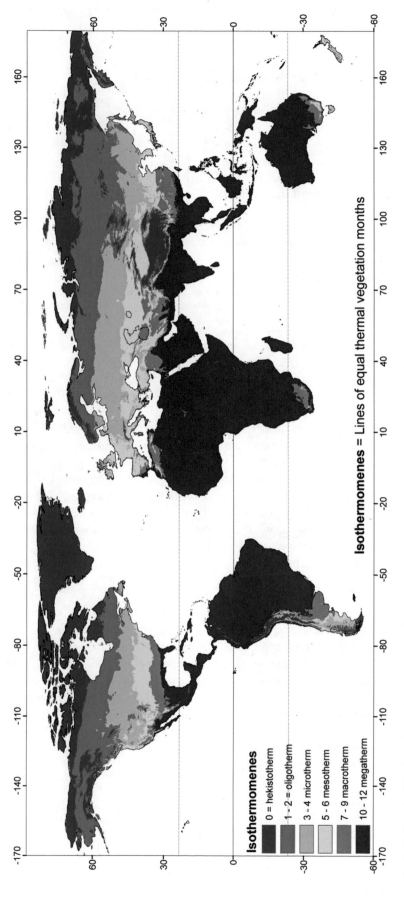

Isothermomes

0 = hekistotherm
1 - 2 = oligotherm
3 - 4 microtherm
5 - 6 mesotherm
7 - 9 macrotherm
10 - 12 megatherm

Isothermomenes = Lines of equal thermal vegetation months

Fig. 2.44 Length of the thermal growing season on Earth

2.2.4   The Water Balance

Water is the sine qua non of all life on earth. Only about 0.001% of the total water on earth is in the atmosphere. Nevertheless, water is extremely important for the climatic processes in the atmosphere and can exist in all three aggregate states (solid, liquid, gaseous). It enters the atmosphere by way of evaporation (Evaporation and Transpiration = Evapotranspiration, ET). Evaporation occurs both over the oceans and through evapotranspiration over the land. The motor for this process is latent heat (see above), which initiates the evaporation process by changing the state of aggregation (from liquid to gaseous state).

Due to the ascent of the air into the atmosphere via the turbulent exchange processes, the gaseous water condenses in clouds (water droplets or ice crystals) by way of cooling. With higher condensation and millionfold enlargement of the water droplets or ice crystals around a condensation nucleus, precipitation occurs (rain, snow or hail ... etc.).

The vapour pressure of water depends on the temperature. The vapour pressure saturation is not linear, but increases steeply with increasing temperature, as shown by the curve of the saturation vapour pressure as a function of temperature (Fig. 2.45). For dry air, for example, with a vapor pressure like point **X** at temperature **Ta** and a vapor pressure of **ea**, the relative humidity can be expressed as a fraction of **Y**. If one reduces the temperature, the relative humidity increases—for the same absolute humidity—e.g. up to 100% at point **Z** at a temperature **Td**; this **Td** is called the dew point. At temperatures below 0 °C, the saturation vapor pressure is very low (ordinate enlarged, Fig. 2.45). It is somewhat lower over ice than over water. If the air cools from **Tb** to **Ti**, the air is still somewhat unsaturated with respect to water droplets in supercooled clouds, for example. This means that at the same temperature these will release water to ice crystals, a significant process for precipitation from clouds (White et al. 1992). Overall, this also means that precipitation can be much more abundant at high temperatures, as in the tropics, than in cold regions. At water vapour saturation, 30 °C warm air contains about 45 mbar of water vapour, i.e. 9 times as much water compared to 0 °C warm air with 5 mbar, or even −20 °C air with only 1 mbar.

Precipitation makes life on Earth possible and is regionally very unevenly distributed (humid and arid areas). In an **arid** region, evaporation predominates in the hydrological water balance. Thus, in contrast to a humid area, no closed, permanent river system will form. Basin landscapes there

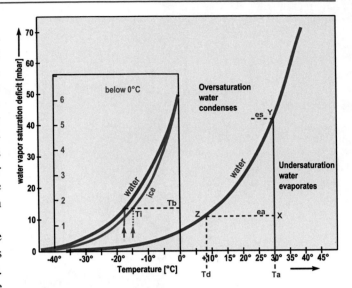

Fig. 2.45 Changes in the water vapour saturation of the air as a function of temperature (modified after Lambrecht, Göttingen)

contain only small terminal lakes (see p. 25) that become saline (Dead Sea, Great Salt Lake in Utah, Aral Sea, Lop-Nor, Hamune-Puzak, Dashte-Nawor) (Fig. 2.46).

In **humid** regions, such basin landscapes (for example Lake Constance, Germany) have an overflow and are filled "to the brim" with water. The internal water cycle of larger regions is thus very different, depending on whether they are humid or arid.

Baumgartner & Reichel (1975) calculated the water balance for the Earth and presented it in maps. A new model of the water balance has been produced by the Max Planck Institute for Meteorology in Hamburg and made available to the public (Fig. 2.47).

According to this, annual precipitation over the world's oceans averages 386×10^3 km^3 and evaporation 469×10^3 km^3, which is about 97×10^3 km^3 more than precipitation. Of the total 469×10^3 km^3 of water evaporated over the oceans, about 40×10^3 km^3 is transported to land. This amount is returned to the oceans by surface and subsurface water flows (40×10^3 km^3). The major portion of precipitation over the mainland comes from the small water cycle (81×10^3 km^3) through evapotranspiration (ET). The sum of ET over the mainland (81×10^3 km^3) and water vapour transport from the oceans to the mainlands (40×10^3 km^3) gives the total annual precipitation (121×10^3 km^3) over the mainland. The measured data of precipitation and

Fig. 2.46 Dashte-Nawor is a dry saline basin in a high plateau in central Afghanistan with clearly formed belts of halophyte vegetation (photo: Breckle)

Fig. 2.47 Scheme of the hydrological cycle on Earth with the measured and modelled proportions of precipitation, runoff and evaporation of various systems on Earth (Source: Max Planck Institute for Meteorology, Hamburg: https://t1p.de/0dqz)

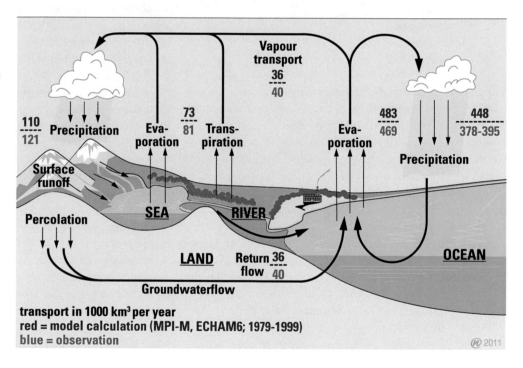

evaporation are in good congruence with the data resulting from the model calculations (Fig. 2.47).

In general, the water conditions of a biogeocenosis, a specific area, an entire landscape section or an entire country can be quantified with the water balance equation. According to this equation, the input variable into the system is precipitation N.

Water balance equation of an ecosystem:

$$N = \pm \Delta W + (E + I + T) + (A + S)$$

N = Precipitation
I = Interception
A = Surface runoff

Fig. 2.48 In Tunisia, farmers increase the distances between the individual trees in olive plantations from north to the south of the country so that a sufficient crop yield is achieved despite the drier climate and lower amount of available water (Photo: Breckle)

E = Evaporation
T = Transpiration
S = Infiltration
ΔW = stored water supply in the system

The discharge can take place in different ways, on the one hand by evaporation **E** (from the soil) and by transpiration **T** (by the plants), in addition there is interception **I** (superficial moistening of the leaves and subsequent evaporation), furthermore discharge is possible by surface runoff **A** and by infiltration **S** into the soil (underground runoff to groundwater). The soil itself, or the entire ecosystem, has a certain water reserve as a storage quantity **ΔW**, which can increase (+) or decrease (−).

Often **E** and **T** together (with **I**) are referred to as evapotranspiration (**ET**). The surplus water that is not released back to the atmosphere by **ET** benefits groundwater and thus the feeding of neighbouring springs, and ultimately the formation of a stream and river system (intact water cycle).

At sites affected by groundwater, allochthonous water can be added to the system, so that in addition to precipitation, upwelling water is also added and the loss variable of leachate is reversed.

In arid areas, most of the water will be lost to **ET**, and there will be no groundwater recharge (arid areas, with interrupted water cycle).

Of particular importance for survival in arid regions is the development of a sufficiently large root system. If the rootable soil volume is large enough, perennial plants may be able to survive several dry years if water is accessible in deeper layers. Some plants with particularly deep roots even seem to be able to lift this water ("hydraulic lift") in such a way that even other plants can benefit from it, as Caldwell et al. (1991) were able to show.

The drier the area, the greater the rooted soil volume of olive trees, for example. In Tunisia, from farmers' experience, the distances between planted trees are much greater in the south than in the north (Fig. 2.48).

For the large-scale landscape differentiation of the Earth into zonobiomes, the determination of the humid and arid months as an expression of the **hygric growing season** over a sufficiently long time series is an effective tool. The hygric growing season is defined by the length of the moisture-determined growing season expressed by the number of humid months (isohygromenes) (Lauer & Rafiqpoor 2002). A month is considered humid if, in it, the precipitation amount at least reaches (N = ET) or exceeds (N ≥ ET) the potential evapotranspiration (ET) of the site flora. The methodological principles of calculating potential evapotranspiration are discussed in detail in Lauer & Rafiqpoor (2002). The water balance in temporally resolved form gives a 12-scale of the number of humid months from perarid (12 arid months) to perhumid (12 humid months). These can be divided into 6 humidity levels of ecological relevance: 0 = perarid, 1–2 = arid, 3–4 = semiarid, 5–6 = semihumid, 7–9 = humid, 10–12 = perhumid. The spatial representation of the lines of equal length of the hygric vegetation period (isohygromenes) gives a differentiated picture of the hygric climate on Earth (Fig. 2.49).

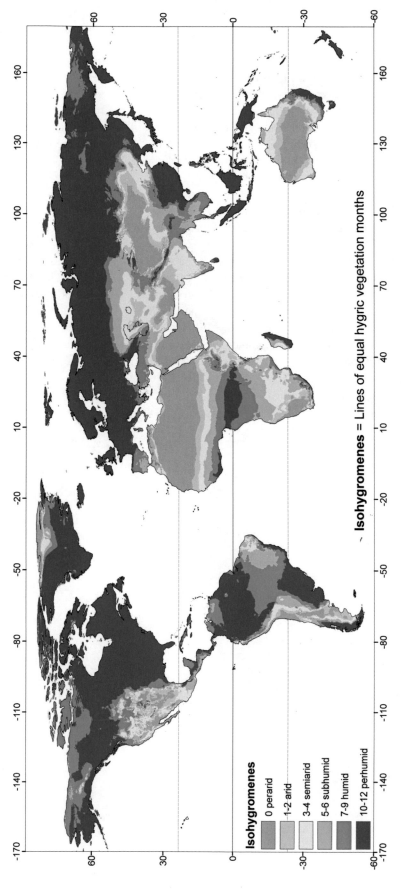

Isohygromenes = Lines of equal hygric vegetation months

Isohygromenes

0 perarid
1–2 arid
3–4 semiarid
5–6 subhumid
7–9 humid
10–12 perhumid

Fig. 2.49 Length of the hygric growing season on Earth (isohygromenes)

2.2.5   The Earth's Eco-Climates (Climate Classification)

Lauer & Rafiqpoor (2002) developed a system of ecoclimate classification in which they used the length of the thermal and hygric growing season on a monthly basis for climate typing of the Earth. They analyzed data from over 2000 climate stations and designed maps for isothermomenes (length of thermal growing season on a monthly basis) (Fig. 2.44) and isohygromenes (length of hygric growing season on a monthly basis) (Fig. 2.49). From the intersection of these two maps, they developed a map of Earth's eco-climate types whose boundaries can be clearly determined empirically (Fig. 2.50). The solar **radiation belts** were used to divide the major climate zones. This is done using the threshold values of the annual daylength variation, which is determined by latitude. Based on this clear mathematical division of the earth, five main climate zones result [tropics (A), subtropics (B), cool mid-latitudes (C), cold mid-latitudes (boreal zone D), polar regions (E)]. They provide the basic framework for the division of the Earth into zonobiomes, which are the basic concept of this book.

From the comparison of the ecoclimate types (Fig. 2.50) and the zonobiomes on Earth (Figs. 3.22–3.27), there is clear congruence in the zonation of the two systems:

ZB I = Megatherm, perhumid climates of the inner tropics

ZB II = Megatherm, humid climate of the outer tropics

ZB III = Megatherm-macrotherm, perarid desert climates of the marginal tropics and subtropics

ZB IV = Macrotherm, humid-semihumid climates of the Mediterranean winter rainareas

ZB V = Macrotherm, perhumid-humid subtropical climates (mostly) of the eastern sides of the continents

ZB VI = Mesotherm, perhumid-semihumid climates of the cool mid-latitudes

ZB VII = Mesotherm, semi-arid and arid climates of the cool mid-latitudes

ZB VIII = Microtherm, perhumid-semihumid climates of the cold mid-latitudes

ZB IX = Oligotherm, perhumid-semihumid climates of the subpolar and polar regions

These zonobiomes can further be differentiated into zonoecotones and biomes according to the degree of continentality and small-scale regional differences, and into different orobiomes according to altitude differences. All these units can be found on the map of the Earth's ecoclimates (Fig. 2.50).

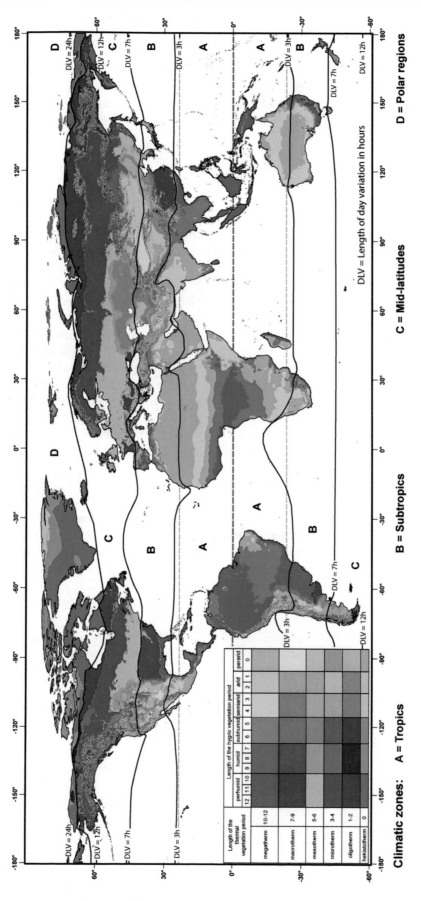

Fig. 2.50 The Earth's eco-climates

Climatic zones: A = Tropics   B = Subtropics   C = Mid-latitudes   D = Polar regions

DLV = Length of day variation in hours

Length of the thermal vegetation period	Length of the hygric vegetation period												
	perhumid		humid			subhumid		semiarid		arid		perarid
	12	11	10	9	8	7	6	5	4	3	2	1	0
megatherm 10-12													
macrotherm 7-9													
mesotherm 5-6													
microtherm 3-4													
oligotherm 1-2													
hekistotherm 0													

2.2.6   Climate Representation: Thermo-Isopleth Diagrams, Ecological Climate Diagrams

Temperature can be recorded in the form of maxima and minima, mean and duration values, and as diurnal and annual variations, and can be represented in diagrams and maps, from which the typical climatic characteristics of the individual zonobiomes can be identified. However, the thermal climate of an area is most easily read off from thermo-isopleth diagrams based on daily and annual fluctuations in air temperature (Troll 1943).

Thermo-isopleth diagrams (Fig. 2.51) provide a quasi-three-dimensional picture of the **thermal** conditions at a climate station. In a thermo-isopleth diagram, the months of the year are plotted on the abscissa and the hours of the day are plotted on the ordinate and connected with lines of equal temperature (to form an isopleth diagram). The graph is extremely informative. Not only the individual values of the diurnal and the annual cycle, but also the fluctuation magnitude (range) and the relationship of the fluctuation values to each other can be made visible through the curve. When comparing individual diagrams, the affiliation to corresponding climate zones is visible at a glance. They are thus suitable instruments for characterising the thermal **homoclimates** (i.e. climates similar in type) on Earth.

In the tropical and bordering tropical, maritime regions, the isolines show a stretching in the direction of the abscissa (Fig. 2.51a). This means that here the seasonal variations remain only slight, since only slightly different temperature values occur at a given hour on each day of the year (**diurnal climate**). The diagrams of extratropical stations, on the other hand, show a stretching of the isolines in the direction of the ordinate (Fig. 2.51c–e), which indicates greater seasonal fluctuations (**seasonal climate**). The dense stratification of the lines, especially in the polar regions (Fig. 2.51e), reflects the rapid change in temperature during the day or in a season, revealing the degree of continentality of an area. Images of stations in the subtropical transition regions (Fig. 2.51b) make it clear that the course of the curve in summer and during the day is more similar to the tropical type, whereas in winter and at night the basic shape of the curve is more indicative of extratropical conditions.

The **hygrothermal** behaviour of the regions can be illustrated most simply with the help of **ecological climate diagrams**. It is a pictorial representation of the overall climate in the area near the ground. However, such a representation must be clearly arranged, i.e. it must contain only the most important data for ecosystems. These are the temperature and precipitation conditions over the course of a year. Almost 9000 climate diagrams from meteorological stations all over the world had been already included in the Climate Diagram World Atlas by Walter & Lieth (1960–1967).

The explanation of some typical diagrams is given in Fig. 2.52. The climate diagrams given there are examples of the nine zonobiomes, each from stations at lower elevations, and with an updated lay-out.

In addition, climate diagrams of orobiomes (OB) are shown in Fig. 2.53. OB I with diurnal climate (Páramo) and the other orobiomes II-IX. Orobiome IX (Vostok, in Antarctica), with a mean annual temperature of −56 °C, is probably one of the coldest stations on Earth.

From the climate diagrams not only the temperature and precipitation values can be seen, but also the duration and intensity of a relatively humid and relatively arid season, as well as the duration and intensity of a cold winter and the possibility of the occurrence of late or early frosts.

The schematic representation of the climate diagrams provides the basis for the assessment of the climate in ecological terms. The indication of the aridity or humidity of the seasons is obtained on the climate diagram by applying the ordinate scale 10 °C $\cong$ 20 mm precipitation. The temperature curve thereby approximately replaces the curve of potential evaporation (whose values are mostly unknown) and can thus be related to the representation of the water balance in comparison with the precipitation curve. The vertical extent of the dotted area, that is, the drought period, is a measure of its intensity, and the horizontal extent is a measure of its duration. The same is true of the humidity area. Gaussen has found the ratio 10 °C $\cong$ 20 mm rain for the Mediterranean region to be particularly approximate with actual weather conditions. For steppe and prairie diagrams, however, it is expedient to use in addition the 10 °C $\cong$ 30 mm scale (Odessa, Fig. 2.52), in order to bring to the representation a dry season which is less extreme than the arid drought season.

The climate diagrams shown in Fig. 2.52 belong to the following zonobiomes:

ZB I (Humid, equatorial diurnal climate): Yangambi on the middle Congo; Bogor on Java

ZB II (Tropical summer rain climate): Harare in Zimbabwe

ZB III (Subtropical desert climate): Cairo on the Lower Nile

ZB IV (Mediterranean winter rain climate): Tunis in Mediterranean North Africa

ZB V (Warm temperate climate): Cheju in the south of South Korea

ZB VI (Temperate, nemoral climate with short cold season): Essen in Germany

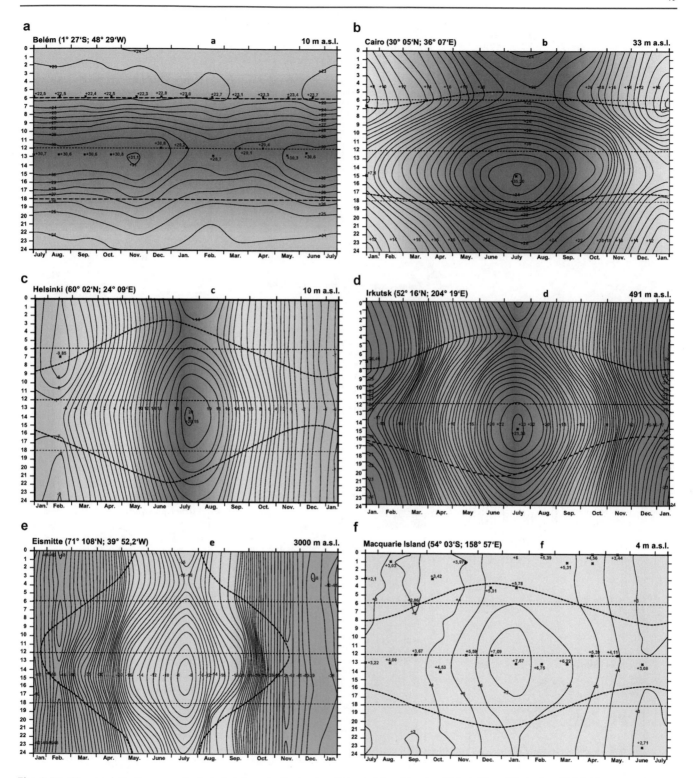

Fig. 2.51 Thermopleth diagrams of selected stations in the tropics (Belem) (**a**), subtropics (Cairo) (**b**), midlatitudes (Helsinki) (**c**), continental region (Irkutsk) (**d**), polar region (Eismitte) (**e**), and South Pacific (Macquarie Island) (**f**) (modified from Troll 1943)

ZB VII (Temperate semi-arid steppe climate with long dry season and low drought): Odessa at the Black Sea

ZB VIIa (Temperate arid semi-desert climate with pronounced drought): Astrakhan on the lower Volga River

ZB VII (rIII) (Extreme arid desert climate with cold winters): Nukuss in Central Asia

ZB VIII (Cold, temperate climate with very long winters): Arkhangelsk in the boreal taiga zone

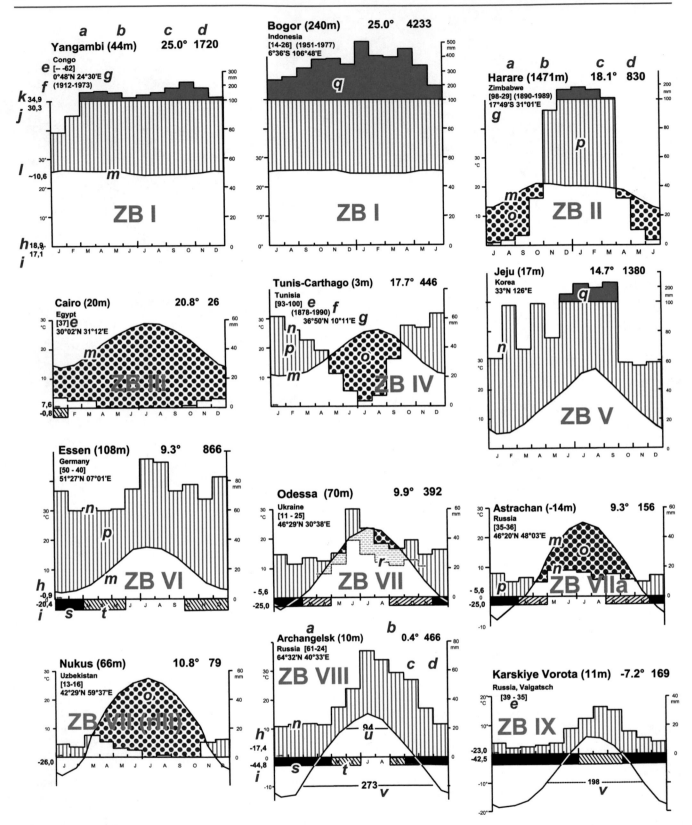

Fig. 2.52 Explanation of the climate diagrams with typical examples, at the same time examples for the different zonobiomes (see below). **Abscissa (horizontal axis):** In the northern hemisphere months from January to December, in the southern hemisphere from July to June (warm season is therefore always in the middle of the diagram. **Ordinate (vertical axes):** Temperature in °C, precipitation in mm. 1 graduation line = 10 °C, or 20 mm precipitation (numbers are often omitted)

The designations and numerical values on the diagrams mean:
a = station (region);
b = height above sea level;
c = mean annual temperature;
d = mean annual rainfall;
e = number of observation years (possibly first number for temperature and second number for precipitation);
f = observation period;
g = geographical coordinates;
h = mean daily minimum of the coldest month;
i = absolute minimum (lowest measured temperature);
j = mean daily maximum of the warmest month;
k = absolute maximum (highest measured temperature);
l = mean daily temperature variation (for tropical stations).
m = curve of mean monthly temperatures;
n = curve of mean monthly precipitation (1 scale division = 20 mm, i.e. in the ratio 10 °C = 20 mm);
o = relative aridity (drought) for the climatic region in question (dotted in red);
p = corresponding to relatively humid season (vertically shaded blue);
q = mean monthly precipitation exceeding 100 mm (scale reduced to 1/10, blue area = perhumid season;
r = precipitation curve, lowered, in the ratio 10 °C = 30 mm, above horizontal yellow dashed area = relative dry season (only for steppe stations);
s = months with mean daily minimum below 0 °C (black) = cold season;
t = months with absolute minimum below 0 °C (diagonally shaded), i.e. late or early frosts may occur;
u = number of days with mean temperatures above +10 °C (duration of the growing season);
v = number of days with mean temperatures above −10 °C.
Not all data are available or given for all stations. If they are missing, the corresponding places in the diagram remain empty.

ZB IX (Arctic tundra climate with July mean below +10 °C): Karskije Vorota (Vaigatsch Island).

The arid season (drought season) shown in the climate diagram is to be regarded only as relatively arid in comparison with the humid season of the climate type in question. The temperature curve, which we use instead of the curve of potential evaporation, is not identical with it, but only runs more or less parallel to it. It is an approximation. The more arid or windy (e.g. Patagonia) the climate in question, the more it lags behind the latter in quantitative terms. More precisely constructed hydroclimate diagrams have been compiled by Henning (1994), and ecological by Lauer et al. (1996) and Lauer & Rafiqpoor (2002). In absolute terms, the greater the aridity of the overall climate, the more arid the season on the ecological climate diagram; that is, an arid season on the climate diagram of a station in the steppe, for example, is not as extreme as that of a Mediterranean station or even one in the Sahara. This is favourable from an ecological point of view, because the drier the climate in which plants are native, the more their sensitivity to drought diminishes. For the species of the tropical rainforest, even a non-perhumid month (less than 100 mm of rain) is relatively dry; the **xerophytes** growing on dry sites in Central Europe would be more likely to be classified as **hygrophytes** in the deserts.

We will include the corresponding diagrams when discussing the vegetation areas, as this will allow us to avoid long tables and thus provide a quick overview.

The climate diagrams are particularly suitable for finding out **homoclimates**, which is extremely tedious when using extensive climate tables.

One only needs to compare the given climate diagram with those in the Climate Diagram World Atlas from areas where homoclimates are suspected. Figure 2.54 shows the homoclimates of Karachi (Pakistan) from zonobiome III (slight transition to ZB II) and of Bombay (very typical of ZB II) in another part of the world. Knowledge of homoclimates is very important for new crop introductions in areas where they are not yet known.

If one looks at climate diagrams of mountain regions, then these are very similar to the diagrams of the surrounding lowlands over the course of the year; however, the temperature curve is more or less lowered depending on the elevation, the precipitation can often be increased. As a result, the aridity usually decreases significantly with increasing altitude. Corresponding mountain stations of the respective zonobiomes are shown as examples in Fig. 2.53 (for more details see p.). If you want to get a quick overview of the climate structure of larger areas, you can use climate diagram maps. These are obtained by inserting the climate diagrams from the Climate Diagram World Atlas in the geographically correct position on large wall maps of individual continents or countries. The clarity is increased if the area of the drought periods in the diagram is marked in red and that of the humid periods in blue. Then the division can be overlooked at a glance. Such climate diagram maps of all continents in large format (black and white) have been published elsewhere (Walter et al. 1975). Here, as an example, we can only bring the climate diagram map of Africa to Fig. 2.55 in small format with only a few climate diagrams (of Africa, the World Atlas contains over 1000 diagrams). Climatic

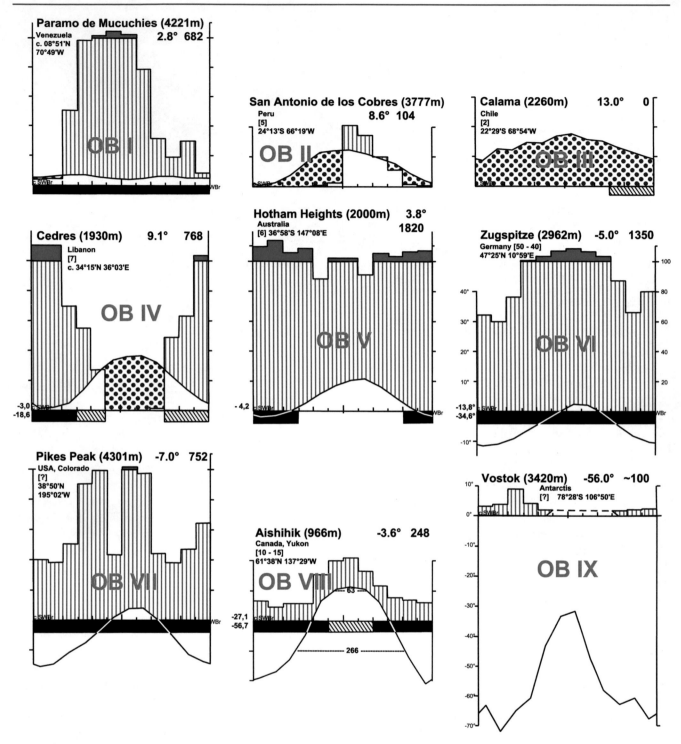

Fig. 2.53 Examples of mountain stations of the different orobiomes: OB I: Páramo de Mucuchies in Venezuela; OB II: San Antonio de Los Cobres in the Peruvian Puna; OB III: Calama in the northern Chilean Desert Puna; OB IV: Cedres in Lebanon; OB V: Hotham Heights in the Snowy Mountains (Australia); OB VI: Zugspitze in the northern Alps; OB VII: Pikes Peak in the Rocky Mountains above the Great Plains of North America; OB VIII: Aishihik in southern Alaska; OB IX: Vostok on the ice cap of Antarctica

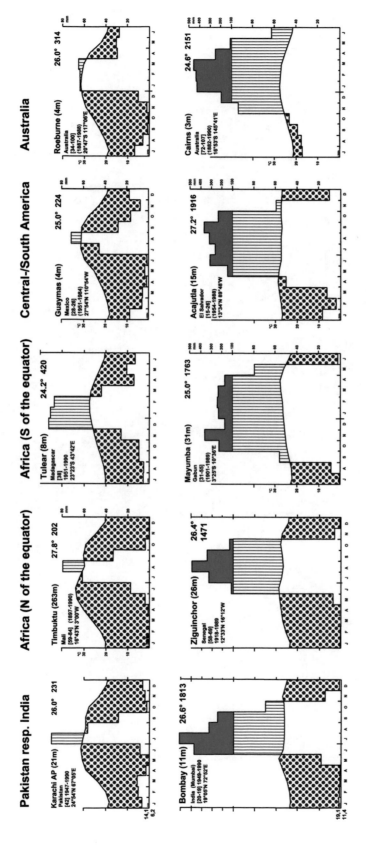

Fig. 2.54 Homoclimates of the two stations Karachi (Pakistan) and Bombay (India) in other continents

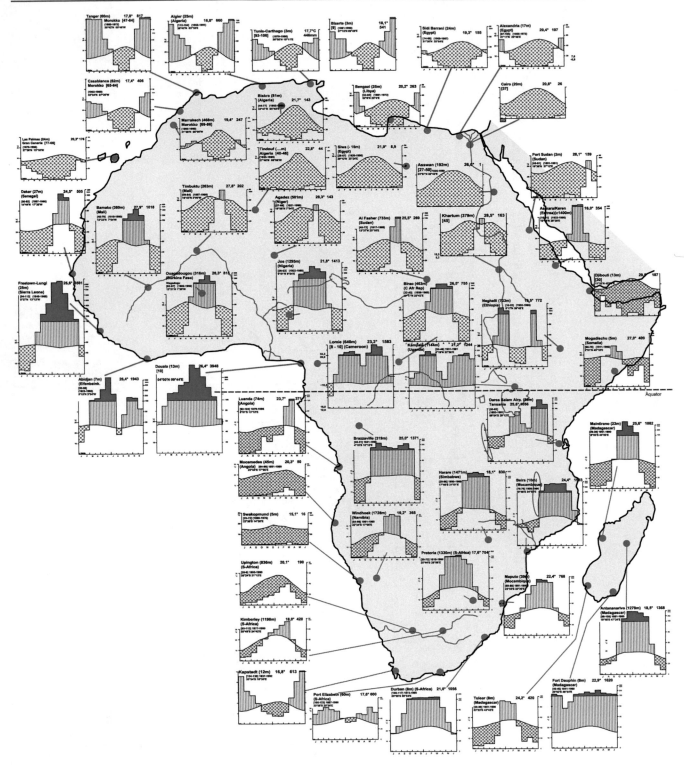

Fig. 2.55 Example of a climate diagram map with 66 stations. Zonobiomes from north to south: IV-III-II-I-III-IV, but north of the equator the east is relatively too dry (monsoon), whereas south is relatively too humid (SE trade winds)

diagram maps of selected regions are also available on the web (www.s.breckle.name/KlimaDiagrKarten/).

References

Adriano D.C. 1986: Trace elements in the terrestrial environment. Springer/New York 533pp.

Barthlott, W., Szarzynski, J., Vlek, P., Lobin, W. et al. 2009: A torch in the rainforest: thermogenesis of the Titan arum *(Amorphophallus titanum)*. Plant Biol. **11** (4): 499-505. Doi: https://doi.org/10.1111/j.1438-8677.2008.00147.x

Baumgartner, A. & Reichel, E. 1975: Die Wasserbilanz. Niederschlag, Verdunstung und Abfluss über Land und Meer sowie auf der Erde im Jahresdurchschnitt. Oldenburg-Verlag, München

Breckle, S.-W. 1976: Zur Ökologie und zu den Mineralstoffver-hältnissen absalzender und nichtabsalzender Xerohalophyten. Habil.-Schr. Bonn, 170 S., Cramer (Diss. Bot.)

Breckle, S.-W. 1986: Studies on halophytes from Iran and Afghanistan. II. Ecology of halophytes along salt gradients. Proceed. Roy. Soc. Edinburgh **89B**: 203-215

Breckle, S.-W. 2002a: Walter's Vegetation of the Earth. The Ecological System of the Geo-Biosphere. Springer Verlag, Heidelberg, 527 p.

Breckle, S.-W. 2002b: Salinity, halophytes and salt affected natural ecosystems. In Läuchli A., Lüttge U. (eds.) Salinity: Environment-Plants-Molecules. Pp. 53-77. Kluwer/Dordrecht

Breckle, S.-W. 2009: Is sustainable agriculture with seawater realistic? In: Aschraf, M., Ozturk, M. & Athar, H.R, (eds.): Tasks for Vegetation Science, Vol. **44**: 187-196

Breckle, S.-W. 2021: Bodenversalzung (Aridität und Fehler bei der Bewässerung, Meeresspiegelanstieg). In: Lozan, J., et al.: Warnsignal Klima: Böden und Landnutzung. Hamburg (in press)

Caldwell, M.M., Richards, J.H. & Beyschlag, W. 1991: Hydraulic lift: ecological implications of water efflux from roots. In: Atkinson, D. (ed.): Plant root growth - an ecological perspective. Blackwell, Oxford

Cheeseman, J.M. 2015: The evolution of halophytes, glycophytes and crops, and its implications for food security under saline conditions. New Phytol. **206**: 557-570

Flowers, T.J., Yeo, A.R. 1995: Breeding for salinity resistance in crop plants: where next? Aust. J. Plant Physiol. **22**: 875-884

Gates, D.M. 1965: Energy, plants, and ecology. Ecol. **46**: 1-13

Gausman, H.M., Allen, W.A. 1973: Optical parameters of leaves of 30 plants species. Plant Physiol. **53**: 57-62

Henning, I. 1994: Hydroklima und Klimavegetation der Kontinente. Münstersche Geogr. Arb. **37**: 144 S.

Hillel, D. 1980: Applications of soil physics. Acad. Press, New York

Ibrahimova, U., Kumari, P., Yadav, S. et al. 2021: Progress in under-standing salt stress response in plants using biotechnological tools. J. Biotechn. **329**: 180-191

Junghans, H. 1969: Sonnenscheindauer und Strahlungsempfang geneigter Ebenen. Abhandlungen des Meteorologischen Dienstes der DDR 85, Berlin.

Kearney, T.H., Briggs, L.J. et al. 1914: Indicator significance of Vege-tation in Tooele Valley, Utah. J. Agric. Res. **1**: 365-417

Kuntze, H., Roeschmann, G. & Schwerdtfeger, G. 1994: Bodenkunde. 5. Aufl., Ulmer, Stuttgart 424 S.

Larcher, W. 1980: Klimastress im Gebirge – Adaptationstraining und Selektionsfilter für Pflanzen. Rheinisch-Westfälische Akad. Wiss., N **291**: 49-88

Larcher, W. 2001: Ökologie der Pflanzen. 5. Aufl., Ulmer, Stuttgart

Lauer, W. 1999: Klimatologie. Das Geographische Seminar. Westermann Verlag Braunschweig

Lauer, W. & Frankenberg, P. 1986: Eine Karte der hygrothermischen Klimatypen von Europa. Erdkunde **40**: 85-94

Lauer, W. & Rafiqpoor, M.D. 2002: Die Klimate der Erde – eine Klassifikation auf der ökophysiologischen Grundlage der realen Vegetation. Erdwissenschaftliche Forschung, Bd. XL. Franz Steiner Verlag, Stuttgart

Lauer, W., Rafiqpoor, M.D. & Frankenberg, P. 1996: Die Klimate der Erde. Erdkunde **50**: 275-300

Lerch, G. 1991: Pflanzenökologie. Akad.-Verlag, Berlin 535 S.

Levitt, J. 1972: Responses of plants to environmental stresses. Vol. 1+2; (1980: 2nd. edit.) Acad. Press, New York 497 + 606 p.

Marschner, H. 1986: Mioneral nutrition of higher plants. Acad. Press Harcourt/London 674 pp.

Munns, R. 2005: Genes and salt tolerance: bringing them together. New Phytologist **167**: 645-663

Popp, M. 1995 Salt resistance in herbaceous halophytes and mangroves. Progress in Botany **56**: 416-429

Scheffer, P. & Schachtschabel, P. 1992: Lehrbuch der Bodenkunde. Enke, Stuttgart 491 S.

Schmithüsen, J. 1976: Atlas zur Biogeographie. Bibliogr. Inst. Mannheim, 80pp.

Schönwiese, C.-D. 1994: Klimatologie. UTB1793, Ulmer, Stuttgart 436 S.

Shepherd JM 2004: The global distribution of precipitation and clouds. Ch.2.4 prepared by the Intern. Aerosol-Precipitation Science Assess-ment Group (IAPSAG) for the Wold Meteor. Org. URL: https://ntrs.nasa.gov/api/citations/20040171853/downloads/20040171853.pdf. Maps download: https://gpm.nasa.gov/data/imerg/precipitation-climatology

Tischler, W. 1984: Einführung in die Ökologie. 3. Aufl. Fischer, Stuttgart, 437 S.

Troll, C. 1943: Thermische Klimatypen der Erde. Petermanns Mitteilungen **89**: 81-89

Waisel, Y. 1972: Biology of halophytes. Acad. Press New York/London 395pp.

Walter, H. 1960: Standortslehre. 2. Aufl., Ulmer, Stuttgart. 566 S.

Walter, H. 1990: Vegetationszonen und Klima. 6. Aufl., Ulmer/Stuttgart 382 S.

Walter, H. & Kreeb, K. 1970: Die Hydratation und Hydratur des Protoplasmas der Pflanzen. Proto-plasmatologia, **Bd. II** C 6, Wien. 306 S.

Walter, H. & Lieth, H. 1967: Klimadiagramm-Weltatlas. Fischer, Jena

Walter, H., Harnickell, E., Mueller-Dombois, D. 1975: Klimadiagramm-Karten der einzelnen Kontinente und die ökologische Klimagliederung der Erde. Fischer/Stuttgart, 365pp. + 10 maps

Wani, S.H., Kumar, V., Khare, T. et al. 2020: Engineering salinity tolerance in plants: progress and prospects. Planta **251**: 76, 29pp.

Weischet, W. 1980: Die ökologische Benachteilung der Tropen. 2. Aufl., Teubner, Stuttgart

White, I.D., Mottershead, D.N. & Harrison, S.J. 1992: Environmental Systems. Chapman & Hall. London 616 p.

Part B: Ecological Basis (Synecology)

3

Contents

© Springer-Verlag GmbH Germany, part of Springer Nature 2022
S.-W. Breckle, M. D. Rafiqpoor, *Vegetation and Climate*, https://doi.org/10.1007/978-3-662-64036-4_3

Photo 5 Tundra with dwarf shrubs like *Loiseleuria procumbens* and *Vaccinium myrtillus, V. vitis-idaea* at snow-poor parts in the tundra (ZB IX) in N Finland (photo: Breckle)

3.1   Environment and Competition

The climate of a site determines its vegetation. But the assumption often made that the distribution of plant species is directly caused by site conditions is almost never correct. These are only of indirect importance by altering the competitiveness of species. Only at the absolute limits of distribution in the arid and cold desert, at the edge of the salt desert, i.e. where the individual plants are isolated, are the site factors (usually a certain extreme factor) directly determining. If we disregard these exceptional cases, plant species can still grow far outside their range if they are protected from competition from other species. For example, the northeastern distribution boundary of beech runs through the Vistula region (river in Poland), but beech still grows in the botanical gardens of Kiev and Helsinki. The Mediterranean evergreen holm oak (*Quercus ilex*) reaches its range northern limit in the southern Rhône valley, cultivated trees still hold out in the botanical gardens of Bonn, Leipzig or Copenhagen.

The natural range limit of a species is reached where changing environmental conditions reduce its competitiveness or competitive strength to such an extent that it can be displaced by other species. It therefore depends above all on the presence of certain competitors (or a certain fauna). For beech, these are hornbeam on the eastern border, oak on the northern border and spruce in the mountains. European beech (*Fagus sylvatica*) is the dominant species of Central Europe's natural forests. It spans a rather broad spectrum of thermal (6.6–13.5 °C mean annual temperature; 16.9–23.0 °C mean temperature of the warmest month), hygric (470–2000 mm annual precipitation) and edaphic site conditions (pH, CEC etc.). Of all the beech species it is the one advanced furthest toward dry climates (Leuschner 2020).

If the northeastern border of beech shows a similar course as the January isotherm of −2°C, or the northern border of oak coincides with the temperature line of 4 months above +10 °C, or the northern border of spruce coincides with the July isotherm of +10 °C, there need be no direct causal connections. At the most, one could conclude that for beech the increasingly colder winter towards the east and for oak and spruce the shorter summer towards the north probably strongly reduces the competitiveness of these species. In future it might be increasingly threatened by climate change related heat waves and longer drought periods in parts of its distribution range.

If we designate as the ecological optimum the conditions under which a species occurs most frequently in nature and as the physiological optimum the conditions under which it thrives best in the laboratory (climate chamber) or in individual culture, these optima often do not correspond (Fig. 3.1).

From the distribution of a species, therefore, one cannot readily determine its physiological requirements. If, for

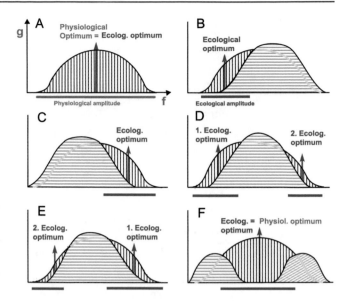

Fig. 3.1 Growth curves (vertically shaded) of a species without (**a**) or under (**b–f**) competitive pressure (horizontally shaded). Ordinate: Growth intensity or substance production; abscissa: Site factor

example, pine is found in the temperate zone under natural conditions only on dry limestone slopes, but also on very dry, acidic sandy slopes or even on very wet acidic moorland soils (Fig. 3.2), this is because it is displaced from the sites more favourable to it by stronger competitors. On the other hand, knowledge of the physiological requirements and adaptations of a species determined in climatic chambers does not yet give us the possibility of predicting or explaining in detail its distribution in nature. Whether the species occupies the site that can be colonized according to its physiological requirements, or only partially, is decided not only by the historical factor, but also by its competitors. In the ecogram, it can be shown for certain ecological factor combinations which species occur dominantly and where their respective ecological optima lie. This is shown for Central European tree species in Fig. 3.2.

Competition is not a direct dependency relationship between species. It can be recognized by the fact that isolated plants develop more luxuriantly than those in a plant community. Inhibition in competition is mostly due to deprivation of light by above-ground organs or of water or nutrients in root competition. Whether certain inhibitors secreted by the plants also play an important role in the competition (allelopathy) is difficult to prove under natural conditions. Only in some cases this seems to be the case. In other cases, there is probably also mutual promotion, in particular through substance exchange via the fungal hyphae network in the soil, which can connect the mycorrhiza of different trees and supply additional nutrients to young seedlings (nursing system). In ecosystems, however, the processes of competition outweigh those of such cooperation.

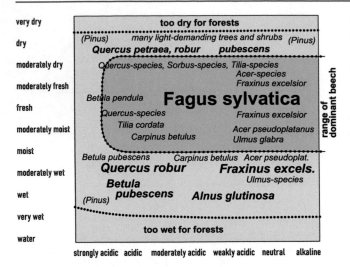

Fig. 3.2 Ecogram of the most important forest tree species of Central Europe of the submontane stage in a temperate-suboceanic climate. The font size expresses approximately the proportion of the tree layer that would be expected as a result of natural competition (modified after Ellenberg 1996)

In competition, a distinction is made between **intraspecific**, which takes place among individuals of the same species, and **interspecific** between different species. The first promotes the survival of the most vigorous individuals and serves to preserve the species. In interspecific competition, one species may gain dominance and displace the other, or an equilibrium may develop in mixed stands, depending on the competitive strength of each partner. In the mountains of Central Europe, for example, it can be observed on the beech-spruce border that beech predominates on southern slopes, spruce on northern slopes, while on eastern and western slopes both more or less balance each other out and form mixed stands. These will also form if the seedlings of one species develop better under alien species than under individuals of the same species, which seems to be true in the tropical virgin forest, perhaps because herbivore and parasite pressure or other inhibiting factors are graded accordingly.

The competitive vigour of a species is a very complicated and difficult phenomenon to measure, especially considering that it can vary greatly with the stage of development. It is weakest in seedlings and young plants and increases with age, especially in trees. It always applies only to very specific environmental conditions. The totality of all morphological and physiological characteristics of a species is important. Biennial species are more competitive than annuals because they begin growth in the second year with greater reserves accumulated during the first year. For the same reason, from the third year onwards, perennial herbs are superior to biennials. Woody species are victorious over perennial herbs if they are not suppressed in the first years of life,

i.e. if they succeed in forming woody axillary organs that rise above the herb layer.

As a result of competition, similar combinations of plant species occur again and again at similar sites in a limited area, which are referred to as plant communities (phytocoenoses). In Central Europe, for example: Beech forests on calcareous soils with their herbaceous flora or alluvial forests, certain types of bogs or reed beds, etc.

In a stable plant community, the species are in a certain ecological balance with each other and with their environment. Together with the animal organisms they form a biocoenosis. The following are decisive for this balance (if one disregards the herbivoral influence of the animals):

1. Interspecific competition
2. The dependence of each species on the presence of others (for example, shade species)
3. The occurrence of complementary species that complement each other spatially or temporally so that almost each ecological niche is filled.

The natural community is thus to a certain extent "saturated" and foreign, introduced species can hardly invade, whereas they have much more of a chance to do so if the equilibrium is disturbed. For this reason, the long-distance transport of seeds plays a significant role in the spread of plants only in the case of areas that have not yet been colonized, for example in the case of young volcanic islands.

The balance of a plant community is not a static but a dynamic phenomenon. Individuals die, others germinate and grow. In the process, there is usually a constant change of place between the individual species. Especially in uninfluenced stands, in moors, and even more so in primeval forests, a constantly changing mosaic of different developmental phases occurs side by side. In primeval forests these processes are obviously very long-term; they lead to the fact that on larger areas all phases are represented. The phases that can be distinguished in this process are interdependent and can merge in different ways with specific cycles (Fig. 3.3).

In terms of quantity, the species composition shows certain or even considerable fluctuations. The species composition does not remain the same either, especially when the external conditions change from year to year, e.g. rainy years are followed by dry periods, etc. As a result, some species are favoured in the competition, then the others are favoured. If the site conditions change permanently in a certain direction, for example if the groundwater level rises slowly over many years, the species combination will also change: Certain species will disappear, others will invade from outside, until finally a new plant community emerges.

If human interventions are carried out in the same way over a long period of time, an anthropogenically induced

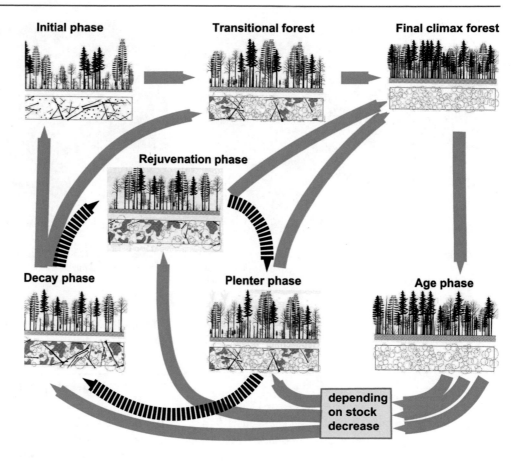

Fig. 3.3 Scheme of the different phases and their transitions in primeval forests, derived from surveys in the Roth forest near Lunz am See (Lower Austria, modified after Zukrigl et al. 1963). Accordingly, the phases and mosaic stands observed in Bialowiecz (primeval forests in eastern Poland and Bielorussia) can also be classified in such a cyclic scheme

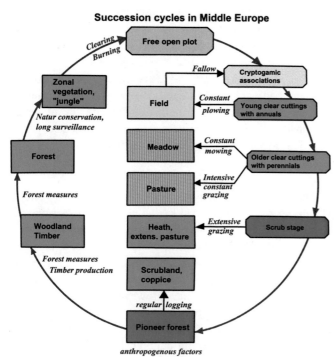

Fig. 3.4 Scheme of the succession cycle of anthropogenic formations in Central Europe, with indication of the main influencing factors

equilibrium develops and plant communities emerge which are referred to as cultural formations in the case of intensive use or as semi-cultural formations in the case of more extensive use. They make up the vegetation of areas densely populated by humans. The essential cultural formations are maintained by certain measures, and the succession sequence is always restarted when there is a change of use (Fig. 3.4).

The sequence of succession is mostly random, depending on which seeds (diaspores) arrive first or in larger quantities on the surface and on the germination conditions of the seeds and the establishment possibilities of the seedlings and their competitive success.

3.2 Pollination and Fertilization (Flowers, Seeds, Fruits)

Since the establishment and maintenance of species depends on their reproduction and this on the pollination and fertilization processes, as well as on the dispersal possibilities of seeds and fruits, it is necessary to briefly discuss their biological basis. However, this does not yet explain the respective distribution of a species, i.e. its current geographical area, because this is the result of a long historical process.

Fig. 3.5 Hummingbird pollination of the flower of a *Stachytarpheta* plant (Verbenaceae) (**a**, photo: H. Breckle) in the New World and of a *Strelitzia* by a nectar bird in the Old World (**b**, photo: Rafiqpoor, S-Africa). The red colour of the flower is an adaptation to the pollination of this plant by animals

Flowering plants are characterized by zoogamous or anemogamous flowers. The mostly coloured flowers of zoogamous species serve to attract insects (entomogamous species), sometimes these are mainly beetles (coleopterogamy) or butterflies (lepidopterogamy), species flowering in the evening especially attract hawkmoths (sphingophilous species), but hymenopterans (bees, bumblebees, wasps) are particularly important as pollinators, also for important crops. In tropical regions, there are some species adapted to bat pollination (chiropterogamous species), whose flowers are usually drooping, pale yellow, and open at night, as well as species with often bright red flowers pollinated by birds (ornithogamous), e.g., by hummingbirds in the New World and nectar birds in the Old World (Fig. 3.5). In the subtropical thorny savannah spots in SE Afghanistan, such species just reach the border of Afghanistan from the Indian subcontinent.

Pollination is the transfer of pollen from one flower to the stigma of another flower. This transfer can take place by wind (anemogamous species): Their flowers are very rich in pollen but inconspicuous and small (grasses). The zoogamous species, as mentioned above, are much more conspicuous. They are sometimes extremely adapted to pollinators as part of **coevolution.** In the case of orchids, only individual pollinia are formed, which are specifically transferred by certain bees or wasps ("registered mail"), conversely, the number of seeds is huge, the dusty seeds are transported by the wind in an untargeted manner ("mail shot").

Pollination is a prerequisite for **fertilization**. It occurs when the pollen tube on the stigma grows out to the ovules in the ovary of the pollinated flower. There, **nuclear fusion** takes place and a diploid zygote is formed, which divides and then continues to grow, ultimately forming an embryo, which is surrounded by more or less nutritive tissue in the seed and protected by the seed coat. From the ovary the fruit grows. But just like the huge number of different structures in the

flowers, the range of variation in fruits and the seeds they contain, is immense.

3.3   Dispersal and Distribution

The mature seeds must be shipped into the environment to maintain the species. This can be done by releasing the seeds from the fruit or by using the fruit itself as a dispersal unit.

Dispersal can take place by wind (anemochorous species), for which the seeds are either fine as dust or they or the fruits develop flight organs (wings, umbrellas). The dispersal can also be done by animals (zoochorous species). In these cases, there are also different ways of attracting animals. Fruits can be eaten, but many seeds are indigestible and after intestinal passage (in birds or also mammals) are even better germinable (endozoochorous species). Other species have adhesive appendages on the fruits, which then remain attached to the fur and are thus spread further (exozoochorous species).

The success of propagation depends on several factors. The number of seeds formed by the plants each year plays a major role. Many seeds that are shipped far compensate for a high loss rate. Few seeds that are targeted to good locations are also a successful strategy; for example, some nuts are purposefully collected and stashed by rodents, where some are then allowed to germinate. However, the rate of loss is generally huge. In old beech forests (*Fagus sylvatica*) in Central Europe, up to or more than 100 million beechnuts fall on 1 ha in mast years in autumn. Depending on winter and herbivore pressure, 10–20% of these germinate. This results in a dense carpet of seedlings. At the end of the first vegetation period only 1–10% are left. In the second and third year more than half of them die due to lack of water, feeding and shading. This still leaves about 5–10 young trees per m^2, which can form a dense stand of shrubby young beeches as soon as they receive a little more light. In the coming years, however, the strong competition will lead to a situation where only a few young trees will be able to grow up to the approximately 300–700 old trees on one hectare (there is no room for more with their broad crowns). These are then every few years a few dozen young trees, this is quite sufficient to maintain the stock.

The **distribution of** a species is the geographical area that the species colonizes. The distribution is very similar for some species, which are then referred to as geo-element to which these species belong (e.g. Mediterranean species, Central Asian species). The range size can be very different. At one extreme are cosmopolitan species; they occur worldwide on all continents. At the other extreme are narrowly restricted occurrences, e.g. only in one valley or only in one mountainous region; these are called **endemics**. However, one must always precisely identify the geographical region one is referring to. One often speaks of country endemics, although of course country borders are usually political and normally do not represent natural borders. Thus, one speaks of endemics of Afghanistan. Of the approximately 5000 species of higher plants in Afghanistan, about 25% are endemics (Breckle et al. 2013). However, directly neighbouring regions (with the same climate, etc., Chitral, Kurram Valley) should be included here (subendemic species).

Most species in Afghanistan have an Irano-Turanian distribution. But also species with an Eastern range (Himalayan, Sino-Japanese) occur in Eastern Afghanistan. In the high mountains, species of the boreal and even arctic region of Euro-Siberia reach into Afghanistan. In the basin of Jalalabad or near Khost even some Saharo-Sindian and Sudanese distributed (subtropical-tropical) species occur, as well as in Southern Afghanistan some Saharo-Sindian desert species.

3.4 Ecotypes and Biotope Change

Many plant species or phytocenoses (plant communities) have a very wide distribution and, if you look at their areas (residential districts) on a map, apparently grow under quite different climatic conditions. This fact may be due to two reasons.

1. The species as a taxonomic unit is often strongly differentiated eco-physiologically, for example with regard to its resistance to cold or drought or its climatic rhythm. Thus, the pine, *Pinus sylvestris,* occurs from Lapland to Spain and eastward to Mongolia, with at most its growth habit showing taxonomically insignificant differences. But the Spanish pine cannot grow in Lapland because it is too sensitive to cold, and the Lapland pine cannot grow in Spain because it needs a long winter dormancy. Therefore, the forester must always pay very close attention to the provenance (origin) of the seed. Most taxonomically uniform species consist of many such ecotypes (races, varieties).

2. The second possibility of a wide distribution is based on a biotope change of the species or phytocenosis, when its range extends into a climatically different area. If, for example, the climate at the northern edge of the range becomes colder, the species is no longer found on the plain, but on the microclimatically warmer southern slopes, i.e. a biotope change occurs which compensates for the climate change, so that the site or environmental conditions for the plants hardly change, i.e. remain relatively constant. This regularity (**law of relative site constancy**) can be observed everywhere: In the southern part of the range, plants move more and more to the northern slopes, into deep moist canyons, or up into the mountains. If the climate becomes wetter, the plants seek out dry limestone or sandy soils. In the dry climate, on the other hand, they are found accordingly on heavy, wet soils or on those with a high groundwater level.

Of course, one must take into account that in the southern hemisphere the northern slopes are warm, and on the equator—the east and west. Similarly, in arid regions, the sandy soils have the most favourable water supply for plants.

This applies not only to water conditions in arid regions, but more generally to all factors that are influenced by climate.

The law of biotope change must also be taken into account in the mountains when determining the elevational belts: Already the differences in elevational limits at different exposures indicate this law. Much more extreme are special niches with intensive irradiation and cold runoff, which allow small stands of trees to grow above the timberline already within the alpine belt. Individual trees were found in the West Pamir in blown-through gorges without cold air dams even at 4000 m NN, and shrubs in the wild terrain even at 5000 m NN; in the Hindu Kush we found such at 5100 m NN in very sheltered niches on southern flanks. On the other hand, in cold-air dolines of the Eastern Alps, forest vegetation is already absent at 1270 m NN, the lowest temperature in Western Europe being measured at −51°C near Lunz (Lower Austria).

Soil factors also play a role. Fragments of alpine vegetation in the Eastern Alps can be found in the middle of the beech belt on dolomite, rocks that are difficult to weather. Special niches are also the avalanche ranges, where the competition of tree species is eliminated, so that the krummholz species of the subalpine belt are able to assert themselves at low altitudes of the forest belt. On such special biotopes one often finds relics of species that formerly had a wider range under different climatic conditions. However, historical evidence for the relict nature of an occurrence should be provided if possible.

3.5   The Historical Dimension

Today's geo-biosphere is closely linked to the history of the Earth. It is the result of a long development of the plant and animal kingdom on the one hand, and of a long geotectonic history of the earth's solid surface on the other. Therefore, in ecology one must always take into account the historical development. The continents did not exist in today's form in former times, also they took a different position to the poles and the equator. This theory of Wegener's continental drift has been further developed today as the theory of plate tectonics. The movements of the land masses are explained by large-scale plate tectonics and convection currents in the Earth's mantle. The movement of the plates of a few centimetres per year leads to very slow changes of the plates in relation to each other. The present position of the plates is shown in Fig. 3.6. Due to the magmatic upwelling areas (for example, opening, extension of the Atlantic Ocean), "submergence" of plate material must occur elsewhere; this occurs in the area of the subduction zones. In their vicinity are usually particularly active volcanic areas, which are important for the evolutionary processes of flora and fauna.

Compared to the shifting continental plates, the atmospheric wind system with the climate zones appears to be relatively a very stable system, which in this form, at least in a comparable form, probably goes back far into the Mesozoic. The climate system as such appears as the more stable, the continents as lithosphere swim under it and are the more variable part in the very long-term overall system of the biosphere (Krutzsch 1992).

Life began in the water. The first land plants are known as fossils since the turn of the Silurian/Devonian. From the fact that NaCl, the main component of sea salt, is not needed by cormophytes and is toxic to all plants except halophytes, one must probably conclude that the ancestors of land plants were freshwater algae, perhaps living in coastal lagoons under humid tropical climates. The halophytes among the angiosperms are young secondary adaptations to saline soils in coastal areas or salt deserts.

The conquest of the land was made possible by large cell vacuoles, which in their entirety, the vacuome, form an inner aqueous medium for the cytoplasm. Around the plasma, the cell wall forms a water-saturated, spongy outer medium that envelops the cell. Toward the outside world, land plants have protected themselves from desiccation by forming a cuticle. The invention of stomata allows controlled CO_2 uptake for photosynthesis, and the root and conduit system provides compensation for transpiration losses (Walter 1967), while also serving as a transport system for mineral nutrients.

Due to the greater isolation of the continents after the formation of the angiosperms in the late Mesozoic, their development took different paths, which led to the formation of six floral kingdoms (Fig. 3.7), which essentially also more or less correspond to the faunal kingdoms.

In the phylogenetically relatively old group of conifers it is evident that the Podocarpaceae and especially the *Araucaria* are found only in the Southern hemisphere, while the large family of Pinaceae and almost all Taxodiaceae have a Northern hemispheric distribution, while the Cupressaceae are found scattered over all continents.

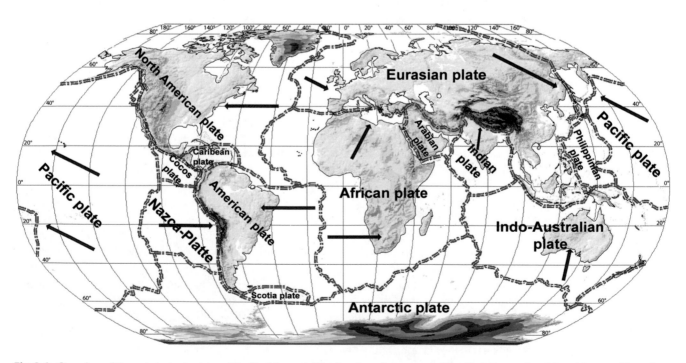

Fig. 3.6 Overview of the main tectonic plates of the Earth's crust. The direction of movement of the plates (arrows) and the mid-ocean ridges along which material is transported from the Earth's mantle to the surface are also indicated

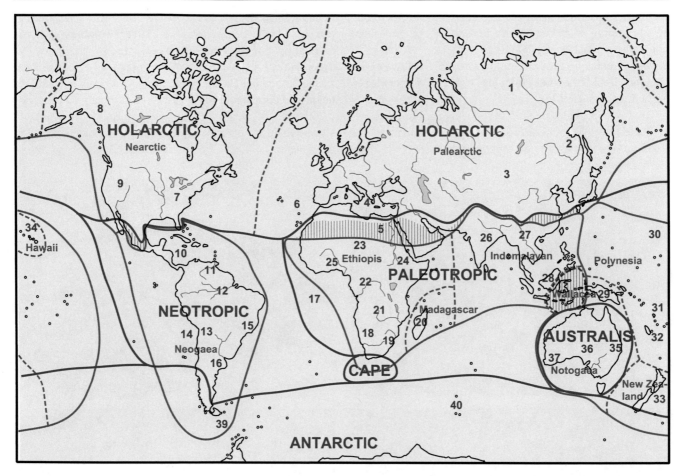

Fig. 3.7 The floral kingdoms of the Earth. On New Zealand and Tasmania, Antarctic as well as Palaeotropical and Australian floral elements occur, respectively

A much stronger differentiation shows the distribution of the flowering plants (angiosperms), the youngest branch of the plant kingdom. Original forms, partly relicts, are still found mainly in Southeast Asia (e.g. New Caledonia). The oldest families of this group of plants are only known from the early Cretaceous, but the flowering plants experienced their main development in the Tertiary, when the Gondwana land mass had already split into the individual continents. In the northern hemisphere this was only the case to a lesser extent; only in the Pleistocene did a final separation between N America with Greenland and Eurasia occur. Therefore, the floristic differences in this area are small, so that these continents are combined into one floral kingdom, the Holarctic. The tropical floras of the so-called New and Old World differ much more. Therefore, they belong to two different floral kingdoms, the Neotropics on the one hand and the Palaeotropics on the other hand. Even less in common are the floras of the southernmost parts of S America and Africa, and of Australia and New Zealand, which lie very isolated. The differentiation led to the formation of three floral

kingdoms: Antarctica, which includes the southern tip of South America and the subantarctic islands, Australis, which is spatially identical with the continent of Australia, and Capensis, the smallest but particularly species-rich floral kingdom at the extreme southwestern corner of Africa (Fig. 3.7).

> **Box 3.1 Plate Tectonics**
> Plate tectonics: From a geotectonic point of view, the current position of the plates is only a specific snapshot. For the understanding of today's distribution of organisms, the former position of the plates to each other and the course of evolution is an important basis.

These six floral kingdoms are not sharply demarcated. Individual floral elements from one floral kingdom can radiate far into the neighbouring one. On New Zealand, one finds both Palaeotropical-Melanesian elements and Antarctic elements, which often interpenetrate like a mosaic. Therefore, the attribution of these islands to one of the two floral kingdoms is a matter of judgement.

The animal regions of the zoologists agree to a large extent with the floral kingdoms, only the Capensis is not distinguished by a special fauna.

The floras provide the building blocks, i.e. the plant species determine the structure of the plant communities that make up the vegetation of the individual areas. If these

building blocks are different, similar life forms can nevertheless arise under certain extreme external conditions; this is known as **convergence**. However, these are more the exceptions. As a well-known example, we cite the stem succulents, which in the arid, i.e. dry regions of America belong predominantly to the family Cactaceae, but in Africa mainly to the genus *Euphorbia* (spurge) (Fig. 3.8). In Australia, on the other hand, there are no succulents at all in climatically similar arid regions, although Australia is

Fig. 3.8 Convergent life forms from the Old World (**a**) *Euphorbia resinifera*, from Morocco (photo: Rafiqpoor); (**b**) *Trichocaulon pedicellatum* (Asclepiadaceae) from SW Africa, (photo: Breckle); (**c**) *Alluaudia procera* from Madagascar (photo: E. Fischer) and from the New World: (**d**) Cacti (*Cereus macrostibas*) from the Atacama Desert S-Peru (photo: Rafiqpoor)

Table 3.1 Endemism on islands (percentage of endemic species of the native flora)

Islands	Degree of endemism (%)
Hawaii	97.5
New Zealand	72
Fiji Islands	70
Juan Fernandez	68
Madagascar	66
Galapagos Islands: in the dry areas	64
Galapagos Islands: in the humid mountain zone only	Only 8–27
Galapagos Islands: in coastal areas	12
New Caledonia	76
Canary Islands	50–55
Offshore islands	0–12

otherwise particularly rich in other convergences not known from the other continents. New Zealand's temperate climate lacks deciduous forests, which are widespread in the Holarctic. The total genetic stock of each flora, determined by historical evolution, is limited, so that the same life forms did not form everywhere. This is particularly true of the Australian floral kingdom, whose vegetation is physiognomically very different from that of other continents; the original mammalian fauna there is also very peculiar.

The Pleistocene left strong traces due to the multiple ice ages, especially in the Northern hemisphere. The flora in Europe became impoverished. Many genera died out, while they are still found in North America and East Asia. There, N-S escape was easier. In Europe, on the other hand, the W-E course of the Alps blocked escape and return.

In parts of the Sahara, the ice ages were temporarily noticeable as rainy, i.e. as pluvial periods, whereas in the tropics they were more noticeable as dry periods.

For this reason, it is essential to consider the historical factor when dealing with the vegetation of zonobiomes that span several floral kingdoms. This is particularly true for zonobiome IV with winter rainfall, which consists of sub-areas in Holarctic, Neotropical, Australis and Capensis. It is convenient to divide this into five vegetation-historical biome groups (Mediterranean, Californian, Middle Chilean, Australis, and Capensis, which differ greatly by floral stock despite similar life forms).

Due to their isolation, islands are also often characterized by strong endemism, i.e. by many species that occur only on them and nowhere else. In percentages of the total flora, the figures given in Table 3.1 are given for the individual islands or island groups.

Endemism is more pronounced the further the islands are from the mainland and the longer they have been isolated, but ocean currents also play a role. Recently many species had been introduced to most of the islands, thus a mixed vegetation is now present.

3.6 Coevolution and Symbioses

Biological systems interact with each other with the result of the evolution of the organisms involved in them (**coevolution**). The expression of the various ecosystems is not understandable without the processes of coevolution in the course of historical development. In many ecosystems the interlocking, i.e. the mutual dependence between certain plants and animals is so close that one must speak of an obligate relationship. It is now certain that in the long evolution of living things, interaction with other organisms has been much more important than with the inanimate world for many radiative events. In Fig. 3.9, the evolution of flowering plants and pollinators is demonstrated.

The radiation of Diptera and Hymenoptera took place simultaneously with the radiation of angiosperms. A.O. Wallace already suspected from the observations of long spurs of tropical orchids (e.g. *Angraecum sesquipedale*) on Madagascar (Fig. 3.10) that butterflies with similarly long proboscises must exist for their pollination. In 1987, the appropriate species (*Xanthopan morgani-praedicta*) was finally found and described for the Malagasy *Angraecum* (Nilsson et al. 1987). This is a result of coevolution, with the long spur evolving through the interaction of nectar predators and pollinators. Thus, coevolution is one of the keys to the origin of new species. Coevolution not infrequently involves linkages between pollinators, herbivores, and certain plant species that change over the course of the year, but this can only be maintained in a large-scale population. In the course of evolution, such close interdependencies have come about through the reinforcement of mutual mechanisms of action. This applies in the same way to numerous relationships between the most diverse organisms. Such a close network of relationships is particularly diverse in those ecosystems that have undergone a particularly long period of evolution (in and since the Tertiary), such as the

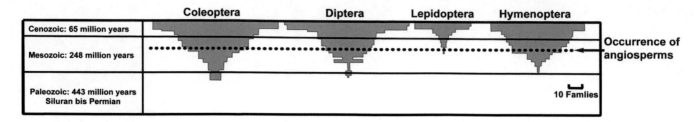

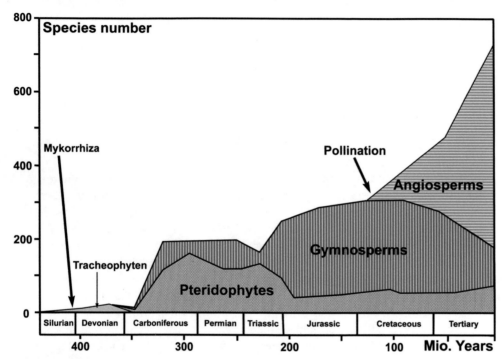

Fig. 3.9 Evolution of flowering plants and their pollinators. Diversity of fossilized insect orders across geologic epochs. A small fraction of fossil insects is documented in Paleozoic deposits. However, the majority evolved with the advent of flowering plants, two-thirds of the fossils of which are documented in deposits from the Mesozoic onwards (below) (modified after Labandeira & Sepkoski 1993 and Niklas et al. 1983)

tropical rainforest. Close functional links between organisms make it more difficult to clearly distinguish the functional compartments in an ecosystem analysis.

The various **symbioses** (a close coexistence in which two partners "parasitize" each other, so to speak: In equilibrium, each provides the other with something vital) are to be emphasized above all. Symbioses that occur ubiquitously are, for example, the various forms of mycorrhiza, to which we will refer in more detail. But also the nitrogen-fixing symbionts, which occur not only on legumes in the form of nodules with *Rhizobium*, but also on a number of other species (for example, *Frankia* on alder roots), improve the competitiveness of the species, or the symbioses even make possible the conquest of certain spaces that are actually

hostile to life, as in the case of lichens, which are thus the dominant primary-producing organisms in Antarctica (Fig. 3.11) or in the nival altitudinal zone of the mountains.

The particularly close interlocking of an extraordinarily large number of different organisms with each other led, in the course of long evolutionary times, to an incredibly diverse network of relationships and to a functional structure in the case of the tropical rainforest, which is in itself very stable under the uniform climatic conditions at the equator. After destruction, however, this feedback network cannot be regenerated within a foreseeable period of time. The secondary forest, which usually has fewer species, therefore has a much wider, looser functional network.

Fig. 3.10 *Angraecum sesquipedale* (right) with the long spur. On the left *Xanthopan morgani-praedicta* with the long proboscis (photo: Barthlott)

3.7 Population Ecology

A population is a group of individuals of a species that occur simultaneously in the same space and are capable of interbreeding. **Population ecology** is concerned with the study of the size (and distribution) of populations and the processes (primarily biological) that determine these parameters (Begon et al. 1996). In particular, the dynamics of populations—i.e. the changes over time in the absolute numbers of individuals and relative proportions of different age or developmental stages in the population—are of great interest, as an understanding of population dynamics facilitates predictions of future population trends.

In highly diverse systems, the question of regeneration of the numerous species is usually difficult or impossible to answer. The fluctuations in the population sizes of the many species involved (seed bank → germinating seed → seedling → sapling → young growth → adult plant; or egg → larva → pupa → imago → etc.) are often not detectable, birth and death rates are only known for a few organisms in their temporal sequence, and even less so are the influencing variables that control population sizes. This is also due to the fact that the input and output of seeds (or diaspores) can be highly variable spatially and temporally, and that, moreover, in some ecosystems some species have a very large seed bank that can be rapidly reactivated years later under changing conditions (for example, meadow becomes fallow land that has been ploughed). This is reflected in the general

Fig. 3.11 In Antarctica, lichens grow on rocks and open ground that protrude from the snow cover in summer (photo: http://is.gd/Oe5qIj)

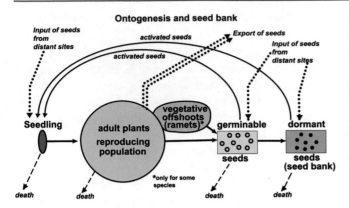

Fig. 3.12 Individual components for the regeneration of a plant species or the maintenance of its population at a given site (modified after Burrows 1990)

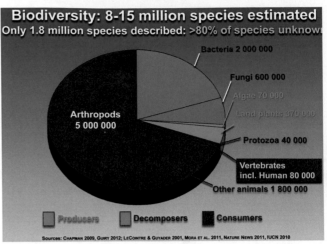

Fig. 3.13 Estimated species numbers. ten million different species are estimated to exist on Earth, but only 1.8 million of these have been scientifically recorded and described: (80% of the species on our planet are unknown (from Barthlott & Rafiqpoor 2016)

scheme in Fig. 3.12 for reproduction in plants. Not infrequently, certain drastic events lead to new developmental impulses of species. The steady development of populations is thus repeatedly interrupted and stimulated anew not so much by periodic as by episodic damaging events (fire, storm, flood).

Periodic events are predictable, occurring regularly (→ winter in ZB VII; the tides on the coast, etc.). Episodic events are not predictable, they occur at irregular, usually longer intervals (→ thunderstorms in ZB III, El Niño, tsunamis, frosts in Brazil's coffee cultivation, etc.).

This is particularly striking in tropical evergreen rainforests, where the stand structure is very heterogeneous, and where gaps of different sizes ("gaps"; Fig. 3.16) are repeatedly torn by branch or tree fall, which are quickly filled by fast-growing species, but where the stand species also rejuvenate. Probably in many more ecosystems than we have thought so far, such episodic events are prerequisites for their long-term maintenance through successive renewal of their structures. However, this then also leads to cyclic renewal of different lengths, which is predominantly stochastic (random) rather than deterministic; individual parts of a biome are younger, others older; the mosaic character and temporal dynamics of natural ecosystems was already characterized decades ago by Aubreville (1938). It is an important principle of maintaining high numbers of species in a dynamic side by side and coexistence.

3.8 Biodiversity

Biodiversity encompasses the different forms of life (plant and animal species, fungi, bacteria), the different habitats in which these species live (e.g. ecosystems such as forests or standing and flowing waters, etc.) and the genetic diversity within the species (e.g. subspecies, varieties, and breeds) (http://is.gd/iirae3). In this way, biodiversity encompasses life existing on earth in all its diversity and is thus also the

basis and potential of all life processes and ecosystem services on our planet (Lozan et al. 2016). Biodiversity is the result of "trial and error" in millions of years of evolution, recently additionally shaped by the influence of centuries to millennia of human use (hunting, gathering, clearing, agriculture, settlement, etc.). Biodiversity can be a measure of the originality and naturalness of ecosystems. Under extreme ecological conditions, however, a completely intact and pristine ecosystem can be species-poor, mainly if only specially adapted specialists can survive (Breckle 2000, 2006). In the case of very high diversity, the question of the regeneration of numerous species can usually not be answered.

Life with its immense biodiversity is the only specific quality of our planet. All the more surprising is the fact that our knowledge of this biodiversity is shockingly low (Barthlott & Rafiqpoor 2016). About 1.8 million different living organisms have been scientifically recorded—but all projections show that at least eight million, but probably far more than ten million different species exist on Earth (Mora et al. 2011).

Based on a rather conservative estimate of ten million species, Fig. 3.13 shows that the various large groups of organisms account for highly unequal shares of global biodiversity. The arthropods, shown in red, account for the largest proportion, with an estimated five million species. In addition to the approximately 1.8 million species, about 90% of all species are unknown. The most species-rich group within the arthropods are the insects (e.g. beetles, hymenopteres, butterflies). Scientifically described ones alone are about 350,000 beetle species from 179 families: What a contrast to the conspicuous vertebrates (e.g. mammals, birds, reptiles), which comprise only about 62,000 species, but are quite well known already because of their body size (knowledge level over 83%) (Barthlott et al. 2014).

Terrestrial plants (higher plants or vascular plants, i.e. flowering plants, gymnosperms and ferns) are a relatively species-poor group, with an estimated 370,000 (Pimm et al. 2014) species. But compared to insects, they are also large and conspicuous creatures that are also "sessile", i.e. do not run away: the trivial reason why the level of knowledge is very high at around 90% (Fig. 3.13).

But anyone concerned with global biodiversity in a constantly changing environment should be largely concerned with arthropods, which after all account for more than half of global biodiversity? Ecosystem-wise, this would be a fatal flaw. Arthropods and the other animals are the consumers within the system. They are all based on the producers, the massive global powerhouse that covers the planet as a worldwide green solar collector: the higher plants.

Figure 3.14 illustrates the ratios in the representation as pyramids, based on the number of species (left, assumption ten million) and estimated mass (right). Consumers, with their high diversity, comprise about 69% of the species, whereas plants comprise only about 5%. However, the latter are the most important structural elements in all terrestrial communities. Our own food, clothing and medical care also rely to a considerable extent on plants (Barthlott et al. 2014).

The most important normative instrument in the field of biodiversity is the Convention on Biological Diversity (CBD), which was adopted at the UN Conference on Environment and Development in Rio de Janeiro in 1992. Currently, 192 countries (including Afghanistan and the European Union) are members of this convention. All UN member states, with the exception of the USA, Somalia and North Korea, have ratified the Convention in an internationally binding manner. The Convention contains three main objectives: (a) protection of biodiversity, (b) sustainable use of its components, (c) equitable sharing of benefits arising from access to genetic resources: The so-called "Access and Benefit-Sharing (ABS)". Biodiversity faces a constant threat despite global efforts, such as the 2002 global treaty to reduce

biodiversity loss by 2010 (the so-called '2010 Target'), which was not achieved; or the Nagoya Summit of the CBD Convention in 2010, which also included a strategic plan, and the so-called '20 Aichi Target'(for more details see Erdelen 2014). Great hopes are currently linked to the establishment of the IPBES (Intergovernmental Science-Policy Platform on Biodiversity and Ecosystem Services) in 2012, which, like the IPCC (Intergovernmental Panel on Climate Change), coordinates global activities in the field of biodiversity. In addition, a paradigm shift in the Green Global Agenda could accelerate our efforts to reduce biodiversity loss (Barthlott & Rafiqpoor 2016).

As regards the protection of biodiversity, the EESC (European Economic and Social Committee) recently stressed that there is no shortage of laws, regulations, political declarations and recommendations in the EU. *The problem is the lack of implementation. This whole judicial framework is not worth the paper on which it is written as long as it is not transformed into real action, it was concluded. The Commission has the tools and means, to encourage the Member States to stick to their obligations. This failure is a sign of the Commission's and Member States' lack of political willingness and cooperation (*www. eesc.europa.eu*). But this can be observed worldwide.

3.8.1 The Uneven Global Distribution of Biodiversity

The extraordinary differences in the number of species per unit area in the different geographical regions of the earth are astonishing. On one hectare of a Siberian taiga, the forest may be formed by only one tree species. On the same area of a rainforest in Amazonian Ecuador, up to more than 300 tree species with BHD >10 cm may grow (Valencia & Balslev 1994) (Fig. 5.37). For comparison, it should be noted that only 40 tree species are native to the entire Federal Republic of Germany (area 375,121 km^2).

In our considerations, we always refer to the relatively well-known land plants. But we would find similar relationships in animals and in marine ecosystems: Comparing a North Atlantic rocky reef with a tropical coral reef leads to similar results (Barthlott et al. 2014). Maps of global terrestrial diversity are best constructed based on plants only. For animals, the state of knowledge is too low (e.g. arthropods), and only a few non-representative groups (e.g. birds or butterflies) have been sufficiently studied; i.e. in general, the determination of biodiversity often encounters great difficulties. It can only be stated for specific groups of organisms, and there are also many different methods and indices etc. (Humphries et al. 1995, Barthlott et al. 2005). The large-scale biodiversity of each region of the world was first compiled in a world map with diversity levels by Barthlott et al. (1996). This map has been continuously

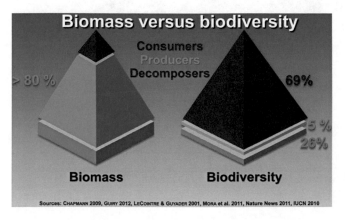

Fig. 3.14 Percentages of producers, consumers and decomposers in terms of their biomass and biodiversity (from Barthlott & Rafiqpoor 2016)

refined and updated by increasing the size of the data base and improving analytical methods (Barthlott et al. 2005, 2007, 2014).

The world map of global biodiversity used here by Barthlott et al. (2014) (Fig. 3.15) is based on analysis of several thousand floras, checklists, and databases (for the method of generating the map, see Barthlott et al. 1996, 1999, 2005, 2007, 2014).

The map shows the spatial global distribution of biodiversity with the best currently achievable high resolution. On this map, areas of high biodiversity are shown in red and those with low biodiversity in light colours (blue → green → yellow). Areas of high biodiversity with more than 3000 species per 10,000 km^2 are concentrated in the tropics and subtropics, especially in mountainous regions. Regions of low diversity are the warm (Sahara, Arabian desert, Atacama desert, etc.) and cold (polar regions, Tibetan plateau, etc.) deserts of the Earth with <100 species per 10,000 km^2. This latitudinal gradient has been known for a long time. The reason is the increasing favourability of hygrothermal parameters towards the equator. A water temperature of more than 26 °C of the sea surface shows a surprisingly good correlation with the highly diverse tropical areas. Where there is too little water (e.g. Sahara) or where edaphic conditions are unfavourable (e.g. nutrient-poor Gran Sabana in Venezuela), species-poor systems can also exist in tropical and subtropical areas. Obviously, however, soil nutrients are not responsible for the development of high biodiversity. On the contrary, ancient, very nutrient-poor areas in Southwest Australia or in the Cape region with extremely poor quartz sands often have an incredible biodiversity.

In the search for causal dependencies for the diversity patterns, a number of principal relationships can be identified. It becomes clear that high biodiversity is by no means linked only to tropical regions: The Caucasus is comparatively richer in species than parts of the Congo lowland rainforest. Here, a second fundamental factor plays the decisive role: The biodiversity of a given space is strongly dependent on habitat heterogeneity (Kreft & Jetz 2007, Kreft et al. 2008), i.e. the diversity of abiotic factors (climate, geology, geomorphology, soils, water availability) within this space, which we subsume under the term 'geodiversity' (Mutke & Barthlott 2005) in contrast to Gray (2004), who subsumes only geological-geomorphological structures and processes under geodiversity. Highest diversities are almost always found in mountainous areas (Agakhanjanz & Breckle 2002). This suggests that higher attention should be paid to the mountainous regions of the world when considering conservation measures on a global scale. They may represent refuges and gene pools for species, especially in the course of global climate change.

If we now combine areas with more than 3000 species per 10,000 km^2 on a map, we obtain a total of 20 centres of biodiversity on Earth (Fig. 3.16). These centres clearly

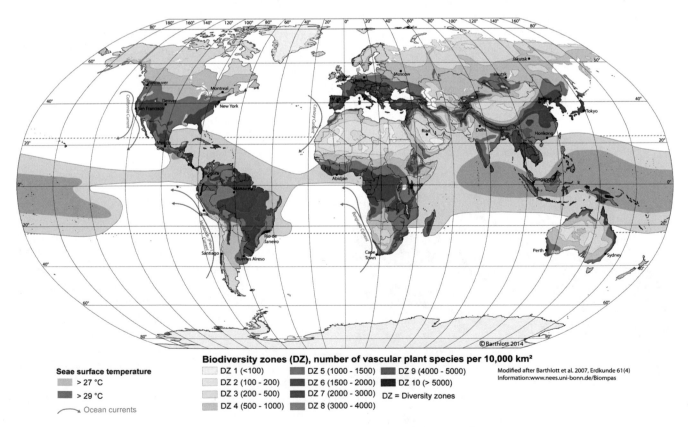

Seae surface temperature
- ▨ > 27 °C
- ▨ > 29 °C
- ⟿ Ocean currents

Biodiversity zones (DZ), number of vascular plant species per 10,000 km²

DZ 1 (<100)	DZ 5 (1000 - 1500)	DZ 9 (4000 - 5000)
DZ 2 (100 - 200)	DZ 6 (1500 - 2000)	DZ 10 (> 5000)
DZ 3 (200 - 500)	DZ 7 (2000 - 3000)	DZ = Diversity zones
DZ 4 (500 - 1000)	DZ 8 (3000 - 4000)	

Modified after Barthlott et al. 2007, Erdkunde 61(4)
Information:www.nees.uni-bonn.de/Biompas

Fig. 3.15 The uneven distribution of global biodiversity: Species numbers of plants per 10,000 km^2 (from Barthlott et al. 2014)

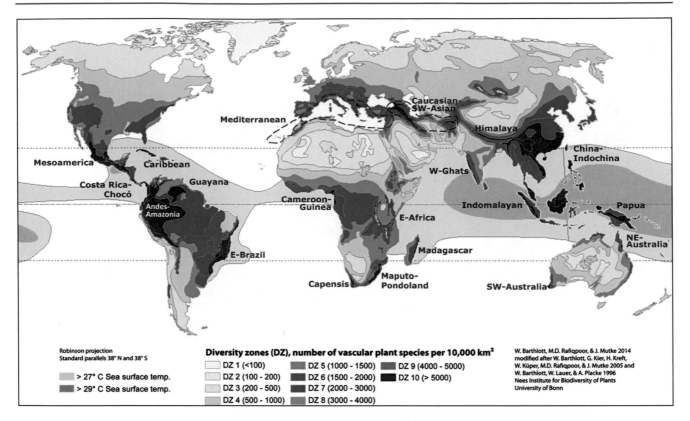

Fig. 3.16 Global centres of biodiversity, each hosting over 3000 species per 10,000 km², largely coincide with mountainous areas of the tropics and subtropics (from Barthlott et al. 2014)

coincide with the mountainous areas of the tropics and subtropics in the Earth's so-called megadiversity countries, which are primarily developing and newly industrializing countries. Afghanistan, with its location in the "Caucasian-SW-Asian" centre of diversity, is one of these megadiversity countries (Barthlott et al. 2014).

Another dimension of biodiversity is the degree of endemism that must be considered for a country or a specific area. Kier et al. (2009) were able to show that, in addition to quantitative aspects, qualitative aspects of biodiversity such as the degree of endemism, i.e. the "specific quality" of an area, play a significant role. A comparison between Hawaii and the federal state of Thuringia in the east of the Federal Republic of Germany would make this clear: The federal state of Thuringia harbours a total of 1570 plant species on an area of 16,200 km², not a single one of which is endemic to this federal state. Hawaii, on the other hand, with the same area as Thuringia (16,600 km²), is home to 1140 native species, 977 of which are endemic to this island. If, in the event of a catastrophe, the entire flora of Thuringia were to be destroyed, there would be no disadvantage for global genetic resources. Such a disaster in Hawaii, on the other hand, would irretrievably destroy a significant part of the genetic resources for mankind. At the same time, this reflects the special role of island systems, but they are not shown on the biodiversity map. Kier et al. (2009) and Weigelt et al. (2013)

have shown elsewhere that oceanic islands comprise only 3% of the land surface but harbour 25% of the known plant species. Among the 10 most endemic areas on Earth, six are islands: New Caledonia, Polynesia-Micronesia, Atlantic Islands, Caribbean Islands, East Melanesian Islands, Madagascar, and Taiwan (Barthlott et al. 2005, 2014).

An interesting but by no means surprising aspect is the contrast between the mega-diversity centres of the tropics and subtropics and the "mega-research centres" of the industrialised nations in predominantly temperate regions. In Fig. 3.17, without claiming to be exhaustive, large research institutions dealing with biodiversity are projected onto the biodiversity map (Barthlott et al. 2014). A North-South gradient is evident.

Afghanistan is not represented in the field of biodiversity research (no red dot). The unequal North-South gradient in research intensity and biodiversity gives rise to a major responsibility for the industrialized nations in terms of "capacity building" to raise awareness of the sustainable use of biodiversity in the megadiversity countries and to protect natural resources for future generations in a sustainable and responsible manner.

Brazil holds the biodiversity record of all countries. The Brazilian rainforest is a determining factor for the world climate. Though Brazil has pledged to stop illegal logging, however, the rainforest in 2020 has shrunk as much as it has

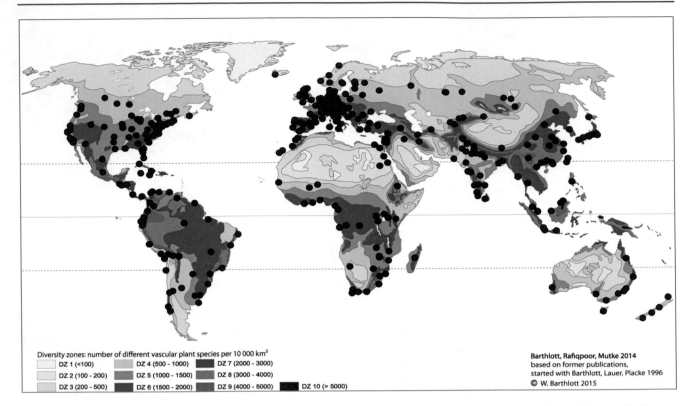

Fig. 3.17 The disparity of "mega-diverse" and "mega-research" countries (red dots) at the global level (from Barthlott & Rafiqpoor 2016)

not since 2008. Lumberjacks, cattle breeders, soil speculators—and an unrepentant government is responsible. The deforestation curve rises and rises. Deforestation in Brazil has been increasing almost steadily for the last 8 years. Now it has reached its highest value in 12 years. Between August 2019 and July 2020, the deforested area in the Brazilian Amazon rainforest was 11,088 km^2—more than the area of Jamaica. The increase compared to the same period of the previous year was 9.5%. The destruction of the rainforests represents a global environmental problem for all of mankind. Burning down the tropical forests not only contributes significantly to the emission of the greenhouse gases CO_2, but also to the irreversible loss of many, quite often still unknown species: This is the real catastrophe! Brazil, which has the largest tropical forests in the world, is increasingly the focus of global attention.

3.8.2　On the Value of Threatened Diversity

The loss of biodiversity since the beginning of the industrial age, but probably only in full force since the 1960s, is alarming (Hammond 1995, Perrings et al. 1997, Duffy 2003). Like the growth of the world population, it appears to be exponential. Despite all international conventions since "Rio 1992" and all political declarations of intent, nothing has fundamentally changed, as the results of "Rio 2012"

shockingly showed (Barthlott et al. 2014). The destruction of habitats, including the clearing of tropical rainforests, continues at an unchanged or even increased rate. At the same time, as a globalization phenomenon, species numbers are increasing locally. The Federal Republic of Germany and the USA have higher diversity than ever before due to invasive species, but globally they are decreasing. Bio-globalization is becoming a central issue. Competitive invasive species also play a major role in studies on the reasons for species extinction (Essl et al. 2008; Klingenstein & Otto 2008; Nehring et al. 2010). A prominent example of the threat and extinction of biodiversity is the extinction of bird roc (Fig. 3.18) in Madagascar (http://t1p.de/uqk1).

The conservation of global biodiversity is only possible with a change in our economic system, and this does not meet with acceptance, especially in industrialized nations, but also in many emerging economies. It is also extremely difficult to convey why we want to maintain a high level of biodiversity (Barthlott et al. 2014). The important ethical aspects should not even be considered here—it is primarily utilitarian considerations that must be brought into the field for conservation (overview in TEEB 2010): Two topics play a role here:

1. The ecosystem value of biodiversity (stability of ecosystems, ecosystems services, for climate, soils, etc.), which can be indirectly expressed in figures and as economic benefits.

Fig. 3.18 Example of the threat to biodiversity on islands: A fossil egg of the elephant bird *(Aepyornis maximus)*, which was native to Madagascar until a few hundred years ago. The elephant birds, which can grow up to 3 m tall and weigh 400 kg, were wiped out after human settlement of Madagascar and survived as the **bird Roc** in "1000 and One Nights" and many other Arabian fairy tales (Photo: Barthlott)

2. The direct use of plants and animals: as food (e.g. rice), medicines, building materials (e.g. wood) or fibre suppliers (e.g. cotton, hemp).

Over 20,000 plant species are used directly by humans. From 100,000 different marine organisms alone, 200,000 extracts are obtained, the pharmaceutical use of which is currently being investigated (http://is.gd/vgiCZ7). Yet over 50% of the diet of billions of people depends on just **four grass species**: Wheat (predominantly Near East), Rice (predominantly Southeast Asia), Maize (predominantly New World) and Millet (predominantly Africa). The diversity of different species and varieties is an important basis for human life. Only one of over 6200 rice varieties studied was resistant to a virus that threatened Southeast Asia's entire rice crop in the 1970s. If all such delicate but also unique apple varieties were replaced in Afghanistan by a slightly higher yielding—

introduced in the early 1970s by Dr. Abdul Wakil—and currently very widespread "Red Delicious" variety, a single virus attack of the newly introduced variety would irretrievably destroy the entire apple crop in Afghanistan. One would then have to look for the resistant native varieties in the remote valleys of Afghanistan to make up for this loss to some extent. The preservation of biodiversity is of high "utilitarian" value for human existence alone.

Every year, new active medical active substances are discovered in plants and animals. For example, drugs against leukaemia and testicular cancer are extracted from the Madagascarian *Catharanthus roseus*. Cultivated and escaped, this plant is also known from eastern Afghanistan (Breckle et al. 2013). Similarly, almost 80% of the new antibacterial agents introduced today are natural products or ingredients derived from them. The same applies to about 60% of the new ingredients introduced in the field of cancer therapeutics during the same period.

2010 was the UN Year of Biodiversity. The countless institutions active in research in the field of biodiversity used this occasion to disseminate information on the protection and sustainable use of biodiversity. Many organizations, institutions, companies and individuals were invited by the UN Secretariat for Biodiversity Conservation (UN-CBD) to participate in this important cause to raise public awareness about the continuing loss of global biodiversity. This key year also saw the publication of the "Field Guide Afghanistan—Flora and Vegetation" with a foreword by the Secretary General of the UN-CBD Dr. Ahmed Djoghlaf (Breckle & Rafiqpoor 2010, p. 7) on the importance and value of this work as a small contribution to Afghanistan as a partner in the association of States Parties that have signed the Convention on Biological Diversity.

The flora of Afghanistan is very diverse due to the high geodiversity of the country. The desert and semi-desert regions of northern and SW Afghanistan fall into diversity level DZ-3 of the world map (Fig. 3.15), with 200–500 species per 10,000 km^2. There is a strong gradient from the lowlands towards the mountains to diversity level DZ-7 (with 3000–4000 species per 10,000 km^2) in the species-rich humid mountain canopies of the Hindu Kush, Kohe Baba and Safed-Koh in Eastern Afghanistan.

Groombridge (1992) has listed species numbers by country. According to this, Brazil hosts about 55,000 species followed by Colombia (35,000 species) and China (30,000 species). India (15,000 species) and Turkey (8000 species) are ranked 12th and 13th respectively. Groombridge (1992) gives an estimated 3500 species for Afghanistan, of which about 30–35% are endemic, i.e. occur only in Afghanistan. About 5–10% await rediscovery according to Groombridge. According to our earlier estimate (Breckle & Rafiqpoor 2010), about 4100 species should occur in Afghanistan, of which about 30% should be endemic. After elaboration of all plant species described to date for Afghanistan (Breckle et al. 2013), about 5000 species now occur in Afghanistan, of which about 25% are endemic. Many parts of Afghanistan have not been targeted for collection. We estimate that a substantial number of plant species could still be newly discovered. According to our findings, Afghanistan with about 5000 species is considered an important hotspot of biodiversity in the Near East after the Mediterranean countries Italy with 5600 species (Pignatti 1982) and Greece with 5700 species (Strid & Kit Tan 1997).

3.9 Zonal, Azonal and Extrazonal Vegetation

Zonal vegetation corresponding to the climate is found only on areas where the typical regional climate is fully effective. Such biotopes are called euclimatopes (Russian: plakor areas). They are flat, slightly elevated areas with deep soils that are neither too permeable (like sand) nor prone to waterlogging.

If we call the vegetation on the euclimatopes zonal vegetation, then, after the biotope change has taken place, it is **extrazonal vegetation** for which the large-scale climate is no longer decisive, but the local conditions. For example, if

forests along rivers extend far into an arid climatic region as gallery forests, these gallery forests are extrazonal vegetation; they are often the basis for irrigated cultivated land (Fig. 3.19). Similarly, the floodplains of rivers and streams (Fig. 3.20) are equally azonal vegetation in forest areas but also in steppes or semi-deserts. The extrazonal vegetation can provide information about the zonal vegetation of a humid or colder zone or of an arid or warmer zone, if the zonal vegetation has been destroyed there.

The term zonal vegetation should only be used in large-scale considerations for the subdivision of the natural vegetation of large areas or entire continents. Only then does the influence of climate become clearly noticeable, and the local differences caused by soil, relief and exposure are less prominent.

On the other hand, under natural conditions, zonal vegetation may be largely absent even over large areas, for example, when groundwater is so high that swamps and bogs cover everything (W Siberia, Sudd swamps in Sudan, tropical swamps in Congo) or in the alluvions of major rivers. A mosaic of vegetation unlike zonal vegetation also grows on extensive lava blankets (Idaho) or on saline soils of wide drainless basins (Aral Sea). In these cases, we are dealing with pedobiomes, or **azonal vegetation,** which is much more strongly influenced by the specific soil properties and on

Fig. 3.19 Gallery forest strip along the broadly meandering (Tajik) Pamir River shortly before the confluence with the (Afghan) Wakhan River (in the background) as an example of azonal vegetation (photo: C. Naumann)

Fig. 3.20 Stream floodplain with dense azonal meadow vegetation north of Almaty (Kazakhstan), surrounded by open forest and steppe (photo: Breckle)

Photo 6 The systematics of the plants begins with collecting them in the field. In this picture is the Norwegian botanist Per Wendelbo (shirtless) and the Danish Botanist Lars Eckberg (opposite Wendelbo) with her Afghan helper during the collecting work in northern Afghanistan (spring 1966; photo: I. Hedge)

which climate has only a weak effect. This does not mean that azonal biotopes would look the same all over the world; they are also influenced by climate, as shown by the very different zonations, for example on seashores.

References

Agakhanjanz, O.E. & Breckle, S.-W. 2002: Plant diversity and endemism in High Mountains of Central Asia, the Caucasus and Siberia. In: Körner, C. & Spehn, E. (eds.): Mountain Biodiversity – A global assessment. Parthenon Publ. Group, Boca Raton, New York etc., Chapter **9**: 117-127

Aubreville, A. 1938: La forêt équatoriale et les formations forestières tropicales africaines. Scientia (Como) **63**: 157

Barthlott, W. & Rafiqpoor, M.D. 2016: Biodiversität im Wandel – Globale Muster der Artenvielfalt. In: Lozán, J.L., Breckle, S.-W., Müller, R. & Rachor, E. (Hrsg.): Warnsignal Klima: Die Biodiversität: 44-50. In Kooperation mit GEO-Verlag. Wissenschaftliche Auswertungen. www.warnsignal-klima.de

Barthlott, W., Biedinger, N., Braun, G., Feig, F. et al. 1999: Terminological and methodological aspects of the mapping and analysis of the global biodiversity. Acta Bot. Finnica vol. **162**: 103-110

Barthlott, W., Erdelen, W. & Rafiqpoor, M.D. 2014: Biodiversity and Technical Innovations: Biomimicry from the Macro-to the Nanoscale. In: Lanzerath, D. & M. Friele (eds.): Concept and Value in Biodiversity. Routledge Studies in Biodiversity Politics and Management, 2014: 300-315. ISBN 978-1-415-66057-0

Barthlott, W., Hostert, a., Kier, G., Küper, W. et al. 2007: Geographic patterns of vascular plant diversity at continental to global scales. Erdkunde **61** (4): 305-315

Barthlott, W., Lauer, W. & Placke, A. 1996: Global distribution of species diversity in vascular plants: towards a world map of phytodiversity. Erdkunde **50**: 317-328

Barthlott, W., Mutke, J., Rafiqpoor, M.D., Kier, G. et al. 2005: Global centres of vascular plant diversity. Nova Acta Leopoldina **92** (342): 61-83

Begon, M., Harper, J.L. & Townsend, C.R. 1996: Ökologie. Spektrum/Heidelberg 750 p.

Breckle, S.-W. 2000: Biodiversität von Wüsten und Halbwüsten. Ber. d. Reinh. Tüxen-Ges. (Hannover) **12**: 207-222

Breckle, S.-W. 2006: Desert and biodiversity – is it area- or resource-related? J. of Arid Land Studies **16**: 61-74

Breckle, S.-W. & Rafiqpoor, M.D. 2010: Field Guide Afghanistan – Flora and Vegetation. Scientia Bonnensis, Bonn, Manama, New York, Florianópolis, 868 S.

Breckle, S.-W., Hedge, I.C. & Rafiqpoor, M.D. 2013: Vascular Plants of Afghanistan – an augmented Checklist. Scientia Bonnensis, Bonn, Manama, New York, Florianópolis, 598 S.

Burrows, C. J. 1990: Processes of vegetation change. U. Hyman/London 551 S.

Duffy, J.E. 2003: Biodiversity loss, tropic skew and ecosystem functioning. Ecology Letters **6** (8): 680687

Ellenberg, H. 1996: Vegetation Mitteleuropas mit den Alpen in ökologischer, dynamischer und historischer Sicht. 5. Aufl., Ulmer, Stuttgart 1096 S.

Erdelen, W.R. 2014: The future of biodiversity and sustainable development. In: Lanzerath, D. & Friele. M. (eds.): Concept and Value in Biodiversity. Routledge Studies in Biodiversity Politics and Management: 149-161. ISBN 978-1-415-66057-0

Essl, F., Klingenstein, F., Nehring, S., Otto, C. et al. 2008: Schwarze Listen invasiver Arten – ein Instrument zur Risikobewertung für die Naturschutzpraxis. Natur und Landschaft **83** (9/10): 418-424

Gray, M. 2004: Geodiversity – Valuing and conserving abiotic nature. J. Willey & Sons Ltd. UK. ISBN 0-470-84895-2

Groombridge, B. (ed.) 1992: Global biodiversity. Status of the earth's living resources. Chapman & Hall/London 585 p.

Hammond, P.M. 1995: The current magnitude of biodiversity. In: Haywood, V.H. and Watson, R.T. (eds.): Global biodiversity assessment. Cambridge University Press: 113-138

Humphries, C.J., Williams, P.H. & Vanewright, R.I. 1995: Measuring biodiversity value for conservation. Ann. Rev. Ecol. Syst. Vol. **26**: 93-111

Kier, G, Kreft, H., Leeb, T.M., Jetz, W. et al. 2009: A global assessment of endemism and species richness across island and mainland regions. PNAS **106** (23): 9322-9327. Doi:https://doi.org/10.1073/pnas.0810306106

Klingenstein, F. & Otto, C. 2008: Zwischen Aktionismus und Laisserfaire: Stand und Perspektiven eines differenzierten Umgangs mit invasiven Arten in Deutschland. Natur und Landschaft **83** (9/10): 407-411

Kreft, H. & Jetz, W. 2007: Global patterns and determinants of vascular plant diversity. PNAS **104** (14): 5925-5930. Doi: https://doi.org/10.1073/pnas.0608361104

Kreft, H., Jetz, W., Mutke, J., Kier, G. et al. 2008: Global diversity of Islands flora from a macroecological perspective. Ecology Letters **11**: 116-127. Doi: https://doi.org/10.1111/j.1461/0248.2007.01129.x

Krutzsch, W. 1992: Paläobotanische Klimagliederung des Alttertiärs (Mitteleozän bis Oberoligozän) in Mitteldeutschland und das Problem der Verknüpfung mariner und kontinentaler Gliederungen (klassische Biostratigraphien - paläobotanisch-ökologische Klimastratigraphie - Evolutions-Stratigraphie der Vertebraten). N. Jb. Geol. Paläont. Abh. **186**: 137-253

Labandeira, C.C. & Sepkoski, J.J. 1993: Insect diversity in fossil records. Science, New Series **261** (5119): 310-315

Leuschner C. 2020: Drought response of European beech (*Fagus sylvatica* L.) . A review. Perspect. Plant Ecol., Evolution and Systematics **47**: 125576

Lozan, J., Breckle, S.-W., Müller, R., Rachor, E. (Hg.) 2016: Warnsignal Klima: Die Biodiversität unter Berücksichtigung von Habitatveränderungen, Umweltverschmutzung und Globalisierung. In Kooperation mit GEO-Verlag. Wissenschaftliche Auswertungen. 354 S.

Mora, C., Tittensor, D.P., Adl, S., Simpson, A.G.B. et al. 2011: How Many Species Are There on Earth and in the Ocean? PloS Biology. DOI:https://doi.org/10.1371/journal.pbio.1001127

Mutke, J. & Barthlott, W. 2005: Patterns of vascular plant diversity at continental to global scale. Biol. Skr. **55**: 521-531

Nehring, S., Essl, F., Klingenstein, F. et al. 2010: Schwarze Liste invasiver Arten: Kriterien-system und Schwarze Listen invasiver Fische für Deutschland und für Österreich. BfN-Skripten **285**. Bundesamt für Naturschutz, 185 S.

Niklas, K.J., Tiffney, B.H. & Knoll, A.H. 1983: Patterns in vascular land plant diversification. Nature **303**: 614-616

Nilsson, L.A., Jonsson, L. & Randrianjohany, E. 1987: Angrecoid orchid and hawmoths in Central Madagascar: specialised pollination systems and generalist foragers. Biotropica **19**: 310-318

Perrings, C., Mähler, K.-G., Folke, C.S. et al. 1997: Biodiversity loss -Economic and ecological issues. Cambridge University Press, Cambridge, UK.

Pignatti, S. 1982: Flora d'Italia I, Edagricola. Bologna

Pimm, S.L., Jenkins, C.N., Abell, R., Brooks, T.M. et al. 2014: The biodiversity of species and their rates of extinction, distribution, and protection. Science **344**: 6187. DOI: 10.1126/ science.1246752

Strid, A. & Kit Tan 1997: Flora of Greece, vol. **1**, Königstein

TEEB 2010: The Economics of Ecosystems and Biodiversity: Mainstreaming the Economics of Nature: A synthesis of the approach, conclusions and recommendations of TEEB. Online available http://is.gd/aL5MWf (accessed 12 Feb 2014)

Valencia, R. & Balslev, H. 1994: High tree alpha diversity in Amazonian Ecuador. Biodiversity and Conservation **3**: 21-28

Walter, H. 1967: Die physiologischen Voraussetzungen für den Übergang der autotrophen Pflanzen vom Leben im Wasser zum Landleben. Z. f. Pflanzenphys. **56**: 170-185

Weigelt, P., Jetz, W. & Kreft, H. 2013: Bioclimatic and physical characterization of the world's islands. Proc. Natl. Acad. Sci. **110**: 15307–15312

Zukrigl, K., Eckhardt, G. & Nather, J. 1963: Standortskundliche und waldbauliche Untersuchungen in Urwaldresten der niederösterreichischen Kalkalpen. Mitt. Forstl. Bundesversuchsanst., Mariabrunn **62**, 244 S.

Part C: Ecological Systems and Ecosystem Biology 4

Contents

Photo 7 Huge and deep rocky slopes with a complicated mosaic of waterfalls and pattern of tropical rainforests (Zonoecotone I/II) near Helbourg, Réunion (photo: Breckle)

4.1 Geo-Biosphere and Hydro-Biosphere

The biosphere comprises the thin layer on the earth's surface in which all life phenomena take place, i.e. on land the lowest layer of the atmosphere, as far as the living organisms permanently reside in it and the plants protrude into it, as well as the rooted layer of the lithosphere, which is called the soil (pedosphere). Besides these, we find life in all waters down to the deep sea. But in this aqueous medium the circulation of matter takes place in a different way from that on land, and the organisms are so different (plankton, for example) that the ecosystems must be treated separately. We therefore divide the biosphere into:

1. Geo-biosphere, comprising the terrestrial ecosystems, and
2. Hydro-biosphere with the aquatic ecosystems that hydrobiology (and also oceanography) deals with.

Today, numerous results from other sciences are required to capture the essential processes in the biosphere. The observation of large-scale ecological relationships is only possible through interdisciplinary evaluation of results; accordingly, ecology has become a very interdisciplinary science today, reaching beyond the original field in biology, as shown schematically in Fig. 4.1.

4.2 The Hydro-Biosphere

The hydrosphere comprises all the solid, liquid and gaseous water occurring on Earth from glaciers and ice caps, oceans, lakes, rivers, soil and groundwater, as well as the water vapour in the atmosphere. It is thus a component of the geo-, litho-, bio-, pedo- and atmosphere.

The earth is covered to 71% by water, nevertheless the hydrosphere is to be treated here in this framework only very briefly.

A quantitative overview of the distribution of water on the globe (Table 4.1) among the various compartments shows that only 3.5% of all water occurs on the entire mainland (Table 4.1). Most of the water is ocean saltwater. The amounts of water that make up the lakes and marshes are also vanishingly small in total. The amount of water in the atmosphere, despite its short residence time of a few days, exceeds that of all the water bound up in living things in the biosphere by a factor of 13.

If only the fresh water is taken into account, $2/3$ of it is bound as ice. Frozen water, i.e. ice, occurs in large quantities in the polar ice caps, in the permafrost soils of the subpolar regions and in the glaciers of the high mountains on Earth. A distinction is made between compact ice, which occurs in various forms in the various compartments (cryosphere), and the less permanent snow, which may be considered separately as chionosphere. The distribution over specific parts of the Earth is given in Table 4.2. Here, the very large imbalance between land and sea is expressed in the

Fig. 4.1 The individual spatial sections on Earth and the field of ecology in the context of other sciences

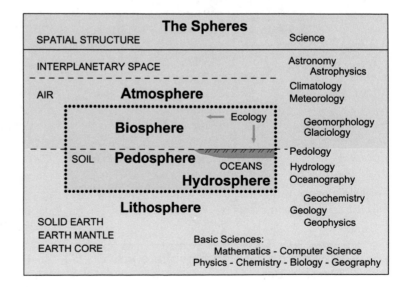

Table 4.1 Quantitative data on the hydrosphere in a global perspective (data from Schönwiese 1994)

Area	Area (10^6 km^2)	Volume (10^6 km^3)	Percentage of the earth's total water volume	Percentage in relation to fresh water
World Wide Sea	361.30	338.00	96.50	–
Mainland	148.40	47.97	3.50	
Groundwater	134.80	23.40	1.70	
Thereof fresh water	(134.8)	10.53	0.76	30.10
Soil moisture	82.00	0.015	0.001	0.05
Polar ice, snow	16.23	24.064	1.74	68.70
Antarctica	13.98	21.60	1.56	61.70
Grassland	1.80	2.34	0.17	6.68
Mountain	0.224	0.041	0.003	0.12
Permafrost	21.00	0.300	0.022	0.86
Freshwater lakes	1.236	0.091	0.007	0.26
Saltwater Lakes	0.822	0.085	0.006	–
Swamps, bogs	2.683	0.0115	0.008	0.03
Watercourses		0.0021	0.0002	0.006
Water of the atmosphere	(510)	0.0129	0.001	0.04
Biologically bound water	(510)	0.0011	0.0001	0.003

Table 4.2 Quantitative data on the cryosphere and chionosphere (after Schönwiese 1994)

Area	Area (10^6 km^2)		Volume (10^6 km^3)	Sea level equivalent (in m)[a]
Land ice	14.44		32.44	81.2
Antarctica	12.2		29.32	73.3
Greenland	1.7		3.0	7.6
Glacier	0.54		0.12	0.3
Permafrost (without Antarctica)				
Constantly	7.6		0.03	0.08
Maximum	17.3		0.07	0.18
Sea ice				
Arctic	Winter	14.0	0.05	–
	Summer	7.0	0.02	–
Antarctic	Winter	18.4	0.06	
	Summer	–	0.02	
Snow				
Northern Hemisphere	Winter	46.3	0.002	Negligible
	Summer	3.7	<0.0001	Negligible
Southern Hemisphere	Winter	0.85	Negligible	Negligible
	Summer	0.07	Negligible	Negligible

[a]Potential rise in sea level with complete melting. Data from Schönwiese (1994)

distribution on the two hemispheres, which is ultimately also reflected again in the distribution of the zonobiomes.

4.3 Division of the Geo-Biosphere into Zonobiomes

Our object of study is only the geo-biosphere, which is the main habitat of man and therefore of particular interest to us. For its subdivision, the large-scale climate lends itself as a primarily independent environmental factor. This is because both soil formation and vegetation depend on it, it is still hardly (or only slightly, but increasingly) changed by humans and can be recorded perfectly everywhere by the increasingly dense network of meteorological stations (on the principles of subdivision, cf. Walter 1976 and Lauer & Rafiqpoor 2002).

From a climatic point of view, a division of the Earth into zonobiomes should be based on an effective climate classification, i.e. one based on the ecophysiological characteristics of vegetation, because such a climate classification relates to natural vegetation, can best circumscribe the orobiomes, and the ecologist is primarily interested in the climate within the geo-biosphere (Breckle 2011). The climate classification also used here (Fig. 2.50) divides the Earth into five main zones according to the length of the thermal and hygric growing season: (A) Tropics, (B) Subtropics, (C) Cool mid-latitudes, (D) Cold mid-latitudes, (E) Polar regions.

Table 4.3 The soil and vegetation types of the individual zonobiomes

Zonobiome (ZB)	Zonal soil types	Zonal vegetation types
I	Equatorial brown loams (ferrallitic soils, latosols)	Evergreen tropical rainforest without seasonal change
II	Red loams, red earths (fersiallitic soils)	Tropical deciduous forest or savannahs
III	Serosemes, syrosemes (grey or red soils, raw soils, saline soils)	Subtropical desert vegetation (rocky landscapes)
IV	Mediterranean brown earth (fossil Terra rossa)	Hardwood vegetation (sclerophylls), (sensitive to soil frost)
V	Yellow or red forest soils, slightly podsolic	Temperate evergreen forest (lauriphylls), (frost sensitive)
VI	Forest brown soils and grey forest soils	Nemoral winter bare deciduous forest (frost resistant)
VII	Chernosemes to serosemes (raw soils)	Steppes to deserts with cold winters (frost resistant), short, hot summers
VIII	Podsole (raw humus bleaching earths)	Boreal coniferous forests (taiga), (very frost resistant)
IX	Humus-rich tundra soils with solifluction (permafrost soils)	Tundra vegetation (treeless)

The climate within the geo-biosphere can be clearly indicated by the ecological climate diagram. It proves to be expedient to further subdivide the very large zone of the mid-latitudes and to combine the subpolar as well as the high-polar to an arctic one. This then results in nine ecological climate zones, which we refer to as Zonobiomes (ZB) in an ecological sense (Walter 1976, Walter & Breckle 1999, Schultz 2008, Olsen & Dinerstein 2002, Wittig & Niekisch 2014), because biome is understood to be a large, climatically uniform habitat within the geo-biosphere. **Humid** refers to a humid (rain-fed) climate, while **arid** refers to a dry (rain-poor) climate. In the case of double designations, the first refers to summer, the second to winter.

The nine zonobiomes (Fig. 2.9):

ZB I Equatorial ZB with diurnal climate, humid tropical ZB

ZB II Tropical ZB with summer rain, humido-arid tropical ZB

ZB III Subtropical ZB with desert climate, hot arid ZB; sparse rainfall

ZB IV ZB with summer drought and winter rain, arido-humid (Mediterranean) ZB

ZB V Warm temperate (oceanic), humid ZB; mild-maritime ZB

ZB VI Typical temperate ZB with short frost period, nemoral ZB

ZB VII Arid temperate ZB with cold winters, continental ZB

ZB VIII Cold temperate ZB with cool summers and long winters, boreal ZB

ZB IX Arctic including Antarctic, with very short summers, polar ZB.

The zonobiomes are the main large units of the biosphere. In the literature, there are many different models with different names. For the large-scale division into ecological units, the zonobiome division has proven itself. The respective boundaries and size of the transitional areas (zonoecotones) is often a matter of subjective opinion. The two polar ice caps—the Arctic and the Antarctic—are joined by zonobiomes more or less in the shape of a belt around the globe. The difference between the northern and southern hemispheres is considerable due to the unequal distribution of the land mass. It is therefore nonsensical to speak of "anti-boreal", a zone that does not exist in the westerly wind belt of the southern hemisphere—where there is no land, only tiny islands—in contrast to the northern hemisphere, where at this latitude there is almost only land mass with the strongest expression of continentality.

Anti-nemoral should be understood only as the zoogeographic region of the shelf areas of the southern half of the southern continents. After all, no one speaks of the North Desert as an "anti-subtropical" desert or the *Nothofagus* forests of Chile as "anti-nemoral" forests.

The differences between the west and east coasts on the continents are responsible for the wedging out of certain belts or for the small-scale occurrence of the zonobiomes due to the warm or cold ocean currents.

The zonobiomes are clearly defined by climate diagram types; moreover, they correspond largely, if not always, to certain zonal soil and vegetation types, as shown in the overview in Table 4.3.

4.4 Zonoecotone

The climatic zones and thus also the zonobiomes are not sharply delimited from one another, but are often connected by very broad transition zones—the zonoecotones (ZE).

Ecotones are, for example, small-scale: a forest with a forest edge is replaced by meadows with a mantle and fringe, or large-scale, for example in E Europe: The deciduous forest gradually merges into the steppe.

In the zonoecotone, both types occur side by side under the same large-scale climatic conditions and are in sharp competition with each other. The decisive factor for the occurrence of one or the other vegetation type is the relief-related microclimate or the soils.

Box 4.1 Ecotones as Transition Zones
Ecotones are ecological areas of tension; transitional areas in which one vegetation type is more or less gradually replaced by another.

Thereby either a diffuse interpenetration of the two types or a mosaic-like arrangement comes out. First, one type is more strongly represented, then the two balance each other out, until the second becomes more and more predominant and the first disappears completely, with which the new zonobiome begins.

The zonoecotones are named according to the zonobiomes that connect them, i.e. we distinguish the zonoecotones: ZE I/II, ZE II/III, ZE III/IV, ZE IV/V, etc.

Triangular zonoecotones can also occur when three zonobiomes meet (e.g. Pannonian Plain: ZE VI/VII/IV). We treat the most important zonoecotones in short sections of their own at the end of the corresponding zonobiomes.

The geographic distribution of the individual zonobiomes and zonoecotones can be seen from the schematic map of the world (Fig. 2.9) or from the maps for the individual continents Figs. 4.22, 4.23, 4.24, 4.25, and 4.26.

4.5 Ecological Systems

On the basis of what we have said so far, we can draw up a scheme for the ranks of both the major and minor ecological units. In addition to the main climatic series, the biomes modified by mountains on the one hand (OB) and by specific soil conditions on the other hand (PB) are thereby marked as corresponding orographic or pedological subsidiary series. This hierarchical scheme of spatial units of ecosystems is used as a basis for the division of the geo-biosphere.

In this context, it should be recalled here once again the very different scales that one has to take care of when characterizing ecosystems (Fig. 4.2). This applies not only to the spatial scales of the structures, but also to the temporal scales. In particular, it is atmospheric or meteorological phenomena that determine the scale size in this context. In Fig. 4.2, the ranks of ecological systems used here are contrasted with those of meteorological processes. Because of the huge differences in scale, such a comparison can only be made on a logarithmic scale. Secondly, this also applies to the time scale on which certain phenomena occur (Fig. 4.3).

For the treatment of the individual biomes, the near-ground air layers with their dynamics and the atmospheric-biospheric interactions play the decisive role, as **Geiger** already pointed out in 1927.

Fig. 4.2 Horizontal-spatial scales in biology, geography and meteorology. Note the logarithmic scale (modified after Schönwiese 1994)

Geography/Ecology		Diameter		Meteorology	
Term	Example	km	m	Term	Example
Zone	Climate zone	100^3		Rossby-Wave	Macro alpha
Region	M Europe	10^3			Macro beta
				ectrop. Cyclone	Meso alpha
Large landscape	Mountain	10^2		tropical Cyclone Foehn	Meso beta
Landscape	City	10		Weather front* Thunderstorm	
Terrain	Valley, basin	1			Meso gamma
				Cumulus-Cloud	Micro alpha
	Mountain slope	0,1	100	Tornado	
	Slope incision		10	Thermals Vortex	Micro beta
Location	Tree		1	Flight turbulence	Micro gamma
			0,1		
	Leaf		0,01	Heat flicker	
Boundary layer	Leaf surface		0,001		

* perpendicular to flow direction (much larger when parallel)

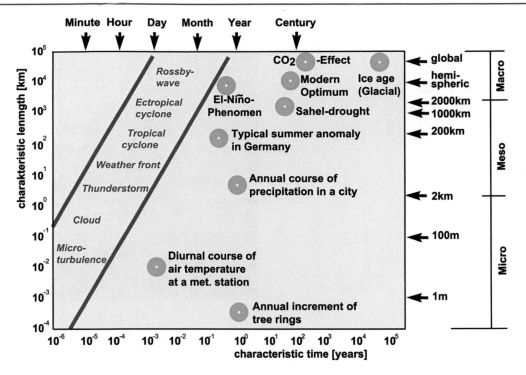

Fig. 4.3 Atmospheric and climatological phenomena in the space-time diagram. Note the logarithmic axes (modified after Schönwiese 1994)

4.6   Orobiomes and Pedobiomes

The geo-biosphere is not only structured horizontally, but also vertically by the mountains. It must therefore be viewed three-dimensionally. The mountains stand out climatically from the climate zones and are therefore treated separately from the zonobiomes. We refer to them as **orobiomes** (OB).

It is characteristic for all orobiomes that the mean annual temperature decreases with altitude. This decrease is about as large per 100 m difference in elevation as that in the Euro-North Asian plain at a distance of 100 km in the direction from south to north. Therefore, the elevational zones in the mountains are about 1000 times narrower than the vegetation zones in the plain from south to north.

Certain similarities of the elevational belts and of the vegetation zones of the higher latitudes are noticeable in Europe and N America on cursory observation, but differences are always present. For, except for the decrease in temperature and the shortening of the growing season with elevation, the mountain climate is different from that on the plains. For example, the length of the day, like the position of the sun, does not change with elevation; on the other hand, the length of the day in summer increases from south to north, while the altitude of the sun decreases at noon. Direct solar radiation increases with altitude, while diffuse radiation decreases. In the plains, the opposite is true in a northerly direction. Precipitation usually increases very rapidly with elevation in the mountains, but is low in the Arctic region.

In addition, the two flanks of a mountain range are almost never symmetrical, but also climatologically different, for example due to foehn effects. As a result, for physical reasons (Fig. 4.4a), the temperature gradient is also no longer the same, because moist-adiabatic and dry-adiabatic heating (cooling) differ energetically. As the air cools due to uplift (and volume expansion as atmospheric pressure decreases), the dew point is eventually reached; this is the condensation level at which clouds form ('cloud forest') and moisture 'rains out'. On the other side of the mountains, when there is only dry-adiabatic heating as the air sinks (with pressure rising), the air is ultimately warmer and drier at the same valley level: Foehn effect (Fig. 4.4a, b), caused by the released heat of condensation (2.26 MJ kg^{-1}, ▶ physical quantities) as the air rises.

Each mountain range within a zonobiome is an ecological unit with a typical sequence of elevational zones (belts), the belts of which are generally referred to as colline, montane, alpine and nival. However, they vary considerably in detail depending on the zone in which the mountain range is located. For example, the elevational belts of mountains in zonobiome I, IV or VI have hardly anything in common.

The further subdivision of the orobiomes is therefore according to the zonobiomes to which they belong. We therefore speak of orobiome I, orobiome II, etc. Furthermore, uni- inter- and multizonal orobiomes (mountains) are

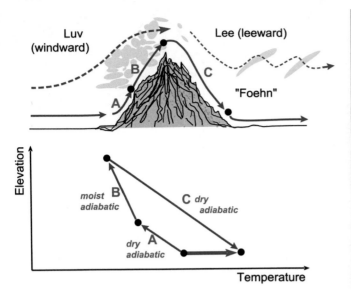

Fig. 4.4 Scheme of the Foehn effect. On the windward side, the air masses are lifted (**a**), then clouds form (**b**), possibly with precipitation. On the lee side (**c**) there is warming, dryness and turbulence. The warming is due to the difference between the moist-adiabatic temperature gradient (**b**) (heat of condensation) during uplift and the dry-adiabatic gradient (**c**, **a**) during sinking (with additional radiation in clear air) (modified after Schönwiese 1994). The photo on the right shows the distinct condensation level on the windward side of the mountains in the Eastern Alps (photo: Breckle)

distinguished, depending on whether they lie within a zonobiome or between two zonobiomes or extend through many, such as the Urals (from IX to VII) or the Andes (from I to IX). Interzonal mountains are the Alps, the Caucasus, or the Himalayas. They are usually sharp climatic boundaries and the elevational belts on the northern and southern margins must be treated separately. In the case of a multizonal mountain range, it is necessary to divide the same according to the zones into separate sections with particular altitudinal sequences. The Andes are both multizonal and interzonal (west and east slopes are different). The elevational belts are also different for inner mountain valleys with low precipitation and continental conditions (intra-mountain elevational belts).

> **Box 4.2 Orobiomes and the Elevational Belts of the Mountains**
> Orobiomes are mountain habitats that are structured according to elevational belts. The individual elevational belts are also referred to as hypsozonal or orozonal vegetation. It is the third dimension that stands out from the associated zonobiome.

> **Box 4.3 Altitudinal and Climatic Zones**
> Altitudinal zones in the mountains are only superficially a short-circuited repetition of the planetary vegetation zones in the plains toward the poles.

Not only do the orobiomes stand out from the zonobiomes, but certain areas with extreme soils and azonal vegetation also behave differently. We refer to them as **pedobiomes** (PB), that is, habitats tied to specific soils. Soils are strongly modified by humans only where soil erosion, i.e. removal of the upper soil layer or the entire soil, has been caused, or where the soil has been processed by agriculture or built over. The large-scale climate affects vegetation unchanged only on the euclimatopes (called "plakor" in Russian), i.e. on flat surfaces with soils that are not too heavy and not too light, so that precipitation does not run off superficially but penetrates into the soil and is retained by it as adhesive water, i.e. does not sink too quickly to the groundwater, it is thus fully available to vegetation. This is not the case with extreme calcareous soils, they are too dry and at the same time too warm biotopes compared to the large-scale climate. On the other hand, the soils may contain harmful substances, such as salts (NaCl, Na_2SO_4), or the soils are extremely poor in nutrients, so that the vegetation also deviates from the normal of the zonobiome. The vegetation of the pedobiome, which is less influenced by the large-scale climate, but much more by the soil and therefore can occur in almost the same formation on the same soils in several zones, we call azonal vegetation.

The pedobiomes are subdivided according to the soils that are typical for them: Lithobiomes (stony soils), Psammobiomes (sandy soils), Halobiomes (saline soils), Helobiomes (marsh or swamp soils), Hydrobiomes (soils covered with water), Peinobiomes (deficient or nutrient-poor soils, from peine in Greek = hunger, lack), Amphibiomes (= alternating moisture soils) and others.

Pedobiomes can often occupy vast areas, for example, the lithobiome of the basalt covers in Idaho (USA), the psammobiome of the S Namib, the Rub-al-Khali in Saudi Arabia, or the Karakum Desert in C Asia with 35,000 km^2, the helobiome of the Sudd marsh on the Nile (150,000 km^2), the marshlands of W Siberia (over 1 million km^2, Fig. 12.19). Their ecology must also be treated separately from that of the zonobiomes.

Box 4.4 What Are Biomes?
Biomes are habitats that correspond to a distinct uniform landscape.

4.7 Biome

By biome (without prefix) we mean the basic unit of large ecological systems.

Biomes are either subunits of zonobiomes (Fig. 4.5) or belong to specific orobiomes or pedobiomes, for example, the central European deciduous forest is a biome of zonobiome VI, Kilimanjaro is a biome of orobiome I, the Salt Desert in Utah (USA) is a biome of pedo-halobiome VII, etc.

In this global overview, biomes are mainly treated as the smallest units of a region.

In Anglo-American literature, the term "biome" is used much more broadly and less sharply defined.

4.8 Small Units of the Ecological System: Biogeocenes and Synusia

If one has made a global classification of the entire land surface of the earth within the nine zonobiomes into the next smaller units (biomes), then one can subdivide these in each case according to the state of knowledge into smaller units.

For the delimitation of small ecological units, it is most appropriate to start from vegetation units. In a limited, landscape-geographically uniform area corresponding to a biome, even slight differences in water and soil conditions are important for the formation of vegetation and thus of ecosystems. This is how the typical ecosystem mosaic of a landscape is formed. It is hardly possible to directly measure the decisive environmental factors, which exhibit constant seasonal changes, and to record their interaction. In contrast, we can assume that the natural vegetation, which is in dynamic equilibrium with its environment, reflects the effect of the environmental factors in an integrating manner. Even small differences in an environmental factor cause a qualitative or at least a quantitative change in the composition of the vegetation cover.

However, since human interventions are now evident almost everywhere, to a greater or lesser extent, caution is called for. It is necessary to carefully distinguish the effects of natural and anthropogenic factors through a critical analysis

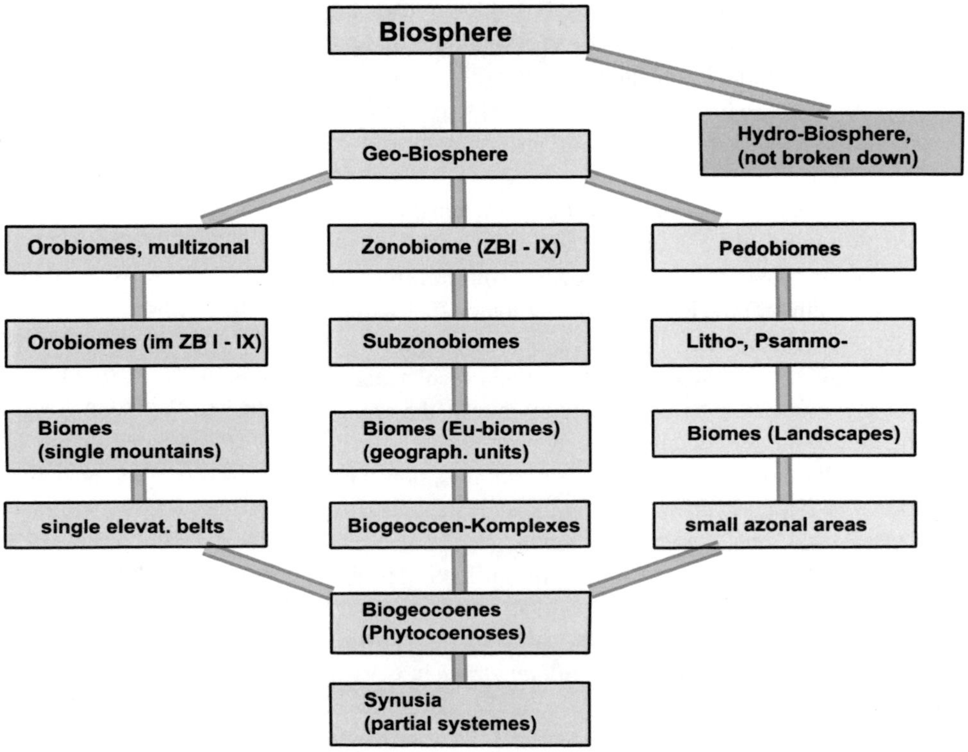

Fig. 4.5 Scheme of the hierarchical organization of the ecological systems of the geo-biosphere

and, in the case of the latter, also to take into account human interventions in the past.

In forest communities, human interventions have an effect even after centuries (clear-cutting, type of regeneration, grazing, litter use, etc.). Although it is often believed that the herb layer in the forest is better suited for the assessment of natural conditions, it depends to a particularly high degree on the composition and structure of the tree layer (shading, higher competitive strength of tree roots, leaf litter) and roots less deeply than the trees, so that only the upper soil horizons are decisive for it. Any change in the tree layer by humans also affects the herb layer. Even the removal of old hollow trees and the trunks decaying on the ground is a serious intervention in the ecosystem.

In densely populated areas, however, we will have to accept the fact that only human-influenced ecosystems are present.

The position of the biogeocene in the size hierarchy of ecological systems is shown in Fig. 4.5.

In an ecosystem, the material cycle, energy flow and phytomass as well as production are mainly determined by the **dominants,** in the forest, for example, by the dominant tree species. Character species that are rare and occur in few specimens may have indicator value for community recognition, but they may not exert the slightest influence on the ecosystem. Therefore, ecosystem research must demand consistency of dominants within an ecosystem type.

Actually, the delimitation of plant communities in the field has to be done after thorough orientation about the prehistory of the individual stands and after exact exploration of the whole area as well as considering the site conditions and the soil profile down to the lower limit of rooting. The ecologist can only examine real (usually heterogeneous) stands and not the abstractly defined associations (plant communities) of plant sociology.

The biogeocene is the basic unit of ecosystems, but not their smallest unit. Within a biogeocene, one can distinguish a number of **synusia.** These are '**working communities'** of species with similar development and ecological behaviour. However, we must not call the synusia ecosystems; for they are only subsystems that do not have their own material cycle. They rather fit into the material cycle of the whole ecosystem, and the production of the synusia is only a small part of the total production of the ecosystem; it is, however, of importance because the turnover in the synusia is usually much faster than in the whole ecosystem.

A typical **example of synusia** are the different species groups with a similar development rhythm and the same demands on environmental factors, such as the **spring geophytes of the deciduous forest** (*Allium ursinum, Corydalis, Anemone, Ficaria* and others), which take advantage of the light phase on the forest floor before foliage emerges (Fig. 4.6), or the herbs that persist during the shade phase in summer, or the herbs with evergreen leaves. Synusia of lower

Fig. 4.6 Spring geophytes in a deciduous forest in the Baybach Valley in the lower Moselle Valley that is not yet green. Before the forest becomes green, the spring geophytes (here *Corydalis cava*) take advantage of the light phase in the forest floor and can complete their entire generative cycle until the forest is fully green (photo: E. Fischer)

plants are the **lichens on the tree trunks** (Fig. 4.7) or the **mosses at the base of the trunks** (Fig. 4.8).

Between the biomes on the one hand and the biogeocenes on the other, there is a large gap that must be filled by units of intermediate rank. These are biogeocene complexes that often coincide with certain landforms and are based on a common genesis, or are linked by dynamic processes. As an example, we mention a biogeocene series on a slope with lateral transport of matter (often with a soil catena, i.e. a sequence of certain interdependent soil types) or lawfully arranged biogeocenes in a river valley or in a basin without drainage, etc. One can also think of biogeocenic complexes with biogeocenes that follow one another in time, as in secondary succession, or biogeocenes that belong side by side to an ecological series that arises in the presence of a constantly changing site factor (decreasing water table or increasing soil depth), etc. The spatial extent of such biogeocenic complexes can be very different. The designations for the individual types diverge greatly. We will content ourselves with using the neutral term, **biogeocene complexes.**

Fig. 4.7 Lichens on a tree trunk as special synusia (photo: Rafiqpoor)

Fig. 4.8 Mosses grow on tree trunks and stones, especially in forests, as the forest floor between the trees becomes covered with a thick layer of dead leaves; eastern slope of Mt. Kinabalu on Saba, Malaysia (photo: Rafiqpoor)

Box 4.6 Real and Theoretical Plant Communities
The plant community ($\rightarrow$ biogeocene as a real spatial unit) is something different from the plant association ($\rightarrow$ theoretically defined construct), the Association as a basic syntaxonomic unit of plant communities—it is a type, the overlying type categories are: Alliance, Order, Class.

Box 4.5 Biogeocene as the Basic Unit of Ecosystems
The basic unit of the smaller ecosystems is the biogeocoenosis (short: biogeocene). It corresponds to a concrete plant community in the range of an association, it is so to speak the walkable ecosystem with for example 20 × 20 m size.

All ecological units are real. The ecologist is only able to carry out his measurements and studies exclusively on real ecosystems and not on abstract units. Only sufficiently extensive real data collection can serve to formulate theoretical models. These summaries, drafted at a desk on the basis of experience gained, must be based on certain assumptions. They will therefore never fully correspond to real ecosystems, but their clarity and comparative view can facilitate our understanding of ecosystems and the processes taking place within them. Provided that they are based on sufficient experience and constant evaluation, they can even enable predictions to be made about future developments (Wissel, oral comm.).

4.9 Ecosystem Biology and the Nature of Ecosystems

Having pointed out the small ecological units, we need to learn more about the principal structures and processes in ecosystems. Here, one often takes as an example a uniform

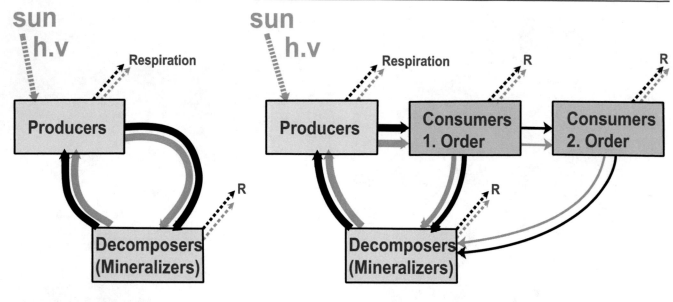

black: material cycling (inorganic and organic)
yellow: energy flow (chemical energy and heat transfer)

Fig. 4.9 Scheme of the simplest ecosystem (left) with material cycle and energy flow (**E**: energy flow; energy = ability to do work); **R**: respiration, respiratory energy; the same with the consumer compartment (right)

deciduous forest stand of zonobiome VI, which has a manageable size and is easily 'walkable'.

If the stand comprises a very specific, limited and homogeneous society, for example a forest, a bog, etc., then it is more appropriately called a biocoenosis. A unit of plants and animals, including the rooted soil and the air layer near the soil, into which the plant organs project, is called a biogeocoenosis (in short: **biogeocene**). The biogeocoenosis 'deciduous forest' describes the static picture, the spatial structure, the organisms. In such a plant community, however, a material cycle and an energy flow constantly take place. Together with the animal organisms and the inorganic environment, the plants form a dynamic structure, an ecosystem, which is not closed in itself, because there is a supply of energy from outside through solar radiation and a supply of matter through precipitation, gas exchange, dust deposition, etc., but at the same time there is also a release of energy in the form of disordered heat energy and matter (water running off or seeping away, through gas exchange, etc.). The dynamic picture of such a spatial section, i.e. the essential structures and processes, is studied by **ecosystem biology**.

The total plant dry matter in a biogeocene is its **phytomass**, and that of animals is its **zoomass**. Together they form the **biomass**. With regard to the role played by each group of organisms in the ecosystem, a distinction is made:

1. **Producers**: They are autotrophic plants that store light energy as chemical energy during photosynthesis by

forming organic compounds from CO_2 and H_2O and extracting mineral nutrients and water from the soil.
2. **Consumers**: These are heterotrophic animal organisms which, as phytophages, use the plants as food and convert a small part of them into animal matter. Predators that eat the phytophages (food chain, food web) also belong to this group.
3. **Decomposers** (mineralizers): They are mostly found in the soil (saprophages, bacteria, fungi) and ultimately degrade all plant and animal remains to CO_2 and H_2O. They mineralize the organic material, thus closing the material cycle.

As the simplest ecosystem, one can imagine the interplay of producers and decomposers (without consumers) (Fig. 4.9). Indeed, there are terrestrial ecosystems in which consumers play only a very minor role.

The total organic matter produced annually by the photosynthesis of plants is called **gross production**; the matter remaining after deducting the amount respired by the plants is called **net** or **primary production**; the matter formed by animal organisms is called **secondary production**. The latter is much smaller. Generally, only a few percent of primary production is consumed by consumers (long cycle, Fig. 4.9); most of it ends up in the soil and is almost completely decomposed by the decomposers (short cycle, Figs. 4.9 and 4.10), producing H_2O, CO_2, and mineral salts. The dead organic matter (litter) is previously crushed by lower animals—the saprophages or mould eaters—during the

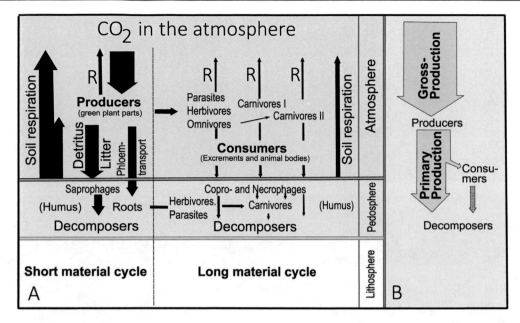

Fig. 4.10 (**a**) Scheme of the short and long metabolic cycles in a deciduous forest biogeocene. (**R**) Respiration (breathing). Thickness of arrows approximates turnover rates. (**b**) Scheme of the energy flow

feeding process. The CO_2 escaping from the soil is called **soil respiration**. The short cycle plays the main quantitative role in terrestrial ecosystems (Fig. 4.10).

The long cycle runs via the consumers, i.e. via the herbivores or phytophages, and via the zoophages or predatory organisms or via the omnivores, which eat both plants and animals. In addition, there are consumers of the 2nd or even 3rd order, but their metabolic rate is vanishingly small (Fig. 4.10). We must also include the parasites of plants among the consumers.

The excreta and corpses of the animals also re-enter the soil where they are prepared by animal organisms (coprophages, necrophages) for decomposition by microorganisms.

Although the long cycle is quantitatively only of minor importance, it plays an even greater role in regulating the balances in the entire ecosystem. Consumers could therefore also be called regulators. As soon as a certain plant species increases disproportionately in the ecosystem, the number of animals consuming it usually also increases. This reduces the population density of the plant species, which then leads to a decrease in phytophages. However, the settling to an equilibrium is rarely constant. Rather, one usually observes cyclic oscillations of population densities, each with typical phase shifts. Such a phase shift exists again between the phytophages and their zoophagous enemies. These regulatory processes, which can be described cybernetically as control loops with feedback, ensure that the ecosystem is kept in a dynamic equilibrium (steady state). It is true that population densities will always fluctuate to a certain extent, but only within certain limits. Such fluctuations are also caused

by changing weather conditions in the individual years, which sometimes favours one plant species over another in the competition.

The interlocking control loops are based, on the one hand, on direct control processes by animals such as pollination (zoogamy) or fruit and seed dispersal (zoochory), and, on the other hand, on food chains that begin with herbivory. The long cycle consists of a whole series of such food chains, which would usually have to be described more accurately as **food webs** and which, despite all fluctuations, give the ecosystem great stability on average. By destroying the predators or by further interventions, man disturbs just these food chains, whereby the whole ecosystem gets into disorder or even collapses (Gigon 1974), or is replaced by another one.

It will continue to be an important task of zoo-ecologists in the future not so much to elucidate the quantitative ratios of secondary production as the various food chains in all their details. For the phytophages and the predators are often strictly specialized in particular species on which they feed. Despite their low density, they have great regulatory importance. A wealth of different adaptations plays a role in this.

Another of many amazing examples from the tropics are the close dependencies between ants and plants. On the one hand, there are the leafcutter ants, which bring pieces of leaves into their nests and grow mushrooms on them (i.e. engage in agriculture), which are their main food source (Fig. 4.11). The close dependence of some ants on *Cecropia* species (of the Neotropics) or *Macaranga* species (of the Palaeotropics) should also be mentioned (Fig. 4.12). The hollow stems of the rapidly growing pioneer trees are colonized by the queen ant through entrance holes preformed by the

Fig. 4.11 A road of leafcutter ants in the lowlands of Tena, Ecuador (photo: Rafiqpoor) and single specimens of these ants transporting a leaf section (taxi!) on a tree trunk, Costa Rica (photo: Breckle)

plant. As the young trees grow, the ant colony increases and continually colonizes new internodes. In addition, protein- or fat-rich food corpuscles are formed on the leaf bases by the plant, which additionally attract the ants. The investment is worthwhile for the plants, because the ants keep the plants free from other herbivores, so to speak as a protective police force.

As can be seen from this, not only the quantitative magnitudes of certain processes, but also the qualitative significance of some processes are quite essential for the stability of natural ecosystems through the interconnectedness of processes.

Parallel to the material cycles, the flow of energy takes place. Solar energy is converted into chemical energy during the photosynthesis of the producers, which is used by themselves, by the consumers and the decomposers for the maintenance of the life processes. In the process, chemical energy is constantly lost as heat in the respiration and fermentations of the microorganisms, until finally, after complete decomposition, it is entirely consumed. This energy flow is shown in Fig. 4.10.

The composition of an ecosystem and its structures can only ever be represented as a model, and there are numerous ways of doing this. Another example is shown in Fig. 4.13, where above all the functional compartments and their interconnections are highlighted.

Box 4.7 Regulatory Importance of Phytophages in Ecosystems
As an unusual example, *Witheringia solanacea* from the tropical rainforest of Costa Rica should be mentioned. The berries contain a natural laxative, so that the birds empty the intestine in less than 10 min and thus distribute the seeds on the forest floor. After this short intestinal passage, 70% of the seeds are still germinable, with longer intestinal passage the rate drops to 20%.

Fig. 4.12 *Cecropia* trees are conspicuous from a distance in the rainforests of the New World because of their shimmering silver leaves. The hollow stems of *Cecropia* are inhabited by ants that live in symbiosis with *Cecropia* (photo: left: Rafiqpoor, right: Barthlott)

Quantitative data of the different ecosystems are helpful for a comparison. Table 4.4 shows essential ecosystem parameters for an oak forest.

The phytomass of forest communities (Table 4.4) is so high because dead wood mass is stored in the stems (heartwood 150 t ha^{-1}). But even without this, the phytomass is more than 1000 times higher than the zoomass.

For the latter, the following figures are given in European forests: Reptiles 1.7 kg ha^{-1}, birds 1.3 kg ha^{-1}, mammals (mainly small species, rodents) 7.4 kg ha^{-1}. Much larger is the mass of invertebrates, especially subterranean (up to 14 kg ha^{-1} TG, 90% dipteran larvae). Figures are available for an American deciduous forest with *Liriodendron* (Reichle 1970) as TG (each in kg ha^{-1}): Aboveground: phytophagous arthropods 2.43, predatory A. 0.61. In the litter: Larger invertebrates 8.42, smaller W. 3.42. In the soil: earthworms (*Octalasium*) 140, smaller invertebrates 2.2.

In the mixed oak forests of E Europe, it was found that after clear-cutting of oaks by caterpillars, the wood increment of ash and lime trees increased due to the better light conditions and overcompensation occurred; in the four years after the caterpillar epidemic, there was a total wood increment of 10%.

Even in a pure pine stand of different ages, compensation occurred over time after infestation with *Dendrolimus pini* by promoting the suppressed and less infested trees. Wood increment in the 2nd year decreased to 76% and in the 3rd year to 56%, but it increased to 150% and 194% in the 4th and 5th years, respectively (Walter & Breckle 1999). Even moderate grazing of grasslands stimulates vegetative growth of grasses to such an extent that total annual production increases when the amount eaten is taken into account (Walter & Breckle 1999). A similar situation probably applies to the material balance in tropical forests when

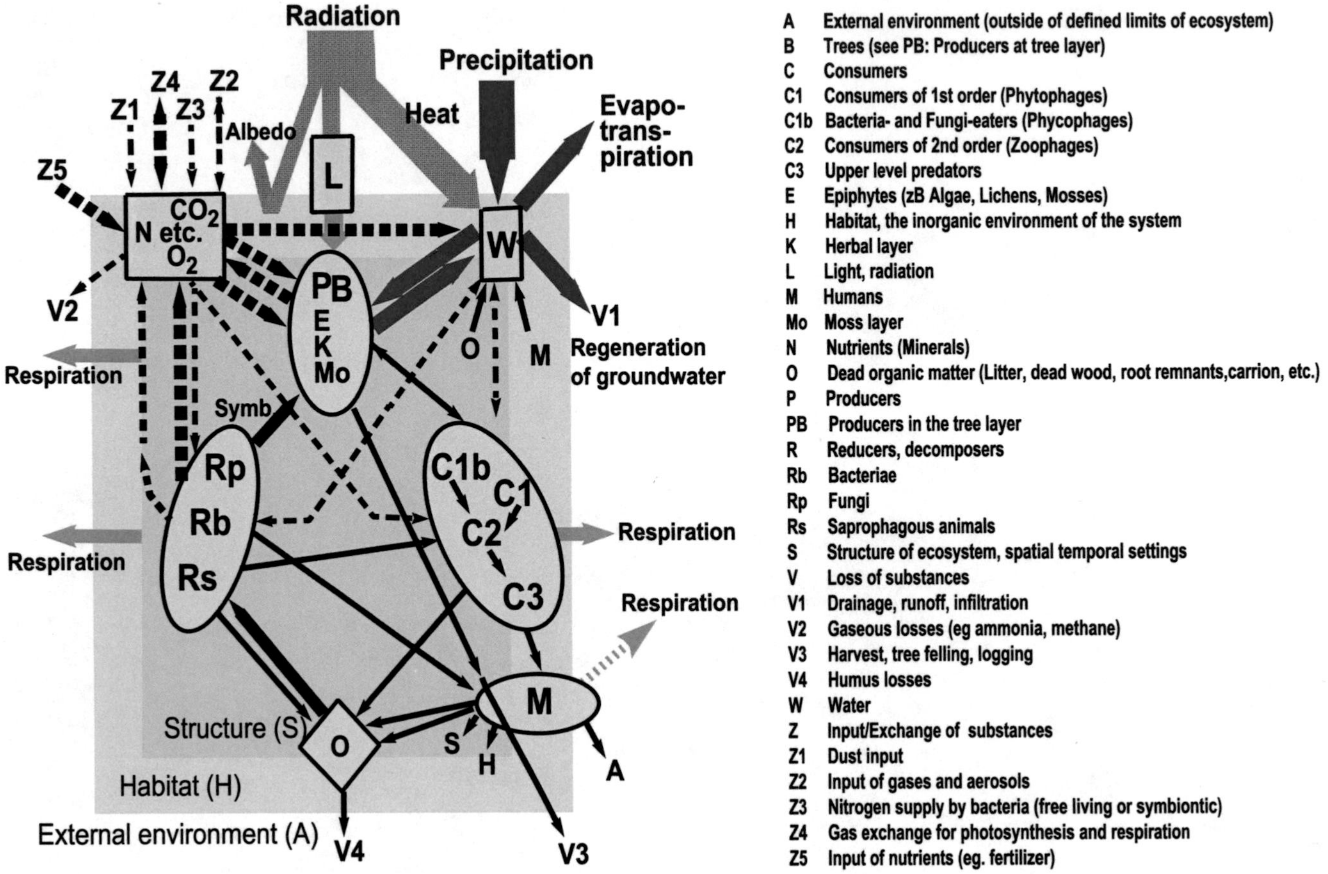
Radiation
Precipitation
Heat
Evapo-
trans-
piration
Z4
Z2
Z1
Z3
Albedo
Z5
L
CO2
N etc.
O2
W
PB
E
K
Mo
V2
V1
O
M
Regeneration
of groundwater
Respiration
Symb
Rp
Rb
Rs
C1b
C1
C2
C3
Respiration
Respiration
Respiration
M
Structure (S)
O
S
H
A
Habitat (H)
External environment (A)
V4
V3
A External environment (outside of defined limits of ecosystem)
B Trees (see PB: Producers at tree layer)
C Consumers
C1 Consumers of 1st order (Phytophages)
C1b Bacteria- and Fungi-eaters (Phycophages)
C2 Consumers of 2nd order (Zoophages)
C3 Upper level predators
E Epiphytes (zB Algae, Lichens, Mosses)
H Habitat, the inorganic environment of the system
K Herbal layer
L Light, radiation
M Humans
Mo Moss layer
N Nutrients (Minerals)
O Dead organic matter (Litter, dead wood, root remnants,carrion, etc.)
P Producers
PB Producers in the tree layer
R Reducers, decomposers
Rb Bacteriae
Rp Fungi
Rs Saprophagous animals
S Structure of ecosystem, spatial temporal settings
V Loss of substances
V1 Drainage, runoff, infiltration
V2 Gaseous losses (eg ammonia, methane)
V3 Harvest, tree felling, logging
V4 Humus losses
W Water
Z Input/Exchange of substances
Z1 Dust input
Z2 Input of gases and aerosols
Z3 Nitrogen supply by bacteria (free living or symbiontic)
Z4 Gas exchange for photosynthesis and respiration
Z5 Input of nutrients (eg. fertilizer)

Fig. 4.13 Simplified schematic representation of the structures and processes in an ecosystem (modified after Ellenberg et al. 1986)

Table 4.4 Important ecosystem parameters of an oak forest of the Belgian Ardennes with a hazel shrub layer (Querceto-Coryletum) and a sparse herb layer (after Duvigneaud 1974)

	Tree layer		3.87
	Shrub layer		1.83
Leaf area index	Total		5.70
Phytomass (t ha^{-1})			
	Aboveground:		260.8
	Thereof	Tree Leaves	3.5
		Twigs and branches	58.3
		Tribes	180.2
		Shrub layer	18.1
		Herb layer	0.7
	Belowground:		55.4
	Total phytomass:		316.2
Primary production per year (t ha^{-1} a^{-1})			
	Aboveground:		15.3
	Thereof	Total litter	6.2
		Lost by herbivores	0.5
		Tree growth	5.9
		Shrub growth	2.1
		Herbal growth	0.6
	Belowground:		2.3
	Entire production:		17.6
Dead organic matter in soil (t ha^{-1})			122

Fig. 4.14 (**a**) Fruits of sunflower (top) and beech (bottom). The seedlings of both plant seeds produce very different amounts of dry matter in the first year due to different growth strategies (photo: Breckle). (**b**) Seedlings of beech in early summer after germination in spring. The two large cotyledons are covered by 2 or 3 primary leaves (photo: Breckle). The density will be strongly diminished by herbivores and competition in the subsequent years

individual trees are cut almost completely bare by leafcutter ants (Wirth et al. 1997).

Primary production is of particular importance for the ecosystem. As production analyses show, the level of primary production depends less on the intensity of photosynthesis, nor on the leaf area index or the total available leaf area, but rather on the assimilate balance of the producers (Walter 1960), i.e. on the way in which the assimilates are used in the course of the growing season (specific investment of organic matter in growth of producing or stabilizing tissues and organs). If they are used productively, in that new assimilating leaves are constantly being formed, then growth increases exponentially. If they are used non-productively to

build up lignifying organs, the benefits of which only become apparent after years, then this corresponds to a long-term strategy. However, this is very different in individual biotopes and depends on the respective life forms.

> **Box 4.8 Influence of Animals on Biomass in a Forest**
> During caterpillar epidemics of the gypsy moth (*Lymantria dispar*) in oak forests, the mass can increase sharply: At 10^6 to 10^7 caterpillars per hectare, their dry mass is 75–150 kg ha^{-1}, with 1–2 t ha^{-1} of dry leaf mass destroyed and 500–1000 kg ha^{-1} of excrement excreted. This throws the whole ecosystem out of balance. But this is true only for forest monocultures of the same age. Deciduous forests usually recover, coniferous forests may even die.

The difference in investment strategy can be illustrated: If, for example, one sows single-seeded fruits of the beech (*Fagus sylvatica*, beechnut) and the sunflower (*Helianthus annuus*) (Fig. 4.14a) under the same conditions in good soil in C Europe, the beech seedling produces only 1.5 g of dry matter in the first year, whereas the sunflower, even in a climate that is not favourable for it, produces about 800 g. This is because it continually forms new large assimilating leaves, whereas the beech seedling is content with two or three small leaves (Fig. 4.14b), and then uses the assimilates to build up a long primary root and a woody stem. It is true that the intensity of photosynthesis in the sunflower is about twice that of the beech, but this does not explain its production, which is 500 times greater. Here, then, the 'compound interest rate' effect of investment in organs of production plays the decisive role. In this the essential types of life differ fundamentally.

For the different forests, the general relationships between net primary production and the main ecological factors, such as temperature and humidity, can be illustrated (Fig. 4.15). Of course, this is only true for a limited range; in both dependencies there are typical upper limits, as shown in Fig. 4.15a, b; in addition, the relationships are not particularly pronounced.

On a global scale, in the case of net primary production (NPP), one can see the high productivity of warm-humid zonobiomes, where the growing season can be more or less year-round. The shorter the growing season and the colder and drier an area, the lower the NPP. The average value of 10–15 t ha^{-1} a^{-1} for the C European beech forest is a relatively high value compared to the rest of the world; most dry areas are far below this value. In the tropics, however, values of up to 25 t ha^{-1} a^{-1} are reached (Fig. 4.16).

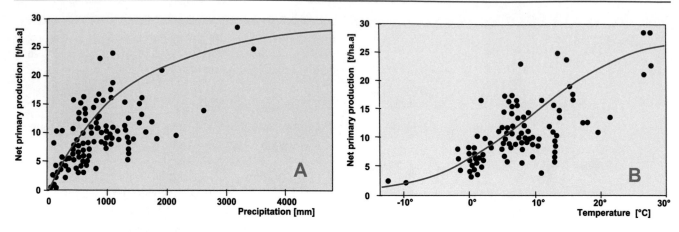

Fig. 4.15 Net primary production of forests as a function of annual precipitation (**a**) and mean annual temperature (**b**) (modified after Ehlers 1996)

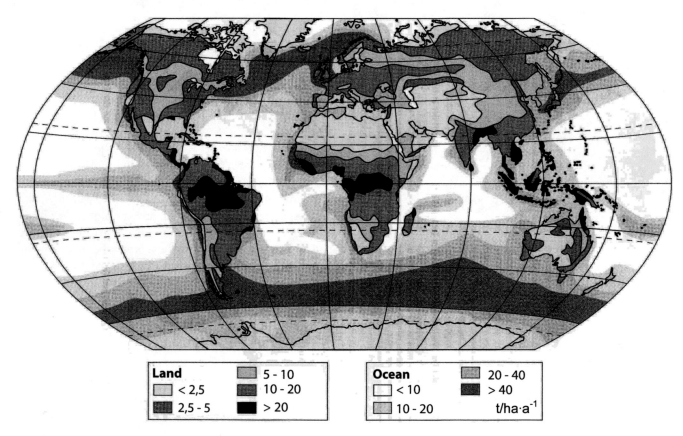

Fig. 4.16 Net primary production ($t_{dm} \cdot ha^{-1} \cdot a^{-1}$) on Earth (after Larcher 2001)

In the tropics, the standing, aboveground phytomass (Fig. 4.17) sometimes reaches well over 500 t ha^{-1} a^{-1}, with the belowground phytomass adding a further 20–30%. In the humid temperate latitudes, the total phytomass is often just as high, and the belowground phytomass is usually even significantly greater than in the tropics. The forested areas of the world usually yield more than 50 t ha^{-1} a^{-1}, but in deserts and semi-deserts the stand stock is often less than 10 t ha^{-1} a^{-1} (Fig. 4.17).

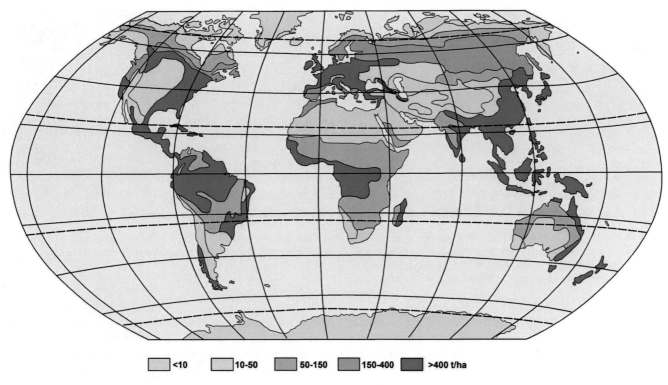

| <10 | 10-50 | 50-150 | 150-400 | >400 t/ha |

Fig. 4.17   Phytomass (t ha^{-1} a^{-1}) on Earth (modified after Schulz 1995; original by Bazilevich & Rodin 1971)

4.10   Highly Productive Ecosystems

Over large areas, NPP in the hot humid tropics reaches average values of up to 25 t ha^{-1} a^{-1}, as shown in the previous section. However, the plant communities of the tall shrubs are also characterized by a particularly high primary production. Like annual plants, tall perennials produce mainly assimilating leaves throughout the growing season and flowering organs and fruits only at the end of the growing season. However, since, unlike a seedling, they have much larger reserves at their disposal when they sprout in the spring, having been laid down the year before, they can build up the richly leafy shoot in a very short time, whereas the seedling of the annuals requires a long start-up time for this, until the leaf area has reached its maximum size. This is why summer cereals need ten weeks to produce the first quarter of the total dry yield, two more weeks for the second quarter, but only one week for the last half (according to the usual exponential growth curve).

In contrast, tall perennials can use almost the entire vegetation period very productively, which explains the very large developed above-ground phytomass and the considerable underground reserves built up in autumn for the next year.

A detailed production analysis of a tall herbaceous stand is shown in Table 4.5. Primary production could be determined by monthly determination of above- and belowground phytomass.

The annual net primary production of about 18 t ha^{-1} a^{-1} is in the same order of magnitude as that of a W European mixed oak forest, but slightly below that of a 50-year-old evergreen *Castanopsis cuspidata* forest in the warm-temperate climate of Japan with an aboveground primary production of 18.3 t ha^{-1} a^{-1} on average.

> **Box 4.9 Influence of Environmental Factors on Productivity**
> Rule of thumb: The higher the temperature, the higher the productivity. The higher the precipitation, the higher the productivity (Fig. **4.15**).

Table 4.5   Production values of a pure stand (in t ha^{-1} a^{-1}) of the adventitious goldenrod *Solidago altissima* in a river floodplain in Japan (Iwaki et al. 1966, cf. Walter 1981); vegetation period from April to October

Growth of the above-ground parts	12.01
Growth of rhizomes and roots	2.94
Parts dead during the growing season	2.83
Total production	17.78

Fig. 4.18 (**a**) Upper Kamchatka River with floodplain. A narrow bear path runs directly along the river bank (grass strip), behind it high herbs of *Filipendula camtschatica* grow in front of the gallery forest of *Salix sachalinensis* with scrawny stems. On the hill there is a forest of *Betula*

ermanii (photo: Breckle); (**b**) High herbs formation with *Campanula*, *Delphinium* and *Heracleum* at the upper forest border in Georgia (photo: E. Fischer)

According to studies by Morozov & Belaya (in Walter 1981), the production of natural giant tall shrubs is even greater on the always moist and nutrient-rich soils of the floodplains on Kamchatka and Sakhalin.

Kamchatka belongs to the subarctic zone with low *Betula ermanii* forests. The growing season is 90–110 days (mean frost-free period only 64 days), mean temperatures are 3.5°, June 10.6°, July 14.3°, August 13.3° and September 7.2 °C in May. Tall herbs reach 3.5 m in height, with *Filipendula camtschatica, Senecio cannabifolium,* and *Heracleum dulce* dominating (Fig. 4.18). According to Hulten (1932), bears sleep in them during most of the day, catching salmon in the river. The maximum standing phytomass reaches $31\,\mathrm{t\,ha^{-1}\,a^{-1}}$ (of which $10\,\mathrm{t\,ha^{-1}\,a^{-1}}$ are underground). Because some of the shoots die during the growing season, primary production is higher than the maximum herbaceous phytomass and probably exceeds 16–$20\,\mathrm{t\,ha^{-1}\,a^{-1}}$ despite the short growing season. The tall herbaceous plants are used as livestock feed in the form of silage. On southern Sakhalin, which is much further south (about 45°N), with a warmer climate and mixed deciduous and coniferous forests, even higher values were obtained. The frost-free period on Sakhalin is 145–155 days and the mean temperature of the warmest month is 18 °C. The tall herbs here grow up to 4.5 m high, and their composition is similar to that on Kamchatka, but they are more heterogeneous, as different species dominate locally. A leaf area index of 13–14 is given for stands dominated by *Filipendula*, and as high as 18–21 when *Polygonum sachalinense* is dominant, which is only possible if the tall shrubs receive additional side light, for example from the river side. This might explain the enormous annual above-ground phytomass produced, reaching $30\,\mathrm{t\,ha^{-1}\,a^{-1}}$ (total phytomass $70\,\mathrm{t\,ha^{-1}}$) when

Polygonum dominates. Thus, small-scale primary production could reach the record value of over $38\,\mathrm{t\,ha^{-1}\,a^{-1}}$.

Whether the primary production of high *Papyrus* stands in the tropics is even higher is not known. It must be borne in mind that in the tropics the respiration losses due to the high night-time temperatures are very great, so that despite the high values of gross production, net production is greatly reduced.

Very lush tall herbs are also known from the western Caucasus Mountains (Fig. 4.18, right) from the subalpine belt (Walter 1974) and not quite as tall in the Alps in the area of the likewise subalpine *Alnus viridis* stands, which indirectly assimilate atmospheric nitrogen, which also benefits the soil. However, exact production values are probably not available in either case.

Tall perennial grasses on moist, nutrient-rich sites also produce a large phytomass annually, for example, about $35\,\mathrm{t\,ha^{-1}\,a^{-1}}$ of phytomass (annual production $18\,\mathrm{t\,ha^{-1}\,a^{-1}}$) is reported for the 2–3 m tall reeds (*Phragmites*) on the lower Amu-Darya.

4.11　Peculiarities of the Material Cycles of Different Ecosystems

With the exception of the narrow shore zone, the littoral, aquatic ecosystems have autotrophic algae floating in the water as producers. They constitute a part of the plankton. By dividing, they can multiply very rapidly. Since they need light for photosynthesis, they only occur in the upper layers of the water bodies. They serve as food for the animal organisms of the micro- and macroplankton, which in turn

Table 4.6 Ratios of phytomass and primary production of terrestrial and aquatic ecosystems

Ecosystem types	Phytomass : Primary production
Terrestrial ecosystems	10–20 : 1
Aquatic ecosystems	1 : 300–400

feed larger animals up to fish and aquatic mammals, as well as birds of prey that take their food from the water. All dead organic waste is mineralized by decomposers in the water or in the mud layer at the bottom of the water bodies.

The phytomass present in the waters is small, yet the primary production may be very high due to the rapid reproduction rate of the algae. Since this primary production serves as food for the animals and is then incorporated to a considerable extent into their body substance (secondary production), the zoomass is very large compared to the phytomass. The ratios, as we have seen, are quite different in terrestrial ecosystems. The corresponding mean ratios are given for comparison in Table 4.6.

In terrestrial ecosystems, much unproductive biomass is accumulated in producers; in aquatic ecosystems, biomass is more accumulated in consumers.

Even the scheme of the deciduous forest ecosystem (Fig. 4.13) is by no means universally valid. Various deviations occur, so that examples must be given below because of their great importance.

Almost all forest trees and most herbaceous plant species (except the Brassicaceae), but especially the Ericaceae and orchids, form a mycorrhiza with fungi, which functionally can be understood as a strong extension and fanning out of the root system. This facilitates the uptake of mineral nutrients from humus-rich soils. The mycorrhizal fungi are also able to supply their host plants with organic substances. This is proven by holosaprophytes among orchids (*Neottia, Corallorhiza* and others), Pyrolaceae (*Monotropa* and others) and other families. In addition, there are certainly hormonal effects. Whether the mycorrhizal fungi also supply organic compounds to the forest trees and the Ericaceae has probably not yet been proven, but could be possible in stands on extremely poor sands with a layer of raw humus. In this case, the short cycle would be even more shortened because the litter does not need to be totally mineralized.

A particularly strange case of an ecosystem without producers was discovered in the dune area of the Namib Mist Desert: The organic mass, which is a prerequisite for the material cycle, is blown into this almost vegetationless dune area by the wind from the neighbouring or distant areas and accumulates on the leeward slope of the dunes or in sandy depressions. It serves as food for the saprophages (beetle species and others), these are eaten by small predators (reptiles and others), which in turn are the food of larger predators. In this way, a rich animal life with very strange adaptations to life in the mobile sand has developed even without plants, that is, an open ecosystem without producers.

4.12 The Importance of Fire for Ecosystems

Well secured is the fact that fire can often replace the decomposers and a very rapid mineralization of the enriched litter is achieved. In this respect, fire also represents a special impact on the material cycle of ecosystems. Natural fires triggered by lightning have always existed, even in the forests of the coal age (Carboniferous). They are typical for areas with a drought, i.e. for all grasslands of the tropics and subtropics, for the steppes of the temperate and cold regions, for the woody vegetation of the winter rainfall areas and for all coniferous forest areas, even without human intervention, and are even necessary for the vegetation if the decomposers are not able to decompose all the dead litter. In Grand Teton National Park (USA), all fires were suppressed for a long time, resulting in a bark beetle disaster in the *Pinus* forests because the beetles were able to proliferate in the enriched dead wood. Since the natural fires are no longer extinguished, the balance in the ecosystem is maintained. However, larger and large-scale fires of varying intensity occur again and again, causing a mosaic of fires in the landscape. Even steppes or prairies (as well as grasslands and savannahs) and natural parks that are completely protected from fire degenerate when litter accumulates that is otherwise periodically mineralized by natural fires. In certain Australian heaths, if the dead organic plant parts do not burn at least every 50 years, the material cycle comes to a halt, because otherwise the mineral nutrients are stored more and more in the accumulating litter, in the large woody fruits of *Banksia*, and also in the hard dead leaves of grass trees. Many *Eucalyptus, Banksia, Grevillea* and *Hakea* species in Australia renew only after fire events. Many annuals also take advantage of open ground, freshly fertilized by ash, after rain and germinate. Many geophytes also sprout suddenly at the same time, and new shoots emerge almost synchronously from many burned stumps (Fig. 4.19).

After a fire, the material cycle is stimulated again by the ash components. The conditions are similar in the large *Protea* stands around Cape Town, in the fynbos, where even shorter fire periods occur naturally, such as on the slopes around Junkershoek, where this was studied. There, on average, a fire occurs every two to three decades, i.e. relatively frequently even under natural conditions. During this time not so much litter and dead matter has accumulated, so that in many places the fires are not too hot and therefore not very devastating. The Cupressaceae *Widdringtonia* can thus always regenerate; it is only competitive enough against

Fig. 4.19 *Macrozamia* in the understory of a tall *Eucalypt* forest north of Melbourne (Australia) with fresh shoots after forest fire (photo: Breckle)

other shrub and tree species when exposed to fire (pyrophyte). The fynbos (Fig. 4.20) thus remains a species-rich mosaic of the most diverse age stages.

Fire is thus very often an important natural environmental factor in maintaining balance in ecosystems. Accurate statistics of forest or grassland fires caused by lightning in the United States are available for the years 1961 to 1970. There were 34,976 = 37% of all fires in the Pacific states, 51,703 = 57% in the Rocky Mountain states, 13,733 = 2% in the Southeastern states, but only 1167 = 1% in the humid Northwest (Taylor 1973).

However, today man-made fires for slash-and-burn etc. have become so devastatingly rampant, especially in the tropics, that thousands of fires can be located in the satellite image every night.

The working group 'Fire Information for Resource Management System' (FIRMS) at the University of Maryland, USA constantly collects comprehensive data from the satellite MODIS, GIS data, Google Earth data etc. about the fire frequency in forests worldwide. NASA creates weekly images of wildfires and makes it available to the public on a platform (for more information ▶ NASA's archive: https://t1p.de/kxqs). We bring to illustrate the scale of forest destruction four world maps of forest fires in Fig. 4.21. From these

Fig. 4.20 Photo top left: Proteoid fynbos at the Cape of Good Hope (South Africa) with sclerophyllous shrubs. Photo top right: *Widdringtonia* (Cupressaceae) stands in the fynbos with numerous Proteaceae bushes in the foreground. Photo bottom left: *Mimetes* *fimbriifolius* (Proteaceae) bushes in full bloom in the fynbos in the Cape region in October 2008. Photo bottom right: *Protea* bushes revived after a fire in the Cape region (photos: Rafiqpoor)

Fig. 4.21 The extent of wildfires in different seasons for 2014 (from top to bottom: spring, summer, autumn, winter) compared using NASA night-time satellite imagery (taken from: https://t1p.de/kxqs)

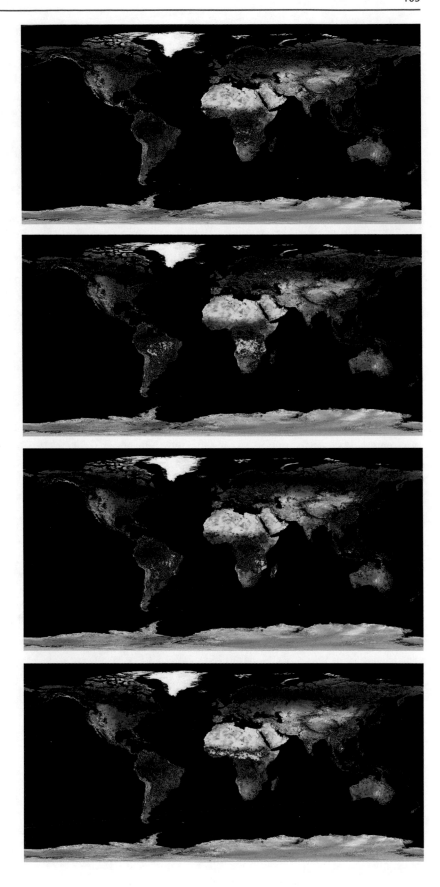

images we can see the seasonal change of forest fires and that the fires are not limited to the forests of the tropics, but also leave equally devastating traces in the forests of the taiga. The smoke particles are distributed throughout the atmosphere, so they contribute a share to the change in radiation absorption and thus to the global climate that is difficult to estimate.

4.13   The Individual Zonobiomes and Their Distribution

The sequence of zonobiomes arranges itself on both sides of the equator, but not quite symmetrically, because the land masses in the southern hemisphere are smaller and the climate is more oceanic as well as cooler. Note here that zonobiomes VI through IX are small-scale in the southern hemisphere. Zonobiome VI and VII are weakly developed in the southern hemisphere, ZB VIII is completely absent, and ZB IX is represented only by the sub-antarctic islands and the southern tip of South America, if one disregards the icy and almost vegetationless Antarctic.

The sequence from the equator to the poles does not always correspond to the numerical order, for example ZB VII in Eurasia is partly inserted between ZB V and ZB VI and represents a very dry variation (Krutzsch 1992 calls these climate facies areas), which even often has a precipitation cycle of ZB IV, but with cold winters and great continentality. The large zonobiomes are usually further

subdivided into subzonobiomes (sZB) due to certain variations.

As a basis for a more detailed discussion of the individual zonobiomes, Figs. 4.22, 4.23, 4.24, 4.25, 4.26, and 4.27 show their distribution on the continents. Additional signatures indicate minor variations within the zonobiomes.

> **Box 4.10 The Importance of Fire in Pyrophytes**
> Plant species that require episodic fire in an ecosystem for maintenance or reproduction are called obligate pyrochorous plants. Their reproduction is tied to fire events.

In Western Europe, the zonobiomes run more from N to S as a result of the influence of the Gulf Stream, whereas in Eastern Europe the normal W-E extension can be seen. They are from N to S: Zonobiome IX (tundra zone) with zonoecotone VIII/IX (forest tundra), zonoecotone VIII (boreal coniferous forest zone), zonoecotone VI/VIII with zonoecotone VI, both of which, however, are tilted to the east (mixed forest and deciduous forest zone) and finally zonoecotone VII (steppe zone). Zonobiomes IX, VIII, and VII find their immediate eastward continuation in Asia (Fig. 4.26). Southern Europe belongs to zonobiome IV (Mediterranean), which is still weakly noticeable in Iran and Afghanistan. Zonobiome III is completely absent from Europe. Only zonoecotone IV/III occupies a small semi-

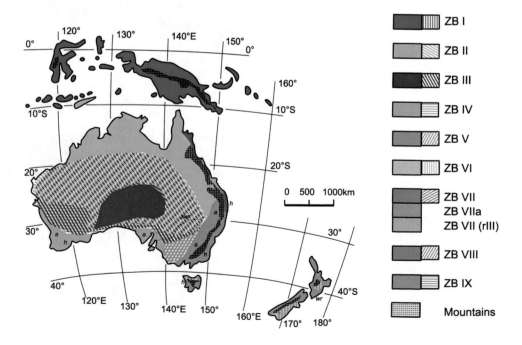

Fig. 4.22 Australia and New Zealand with zonobiomes I–V

Fig. 4.23 North and Central
America with zonobiomes I–IX

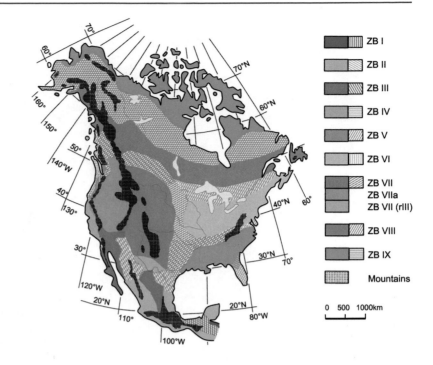

Fig. 4.24 South America with
zonobiomes I–VII and IX

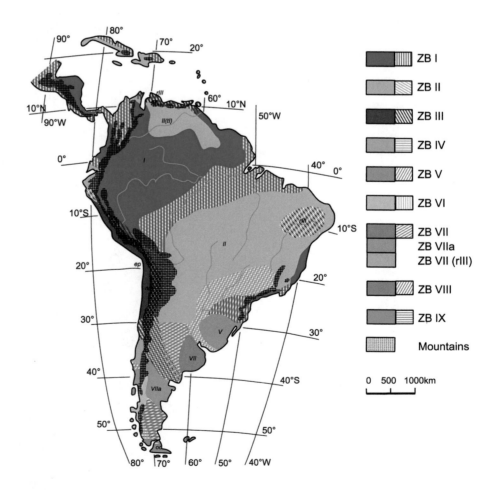

Fig. 4.25 Africa with zonobiomes I–V

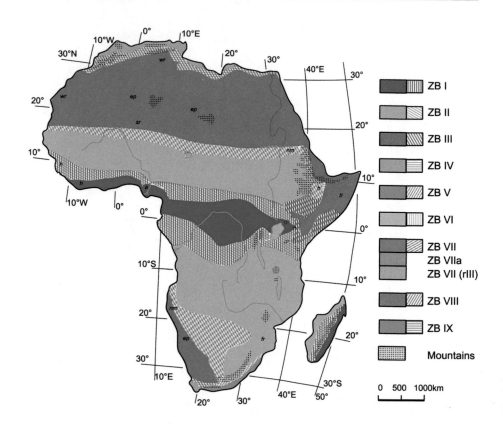

Fig. 4.26 Europe with zonobiomes IV–IX, plus the Near East

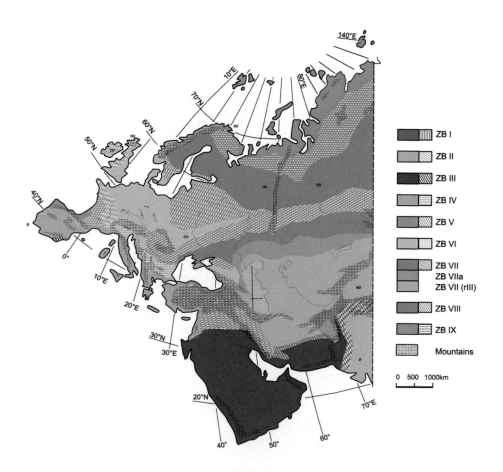

Fig. 4.27 Asia with zonobiomes I–IX (see Fig. 4.26 for the Near East)

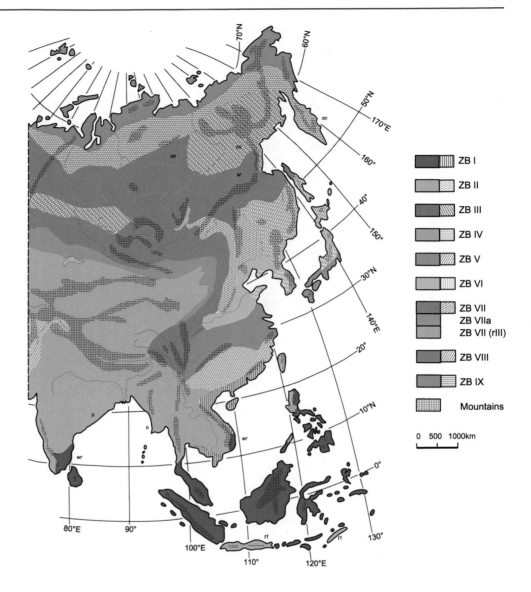

ZB I
ZB II
ZB III
ZB IV
ZB V
ZB VI
ZB VII
ZB VIIa
ZB VII (rIII)
ZB VIII
ZB IX
Mountains

0 500 1000km

desert area in SE Spain, the driest part of Europe. In C Europe, the zonation is strongly modified by the Alps and the other mountains. The zonation in mountainous Balkan Peninsula is also very complex.

In Figs. 4.22, 4.23, 4.24, 4.25, 4.26, and 4.27 the major ecological subdivision of the continents (modified after Walter et al. 1975) is shown. Signatures I–IX correspond to zonobiomes (ZB); zono-ecotones (overlapping hatchings) and mountains (gridded) are shown between them.

References

Bazilevich, N.I. & Rodin, L.E. 1971: Geographical regularities in productivity and the circulation of chemical elements in the earth's main vegetation types. Soviet Geogr. - Transl. American Geogr. Soc., New York

Breckle, S.-W. 2011: Vegetationszonen und Klima – gestern, heute und morgen. In: Anhuf, D., Fickert. T., Grüninger, F. (eds.): Ökozonen im Wandel. Passauer Kontaktstudium Geographie **11**: 27-36

Duvigneaud P 1974: La synthèse écologique. Population, communautés, écosystèmes, biosphère, noosphère. Paris. 296 pp.

Ehlers, W. 1996: Wasser in Boden und Pflanze. Ulmer, Stuttgart 272 S.

Ellenberg, H., Mayer, R. & Schauermann, J. 1986: Ökosystemforschung, Ergebnisse des Sollingprojektes 1966-1986. Ulmer, Stuttgart 507 S.

Gigon, A. 1974: Ökosysteme. Gleichgewichte und Störungen. In: Leibundgut, H. (Hrsg.) Landschaftsschutz und Umweltpflege. Huber, Frauenfeld: 16-39

Hulten, E. 1932: Süd-Kamtschatka. Vegetationsbilder **23**. Reihe, Heft 1/2, Jena

Iwaki, H., Monsi, M. & Midorikawa, B. 1966: Dry matter production of some herb communities in Japan. 11th Pacific Science Congress, Tokyo

Krutzsch, W. 1992: Paläobotanische Klimagliederung des Alttertiärs (Mitteleozän bis Oberoligozän) in Mitteldeutschland und das Problem der Verknüpfung mariner und kontinentaler Gliederungen (klassische Biostratigraphien - paläobotanisch-ökologische

Klimastratigraphie - Evolutions-Stratigraphie der Vertebraten. N. Jb. Geol. Paläont. Abh. **186**: 137-253

Larcher, W. 1994, 2001: Ökologie der Pflanzen. 5. Aufl., Ulmer, Stuttgart

Lauer, W. Rafiqpoor, M.D. 2002: Die Klimate der Erde – eine Klassifikation auf der ökophysiologischen Grundlage der realen Vegetation. Erdwissenschaftliche Forschung, Bd. XL. Fran Steiner Verlag, Stuttgart

Olsen, DM. & Dinerstein, E. 2002: The Global 2000: Priority ecoregions for global conservation. Annals Missouri Bot. Garten **89**: 199-224

Reichle, D.E. 1970: Analysis of temperate forest ecosystems. Ecol. Stud. **1**, 304 p.

Schönwiese, C.-D. 1994: Klimatologie. UTB1793, Ulmer, Stuttgart 436 S.

Schultz, J. 2008: Die Ökozonen der Erde. 4. Aufl. Ulmer, Stuttgart

Taylor, A.R. 1973: Ecological aspects of lightning in forests. Ann. Proc. Tall Timber Fire Ecol., Tallahassee **13**: 455-482

Walter, H. 1960: Standortslehre. 2. Aufl., Ulmer, Stuttgart. 566 S.

Walter, H. 1974: Die Vegetation Osteuropas, Nord- und Zentralasiens. Vegetationsmonographien, Fischer, Stuttgart. 452 S.

Walter, H. 1976: Die ökologischen Systeme der Kontinente (Biogeosphäre). Prinzipien ihrer Gliederung mit Beispielen, Fischer, Stuttgart 131 S.

Walter, H. 1981: Höchstwerte der Produktion von natürlicher Riesen-Staudenvegetation in Ostasien. Vegetatio **44**: 37-41.

Walter, H. & Breckle S.-W. 1999: Vegetationszonen und Klima. 7. Aufl., Ulmer/Stuttgart 382 S.

Walter H, Harnickell E, Mueller-Dombois D 1975: Klimadiagramm-Karten der einzelnen Kontinente und ökologische Klimagliederung der Kontinente. Fischer/Stuttgart. 36 pp.

Wirth, R., Beyschlag, W., Ryel, R.J. et al. 1997: Annual foraging of the leaf-cutting ant *Atta colombica* in a semi-deciduous forest in Panama. J. Trop. Ecol. **13**: 741-757

Wittig, R. & Niekisch, M. 2014: Biodiversität: Grundlagen, Gefährdung, Schutz. Springer-Spektrum. Springer Verlag Berlin, Heidelberg

Part II

Special Part

Part D: ZB I—Zonobiome of the Evergreen Tropical Rainforest or of the Equatorial Humid Diurnal Climate

5

Contents

© Springer-Verlag GmbH Germany, part of Springer Nature 2022
S.-W. Breckle, M. D. Rafiqpoor, *Vegetation and Climate*, https://doi.org/10.1007/978-3-662-64036-4_5

Photo 8 Dipterocarpaceous tree giants with their enormous heights of over 60 m in the evergreen rainforests of SE Asia (Zonobiome I), e.g. in Borneo, Saba, tower above the irregular canopy of the forests (photo: Rafiqpoor)

5.1 Typical Features of the Climate in ZB I

The entire tropics are generally radiation surplus areas of the earth because of the equally favourable radiation conditions in the area between the tropics. They give off energy to the high latitudes through atmospheric circulation. Nevertheless, there are radiative climatic differences between the humid and arid tropics. The latter, especially the extensive desert regions (e.g. the Sahara) at the transition to the subtropics, have a negative radiation balance because of the maximum nocturnal radiation emission. Despite these differences, the uniform irradiation conditions make the tropics to be primarily the Earth's **heat belts**. They are an area of **thermal uniformity** with 12 thermal vegetation months (Fig. 5.1a) and a zone without noticeable temperature seasons at all elevations from sea level to the summit regions of the high mountains (Lauer 1975, Breckle 2004).

The clearly pronounced diurnal variation in temperature is the result of the greater temperature contrasts between day and night (**diurnal climate**). The decrease in temperature with altitude is continuous from the **warm tropics** of the lowlands to the **cold tropics** of the high mountains.

One can contrast the thermal tropics with the **hygric tropics**, since the seasons can be divided by rainy and dry seasons of different length and intensity and from year-round humidity to year-round aridity. As rainfall amounts and temporal duration decrease from the interior to the marginal tropics, **humid tropics** can be distinguished from the **arid tropics** (Fig. 5.1b) (Lauer 1975, 1999).

In **Zonobiome I**, the monthly mean temperature often varies by only 2–3 K. But the diurnal fluctuations of temperature can reach over 9 K on sunny days (**diurnal climate**); on cloudy days they are insignificant at only 2 K. Accordingly, the humidity changes only slightly (Fig. 5.2). Frost never occurs, only in the high mountains, but even there a tropical diurnal climate prevails year in, year out.

Bogor (Buitenzorg) on Java, for example, has a distinctly permanently humid rainforest climate. The monthly average temperature there varies only between 24.3 °C (February) and 25.3 °C (October), the average annual precipitation is 4370 mm, the rainiest month has 450 mm of rain, the less rainiest 230 mm. A month with less than 100 mm of rain is already considered relatively dry in this rainy zonobiome and is an exception. Only on the Malay Peninsula and in Indonesia are larger areas found that are wet all year round; in the Amazon Basin it is only a sub-area along the Rio Negro, and in C America it is a few small, rain-facing mountain areas. In the Congo Basin, two periods of lower rainfall are mostly noticeable (Fig. 5.3).

In southern India, too, there are always one or two drier periods per year. Bogor (Buitenzorg) on Java has a distinctly permanently humid rainforest climate (Fig. 2.52).

> **Box 4.1 The Tropics as Areas with Diurnal Climates**
> Zonobiome I (the tropical rainforest) has a distinct diurnal climate: The diurnal amplitudes of the air temperature are considerably larger than the annual amplitudes of the monthly mean values.

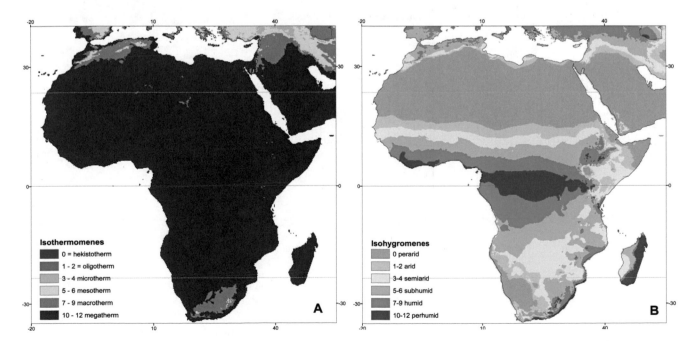

Fig. 5.1 The hygrothermal character of the climate of the tropics, using Africa as an example: thermal uniformity (**a**), hygric differentiation (**b**)

Fig. 5.2 Diurnal variation of weather factors in Bogor (Java) during the rainy season (compare sunny 12th February, when humidity dropped to almost 50%, with hazy 14th February). Figures for rain indicate absolute rainfall amounts in mm (data after Stocker, modified after Walter 1990)

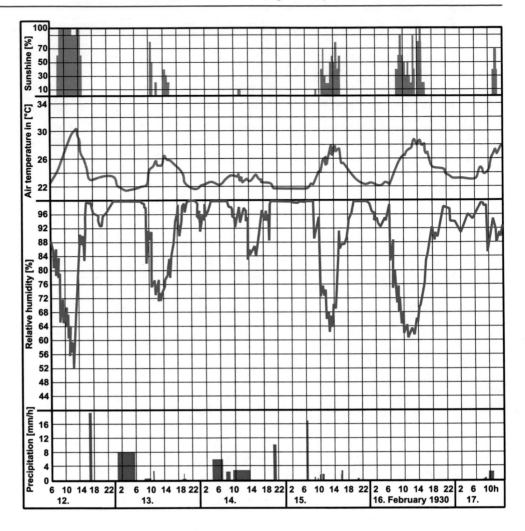

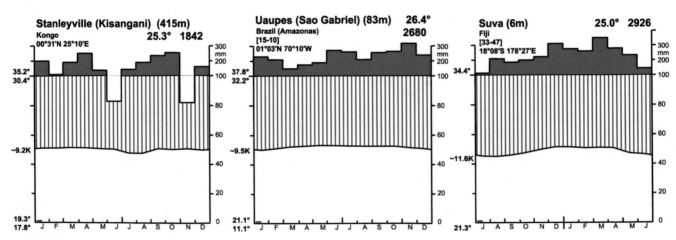

Fig. 5.3 Climate diagrams of stations in the tropical rainforest region: Congo, Amazon Basin and New Guinea

The rains fall mostly in the afternoon as short, heavy downpours, in the evening hours the sun shines again. Its radiation, when it is at its zenith, is very strong. As a result, leaves directly exposed to the radiation (e.g. in the canopy) heat up by several degrees (up to 10 K) above the already very high air temperature. Therefore, high water vapour saturation deficits occur at the leaf surface even in vapour-saturated air (Fig. 5.4). Excess temperatures of 10 to 15 K have been measured on unshaded *Coffea* leaves on clear days in Kenya. Clear days are not so rare even in permanently

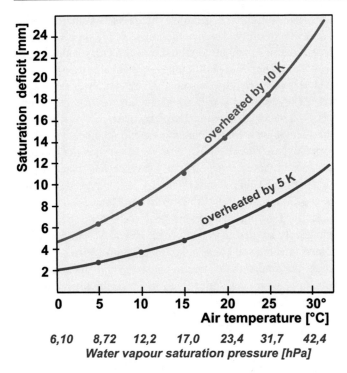

Fig. 5.4 Curves of saturation deficits in mm Hg at the leaf surface by overheating of 5 and 10 K, respectively, as a function of air temperature in water vapor-saturated air

humid Bogor (Buitenzorg) (Fig. 5.2). When this occurs, humidity drops to nearly 50% and temperature rises above 30 °C, increasing the saturation deficit to nearly 6 kPa with leaves overheated by 10 K, meaning that leaves are exposed to extreme dryness for hours at a time even in the wettest tropics. In contrast, humans, who have their own body temperature, constantly experience the air as muggy.

Researchers who worked in the jungle for years emphasize that even in the perhumid area on Borneo there are always weeks without rain, which means a dry period for the jungle trees. The long-term monthly rainfall averages do not indicate this. This is equally true for Amazonia.

It is therefore understandable that the leaves have high resistance to transpiration and also a very thick cuticle. They are leathery, but not completely xeromorphic (compare the gum tree *Ficus elastica*, *Philodendron*, *Anthurium* and others); they can strongly limit their transpiration when stomata are closed and permanently maintain a high hydrature of the plasma. They are often lauriphyllous, but not sclerophyllous. The cell sap concentration is usually only 1.0 to 1.5 MPa, and it is significant that some of these species, as houseplants, tolerate well the dry air in heated living rooms.

Conditions are different for the species that grow in the forest shade. Inside the rainforest, the microclimate is much more balanced, especially on the ground, which receives almost no direct sunlight. Here, temperature fluctuations almost cease, and the air is constantly saturated with water vapor. With the high humidity, even with the slight nightly cooling, dew regularly falls on the treetops. It drips off and wets the leaves of the lower layers. Important for the forest plants are the light conditions. The irregular contour of the canopy and the highly reflective, leathery leaves allow light to penetrate deep into the forest interior, but at ground level the average intensity is very low. However, the brief sunspots on the ground play an important role in the light yield. Depending on the structure of the forest, on average 0.5 to 2% of the daylight (as in our deciduous forests), and more rarely only 0.1%, reaches the herb and ground layer. If the numerous gaps in the stand, which cause a very heterogeneous structure, are included, i.e. if the light yield is integrated over a larger area, values well above 2% are obtained; on average, more than just 2% of the light reaches the ground. This is due to the very non-uniform structure of the stands. However, the individual plant sometimes receives less than 1% of the light. Some of the very tender herbs have bluish reflecting undersides.

5.2 Soils and Pedobiomes

If we disregard the young volcanic soils and the alluvions, the soils of the rainforest areas are mostly very old. They often go back to the Tertiary. Weathering penetrates several meters into the depth of siliceous rocks. A leaching of the bases and the silicic acid takes place; what remains are the sesquioxides (Al_2O_3, Fe_2O_3), i.e. a lateritization occurs and red-brown to yellow-red loams (ferrallitic soils or latosols) without visible division into horizons are formed. If we compare the great diversity of soil types, we find that about $2/3$ of the soils in the tropics belong to the oxisols and ultisols; these are soils with only very moderate to very low fertility. About 7% of the tropical soils are alluvial terraces rich in quartz sand or other highly weathered, leached surfaces (psammente or spodosols) with extreme nutrient poverty. Only on about 20% of the tropical soils can arable farming be practiced with current methods; these are the younger volcanic soils (alfisols) and the rich alluvial areas in large river plains (fluvente, aquepte).

The decay of the litter is very rapid. The wood is destroyed by termites, which are not noticeable in the jungle because their burrows are underground. Usually, under a very thin layer of litter and dark humus (1 to 3 cm), the reddish-brown soil immediately appears. The typical soils are found on slightly sloping ground, because on flat surfaces, waterlogging with siltation easily becomes noticeable with the large amounts of rain. The soils are generally very poor in nutrients and acidic (pH = 4.5 to 5.5). This seems to contradict the outwardly lush vegetation. But almost all the nutrients needed by the forest are contained in the above-ground phytomass, almost nothing in the soil. Annually, some of the biomass dies, is rapidly mineralized, and the released nutrients can be immediately reabsorbed by the roots (always via mycorrhiza). Therefore, despite the high rainfall, there is almost no loss of nutrients through leaching. This is shown by the fact that the water in the streams of these areas has an electrical conductivity almost equal to that of distilled water. At most, it is slightly brown in colour due to humus brine (colloidal dispersion).

Nutrients are even reabsorbed before the mineralisation of the litter. In the lowland rainforest near Manaus on the Amazon, the suction roots of the trees on very poor sandy soils have a mycorrhiza at a depth of only 2 to 15 cm. Through the fungal hyphae, this is directly connected to the litter layer; through the fungi, the trees can receive nutrients in organic form directly from the litter (short-circuited cycle), similar to saprophytic flowering plants. This prevents leaching of nutrients by rain and thus loss from the ecosystem. The amount of leaves falling daily ranges from 4.5 to 12.6 g of dry matter per m^2. The leaf turnover ranges between 0.9 and 2.2 years.

As a result of the rapid circulation of substances, the primeval forest may persist for thousands of years on the same soil. But as soon as it is cleared and all the wood is burned or removed, there is a heavy leaching of the entire nutrient capital suddenly mineralized by the fire. Only a small part is adsorbed by the soil colloids and can be utilized by cultivated plants for only a few years.

After abandoning the crops, as is the case with shifting cultivation, a secondary forest grows up, which, however, by no means reaches the lushness and diversity of the primeval forest. After the forest has been cleared again in shifting cultivation, nutrients are again lost through leaching until, after repeated use, only bracken (Pteridium) or Gleichenia species are able to thrive. If these areas are burnt down, they often become grassed over by alang-alang grass (Imperata) or other undemanding species that are of no value for grazing.

On completely degraded areas, virgin forest can re-emerge when soil erosion removes the entire soil down to the bedrock, which then weathers and a new primary succession begins, which of course requires considerable time and corresponding diaspores inputs from the surrounding area. If, on the other hand, the parent rock is very poor in nutrients from the outset, for example if it is weathered poor sandstones or alluvial sands, the nutrients are only sufficient for the establishment of very poor tree or heath stands or light savannahs. These are pedobiomes, specifically **peinobiomes** (deficient or nutrient-poor soils), which can cover very wide areas. They grow on podsol soils with 20 cm thick raw humus (pH = 2.8) and bleaching horizons or even on peat soils. These are known from Thailand and Indomalaya, as well as from Guyana (Humiria bush, Eperua forest) and the Amazon lowlands in the Rio Negro basin, which carries 'black water' (rich in humic acid colloids). They are also reported in Africa for the Congo Basin and the heaths on Mafia Island off the coast of Tanzania. However, peat soils have been most extensively studied in NW Borneo. Extensive (14,600 km^2) domed forest raised bogs (helobiomes) with Shorea alba and others are found there, starting just beyond the mangrove boundary, with peat deposits up to 15 m thick (pH around 4.0). Heath forests (Agathis, Dacrydium, and others) on raw humus soils with Vaccinium as well as Rhododendron also occur there. The total area of tropical podsol soils is estimated at seven million hectares.

At the other extreme are the tropical calcareous soils, or **lithobiomes**, which are associated with very striking relief forms and have been described from Jamaica and Cuba. In the humid tropical climate lime is readily dissolved. Clints or grikes (karren) are formed, the softer limestone disappears, and only the hardest parts remain as razor-sharp ridges (Fig. 5.5). The whole area karsts and is dissected into a network of ridges (the remains of the former plateau surface) by sinkholes, which form as collapse funnels, are round and sometimes up to 150 m deep. If the erosion goes even further, as in Cuba, then only single, unconnected towers or cone karst mountains with almost vertical walls remain, like the 'Mogotes' ("organ" mountains or cockpit karst) in Cuba or the 'Moros' in N Venezuela. The soil of the sinkholes is filled with bauxitic red earth, on which a moist evergreen forest develops. The limestone ridges, on the other hand, form a very heterogeneous site, depending on whether alkaline soil (pH = 7.7) can accumulate in individual depressions or not. Therefore, one usually finds a very interesting flora with representatives from rain forest to cactus desert. In the mentioned areas the climate is characterised by little over 1000 mm rainfall. In NW Madagascar lush tropical rainforests are common outside the karst areas. On the karst landscape itself, the vegetation cover is sparsely developed with mostly deciduous tree species (Fig. 5.5).

We will come back to the halobiomes (mangroves) later.

Fig. 5.5 Impressive karst landscape with clint (karren) formation on Jurassic limestones at the NW tip of Madagascar with an annual precipitation of >2000 mm. Forest vegetation grows in sinkholes with soils (photo: E. Fischer)

5.3 Vegetation

5.3.1 Structure of the Tree Layer, Flowering Periodicity

The most striking feature of the tropical rainforest is the large number of wood species that make up the tree layer (Homeier et al. 2010). One often finds over 100 up to 300 species per hectare with a BHD of >10 cm (Valencia & Balslev 1994). But there are also forests with only a few tree species: Indomalaya is often dominated by Dipterocarpaceae and in Trinidad the upper tree layer is formed by *Mora excelsa* (Fabaceae). The floristic differences between the forests of S America, Africa and Asia are very large (Fig. 5.6). The forest types are correspondingly diverse; but we can discuss only those features that are more or less common to all. Palm trees are almost entirely absent from the African rainforests, but are common in the C and S American ones (especially on wet sites). The tree layer reaches a height of 50 to 55 m, occasionally 60 m.

Sometimes one distinguishes three tree layers, an upper, middle and lower one; but mostly a stratification is not recognizable. The upper tree layer is not closed; there are individual giants that protrude above the other trees. Only the middle or lower layer forms a dense canopy of leaves; in this case, for lack of light, the lower trunk space is quite free, so that locomotion is easily possible. But the structure of the forest varies greatly in detail; one must be cautious in making generalizations. Examples of profiles of the stand structure will illustrate this (Figs. 5.7, 5.8, and 5.9).

In terms of tree shape, the trunks are generally slender and thin-barked, the crowns set high and are relatively small and irregular in outline, corresponding to the dense stand. The age of the trees is difficult to determine, as annual rings are usually absent. Estimates based on increment measurements gave 200 to 250 years for the thick old trees. The root depth is greater than assumed. 21 to 47% of the roots are in the upper 10 cm, most of the rest below that to 30 cm depth, but 5 to 6% go as deep as 1.3 to 2.5 m (Hüttel 1975).

The root mass was determined to be 23 to 25 t ha^{-1} (49 t ha^{-1} according to another method). The large tree giants

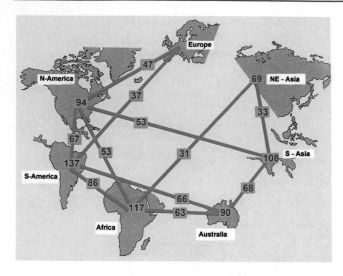

Fig. 5.6 Number of flowering plant families in each continental region (number in each region) and percent similarity (number among major regions), excluding cosmopolitan plant families (modified from Terborgh 1991)

achieve their stability through powerful board roots (Fig. 5.10), which can reach up to 9 m up the trunk in the shape of a pillar and run radially outwards from the base of the trunk with only a small thickness; many roots grow vertically into the soil from their base (Vareschi 1980), which can become stilt roots (Fig. 5.11) when the soil is washed out.

The leaves of trees are larger the more humid and warmer the climate, but the leaves exposed to the light are always much smaller in one and the same tree species. In the East African rainforest, for example, a ratio of 8:1 (largest leaf 48 × 19 cm, smallest 16 × 7 cm) occurs in *Myrianthus arboreus,* and even 28:1 (largest leaf 162 × 38 cm, smallest 22 × 10 cm) in *Anthocleista orientalis.* Both are trees of the lower tree layer. However, in *Elaeagia auriculata* in the mountain forest of Costa Rica, a species that always has very large leaves, the differences are smaller.

Bud protection is not necessary for trees of the rainforest. The young leaf plants are sometimes enclosed by hairs, mucilage or juicy scales or by specially formed stipules.

Fig. 5.7 Rainforest profile through the Shasha Protected Forest (Nigeria). The strip of forest shown is 61 m long and 7.6 m wide. All trees taller than 4.6 m are plotted. The letters indicate different tree species (data after Richards, from Walter 1973)

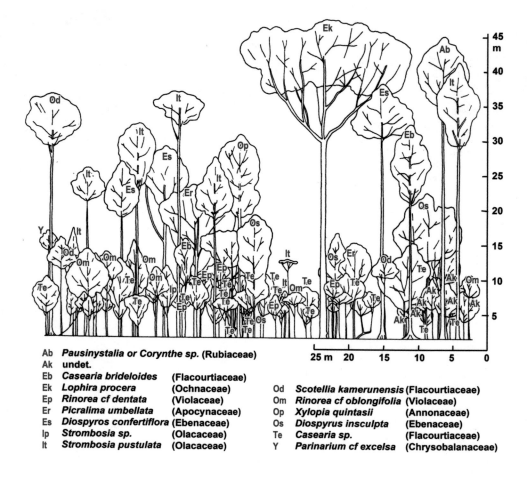

Ab	*Pausinystalia or Corynthe sp.* (Rubiaceae)	
Ak	undet.	
Eb	*Casearia brideloides*	(Flacourtiaceae)
Ek	*Lophira procera*	(Ochnaceae)
Ep	*Rinorea cf dentata*	(Violaceae)
Er	*Picralima umbellata*	(Apocynaceae)
Es	*Diospyros confertiflora*	(Ebenaceae)
Ip	*Strombosia sp.*	(Olacaceae)
It	*Strombosia pustulata*	(Olacaceae)
Od	*Scotellia kamerunensis*	(Flacourtiaceae)
Om	*Rinorea cf oblongifolia*	(Violaceae)
Op	*Xylopia quintasii*	(Annonaceae)
Os	*Diospyrus insculpta*	(Ebenaceae)
Te	*Casearia sp.*	(Flacourtiaceae)
Y	*Parinarium cf excelsa*	(Chrysobalanaceae)

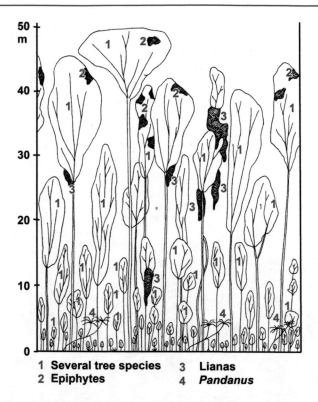

1 Several tree species   3 Lianas
2 Epiphytes   4 _Pandanus_

Fig. 5.8 Schematic profile through dipterocarpaceous rainforest on Borneo, length 33 m, width 10 m, herbs absent (after Walter 1973)

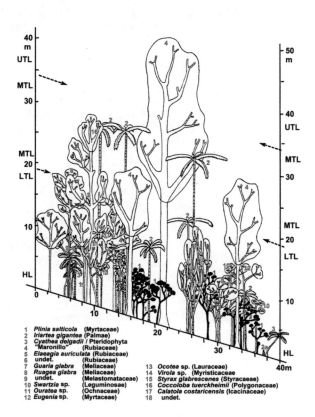

1   _Plinia salticola_   (Myrtaceae)
2   _Iriartea gigantea_ (Palmae)
3   _Cyathea delgadii_ / Pteridophyta
4   "Maronillo"   (Rubiaceae)
5   _Elaeagia auriculata_ (Rubiaceae)
6   undet.   (Rubiaceae)
7   _Guaria glabra_   (Meliaceae)
8   _Ruagea glabra_   (Meliaceae)
9   undet.   (Melastomataceae)
10 _Swartzia_ sp.   (Leguminosae)
11 _Ouratea_ sp.   (Ochnaceae)
12 _Eugenia_ sp.   (Myrtaceae)
13 _Ocotea_ sp. (Lauraceae)
14 _Virola_ sp.   (Myristicaceae
15 _Styrax glabrescens_ (Styracaeae)
16 _Coccoloba tuerckheimii_ (Polygonaceae)
17 _Calatola costaricensis_ (Icacinaceae)
18 undet.

Fig. 5.9 Schematic profile through tropical montane rainforest in the Sierra de Tilaran (Costa Rica) (from Sprenger & Breckle 1997)

Fig. 5.10 The development of board roots in the world's tropical rainforests is an important strategy for the large rainforest trees: they provide stability (photo: E. Fischer, Makokou-Ipassa rainforest, Gabon)

Although the growing conditions are permanently favourable, the growth of the shoots takes place in stages. The sprouting branch ends often show the appearance of **shaking leaves (nodding foliage)** (Fig. 5.12). During the rapid extension growth, no supporting tissue is formed at first, so that the young shoots droop downwards with the leaves; they are white or bright red in colour initially and only turn green later when they become stronger. Rapid differentiation of the leaf tip leads to the formation of a drip tip in some species (Fig. 5.13). It is found in 90% of the species in the understory in Ghana. Experiments in the forest showed that leaves with drip tips were dry within 20 minutes after rain, but those without still remained wet after 90 minutes (Longman & Jenik 1974).

A special problem is the periodicity of the development and growth of plants in the ever-humid tropics without annual variation of temperature. That a periodicity of shoot growth can be observed, has already been mentioned. Many trees have growth rings in their wood, usually several per year, which are therefore not annual growth rings. A similar thing is true of the rhythmicity of flowering. With the always uniform external conditions, periodic phenomena are usually not confined to a particular season.

Fig. 5.11 From the trunk base of some rainforest trees arise many roots growing vertically into the soil, which can become stilt roots when the soil is washed out (photo: Breckle, rainforest in Rancho Grande, Venezuela)

Fig. 5.12 The process of 'shaking leaves' is the distinctive feature of some rainforest trees (photo: Breckle)

In Malaysia, in wet weather, the old leaves are said to fall off after the young ones sprout, and in dry weather, they fall before. In this way, deciduous wood species have arisen in the climate zone with a drought. A tree may even be leafless for a short time. Then it can be observed in individuals of the same species that leafy and non-leafy trees stand side by side. In others, even the branches of one and the same tree behave differently, i.e. they do not sprout at the same time. The same applies to the flowering period. Different individuals of the same species or different branches of the same tree flower at different times (Fig. 5.14).

In all these cases, therefore, it is an autonomous periodicity which is not linked to the twelve-month period. Periods of two to four months, of nine months, but also of 32 months occur. As a consequence, there is no general flowering period in the rainforest. There are always only individual trees in flower and their blossom is only slightly noticeable in the prevailing greenery, however beautiful and large the blossoms may be. European species of trees (beech, oak, poplar, apple, pear, almond) have been planted in tropical mountains without seasons. The general experience was that at first they retained their annual periodicity of leaf fall, budding, and flowering. Over time,

Fig. 5.13 Some rainforest trees form drip tips in their leaves. This causes the leaves to dry more quickly after a rain event (photo: Breckle)

disturbances in inflorescence development occurred, the individual branches reacted differently, and finally one could see all seasons on a tree, i.e. leafless, sprouting, flowering and fruiting branches.

Fig. 5.14 Seasonality is extinguished on a coffee tree with flowers, green non-ripe and red ripe fruits (photo: http://bit-Do/brNss)

Fig. 5.15 The new shoots of the mango tree in a mango plantation in the vicinity of Guayaquil, Ecuador, are conspicuous by their bright red colouring from afar (photo: Rafiqpoor)

Central European species are mostly long-day plants, i.e. they only flower when exposed to long days (>14 h sunlight), as in summer in the temperate zone. Therefore, they generally do not flower in the tropics, but lower temperatures may substitute for long-day: *Primula veris* grows only vegetatively in Indomalaya at 1400 m altitude, and flowers and fruits abundantly at 2400 m altitude. *Fragaria species* do not flower at low altitudes in the tropics, but form many stolons; in the mountains they only flower and fruit, while stolon formation is suppressed (for example in Sri Lanka) (Zeller 1973).

Pyrethrum plantations are found in Kenya at altitudes of 1500–2500 m, where the flowers are harvested that do not develop at lower altitudes. However, the endogenous rhythmicity of these plants immediately adapts to the climatic rhythmicity as soon as one is present, for example even in the humid tropics with only a short, slightly drier season, which, by the way, is the case in many areas, so that a 'seasonal generator' is usually present after all. In the case of the mango tree, which is cultivated everywhere in the tropics, the individual light sprouting branches in the otherwise very dark crown are particularly striking (Fig. 5.15). However, as soon as there is a distinct dry season, the sprouting and flowering of all branches and trees adapt to it. The teak or djatti tree *(Tectona grandis)* never becomes bare in W Java, which is always moist, while in East Java it

sheds all its leaves during the dry season. But even in the humid tropics there are species, such as the dove orchid *(Dendrobium crumenatum)*, which blooms in a larger area on one and the same day. It does form the buds, but for them to unfold, a sudden cooling as a signal is necessary for synchronization, for example, after particularly strong thunderstorms. The coffee tree also opens buds only after a short drought. Bamboo species often develop reproductive organs only after a drought year, then they all flower synchronously and die afterwards. In the very uniform climate just certain species are very sensitive to small deviations in the weather.

> **Box 4.2 Day Length in the Tropics**
> The tropics differ from the temperate latitudes in that the days are constantly short, with about 12 h of daylight.

A common phenomenon in tropical tree species is Cauliflory, i.e. the formation of inflorescences on old wood, for example on the trunk (Fig. 5.16). It is found in about 1000 tropical species. It occurs in lower storey tree species and often in those that are chiropterogamous or chiropterochorous, that is, in which bats or flying foxes are the pollinators of the flowers or the dispersers of the seeds. They are particularly comfortable flying to cauliflorous flowers and fruits. Cauliflory also occurs in the carob tree *(Ceratonia siliqua)* and the Judas tree *(Cercis griffithii* in Afghanistan, *C. siliquastrum)*, which are now widespread in the Mediterranean.

Fig. 5.16 Cauliflory is one of the special features especially of the trees of the humid tropics. Here the flowers grow on the old stems of the trees (photos: **a** *Ixora cauliflora*, New Caledonia, Breckle; **b** *Cola* spp., Cameroon, Rafiqpoor; **c** Cacao, Ivory Coast, Barthlott; **d** *Vitex* spec Madagascar, E. Fischer)

5.3.2 Mosaic Structure of the Habitats

A difficult question to study is the regeneration of virgin forest stands. When a giant tree falls, a large gap forms in the forest. If a large branch falls, there is a smaller gap. In these gaps, fast-growing species of the secondary forest (balsa = *Ochroma lagopus* and *Cecropia* in Central and S America, *Musanga* and *Schizolobium* in Africa, *Macaranga* in Malaya) often develop first. *Ochroma* forms annual shoots 5.5 m long with light wood, *Musanga* 3.8 m and *Cedrela* 6.7 m long. These trees are then gradually displaced again over time by the species of the upper tree layer (Fig. 5.17).

It has been found that among the tree species of the primeval forest there is often a lack of its own offspring, and from this it has been concluded that the primeval forest is composed in a mosaic fashion, that is to say that each tree species is replaced by another during regeneration and can only take the same place again after several generations. The cause can be very different. Often, however, it is due to herbivore or parasite pressure in close proximity to the parent tree. The seeds, seedlings or saplings are then exposed to a higher feeding pressure near the parent tree than at a greater distance; conversely, the number of seeds naturally decreases at a greater distance. It is therefore not uncommon for a maximum of the best establishment of seedlings to form at a certain distance (Fig. 5.18). However, other factors often also play a role; generalized, this can be characterized by the degree of disturbance.

This is explained in Fig. 5.19, extrapolated also to animal communities that react differently. However, it must be emphasized that other processes also play a role beyond this. The model is therefore not valid for all tropical forests.

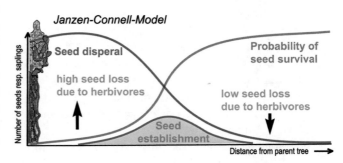

Fig. 5.18 The spatial occurrence of young growth often follows the Janzen hypothesis; it describes the interplay of disturbance (e.g. herbivory) and seed set as a function of distance from the parent tree. At intermediate distance, the best establishment is often observed, ultimately leading to a change of place of the species from generation to generation

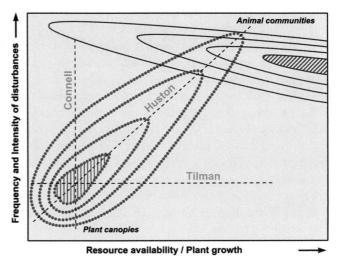

Fig. 5.19 High biodiversity in plant and animal communities have different maxima depending on resource availability or maximum plant growth and disturbance frequency. Subsequently, different patterns of lines of equal species density (isotaxes) emerge. Species density in plants depends on the frequency or intensity of disturbance in the stand (Janzen hypothesis 1978) or on a certain maximum growth rate and resource availability (Tilman's hypothesis 1982), summarized in the hypothesis of Huston (1980). Animals, including arthropods have their greatest species richness in disturbed habitats, especially when resource availability is high (modified from Begon et al. 1996)

Fig. 5.17 The formation of a 'gap' (top) and competitive growth until the gap is closed (bottom) (after Tomlinson & Zimmermann 1976)

However, the underlying dynamics of dispersal and establishment usually result in the individual species changing their place almost completely from generation to generation in a complex mosaic.

Something similar has been observed in meadows of temperate latitudes, in undisturbed primary forests of the taiga and in primeval forests in eastern Poland.

> **Box 4.3 The Regeneration of Trees in Tropical Rainforests**
> In tropical forests, rotation or cyclic rejuvenation and regeneration of tree species occurs.

> **Box 4.4 The Mosaic Structure of Plant Communities**
> The cyclical change of species and heterogeneous mosaic formation is a generally valid principle for all species-rich, original plant communities in a dynamic equilibrium. This explains why none of the species achieves absolute dominance in competition, but species-rich mixed stands are the rule in the long term.

For tropical forests, attempts have been made to explain the 'rotation' of species by the herbivore hypothesis of Janzen (1978). Only in the vicinity of old trees will there be a sufficient number of seeds, fruits and young plants at certain times, so that there the reproduction of herbivores, the extensive occurrence of parasites, the inhibition by mycorrhizal fungi or other factors can considerably limit the density of young growth. In the case of palms, it has been observed that the falling of leaves up to 10 m long and weighing many kilograms slays and crushes many young plants. These processes result in a reduction in the density of individuals of the seedlings and saplings, which is particularly severe in the vicinity of the old tree. Only at a certain distance from the old tree does a density maximum then form under certain circumstances. This has indeed been found on many occasions.

For somewhat species-poorer montane tropical oak forests in Costa Rica, Kapelle (1990) has carried out detailed investigations of the succession sequences. From the transects in Fig. 5.20, on the one hand, the great heterogeneity of the stands with their only very indistinct stratification, the uneven upper canopy layer, but also the 'clumping' of certain species (Fig. 5.20) can be seen, which then grow up in the further course, in addition to many smaller or larger gaps in the stand ('gaps'). The number of stems per hectare (from 3 cm BHD) decreases only slightly at the beginning. The thinning process during late succession is a sign of particularly strong competition, during which some stems become dominant, out-competing the others.

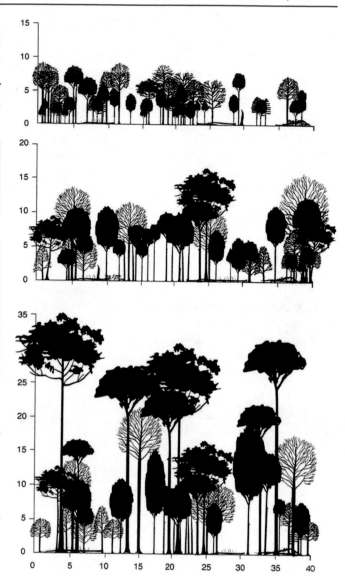

Fig. 5.20 Profile diagrams of three transects of tropical subalpine oak forest from the Cordillera de Talamanca (Costa Rica), comparing different aged secondary forest phases and primary forest structure (after Kapelle 1990)

5.3.3   Herb Layer

About 70% of all species found in the rainforest are phanerophytes, i.e. trees. They are also absolutely dominant in terms of mass. The shrub and herb layers are difficult to separate, because herbs can grow several metres high, such as bananas, heliconias (Fig. 5.21), Scitamineae and others. Undergrowth is often absent even in relatively good light conditions at ground level, perhaps due to competition for nitrogen or other nutrients from tree roots near the surface. Low herbs must make do with little light. They can withstand this even as houseplants under very low light (*Aspidistra, Chlorophytum, Saintpaulia* = Usambara Violet).

Fig. 5.21 The various *Heliconia species* (**a**) can form the herbaceous layer of rainforests in the Amazon basin of Ecuador (photo: Rafiqpoor) and the *Musa* species (**b**) (photo: Breckle) in Asia together with other herbaceous plants

Curious is the frequent occurrence of velvety dull leaves or variegation, with white or red patches or metallic shimmer.

At high humidities, guttation plays a major role, accordingly the hydrature of the plasma is very high (cell sap concentration only 0.4 to 0.8 MPa). In the ferns with less efficient pathways, the cell sap concentration is 0.8 to 1.2 MPa. Heterotrophic flowering plants, saprophytes or parasites occur but play only an insignificant role. There are certainly many different synusia depending on light and water conditions, but corresponding investigations of the linkage of the subsystems are hardly available yet. Typical synusia of different types are the groups and life forms mentioned below.

5.3.4 Lianas

In the dense tropical jungle, the battle of the autotrophic plants is above all about light. The higher a tree is, the more light its leaves receive, the higher the production of organic matter can be. But to reach to the light in the tree layer, a trunk must first be formed over the course of many years, which requires a substantial investment of organic matter. The lianas and epiphytes arrive at the favourable enjoyment of light in a simpler way. The former do not form a rigid stem, but use the trees as a support for their rapidly growing flexible shoots (Fig. 5.22).

The epiphytes, on the other hand, relocate their germination site from the beginning to the upper branches of the trees, which serve them only as a support (Fig. 5.23).

The attachment of the lianas to the supporting trees takes place in various ways: In the **splay climbers** (Fig. 5.24), it is splaying branches that grow into the branch system, with slippage prevented by thorns or spines, for example, in the climbing palm *Calamus* (Rotang), *Smilax,* or the *Rubus* lianas. The **root climbers** (Fig. 5.25) form roots that adhere to the cracks in the bark or encircle the trunk (many Araceae).

The **climbers** (Fig. 5.26), also called petiole climbers, develop small corkscrew-like coiled holding organs with which they wrap around the branches of the host plant (the climbing aid). The climbing aid for the climbers must not exceed a diameter of 8 mm, otherwise the wrapping is no longer possible. The **bindweed** (Fig. 5.27) has rapidly growing, twining branch tips with very long internodes or tendrils on which the leaves initially remain undeveloped. To grow, the lianas need light. They therefore develop in the clearings of the forest and grow upwards at the same time as the trees; in the process they reach the canopy in time.

Fig. 5.22 Lianas in the tropical rainforest of Arroyo Blanco, Dominican Republic (photo: Breckle)

Fig. 5.23 Numerous bromeliads growing epiphytically on trees in a mountain rainforest on the eastern slope of the Andes of Ecuador (photo: Breckle)

Fig. 5.24 *Calamus lianas* belong to the 'scrambling lianas'. The specimen here is growing through a stand in the rainforest of Cameroon (photo: Rafiqpoor)

The tropical lianas, unlike those of the outer tropics, are long-lived. Their axial organs possess secondary thickness growth; but since they must remain flexible to follow the movements of the supporting trees, no compact woody body is formed, but a woody part fissured into individual strands by parenchyma tissue and broad medullary rays (anomalous thickness growth). The vessels are very large on the cross-section, therefore easily visible to the naked eye. They have no transverse walls, so that the crown of the liana can be supplied with sufficient water regardless of the small diameter of the flexible stem. When the leaves serving as support die and rot, the lianas nevertheless remain attached to the canopy of other trees, and the liana stems hang down freely like ropes. Often they slip off partially and then lie with the lower end in loops on the ground. The shoot tip, however, works its way back up. If this is repeated several times, the liana stem can reach a great length. In the case of *Calamus* (Fig. 5.24), a total length of 240 m was measured!

Large clearcuts are particularly favourable for liana development. Lianas are therefore much more numerous in secondary forests than in pristine virgin forests, where they cover more of the forest edges. 90% of all liana species are restricted to the tropics; in C America, 8% of all species are lianas. The fact that lianas are mainly restricted to the humid tropics is probably related to water recharge. In dry climates, strong suction stresses (deep water potentials) develop in the leaves, causing the long water filaments necessary for water conduction to break by overcoming cohesion in the wide vessels. Even in temperate climates, woody lianas are most common only in moist riparian forests. Here there are only a few woody lianas: The root-climbing ivy *(Hedera helix)*, spreading and climbing woodland vines *(Clematis vitalba)* and grapevines *(Vitis silvestris)*, and the twining *Lonicera* species. The blackberry species *(Rubus* spec.) do not rise high

Fig. 5.25 Two Araceae (**a**, **b**) as root climbers in a rainforest in Ecuador (photos: Rafiqpoor)

Fig. 5.26 *Flagellaria* from New Caledonia is a good example of the rank climbers (photo: Breckle)

Fig. 5.27 The bindweed wraps itself around the host tree and climbs upwards to reach the light. In C European forests, as shown here, *Hedera helix* can take over this task (photo: Breckle)

above the ground in Europe, whereas in New Zealand they grow as thick as an arm and reach the tops of trees.

5.3.5   Epiphytes, Hemi-Epiphytes and Strangler

For the tropical rainforests, the epiphytic ferns and flowering plants are considered particularly characteristic. But this is true only for those forests where wetting water (mist, fog) is often available; high humidity is therefore not sufficient. There are many types with interesting adaptations (Fig. 5.23). In Liberia 153 species have been ecologically studied (Johansson 1974).

Germination high up on the branches of the trees can take advantage of the favourable light conditions, but the more difficult is the water supply; it lacks the permanent water reservoir of the soil, from which water is absorbed. The epiphytic site can be compared to a rocky site. In fact, epiphytes can usually grow just as well on rocks if they have favourable light conditions. Water uptake in general is possible for epiphytes only during rain. Therefore, wetting frequency is more important to them than absolute rainfall. The frequency of rainfall is greater on mountain slopes, where ascending air masses cause upslope rains, than in the lowlands; for this reason, montane forests are usually richer in epiphytes, especially the cloud forest, where it drips constantly from the leaves (Fig. 5.28).

In order to be able to survive longer intervals between rain showers, the epiphytes must either endure temporary desiccation without undergoing damage - this is the case with many epiphytic poikilohydric ferns, or they must store water in their funnels (bromeliads), or like the succulents of the dry regions; a number of cacti, for example, have switched to the epiphytic way of life (*Rhipsalis, Phyllocactus, Cereus* species) (Fig. 5.29). Like the succulents, the epiphytes release water very economically. Leaf tubers as water reservoirs are possessed by many orchids, woody tubers by some Ericaceae, succulent leaves have developed by most orchids, but also by Bromeliaceae, *Peperomia,* and others. Special devices for the rapid absorption of water during wetting by rain are the aerial roots of the orchids with the velamen absorbing the water, and the sucking scales of the Bromeliaceae, which take up the water from the funnels formed by the bases of the leaves, which collect the rainwater, or hold it capillary through the dense scaling of the leaves and then suck it up.

The roots are in the epiphytic Bromeliaceae only adhesive organs (Fig. 5.30) and are completely absent in *Tillandsia usneoides,* which is reminiscent of bearded lichens, as well as other Tillandsias etc. Special hollow organs, partly inhabited by ants, are formed by *Myrmecodia, Hydnophytum, and Dischidia* species. Ferns that cannot tolerate desiccation can form their own soil by accumulating falling litter and detritus between the funnel-shaped leaves *(Asplenium nidus)* or with the help of overlapping special niche leaves *(Platycerium)* (Fig. 5.31). A humus-rich, water-containing soil is thus formed into which the roots grow. But this can also be observed in many other species.

In a forest densely populated by epiphytes, the epiphytic humus can amount to several tons per hectare. In this way, a new biotope is created high above the ground, which can even be considered as an almost closed ecosystem. It is only now that attempts are being made to supplement the previous rather destructive (helicopters, balloon nets, etc.) or inadequate methods ('canopy walk ways', climbing techniques, etc.) in the exploration of this ecosystem by using new techniques (Fig. 5.32).

Nitrogen and nutrients are supplied to the canopy by dripping water and dust. Ants can settle and build their nests. They drag in seeds that germinate and grow into

Fig. 5.28 The cloud forests (**a**) in the montane altitudinal zone of the tropical mountains are rich in epiphytic flowering plants (**b**). In these forests, although the amount of precipitation is somewhat lower compared to cloud forests, the always high humidity constantly provides the epiphytes with sufficient water (photos: Rafiqpoor, Mt. Kinabalu)

Fig. 5.29 Examples of some epiphyte types from the tropical cloud and fog forests: (**a**) *Guzmannia* spec. (Bromeliaceae) (photos: Rafiqpoor); (**b**) *Epiphyllum phyllanthus* (Cactaceae); (**c**) *Rhipsalis aff. Crispata;* *Schlumbergera orsichiana* (Cactaceae), (**d**) *Rhipsalis pilocarpa* (Cactaceae) (photos: Barthlott)

flowering plants. Such 'flower gardens' or 'ant gardens' (Fig. 5.33) are described from South America. They also harbour a special fauna and microflora; mosquito larvae, aquatic insects and protists live in the funnels of the Bromeliaceae, which often reach considerable dimensions (phytotelmae). In addition, there is an enormous diversity of insect species.

Fig. 5.30 (**a**) *Usnea barbata* (yellowish) together with a bromeliad species on a tree in the cloud forests of the lower Charazani Valley (Bolivia) (photo: Rafiqpoor); (**b**) Another tree is completely overgrown with *Tillandsia usneoides* (photo: Breckle)

Fig. 5.31 Dead organic material has accumulated in the funnel of *Asplenium nidus* on a tree trunk in the rainforest of Ecuador. This provides the epiphytic plant with the necessary nutrients and also retains rainwater (photo: Rafiqpoor)

It should be mentioned that the insectivore *Nepenthes* (pitcher plant) (Fig. 5.34) can also grow epiphytically, as can various *Utricularia* species. Epiphytes are dispersed by spores (ferns), by dusty seeds (orchids) or by diaspores with membranous appendages for wind dispersal, or by berry fruits (Cactaceae, Bromeliaceae) that are eaten by birds, so that the seeds with the excrement easily reach tree branches far away. Many epiphytes can survive a

Fig. 5.32 Different techniques for recording epiphytes in the canopies of tropical rainforests. (**a**) and (**b**) in Gabon (photos: Szarzynski); (**c**) in Mt. Kinabalu Malaysia (photo: Rafiqpoor) and (**d**) in Venezuela (photo: Barthlott)

prolonged dry period, for example orchids, some moving in completely, or densely scaled tillandsias, poikilohydric ferns and others. They also occur in dry tropical forests. Coutinho (1982) found diurnal acid metabolism (CAM) in some epiphytes in Brazil, that is, the uptake of CO_2 at night when the stomata are open and the binding as organic acid (usually malate). The latter is then degraded during the day and assimilated immediately with the stomata closed. This is a process by which water loss by transpiration during the day is avoided, and is frequently found in succulents of arid regions. Medina already examined the Bromeliaceae in this respect in 1974.

Mosses and Hymenophyllaceae (skin ferns) require permanent wetness and are therefore the typical epiphytes of the Páramo, as are the epiphyllous species.

Fig. 5.33 On the rainforest trees in the Amazon lowlands of Ecuador, ant gardens represent small ecosystems (photo: Rafiqpoor)

More interesting are the strangler trees, of which the many strangler figs *(Ficus species)* are the best known. However, there are such strangler trees in many different families, for example the *Clusia* species (Guttiferae) in S America, *Metrosideros* (Myrtaceae) in New Zealand, Hawaii and others more. These species germinate as epiphytes in a branch fork and initially form only a small shoot, but a long root that grows rapidly down the trunk of the supporting tree, entwining it in a net-like manner.

Only when the root has reached the ground does the shoot grow up; at the same time the roots thicken more and more, forms interconnections (anastomoses) and prevent the secondary growth of thickness of the bearing tree, that is, the tree is strangled; it then dies and its wood decays. The strangler's root network closes to form a true trunk that supports a broad crown (Fig. 5.35b). Such tree like structures can reach huge dimensions, and it is not obvious that they began their existence as epiphytes. The developmental strategy of the strangler from germination to complete strangulation of the host tree is shown in a diagram (Fig. 5.35b). Palms without secondary thickening growth are not strangled and remain alive longer until eventually the strangler crown shades their leaves too much. In temperate latitudes, only ivy *(Hedera)* is known to strangle (Fig. 5.27).

Hemi-epiphytes occupy an intermediate position between lianas and epiphytes. Many Araceae germinate on the ground and then grow upwards as lianas (Fig. 5.25), usually as root climbers. In time, the lower part of the stem dies, and they are then epiphytes, but they can remain connected to the soil by aerial roots.

Fig. 5.34 The *Nepenthes* species on Mount Kinabalu (**a**, photo: Rafiqpoor) live both terrestrially and epiphytically in the rainforests at different altitudes. An endemic *Nepenthes* species (**b**) *Nepenthes* *vieillardii*, (photo: Breckle) also occurs in New Caledonia, partly growing on extreme sites such as heavy metal soils

Fig. 5.35 (a) Area consisting of a single fig tree that looks like a forest (photo: Barthlott). (b) Schematic representation (after Barthlott's lecture "Vegetation of the Earth") of the development of a strangler tree (red) from the initial phase to the extinction of the host tree (green) and its complete replacement by the epiphyte (full red), which then continues its life in place of the host tree

5.3.6 Epiphyllic Plants

Epiphyllic plants grow on the surface of leaves of other plants. These are microscopic algae (Cyanophyceae and other bacteria; *Azotobacter*, which can bind N, green algae), yeasts and fungi, lichens and mosses, (especially liverworts,

but also other mosses), then *Selaginella*, even small seed plants growing on leaves occur (Fig. 5.36).

Epiphylls occur mainly in the particularly humid tropical rainforest. The illumination, the wettability of the leaves and their longevity are fundamental for the colonization of the leaves by epiphylls. The leaves suffer additional light loss as a result (Fig. 5.36). However, some epiphylls even may grow into the leaf tissue.

5.3.7 Biodiversity

Central European forests usually have only five to ten tree species, of which one or two are dominant, i.e. they account for more than 90% of the stems. Corresponding temperate forests in N America or E Asia are not quite so species-poor, but there are still only 15–40 tree species per hectare. In the tropics, the number of species is incomparably greater. On the island of Barro Colorado in Panama, about 1400 higher plants occur in a 15 km^2 research reserve, including 365 tree species. In the mountain rainforest in the Biological Reserve N of San Ramón in Costa Rica, there are 94 tree species per hectare alone (breath height diameter, DBH 10 cm and larger), belonging to very different families (Wattenberg & Breckle 1995), and in Ecuador's Yasuni National Park about 300 tree species with >10 cm BHD per ha (Valencia & Balslev 1994), these are still the 'diversity records' today. Over a third of these species are represented by only a single stem, meaning that the minimum area for recording the species assemblage is thus much higher than 1 ha (Fig. 5.37); it cannot be determined. Also from other places, for example in Peru (Yanamono area), almost 300 tree species have been described from one hectare. There, 63% of the species are represented by only one stem per hectare.

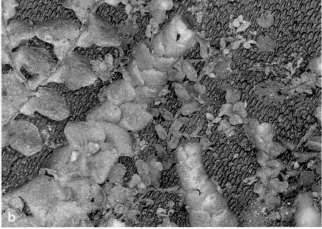

Fig. 5.36 (a) Coating of epiphylls on a large leaf of *Cyclanthus*, composed of various blue-green algae, green algae, mosses and lichens, Hymenophyllaceae, *Selaginella* and even a B*egonia*; in primary forest (Reserva Biol. San Ramón, Costa Rica) (photo: Breckle); (b) Part of a leaf with the epiphyllous liverwort *Aphanolejeunea* in Gisakura, Nyungwe National Park, Rwanda (photo: E. Fischer)

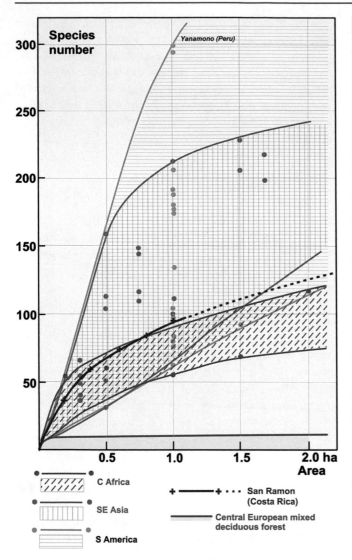

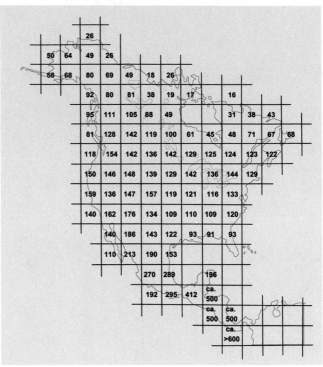

Fig. 5.38 The number of breeding bird species in N America in grid areas of 0.31 million km². Despite the tiny area, more land bird species breed in Costa Rica than in the USA and Canada combined, cf. with Fig. 5.39 (after Terborgh 1991, from MacArthur 1972)

Fig. 5.37 Increasing species number of tree species (≥10 cm DBH) with study area in montane rainforest in the Sierra de Tilaran (Costa Rica). There is no minimum area. When the area is increased from 1 to 2 ha, 30 new tree species are added: dashed red line, and for the tropical rainforest in C Africa (brown area), SE Asia (green area) and S America (blue area) (modified after Wattenberg & Breckle 1995, partly after Terborgh 1991)

However, the species assemblage of larger areas is still hardly known, as it requires years of effort to identify all species.

The tendency for the number of species per area to increase towards the equator applies not only to higher plants or trees, but also to reptiles, amphibians and birds, insects, etc. (exceptions: salamanders and aphids). For birds, MacArthur (1972) has shown the great difference in species numbers on a map of North and Central America (Fig. 5.38). Thus, tiny Costa Rica has more breeding land-bird species than the USA and Canada combined, although the land mass is only a small fraction (Fig. 5.39).

The greater structural diversity of tropical rainforests, the closer interconnectedness and tighter network with many

more different food sources, the year-round activity of the organisms, their closer intermingling, narrower niches and specialization and the huge variety of mutual interdependencies (symbioses) that are possible as a result is one possible explanation for the higher diversity.

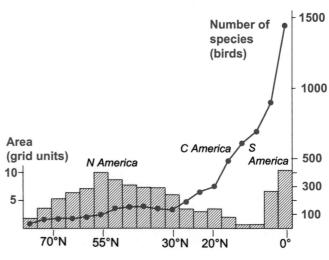

Fig. 5.39 The north-south gradient of landbird species in northern America (dot symbol) and, in comparison, the land area along latitudes to the equator (modified after Reichholf 1990)

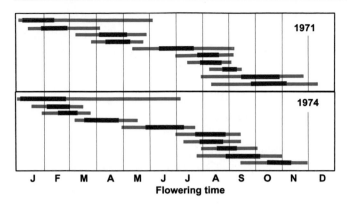

Fig. 5.40 The flowering times of heliconias in the rainforest in Costa Rica are distributed throughout the year and are similar in the individual years. They ensure a constant supply of nectar to the hummingbirds (modified after Terborgh 1991)

An important fact is the close functional interconnectedness of very many organisms. The temporal restriction of the flowering times of *Heliconia* species throughout the year can be taken as a relatively simple example (Fig. 5.40). Thus, for the various hummingbird species (Fig. 5.41), a food source is almost always available. However, this close web of relationships between several heliconias and several hummingbirds requires sufficiently large areas. If these are isolated too much, then at one point the network of relationships breaks down with far-reaching consequences for the other hummingbirds and in turn for other *Heliconia*.

Box 4.5 Diversity of Tropical Rainforests
In the tropics, up to c.300 different tree species with a BHD of ≥10 cm per hectare occur; in the whole of Europe, north of the Alps to the Urals, a total of barely 50 tree species are native.

Fig. 5.41 *Hummingbird* moving its wings so fast that it appears to hover almost still in front of the nectar plant (photo: Barthlott)

During **glacial periods**, rainforest areas were sometimes drier than today, deserts wetter: **pluvial periods**. In earlier epochs, the extent of the **Amazonian** rainforests was probably severely restricted and probably fragmented into retreat areas. The present-day rainfall distribution is given by the map in Fig. 5.42. It must be assumed that areas that now receive more than 3000 mm of annual precipitation also received sufficient rainfall (more than 2000 mm) to maintain closed tropical forests about 15,000 years ago.

These areas coincide quite well with certain retreat areas where both bird life, butterfly fauna, lizards, but also flowering plants are particularly rich in endemic species (Fig. 5.43). From this evidence, it can be concluded that the rainforests alternately shrank and expanded again, and that the foci of species richness and endemism correspond to sites that were permanently covered by rainforest. In between, large areas were probably covered with drier, seasonal rainforest. However, these changes occurred very slowly, whereas today's anthropogenic destruction is occurring at such a rapid speed, that organisms cannot adapt.

In the African forests, for example in Upper Guinea, Cameroon/Gabon and in E Zaire, endemism-rich retreats have also been identified. Only there have forests rich in tree species become known, in which up to 140 tree species per hectare occur, while in all other regions in Africa the number is always below 100, in Nigeria for example, only 23.

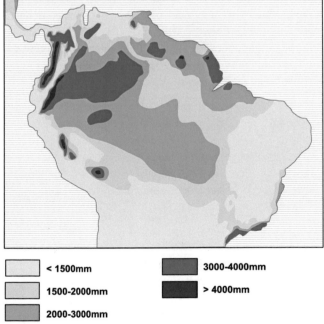

☐ < 1500mm	☐ 3000-4000mm
☐ 1500-2000mm	■ > 4000mm
☐ 2000-3000mm	

Fig. 5.42 The present-day distribution of annual precipitation in tropical S America. Areas with more than 3000 mm per year probably also received sufficient rainfall 15,000 years ago (at least more than 2000 mm) so that closed rainforests survived there (modified after Simpson & Haffer 1978)

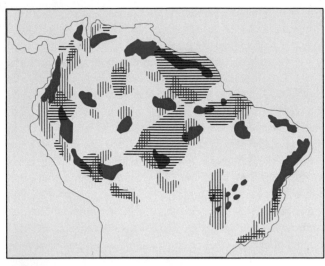

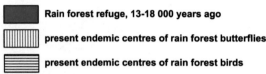

Rain forest refuge, 13-18 000 years ago

present endemic centres of rain forest butterflies

present endemic centres of rain forest birds

Fig. 5.43 The distribution of former rainforest refugia (c. 10,000 years ago) and present-day endemism in butterflies and birds (modified after Brown & Ab'Saber 1978)

The vegetation history of the Malayan rainforests is somewhat different. There, during the Pleistocene, a large part of the shelf sea was covered with rainforest. Presumably, the perhumid rainforests above sea level today have been preserved, which would explain their extreme species richness (with up to 180 tree species per ha) and the lack of geographic-geological evidence of earlier seasonal climates. Mt. Kinabalu in North Borneo has as many fern species as the entire African continent.

> **Box 4.6 The Functional Networks and Diversity of Life Forms**
> The preservation of the functional network with the immense diversity of tropical life forms requires much larger protected areas than in temperate latitudes.

5.4   Different Types of Vegetation in Zonobiome I Around the Equator

For zonobiome I, climate diagrams with a perhumid diurnal climate are typical (Fig. 5.3), which has two equinoxial rainfall maxima coinciding with the zenith position of the sun around noon. However, such a climate is not present everywhere in the equatorial zone. Areas with humid monsoon winds (Guinea, India, SE Asia) show only a particularly pronounced rainfall maximum in summer, but a short dry

season or even drought is noticeable (tendency to ZB II). The vegetation still consists of rainforests, but leaf fall and flowering are clearly bound to a certain season. One speaks of **seasonal rainforests**. On the Gold Coast (around Ghana), which is not hit by the monsoon, it still exist two rainfall maxima with droughts in between, similar to E Africa, where the monsoon winds are dry and the rain falls in the time of the wind change, distinguishing a major and a minor rainy season. In Somalia, the rainfall decreases so much that in some cases no humid season can be seen on the climatic diagram and the vegetation becomes desert-like: It is a zonoecotone I/rIII.

The trade winds also change the character of the climate, especially on the eastern sides of the continents. The SE-trade wind is humid and produces a rainforest climate in SE-Brazil, eastern Madagascar, and NE-Australia from the equator to beyond 20° S, with only one rainfall maximum. In contrast, the NE-trade wind in the south of the Caribbean Sea brings rain only to the mountains during wind jams at obstacles like mountains. As a result, Venezuela, with its many mountain ridges, has very diverse climates and vegetation (Fig. 5.44). The situation is similar in mountainous Costa Rica.

Venezuela is located between the equator and 12° N. All altitudes are present from sea level to the glaciated Pico Bolivar (5007 m). The northern half of the country is under the influence of the strong trade winds from November to March; it rains in the lowlands only during the windless seven summer months, with rising air masses and frequent thunderstorms. Only in the south of the country, in the Amazon basin, does no month have less than 200 mm of rain. Annual rainfall varies from 150 mm on the island of La Orchila to over 3500 mm in the south. In the mountains, on the windward side, rainfall increases rapidly up to the condensation level and decreases again above it. At the same time, temperatures decrease on average by 0.57 K per 100 m increase in altitude. The interior valleys, which lie in the rain shadow, are very dry (Fig. 5.45). The change in vegetation from north to south with increasing rainfall amounts as well as the elevational belts are shown schematically in Fig. 5.46.

In the driest parts, a semi-desert with cacti dominates (Fig. 5.47). The succulents store so much water that they easily survive a dry period of half a year or longer. If the rainfall increases slightly, thornbushes and ground bromeliads take hold. Impenetrable thickets develop, which correspond to the Caatinga in the dry area of NE-Brazil or the dry formations of the dry valleys of Ecuador. If the rainfall reaches 500 mm per year, the briars with umbrella crowns (*Prosopis, Acacia*) predominate. They are joined by *Bursera, Guaiacum, Capparis* and *Croton* species, as well as *Agave, Fourcroya* and others.

Peireskia guamacho (Fig. 5.48), the tree-shaped Cactaceae, which still has true leaves and is probably close to the ancestral form of the cacti, also occurs in the Caatinga. During the dry season these woody plants are leafless. The

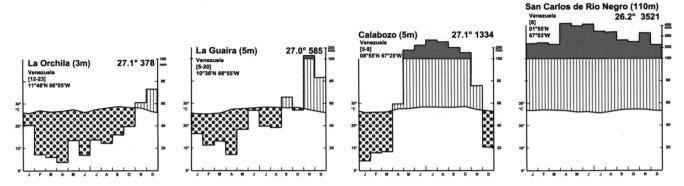

Fig. 5.44 Climate diagrams along a north-south profile through Venezuela (after Walter & Medina 1971). **1** offshore island, **2** coastal station, **3** typical trade wind climate (rainy season 7 months), **4** ever humid climate in the Amazon basin

Fig. 5.45 North-south sequence of vegetation formations in Venezuela, arranged climatically according to climatic diagrams in Fig. 5.44. (**a**) Sand desert vegetation on the north coast of Venezuela with *Melochia, Suriana, Jatropha, Sporobolus* and many other species, among others; (**b**) Semi-desert north of Mérida with columnar cacti and numerous tillandsias; (**c**) Wet savannah with numerous palms in the Sierra d'Avila near Caracas; (**d**) Tropical rainforest in Rancho Grande with epiphytes, lianas, stilt-root palms, Araceae, Melastomataceae and very many other species (photos: Breckle)

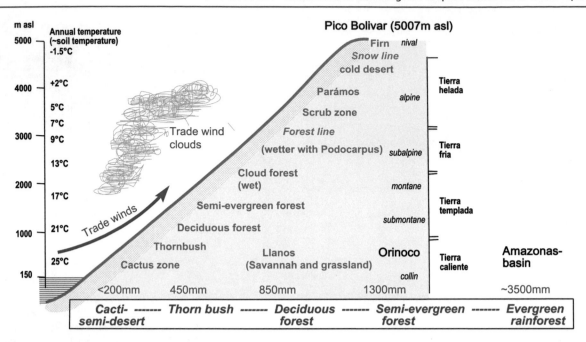

Fig. 5.46 Schematic representation of the vegetation zones in Venezuela from north to south, showing the annual precipitation in mm and the altitude levels as well as the mean annual temperature in °C (left)

Fig. 5.47 Cactus thornbush semi-desert with *Cereus* species on the wind-exposed slope of the coastal cordillera of Venezuela (photo: Breckle)

Fig. 5.48 *Peireskia guamacho* (Cactaceae) has the shape of a deciduous tree (**a**) with splendid flowers (**b**); its trunk, however, resembles cacti (photos: Barthlott)

cactaceous semi-desert and the thorn bush are only used as goat pasture.

If the rainfall increases further, the number of different tree species increases and true deciduous forests begin, which are very rich in species. The tree layer becomes 10–20 m high, only the Bombacaceae *(Ceíba)* and Malvaceae (Fig. 5.49), with thick trunks serving as water reservoirs, and the beautifully flowering *Erythrina* or *Tabebuia* species

Fig. 5.49 (**a**) *Adansonia fony* (Malvaceae; Bombacoideae), SW-Madagascar, (**b**) *Cavanillesia arborea* (Malvaceae) in Bahia State, Brazil, are two examples of trees that store water in their bottle-like trunks (photos: Barthlott)

Fig. 5.50 *Tabebuia* a yellow-flowered species from the seasonal tropical forests of Ecuador, flowering in the dry season when the tree is still leafless (photo: Breckle)

(Fig. 5.50) rise above this. During the dry season, such a forest looks much like a deciduous forest in winter in temperate latitudes. However, some tree species already begin to flower during this season. A distinction is made between, among others, dry tropical deciduous forests and moist ones with a precipitation level of up to 2000 mm. The latter reach a height of over 25 m and contain forestry valuable woods, such as *Swietenia* (mahogany), *Cedrela* and many other species.

The deciduous forests are sometimes cleared for the plantation of coffee crops under shade trees. Sugar cane, maize, pineapple and many other crops can also be grown here. Cattle pastures can be established after sowing *Panicum maximum*. The forests are poor in lianas, but epiphytes (drought-resistant ferns, cacti, Bromeliaceae and orchids) are common.

In even rainier areas with an even shorter dry season, the semi-evergreen forest occurs, in which only the lower shrub and tree layer consists of evergreen species. Finally, with even more rain, the tropical evergreen rainforest commences (Vareschi 1980).

A peculiarity of **Venezuela** is that in the area of the llanos of the Orinoco basin, which extend far into Colombia, instead of the deciduous forests suddenly appears a grassland with interspersed small forest stands or individual small trees. It is savannah or pure grassland. Climatically it is an area of deciduous forests. The grass that is now pasture is burned regularly, but here fire cannot be the primary cause of the absence of forests, rather it is the soil. We shall return to the special soil conditions in this area. Not due to the climatic, but also due to the edaphic (pedobiome) conditions or due to the relief it exist also still the following vegetation formations in Venezuela: the mangroves, on the sea coast and in the estuaries, the beach and dune vegetation, the freshwater swamps and the aquatic plant communities as well as the floodplain forests and the vegetation of dry shallow rocky soils.

Deciduous forests are an extrazonal occurrence in Venezuela, due to the dry trade winds, and are discussed in more detail in the context of ZB II.

The orobiomes must also be treated separately, since the elevational belts of the orobiome have certain peculiarities. Very diverse is also the equatorial mountainous East Africa.

5.5 Orobiome I: Tropical Mountains with Diurnal Climate

5.5.1 Forest Belt

In many tropical areas, mountains or volcanoes rise from the lowland rainforests (Fig. 5.51).

The tropical orobiomes are the particular habitats where the main mass of the population lives, especially in the Latin American Andean countries (Mexico, Venezuela, Colombia, Ecuador, Peru, Bolivia), finding a favourable livelihood due to the diversity of habitats along the elevational belts. The climates of the tropical orobiomes (high mountains) are characterized by an altitudinal gradient of temperature and precipitation. They are associated with an altitudinal gradient of vegetation. The thermal altitude gradient is 0.5–0.7 K/ 100 m. The gradient of heat with elevation is not only of climatological but also of vegetation-geographical and economic relevance, because it results in a vertical division of the zones of life both for plants and animals and for man and his economic activity. Since temperatures start from a high level at sea level (27 °C on average), the vertical extent of the

habitat in the orobiomes of the tropical zonobiomes is very extensive. The mean altitudes between 1000 and 3000 m also correspond to the most favourable temperature interval between 24 and 12 °C for the human organism.

It is therefore not surprising that in Latin America some of the mega-cities (Mexico City, Mérida, Bogotá, Quito, La Paz) have developed in the mountains. Beyond the forest and tree line, i.e. in the subnival elevation range between 4000 and 5000 m, there is hardly any snow cover that lasts more than a day. Here, frosts are also tied to the rhythm of the diurnal periods (nocturnal cooling, diurnal warming).

The thermo-isopleth diagram of Quito shows (Fig. 5.52) that the elevational areas within the tropics must be classified climatologically as tropical, since the isotherm of the seasons persists up to the summits of the high mountains. The annual variation in temperature is only 0.4 K at the inner tropical station of Quito at 2850 m altitude. Towards the tropics, this fluctuation increases successively, and the annual fluctuation also gradually becomes noticeable (Mexico City: daily fluctuation = 16 K, annual fluctuation = 13.5 K). Lauer (1995) has illustrated the structure of the tropical orobiomes in a diagram for South America (Fig. 5.53).

Precipitation also has a characteristic vertical distribution in tropical orobiomes, where a continuous thermal decrease with elevation is contrasted by a mostly discontinuous gradient of precipitation. Although the absolute water vapour content of the air decreases exponentially with altitude during convective processes, the multiple attainment of levels of complete water vapour saturation (100% relative humidity) also leads to the formation of multiple condensation levels with corresponding precipitation events at different elevational belts of the tropical mountains (Lauer 1975).

Fig. 5.51 The extinct volcano Mt. Kinabalu in Saba (**a**) and the Antisana (Eastern Cordillera Ecuador, (**b**) are two good examples of mountains (orobiomes) rising from the rainforests of the tropical lowlands. On these mountains, all elevational belts of climate and vegetation are formed according to physical specifications (photos: Rafiqpoor)

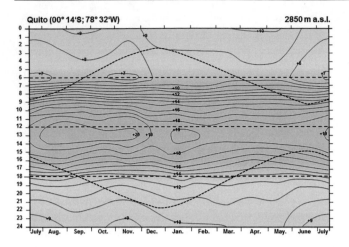

Fig. 5.52 The thermo-isopleth diagram of Quito (modified after Troll 1943)

Figure 5.54 shows a synthetic picture of the elevational belts of climate and vegetation on the eastern slope of the Eastern Cordillera of Ecuador, with a level of maximum precipitation at about 1500 m NN (Fig. 5.55) and a second, but somewhat weak, level of condensation in the area of fog forests at 3200–3800 m NN. In the cloud forest area at about 1500 m, epiphytic plants reach their maximum abundance and diversity (Fig. 5.56). In the **tierra fría**, where the second condensation level as the so-called **ceja de la montaña** (eyebrow of the forest, Fig. 5.27) allows the formation of fog forests, the branches and twigs of the trees are covered with epiphytic mosses and lichens (Fig. 5.57). The forest here appears more humid than in the cloud forest further down, because here fine droplets of water float in the water-vapour-saturated fog atmosphere and only precipitate when they come into contact with objects. The plants comb the moisture out of the water-saturated fog (Fig. 5.55). Passing through the

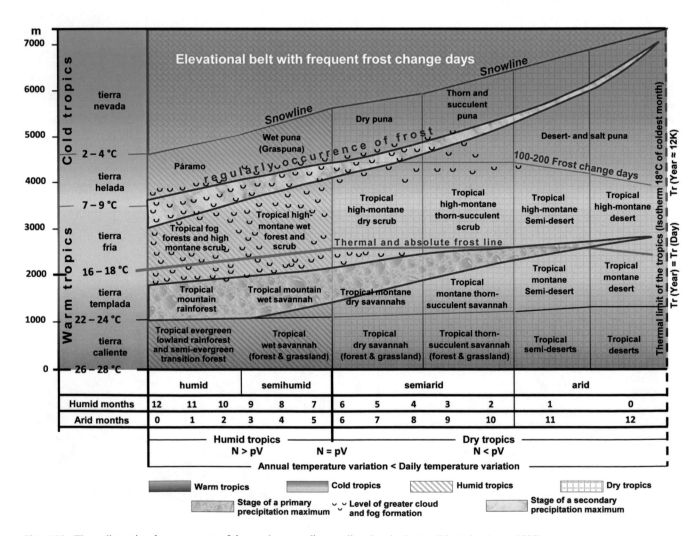

Fig. 5.53 Three-dimensional arrangement of the tropics according to climatic criteria (modified after Lauer 1995)

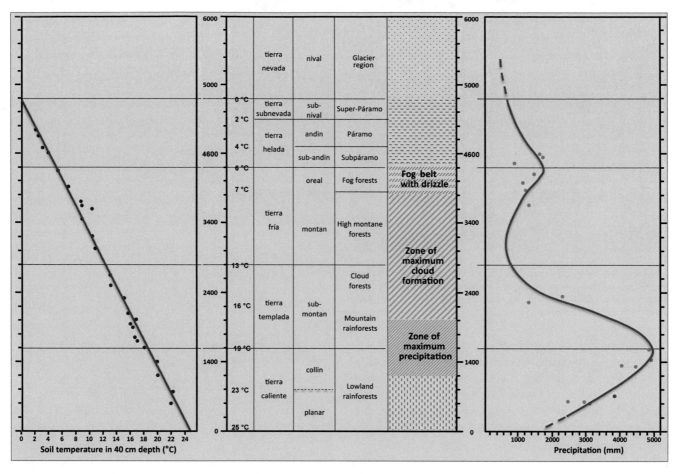

Fig. 5.54 Scheme of the three-dimensional arrangement of the elevational belts of climate and vegetation on the humid eastern slope of the Eastern Cordillera of Ecuador (modified after Lauer 1995)

Fig. 5.55 Formation of the first condensation level at about 1500 m asl in the altitudinal belt of tropical mountain rainforests on the western slope of the Western Cordillera of Ecuador above Machala (photo: Rafiqpoor)

Fig. 5.56 Abundance and frequency of epiphytic flowering plants are highest in cloud forests in the area of the first condensation level (Fig. 5.55). The picture shows an epiphyte survey in Gabon (photo: Barthlott)

Fig. 5.57 In the tropical fog forests, the branches and twigs of the trees are covered with mosses and lichens. Fine water droplets float in the air and are combed out by the plants when touched (Mount Kinabalu, Saba, Borneo) (photo: Rafiqpoor)

Fig. 5.58 Exotic crested-leaf plants (*Espeletia*) grow in a sea of grass in the altitudinal zone of the Páramo in the Andes (here Páramo del Angel in Ecuador) (photo: Rafiqpoor)

In N Venezuela, the following altitudinal sequence occurs:

Firn surfaces (glaciers)
--------------------*Climatic snow line*--------------------
Cold desert
Andine (alpine) level (Páramos)
---------------------------*Potential timber line*---------------------------
High montane forests with many *Podocarpus*
Cloud forests
Semi evergreen forests
Deciduous forests
Thornbush
Cactus semi-desert

dripping wet fog, clothes become wet without rain. Above the timberline, the equatorial orobiome contains **Páramo**, an exceedingly exotic-looking vegetation formation of grasses (Poaceae), espeletias (Asteraceae) (Fig. 5.58), bromeliads (*Puya*), etc. In the nival belt, glaciers occur, but with short tongues, as the year-round constant frost line is quickly reached here (Fig. 5.59).

The mountains thus often have very different altitudinal belts. If the trade wind meets a mountain ridge at right angles to the direction of the wind, the cooling of the air masses forced to rise causes condensation, i.e. cloud formation and upslope rain. As the strength of the trade wind diminishes in the late evening, the nights and early mornings are clear; during the rest of the time the cloud cover is at a certain height, so that this elevational belt is shrouded in fog during the day. In addition to the rising rains, condensation of the fog droplets on the branches of the trees and the lack of transpiration are present here because the atmosphere is saturated with water vapor (Fig. 5.57).

The extremely humid and, due to the altitude, also cooler climate causes the development of the hygrophilous, tropical Páramo, which is characteristic of all tropical mountains exposed to the winds. The sequence along increasing elevation is determined by the increasing precipitation level, while the decreasing temperature only becomes clearly noticeable above 2000 m asl.

The cool cloud forest, which is always dripping wet, differs from the hot tropical rain forest in the large number of tree ferns and epiphytic mosses that hang down from all the branches, as well as in the Hymenophyllaceae (filmy ferns) that cover all the branches and trunks. In the high montane forest, which is often above the cloud cover and not so humid, the epiphytic lichens are more predominant.

Due to the upslope rains, precipitation on the mountain slopes, as far as they are not in the lee, increases with altitude. Any dry season that may occur in the lowlands becomes shorter or disappears with altitude. The montane forests are therefore particularly luxuriant and rich in epiphytes, which are frequently wetted. As the slopes in the tropics are usually very steep, the soils are well drained, and the marshiness of the lowlands is absent. The decrease in temperature is at first scarcely noticeable. Eventually the cloud level is reached to which the Páramos are attached with maximum moisture.

Fig. 5.59 Glaciers exist in the nival altitudinal zone of the humid tropics, but they form only short tongues, as here on Cotopaxi volcano (5600 m asl) in Ecuador (photo: Breckle)

The more humid the air at the foot of the mountains, the lower is the cloud cover; in a climate with rainy season and dry season, the position of the clouds is higher in the dry season. Cloud forests may occur between 1000 and 2500 m and even higher and may have different temperature conditions, which causes floristic differences. The height of the tree layer also decreases upwards in the mountains.

More upwards in the fog forests only wind-formed, low trees occur. With increasing altitude, the number of heat-loving epiphytic flowering plants also decreases, but that of ferns, lycopodia and above all Hymenophyllaceae and mosses increases. The ground is often covered with a bright green carpet of *Selaginella* species. In many tropical mountains, the wettest elevational belt is characterized by palms (S-America) or dense stands of bamboo (E-Africa). Soils change with altitude: The red loams of the lower belt change to more yellowish or brown ones; at the same time a gauze horizon forms and the clay content decreases. Still higher, a slight podsolization becomes noticeable and finally true podsols with raw humus and bleaching horizon develop (Fig. 5.60); in the perhumid cloud level, gley soils can be found.

Fig. 5.61 Oak forest at 2500 m asl north of Cerro de la Muerte (Central Costa Rica) with small *Puya* bogs in the middle. Oaks dominate, with some other tree species also admixed. However, the forest line is about 800 m higher (photo: Breckle)

Fig. 5.60 Mountain podsol under a dense grass layer of *Calamagrostis effusa* (Poaceae) over Cangahua in the Páramo de Papallacta, Eastern Cordillera Ecuador (photo: Rafiqpoor)

Weinmannia, Myrrhodendron (shrubby Apiaceae), etc., are then replaced by bamboo (*Chusquea* species) as a sign of anthropogenic encroachment.

From the Central Cordillera in Costa Rica (Sierra de Talamanca), which rises to almost 3800 m (at Chirripó), very detailed studies of montane forests are available from Kapelle (1990). In most cases, oak species (Fig. 5.63) and bamboo *(Chusquea)* dominate, so that characterization according to dominant species is possible. Species diversity decreases with increasing elevation; for the woody species, this also results in a considerable change in importance, with the Rubiaceae, for example, decreasing from 2000 m with 31 species to two species at 3200 m (Table 5.1).

5.5.2   Forest Line

Above the cloud level (mostly above 2500–3000 m) precipitation decreases rapidly. If the forest extends even higher up the slope, the foliage of the trees becomes smaller and more xeromorphic. In Venezuela, conifers occur, namely *Podocarpus* species, which have no needles but hard, narrow, leaf-like structures. The mosses are replaced by bearded lichens (*Usnea barbata,* Parmeliaceae). Finally, the forest line is reached, which changes into a scrub zone and is lower in the tropics than in the subtropics. From the Andes of Venezuela, an altitude of 3100 to 3250 m asl is recorded, from Costa Rica 3200–3300 m (Fig. 5.61); in Venezuela, shrublands sheltered by rocks are still found at 3600 m asl; in N Ecuador, the forest boundary is about 4100 m (Lauer et al. 2001), in S Ecuador it is slightly lower (about 3500 m).

The shrub zone is narrow, but the shrubs also become lower further up (Fig. 5.62); in Costa Rica, *Escallonia,*

Fig. 5.62 The elevational belt of the scrub Páramo in Ecuador lies between 3700–4100 m asl and is composed of various species of the genera *Bacharis, Loricarya, Chuquiraga, Ribes* etc. They are accompanied by horst grasses *(Calamagrostis effusa, Festuca subulifolia,* etc.) (photo: Rafiqpoor)

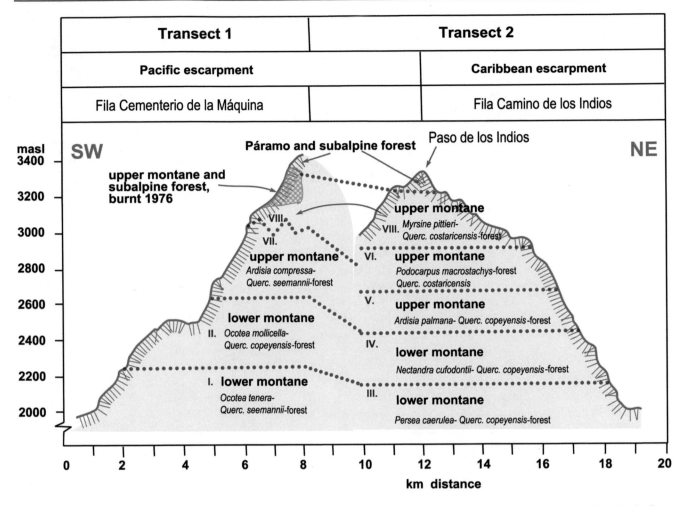

Fig. 5.63 Schematic mountain profile of the upper belts with montane-subalpine tropical oak forest in the area of Chirripó National Park (Costa Rica) on both mountain flanks (modified after Kapelle 1990)

The question of which factors are decisive for the tree line in the tropics is difficult to answer. The precipitation decreasing again from a certain height upwards made it seem possible that it is a drought limit. On the other hand, it could also be a frost line, because frosts can already occur at this altitude. However, studies in Venezuela (Walter & Medina 1969) and Ecuador (Lauer et al. 2001, Bendix & Rafiqpoor 2001), as well as worldwide, (Körner 2012) make it likely that the soil temperature is of decisive importance, although a wide variety of factors always interact in such phenomena. The diurnal climate in the equatorial zone means that temperature fluctuations penetrate very little deep into the soil. When the soil is shaded, the temperature at a depth of about 30 cm is constant throughout the year (Walter & Medina 1969, Körner 2007) and equal to the mean annual air temperature calculated by meteorologists on the basis of their measurements. With a few pricks of the spade and a thermometer, it is thus possible to determine the annual

temperature at any point in the tropics in a few minutes (Lauer 1982, Lauer et al. 2001).

In dense forests, the temperature is already also constant just below the surface. It is decisive for the root system. Körner (2007) mentions about 6.5 °C as an almost worldwide value for the temperature limit at tree lines. Although we do not know the temperature minima for the root growth of tropical trees, it is known that the enzymes that are decisive for the metabolic processes taking place in roots have a temperature minimum in tropical species that is far above 0 °C; tropical species can therefore already 'catch cold' at temperatures above freezing point and slowly die. *Ceíba* seedlings only grow at temperatures above 15 °C. If we assume that at the roots of trees at the tree line the temperature minimum is 6–8 °C, this would just correspond to the soil temperature in Venezuela at the tree line. The latter is formed by typical tropical species, holarctic species being completely absent. If our assumption is correct, this would

Table 5.1 The plant families with woody species, arranged according to their number of species (indicated in parentheses), from five different elevations of montane oak forests of the Sierra de Talamanca in Costa Rica (after Kapelle 1990)

2000 m	2300 m	2600 m	2900 m	3200 m
Rubiaceae (31)	Lauraceae (20)	Ericaceae (14)	Ericaceae (9)	Asteraceae (11)
Lauraceae (27)	Melastomataceae (14)	Melastomataceae (11)	Rosaceae (9)	Ericaceae (9)
Melastomataceae (27)	Asteraceae (12)	Myrsinaceae (11)	Poaceae (8)	Rosaceae (6)
Asteraceae (15)	Myrsinaceae (12)	Loranthaceae(9)	Asteraceae (6)	Clusiaceae (5)
Myrsinaceae (14)	Araliaceae (11)	Poaceae (9)	Clusiaceae (6)	Poaceae (5)
Araliaceae (11)	Ericaceae (11)	Araliaceae (8)	Cunoniaceae (6)	Cunoniaceae (3)
Solanaceae (11)	Rubiaceae (10)	Asteraceae (8)	Loranthaceae (6)	Scrophulariaceae (3)
Ericaceae (10)	Solanaceae (10)	Lauraceae (8)	Araliaceae (5)	Clethraceae (2)
Euphorbiaceae (9)	Rosaceae (9)	Rosaceae (8)	Lauraceae (5)	Lauraceae (2)
Piperaceae (9)	Fagaceae (6)	Solanaceae (8)	Myrsinaceae (5)	Loranthaceae (2)
Rosaceae (9)	Poaceae (5)	Cunoniaceae (7)	Solanaceae (5)	Melastomataceae (2)
Loranthaceae (7)	Celastraceae (5)	Rubiaceae (7)	Caprifoliaceae (4)	Rubiaceae (2)
Myrtaceae (7)	Cunoniaceae (5)	Aquifoliaceae (4)	Aquifoliaceae (3)	
Poaceae (7)	Loranthaceae (5)	Caprifoliaceae (4)	Fagaceae (3)	
Clusiaceae (6)	Aquifoliaceae (4)	Chloranthaceae (4)	Melastomataceae (3)	
Moraceae (6)	Acanthaceae (3)	Fagaceae (4)	Rubiaceae (3)	
Celastraceae (5)	Caprifoliaceae(3)	Myrtaceae (4)	Clethraceae (2)	
Cyatheaceae (5)	Chloranthaceae (3)	Celastraceae(3)	Myrtaceae (2)	
Fagaceae (5)	Clusiaceae (3)	Clethraceae (3)	Polygalaceae (2)	
Smilacaceae (5)	Cyathaeceae (3)	Clusiaceae (3)	Rhamnaceae (2)	
Urticaceae (5)	Myrtaceae (3)	Loganiaceae (3)	Rutaceae (2)	
Cunoniaceae (4)	Onagraceae (3)	Rutaceae (3)	Scrophulariaceae (2)	
Flacourtiaceae (4)	Rhamnaceae (3)	Symplocaceae(3)	Symplocaceae (2)	
Mimosaceae (4)	Rutaceae (3)			
Theaceae (4)	Theaceae (3)			
14 families (3)	11 families (2)	14 families (2)		
17 families (2)	35 families (1)	23 families (1)	25 families (1)	22 families (1)
26 families (1)				
82 families	**71 families**	**60 families**	**48 families**	**34 families**
349 Species	**226 Species**	**197 Species**	**125 Species**	**74 Species**

also explain the higher position of the tree line in the subtropics. There an annual variation of temperature already occurs, so that during the summer season the soil warms up considerably above the annual temperature, and the tree species can take advantage of this favourable season, which is lacking in the diurnal climate.

5.5.3 Andine (Alpine) Belt

The Andean elevational zone of the humid tropics is called the **Páramo**. It is almost permanently humid and foggy, inhospitable and cold. In the inner tropics of Ecuador, the elevational belt of the Páramo follows above the upper timberline in about 3500–3700 m. Year-round low temperatures with high interdiurnal variability, night frosts, constant cloud cover, frequent fog, occasional short snowfalls and reduced evapotranspiration are the climatic-ecological characteristics of the Páramo. On the frequent foggy days, the temperature level does not increase significantly. This results in a balanced day/night difference of air and soil temperature mostly excluding night frost up to about 4200 m asl. On such days there is practically no evaporation. During a 14-day period of bad weather from 12 to 25 November 1988, only 1.8 mm of evaporation was recorded in the Páramo de Papallacta (Ecuador) at 4200 m using a Piche tube (Table 5.2).

Table 5.2 Measurements of evaporation using a Piche tube in the Páramo de Papallacta (Ecuador) at 4100 m asl between 12–25 November 1988 (data from Lauer et al. 2001)

Date	Evaporation (mm)	Weather situation
12–19 Nov 1988	1.4	Cloudy, fog
19–20 Nov. 1988	0.0	Rain, fog
20–22 Nov. 1988	0.3	Rain, fog
22–25 Nov. 1988	0.1	Rain, fog
Total	1.8 mm in 14 days	

Table 5.3 Altitudinal belts of climate and vegetation in the Páramo de Papallacta, Eastern Cordillera of Ecuador (after data from Lauer et al. 2001)

Elevation asl (m)	Climate		Vegetation
	Elevational belt	Temperature (°C)	
>4800	Tierra nevada	< 0°	**Glacier region**
4800	Tierra subnevada	1°	**Frost debris belt** (Super Páramo) single *Werneria crassa* **cushions**
4600		2° Third condensation **level**	**Cushion plant Páramo** with *Distichia muscoides, Xenophyllum humile, Plantago rigida, Azorella compacta, Gentiana sedifolia*
4200	Tierra helada	5°	**Dwarf shrub Páramo** with *Loricaria ilinissae* potential forest boundary
4100		6°	**Genuine Páramo** with *Polylepis* forest islands dominated by *Polylepis pauta, P. incana, Gynoxis acostae, G. halii, Escallonia myrtilloides, Hesperomeles heterophylla, Plantago rigida* present forest boundary
3700		7,5°	**"Ceja de la montaña"** with *Miconia salicifolia*
3500	Tierra fría II	Second condensation **level**	**Fog forests** of *Tournefortia fuliginosa, Miconia latifolia, M. bracteolata, Senecio onae, Vallea stipularis* with ferns, and many epiphytic mosses and lichens
3100	Tierra fría I	10°	**High-altitude evergreen forests** with numerous *Miconia* species
2100 1000	Tierra templada	16° First condensation level 22°	**Evergreen mountain forests** with *Geonoma, Prestoea, Nectandra, Cestrum, Solanum, Boehmeria,* and numerous epiphytic flowering plants (orchids, bromeliads)

The altitudinal distribution of the Páramo de Papallacta (Eastern Cordillera of Ecuador) is shown in Table 5.3. The Páramo spreads in the humid Eastern Cordillera of Ecuador above the present forest line from 3700 m following the Ceja forest, consisting mainly of *Tournefortia, Miconia, Senecio, Vallea, Podocarpus,* ferns, and numerous epiphytic mosses and lichens. The grass Páramo between 3700 and 4100 m consists mainly of *Festuca subulifolia* and *Calamagrostis intermedia* (Fig. 5.64e) and is interspersed with small forest islands of *Polylepis, Gynoxys, Escallona, Hesperomelis,* etc. Lauer et al. (2001) suggest that the entire Graspáramo may once have been completely covered with a dense, medium-high forest, and only after settlement and anthropogenic deforestation with fire-clearing has it been cleared and converted to open landscape. The potential forest line is marked at 4100 m by the *Polylepis-Gynoxys* groves (Fig. 5.64f) (Keßler 2002).

Above this, between 4100 and 4200, a belt of shrub-grass Páramo begins, up to about 100 m wide, dominated by *Loricarya ilinissae, Diplostephium rupestre,* and *Chuquiraga jussieui,* together with *Festuca subulifolia* (Fig. 5.64c). Between 4200 and 4600 m, the altitudinal belt of cushion Páramo is formed. Here the surface is almost

completely covered with cushion-forming *Xenophyllum humile, Plantago rígida, Azorella compacta,* and *Huperzia crassa,* interspersed with small shrubs of *Diplostephium rupestre* and *Gentiana sedifolia,* etc. (Fig. 5.64d). This stage eventually becomes more patchy in the upper part due to the action of daily frost change (Fig. 5.64a). Finally, the 'super-Páramo' belt is dominated by frost-patterned soils (patterned grounds) with single specimens of *Werneria crassa* and *Azorella compacta* (Fig. 5.64b), very sharply demarcated upwards against the nival belt, which has almost no higher plants because of the continuous frost in the diurnal climate. The elevational belts are asymmetrically structured on the two sides of the mountains because of different hygric endowments (east side humid, west side dry).

In the Páramos of Venezuela, there is very little rain during the trade wind season (November to March). One can experience a whole week of cloudless skies in January. The cloud cover is lower. The hourly temperature values for days during the rainy season or during the dry season reflect the lack of radiation, or strong radiation (February 10) or strong radiation at night (February 12) (Fig. 5.65). The coldest day of 1967 almost immediately followed the warmest. During the dry season, the air at 3600 m usually

Fig. 5.64 Images of elevational belts of vegetation in the Páramo de Papallacta, Eastern Cordillera Ecuador. (**a**) A double cushion of *Xenophyllum* and *Azorella* surrounded with frost-patterned soils; (**b**) Single cushion plants in superpáramo with mainly frost-patterned soil forms; (**c**) Scrub páramo with *Loricarya, Diplostephium* and single *Gynoxys* shrubs; (**d**) Cushion páramo with *Xenophyllum, Azorella* and *Lycopodium;* (**e**): *Calamagrostis* grass páramo, (**f**) Grass páramo with *Polylepis* forest islands (photos: Rafiqpoor)

warms to 10 °C during the day, while it freezes at night. The plants, of course, are exposed to much more extremes than the thermometer in the hut. But this constant change of frost does not harm the plants, just at this time of the year is the main flowering. About this time also the upper layers of soil, in which the Páramo plants are rooted, warm up during the

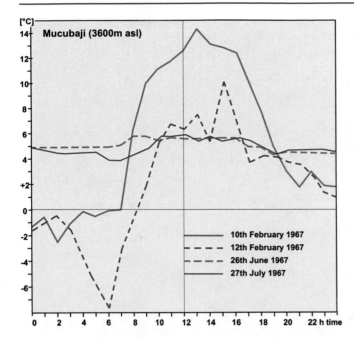

Fig. 5.65 Diurnal variation of temperature at the meteorological hut (Páramo belt at 3600 m NN in Venezuela) on 26 June as well as 27 July during the rainy season (variation only 1.6 and 2.0 °C, respectively) and on 10 February (hottest day) as well as on 12 February (coldest day) during the dry season with a variation of 17.0 and 17.5 °C, respectively, a t-maximum of 14.5 and a t-minimum of −7.5 °C

day above the annual temperature. Rocky sites seem to be more favourable than wet soils. The annual temperature was determined by measurements in the soil: at an elevation of 3600 m 5.0 °C (corresponding to the meteorological data), at 3950 m 3.9 °C, at 4250 m 2.0 °C and in the firn snow at 4765 m −1.5 to −3.5 °C.

As the temperature decreases, the plants are forced to root shallower and shallower. Thus, the plant cover becomes more and more open until finally a vegetationless belt develops below the firn and snow zone. This belt of cold desert with frost-sheltered and -patterned soils (Fig. 5.66) due to permanent frost-change days is characteristic of the tropical mountains. The altitudinal limit of occurrence of vascular plants in tropical mountains is much sharper than in temperate mountains. In Bolivia, it is almost exactly 5200 m, e.g. at the Chacaltaya. At higher latitudes (e.g. in the Alps) plants can take advantage of the most favourable season for growth even in the nival belt in places not covered by snow. The soil of the Páramos is moist even during the dry season, so that the vegetation does not suffer from drought and has a hygromorphic appearance. In Colombia, in addition to Páramos with a dry season, permanently wet soils with cushion plants, dwarf bamboo, grasses and mosses have also been studied.

The flora of the Páramos in S-America, Africa and Indonesia is very different and each area has its own peculiarities. However, it is striking that besides the plants pressed to the ground, there are also tall plants, mostly Compositae (Asteraceae), with a proper trunk and crested large leaves that have a thick white hair felt. In the Andes these are *Espeletia* (27 species) (Fig. 5.67a, b), in the equatorial African regions the tree *Senecio* species (Fig. 5.67c), in Indonesia *Anaphalis* species. In addition to the crested tree form, the woolly candle forms of *Lupinus* (Fig. 5.67g), *Lobelia* (Fig. 5.67h), and *Puya* species (Fig. 5.67d–f) should also be mentioned as special life forms. The many *Helichrysum* species on Kilimanjaro, Mt. Kenya or Mt. Elgon, which occur up to over 4400 m, are also very heavily hairy. That this pubescence (Fig. 5.67f) serves as thermal insulation and thus as protection against sudden extreme fluctuations in leaf temperature seems probable. On radiating days at these elevations, the passage of a cloud always results in a fall in temperature. The upper limit of vegetation, which is usually very sharp, is at about 4400–4600 m and is likely to coincide with an annual temperature of about +1 °C. At this elevation frost occurs once a day.

It is particularly strange, however, that in the Andes of Venezuela, even in the middle of the alpine belt at an altitude of about 4200 m, i.e. at an annual temperature of 2 °C, small stands of trees of the Rosaceae *Polylepis* occur. They are always attached to steep boulder slopes in east or west exposure, irradiated by the sun in the morning or afternoon, respectively. The rooting depth of *Polylepis* can reach 1.5 m.

The explanation for this occurrence of trees 1000 m above the present timberline is that block piles have particularly favourable temperature conditions. When insolation occurs, the layer of air near the ground above the block pile heats up very strongly; the cold air in the block pile is specifically heavier and should flow out in the lower part of the block pile, whereby the warm air in the upper part should be sucked in. This explanation is supported by the fact that the lower part of the block dump is not forested and is often completely bare.

Accurate temperature measurements at block piles of Mexican mountains have shown the same phenomenon. Numerous such high-altitude *Polylepis* forest plots are also found in Ecuador. They are now also discussed in connection with long-lasting historical fire clearing events (see above and Keßler 2002).

Somewhat less humid is the elevational sequence at the African volcanoes (Mt. Elgon, Mt. Kenya, Kilimanjaro), which rise from a humid savannah zone. Soil temperatures at the timberline (with *Hagenia, Podocarpus* species) are

Fig. 5.66 Comb ice (**a**, **b**), striped soils (**c**), cellular soils or miniature polygons (**d**), and soil buds or clay patches (**e**) as small frost pattern soil forms in the subnival altitudinal zone of tropical mountains, using examples from the Andes in Bolivia (photos: Rafiqpoor)

similar to Venezuela. *Erica arborea* plays a major role in the lower alpine belt, with crested *Senecio* (Fig. 5.67) and candle *Lobelia* above. Interspersed, however, are bogs with water-logged soils, on which Cyperaceae, *Alchemilla* or *Lachemilla*

species, Gentianaceae and the southern hemispherically widespread thickish *Huperzia saururus* (*Lycopodium crassum*) (Lycopodiaceae) occur (Fig. 5.68).

Fig. 5.67 Convergent life forms in different Páramos. (**a**) *Espeletia hartwegiana* in the Páramo de Mucubaji, Venezuela (photo: Breckle), (**b**) *Espeletia hartwegiana* in the Páramo del Angel, Ecuador (photo: Rafiqpoor); (**c**) *Senecio keniodendron* in the afroalpine belt of Mt. Kenya in the Teleki Valley (photo: Breckle); (**d**) *Anaphalis triplinervis* representing the Páramo of Java (photo: BotGart Berlin-Dahlem: http://bit.ly/2maGin1); (**e, f**) *Puya clava-hercules* in the Páramo del Angel, Ecuador (photos: Rafiqpoor). (**g**) *Lupinus humilis* in the Páramo de Pichincha, Ecuador (photos: Rafiqpoor); (**h**) *Lobelia deckenii* ssp. *kenyensis* (photo: Breckle). All pictures (except **d**) at about 4200 m asl

Fig. 5.68 *Huperzia saururus (Lycopodium crassum)* (**a**), shimmering red from a distance, in the elevational belt of the cushion Páramo (**b**) in Papallacta is one of the dominant life forms in the humid Páramos of the Eastern Cordillera of Ecuador (photos: Rafiqpoor)

5.6 The Biogeocoenes of the Zonobiome I as Ecosystems

The tropical evergreen rainforest is one of the most complicated plant communities. The individual biogeocoenes are still largely unknown, probably they cannot be plausibly delimited at all. The difficulty of dividing them into ecosystems is thus extraordinarily great.

The luxuriance of the vegetation and its high biodiversity tempt one to assume a very large primary production. The first estimates were 100 t ha^{-1} a^{-1} (dry weight), but they were much too high. It must be borne in mind that the phytomass in the tropical jungle is characterized by a very high water content (75–90% for herbaceous parts). The green leaves can assimilate CO_2 all year round, but the respiration losses at night are also particularly high due to the high temperature. The phytomass of wood and leaves is two to three times higher in tropical forests, but the costs of maintaining this mass, the respiration losses, are four times higher in the wood and six times higher in the leaves. Tropical forests are forced to greater metabolic turnover at the high temperatures, so relatively less can be invested in the production of wood.

Of very great importance for the matter production of a biogeocoene is the leaf area index (LAI), i.e. the ratio of the total leaf area of a stand to the soil surface covered by the stand. It was very low in the Ivory Coast. But this experimental plot cannot be considered representative. Although the gross production is very high, 75% of the organic matter produced is lost again through respiration, whereas in the central European beech forest only 43% of the gross production is lost.

We therefore understand that the annual primary production of the tropical virgin forest in this case was not higher than that of a well-managed beech forest in Central Europe:

Tropical virgin forest 13.4 t ha^{-1} a^{-1}
Beech forest 13.5 t ha^{-1} a^{-1}

Timber yields in forest plantations in the tropics reach 13 t ha^{-1}, which is only about twice as high as in a good European beech forest, which is also due to the twice as long vegetation period. A forest studied in Thailand with 2700 mm rainfall and an annual temperature of 27.2 °C has an above-ground phytomass of 325 t ha^{-1}, which should correspond to a total phytomass of 360 t ha^{-1}. It still increased by 5.3 t ha^{-1} per year during the 3 years of observation. They were: LAI = 12.3.

> **Box 4.7 Turnover Rates in Tropical Forests**
> Tropical forests have high turnover rates due to the permanently high temperatures. However, the very high productivity is more than compensated by particularly high respiration losses.

One must distinguish three phases in virgin forests, which form a small-scale mosaic: A **juvenile phase** with stand regeneration and positive phytomass increase, an **optimal phase** with maximum phytomass remaining unchanged, and an **aging phase** with decreasing phytomass. The Ivory Coast stand was probably a light juvenile phase. But all three phases usually occur mixed as a mosaic.

According to the available data, the following mean values can be given for the optimum phase of a lush tropical rainforest in ZB I:

Total phytomass 350 to 450 t ha^{-1} and, with a leaf area index of 12 to 15, a gross production of up to 120 to 150 t ha^{-1} per year, which would correspond to a primary production of 30 to 35 t ha^{-1}, with 10–12 t ha^{-1} on litterfall.

Soil respiration is approximately equal to the amount of litter, but a substantial part of the primary production is probably already mineralized above ground (standing dead trees, epiphytes). Litter returns 106 kg ha^{-1} of nitrogen annually to the soil in the Amazon, but only 2.2 kg ha^{-1} of phosphorus. The depletion of secondary forests is likely to be mainly a phosphorus problem, especially since P is rapidly bound to Fe and Al in the soil and is then no longer available to plants. Nitrogen is also steadily supplied from the atmosphere during the frequent heavy thunderstorms.

5.7 Fauna and Food Chains in the Zonobiome I

Only a few remarks can be made here. The organismic diversity and also the gaps in knowledge are still very large. The interrelations of organisms in the tropical rainforest are very close, and the resulting sensitivity to interventions has already been pointed out.

There are now many generally understandable books and extensive literature, also with comprehensive overviews of the tropical rainforest and its fauna. In each case, the close interconnection of the organisms is emphasized. A few examples of general works should be mentioned (Terborgh 1991, Richards 1996, Scholz 2003, Germanwatch 2014, Reichholf 2011, etc.).

For the animal world, as well as for many plants, it is characteristic that the canopy space is an important action space. More than half of the mammals live in the treetops and have a prehensile tail; very large is the number of birds, again with a focus of activity in the canopy area. The number of species can only be roughly estimated so far and especially the number of invertebrates above and below ground is actually unknown so far, as well as the functional relationships.

We know the most about bird species, e.g. their functional guilds in bird communities. As an example in comparison with a region from the temperate latitudes of the USA (ZB VI), such an overview of the guilds is given in Table 5.4. It becomes clear that the number of species on the one hand, and the number of guilds on the other hand, is considerably larger in the tropics. In temperate latitudes some guilds are completely absent. It is also easy to see that seed and fruit dispersal or destruction by birds is of great importance for the regeneration and thus future structure of the forest.

Furthermore, termites and ants are of particular importance for ecosystem processes; they do turn over quite a bit of biomass, but their zoomass, despite their diversity, is not large.

Typical canopy animals in the Neotropics are the sloths (*Cholopus, Bradypus),* whose lifestyle has been studied in detail (Montgomery & Sunquist 1975). The total zoomass of the animals was 23 kg ha^{-1}, and the leaf mass eaten annually

Table 5.4 The bird communities of a temperate forest (Congaree floodplain, USA) and of a tropical rainforest (Peru), broken down by guilds (functional units) (from Terborgh 1991)

Bird community guild	Number of species in the tropical rainforest Peru	Species count in midlatitude Congaree floodplain USA
Scavenger	2	1
Mammalian predator	7	1
Bird predator	4	1
Other birds of prey	7	1
Owls	5	2
Nightjar	1	0
Terrestrial seed eaters	5	2
Arboreal seed eaters	8	0
Terrestrial fruit eaters	3	0
Arboreal fructivores	18	1
Nectarivore	8	1
Terrestrial insectivores	10	2
Woodpeckers	8	5
Leaf-scanning insectivores	9	1
Leaf peepers	19	15
Failures undertaking insectivores	27	3
Insectivores hunting in the airspace	4	1
Ants following insectivores	6	0
Dead leaf scavenging insectivores	7	0
Climbing insectivores	7	0
Fruit eaters, predators	6	2
Tree-dwelling substrate-feeding fruit and insectivores	12	1
Tree-dwelling, foraging fruit and insectivores	13	0
Fruit, insect and nectar eaters	11	0
Total	197	40

was 53 kg ha^{-1}; this represents 0.63% of leaf production. The excrement decomposes slowly and provides a reserve of nutrients in the soil.

Leafcutter ants *(Atta)* (Fig. 4.11) exert a particularly strong influence through selective infestation (Haines 1975). They increase light enjoyment in the stand by up to 7%. They haul their material from tree species in the secondary forest up to 180 m to the 10-m-diameter underground nest, where they establish fungal gardens on the cut leaves. The cut leaf area can reach 4000 m^2. The mushrooms provide the food for the ants. So, they are perfect states with 'microbiological agriculture'. The number of other ant species living there on a single tree not infrequently exceeds the number of all ant species in a temperate country in Central Europe (Wilson 1988).

5.8   Man in the Zonobiome I

The tropical rainforest on poor soils is hostile to settlement and is mostly avoided by humans. It is often the refuge of native tribes. In Africa these are the pygmies, in Latin America the original Indian tribes. In Southeast Asia, too, remnants of the original inhabitants still live. In contrast, the former primeval forest areas on nutrient-rich, young volcanic soils are now densely populated cultivated land (Java, Central America and others). Only there is reasonably sustainable agriculture possible. On all poorer soils, clearing leads to catastrophic nutrient losses. The 'ecological disadvantage of the tropics' (Weischet 1980) is particularly evident here. Cleared areas are worthless after a few years and fall victim to erosion or cover themselves with worthless *Gleichenia* or *Imperata* thickets.

Yet today there are good reasons to realistically assess the economic value of tropical rainforests. Even without the completely inestimable genetic resources of an incredible biodiversity that has not yet been clearly recognized, the tropical rainforest is always worth much more than the wood standing on it, as the simple calculation in Table 5.5 shows. What an appalling overexploitation deforestation represents is clear from these figures. At the same time, the loss of biodiversity cannot be calculated; the enormous abundance and value of possible secondary constituents in such calculations is not taken into account.

Deforestation has accelerated at an incredible rate in recent decades. Costa Rica should be cited as an example: The forest area there has decreased alarmingly in just a few decades (between 1940 and 1987) (Fig. 5.69). It was not until after 1987 that the forested area increased again due to effective conservation measures and an increase in environmental awareness among the population (FONAFIFO 2012: http://bit.ly/2vnPtau).

Today in Costa Rica, although 21% of the country's land is under protection (national parks, reserves, etc.), the pressure on these areas is great, since hardly any other forest with larger wood reserves is available. In the Dominican Republic

Table 5.5 Wood value of marketable logs per hectare on an experimental area in Amazonia (rainforest of Misana on the Rio Nanay in Peru) in the case of irreversible one-time deforestation and compared with the annual yield and market value of fruits, raw rubber, resins and other continuously usable products per year (Peters et al. 1989, Reichholf 1990)

	Unique wood value	Ongoing use (per year!)
Number of species	27	12
Wood volume (m^3)	94	–
Wood value ($)	1001	–
kg Raw products	–	160
Number of fruits	–	5500
Market value ($)		698

on Hispaniola, the primary forest area has fallen from over 70% to less than 6% in half a century, and in Haiti on Hispaniola everything has been deforested.

The forecasts for the preservation of the rainforests foresee dire things. Of the approximately six million square kilometres of humid tropical forests that still exist today, all will have been cleared by 2040 according to current deforestation rates; according to other forecasts, this catastrophic state will occur as early as 2025, as the rate of deforestation will increase even further due to overpopulation and impoverishment (Fig. 5.70). This is not just a regional or national problem, but a global one. Even if in the USA or in Central Europe, too, almost all native and primary forests and prairies have been destroyed and replaced by monotonous forests or maize fields, the endowment of the landscapes and the climate are so favourable there that it is possible to establish an efficient agriculture and forestry industry and to hope for a largely sustainable use. In the tropics this is quite different (Box 4.8). It is worse, rather than better.

New data by FAO on all tropical rainforests indicate a total area of 13.4 million km^2. The net loss of forest area for the three rainforest regions was 54,000 km^2 annually between 2000 and 2010. Of this, the Amazon Basin accounted for 36,000 km^2, Southeast Asia 10,000 km^2 and the Congo Basin 7000 km^2 (see also Homeier 2021).

Global Forest Watch (GFW) has calculated that in 2018, about 120,000 km^2 of forest were lost in the tropics, including 36,000 km^2 of humid tropical rainforests, namely primary forests, i.e. the largely untouched forests. In 2019, the area of lost primary rainforests increased to 37,700 km^2, according to GFW. A total of 121,500 km^2 of forest was lost in the tropics this year. So again, it is worse, rather than better.

Box 4.8 Protecting Tropical Rainforests
The close coupling with the global cycle, the unique, incredibly high biodiversity with its corresponding irreplaceable genetic resources, the sensitive soils, the irreversible damage to landscapes when deforestation occurs, all call for an immediate, global effort to save the rainforests.

Fig. 5.69 Decline in percent forest cover in Costa Rica by 1987 as a result of deforestation and regrowth of forested areas by 2010 as a result of conservation measures in recent years (data from FONAFIFO 2012; http://bit.ly/2vnPtau)

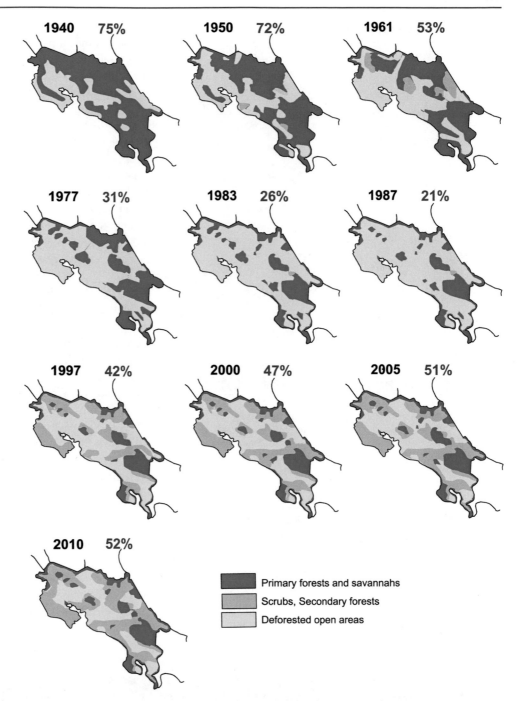

Primary forests and savannahs

Scrubs, Secondary forests

Deforested open areas

Today, thousands of fires burn every night (Fig. 4.20). The smoke still inhibits the greenhouse effect. Slash-and-burn must not be allowed to continue. Natural fires almost do not occur in the rainforests in ZB I.

It is getting drier and drier in the Amazon in recent years, more and more fires are deliberately set in order to gain land, and more and more often these fires get out of control; whereby out of control fires also create new acreage for invading colonists, according to WWF. In September 2020 there were over 32,000 different sources of fire in the Brazilian Amazon alone.

Clearance and fire clearance on a large scale only know the value of the forest, which comes from its wood, not really its natural resources – above all, its area as arable land. The year 2021 is not over yet, but one can extrapolate already a record deforestation. It is assumed that the deforestation rate is a third higher than the year before. This disaster entails: probably more than 200,000 fires by the end of the year 2021.

Area of rain forest [Mio km²]

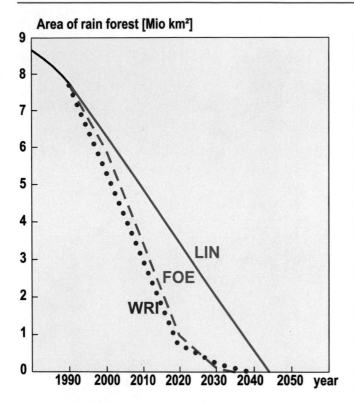

Fig. 5.70 The humid tropical rainforest is constantly losing area. The straight line shows the prognosis for constant deforestation, i.e. for a constant annual area cleared from 1990 onwards. The other two curves are based on projections by various organizations and take into account the still increasing demand (after Terborgh 1991)

5.9   Zonoecotone I/II: Semi-Evergreen Forest-Thorn Savannah

The zonoecotone between ZB I with evergreen rainforest and ZB II of the tropical summer rainfall area with deciduous forests is the semi-evergreen tropical rainforest, i.e. a transition zone with diffuse mixing of the two vegetation types.

In small areas, this is also locally formed as a vegetation mosaic, sometimes a patchwork of the most diverse vegetation types, modified according to groundwater, soil structure, water and nutrient availability. On very poor sandy soils in Venezuela and Guyana, the periodically flooded Igapo forest grows as a pedobiome near the river. Higher up on sandy soils follows the caatinga, which becomes a low caatinga or even a puny 'bana' when sandy soils are thick, although rainfall is high (3300 mm). However, the sandy soil has no nutrients and the water storage capacity during the dry season is very low. Viewed over a larger area, the following series can be identified with decreasing annual precipitation and increasing duration of the dry season in Venezuela (Figs. 5.44 and 5.45, respectively):

Evergreen rainforest - Semi-evergreen forest - Deciduous forest.

Within the equatorial climatic zone this series is seldom observed, such a gradation of rainfall as occurs in Venezuela being an exception. This series, however, can be generally observed as we move from the equator toward the tropics (of Cancer or Capricorn); for we thereby enter more and more into the tropical climatic zone of cenital summer rains, the absolute amount of rain constantly decreasing, the rainy season shortening, and the dry season lengthening. The difference with Venezuela is that in this the annual variation of temperature becomes gradually perceptible and more and more marked, the dry season being the cool season. However, since the latter is a dormant season for vegetation, temperature differences do not play a significant role for vegetation.

It has already been mentioned that in the very humid tropical area, when a short dry season occurs, the endogenous rhythmicity of the tree species adapts to the climatic rhythmicity. The general character of the forest does not change, but many tree species lose their leaves at about the same time, or sprout and flower at the same time. The vegetation thus exhibits a clearly synchronized seasonal aspect sequence (seasonal rainforest).

Fig. 5.71 A semi-evergreen tropical rainforest with *Erythrina* trees, conspicuous from afar by the red colour of its flowers. This forest is native to the leeward sides of the Ecuadorian Coastal Cordillera (photos: Rafiqpoor)

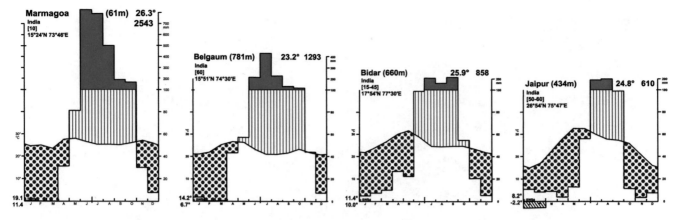

Fig. 5.72 Climate diagrams of Indian stations in the area of the evergreen, semi-evergreen, wet and dry monsoon forest

If the duration of the dry season increases further, the forest type changes: the uppermost tree layer is formed by deciduous tree species; in S America these are the large, thick-stemmed Bombacaceae and beautifully flowering *Erythrina* species (Fig. 5.71), while the lower layers still consist of many evergreen species. We therefore speak of the semi-evergreen tropical forest.

If the precipitation decreases further and the dry season is prolonged even more, then all tree species shed their leaves, so that the forest is bare for a shorter or longer period of time, i.e. it is a moist or dry deciduous tropical forest. This sharp transition is realized from central Costa Rica to the northwest (Guanacaste) and in Ecuador at the Andean western slope near Loja in a very short distance.

The climate diagrams for corresponding forest types in India, where this transition can be observed particularly well in the area of monsoon rainfall in summer, are shown in Fig. 5.72.

The question arises as to what determines the structure of the forest, the amount of precipitation or the duration of the dry season. The diagram Fig. 5.73 shows that both factors are ecologically important. One must not consider either factor alone. From the course of the boundary lines (slope) one can see that for the moist forest types the duration of the drought period is more important, whereas for the dry types the amount of rainfall is more significant.

In Africa the above-mentioned series is not so clearly observable. Due to increased settlement and the practice of shifting cultivation, it is precisely the area of semi-evergreen forests and moist deciduous forests that has been largely cleared. These forests are easier to clear than the rainforests, because they could be burned down during the dry season in the past; however, the rainfall is still high enough that one can expect an annual crop yield when cultivating the land.

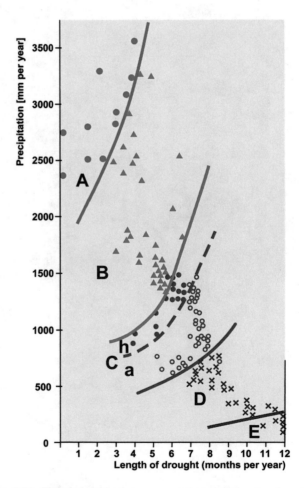

Fig. 5.73 The relationships between annual precipitation (*y*-axis, ordinate) and duration of drought in months (*x*-axis, abscissa) in India for different types of forest vegetation. **A** evergreen and **B** semi-evergreen tropical rainforest, **C** monsoon forest (**h** wetter: more humid, **a** drier: more arid), **D** savannah (thorn bush forest), **E** desert (after Walter, from a work commissioned by UNESCO)

Photo 9 *Rafflesia arnoldii* in the tropical rainforest of Borneo (ZB I) attracts flies as pollinators, growing as parasite on liana roots (photo: Rafiqpoor)

Photo 10 Tropical rainforest (ZB I) on the slope of the Ambohitsitondroina Mountains on the Masoala Peninsula in NE Madagascar (photo: E. Fischer)

Photo 11 Submontane tropical rainforest (ZB I) in the morning fog in the Sierra de Tilarán in Costa Rica (photo: Breckle)

Photo 12 Mangrove trees mirror in shallow sea water ditches on Isla de Coco in Southern Costa Rica (Pedobiom I) (photo: Breckle)

References

Begon, M., Mortimer, M. & Thompson, D.J. 1996. Population Ecology: A Unified Study of Animals and Plants. Third Ed. Blackwell Science Ltd., Oxford, UK, 247 p. ISBN 0-632-03478-5

Bendix, J. & Rafiqpoor, M.D. 2001: Studies on the Thermal Conditions of Soils at the Upper Tree Line in the Páramo of Papallacta (Eastern Cordillera of Ecuador). Erdkunde 7: 257-276

Breckle, S.-W. 2004: Flora, Vegetation und Ökologie der alpin-nivalen Stufe des Hindukusch (Afghanistan). In: Breckle, S.–W., Schweizer B., Fangmeier, A. (eds.): Proceed. 2nd Symposium AFW Schimper–Foundation: Results of worldwide ecological studies. Stuttgart–Hohenheim: 97–117

Brown, K.S.J. & Ab'Saber, A.N. 1978: Ice age refuges and evolution in the neotropics: correlation and paleoclimatological, geomorphological and pedological data with modern biological endemism. Paleoclimas (Sao Paulo) 5: 1-30

Coutinho, L.M. 1982: Ecological effect of fire in Brazilian Cerrado, 273-291. In: Huntley, B. J. & Walker, B.H. (eds.) s. there

FONAFIFO (Fondo Nacional de Financiamiento Forestal) 2012: Ministerio de Ambiente, Energía y Telecomunicaciones. Estudio de cobertura forestal de Costa Rica 2009-2010. 26 pp.

Germanwatch 2014: Die Bedrohung der tropischen Regenwälder und der internationale Klimaschutz. Arbeitsblätter zum globalen Klimawandel. http://bit.ly/2eSyeXe

Haines, B. 1975: Impact of leaf-cutting ants on vegetation development at Barro Colorado Island. Ecol. Stud. 11: 99-111

Homeier J 2021: Die Vernichtung der tropischen Regenwälder. In: Lozan J, Breckle S-W, Graßl H, Kasang D (eds): Warnsignal Klima: Boden und Landnutzung. 184-189

Homeier, J., Breckle, S.-W., Günter, S., Rollenbeck, R.T. et al. 2010: Tree diversity, forest structure and productivity along altitudinal and topographical gradients in a species-rich Ecuadorian montane rain forest. Biotropica 42: 140–148

Hüttel, Cl. 1975: Root distribution and biomass in three Ivory Coast rain forests plots. Ecol. Stud. 11: 123-130

Janzen, D.H.1978: Seeding patterns of tropical trees. In: Tomlinson, P.B. & Zimmermann, M.H. (eds.): Tropical trees as living systems. Cambridge Univ. Press: 83–128

Johansson, D. 1974: Ecology of vascular epiphytes in West African rain forest. Acta Phytogeogr. Suecica 59: 129 p.

Kapelle, M. 1990: Ecology of mature and recovering Talamancan montane Quercus forests, Costa Rica. Acad. Proefschrift, Amsterdam 270 p.

Keßler, M. 2002: The "Polylepis-Problem": where do we stand? Ecotropica 8: 97–110

Körner, Ch. 2007: Alpine ecosystems. Encyclopedia of Life Sciences, John Wiley

Körner, Ch. 2012: Alpine treelines. Springer-Verlag, Basel

Lauer, W. 1975: Vom Wesen der Tropen. Klimaökologische Studien zum Inhalt und zur Abgrenzung eines irdischen Landschaftsgürtels. Abh. d. Akad. d. Wiss. u. d. Lit. Mainz, Math.-nat. Kl., Nr. 3. Franz Steiner Verlag, Stuttgart.

Lauer, W. 1982: Zur Ökoklimatologie der Kallawaya-Region (Bolivien). Erdkunde 36: 223-248.

Lauer, W. 1995: Die Tropen – Klimatische und landschaftsökologische Differenzierung. Rundgespräche der Kommission für Ökologie. Bay. Akad. der Wiss.: "Tropenforschung", Bd. 10: 4360.

Lauer, W. 1999: Klimatologie. Das Geographische Seminar. Westermann Verlag, Braunschweig

Lauer, W. & Rafiqpoor, M.D. 2002: Die Klimate der Erde – Eine Klassifikation auf der Grundlage der ökophysiologischen Merkmale der realen Vegetation. Erd. Wiss. Forschung Bd. XL. 271 p. Franz Steiner Verlag. Stuttgrat.

Lauer, W. Rafiqpoor, M.D. & Theisen, I. 2001: Physiogeographie, Vegetation und Syntaxonomie des Páramo de Papallacta (Ostkordillere Ecuador). Erdwissenschaftliche Forschung, Bd. 39, Franz Steiner Verlag, Stuttgart

Longman, K.A. & Jenik, J. 1974: Tropical forest and its environment (Ghana), Thetford, Norfolk, 196 p.

MacArthur, R.H. 1972: Geographical ecology: patterns in the distribution of species. Harper & Row, New York

Medina, E. 1974: Dark CO_2-fixation, habitat preference and evolution within the Bromeliaceae. Evolution 28: 677–686

Montgomery, G.G. & Sunquist, M.E. 1975: Impact of sloths on neotropical forest. Energy and nutrient cycling. Ecol. Stud. 11: 69–98

Peters R, Gentry AW, Mendelsohn RO 1989: Valuation of an Amazonian rainforest. Nature 339: 655-656

Reichholf, J.H. 1990: Der unersetzbare Dschungel. Leben, Gefährdung und Rettung des tropischen Regenwaldes. BLV, München 207 S.

Reichholf, J.H. 2011: Der Tropische Regenwald: Die Ökobiologie des artenreichsten Naturraums der Erde. Fischer Taschenbuch Verlag, Frankfurt.

Richards, P.W. 1996: The tropical Rain Forest: An Ecological Study. Cambridge Uni. Press, 450 p.

Scholz, U. 2003: Die feuchten Tropen. Das Geographische Seminar. Braunschweig.

Simpson, B.B. & Haffer, J. 1978: Speciation patterns in the Amazonian forest biota. Ann. Rev. Ecol. Syst. 9: 497–518

Sprenger, A. & Breckle, S.-W. 1997: Ecological studies in a submontane rainforest in Costa Rica. Bielefelder Ökologische Beiträge 11 (Contributions to tropical ecology research in Costa Rica): 77-88

Terborgh, J. 1991: Lebensraum Regenwald, Zentrum biologischer Vielfalt. Spektrum Akad. Verl., Heidelberg 253 S.

Tomlinson, P.B. & Zimmermann, M.H. 1976: Tropical trees as living systems. Cambridge Univ. Press, UK

Troll, C. 1943: Thermische Klimatypen der Erde. In: Petermanns Mitteilungen 89: 81-89

Valencia, R. & Balslev, H. 1994: High tree alpha diversity in Amazonian Ecuador. Biodiversity and Conservation 3: 21–28

Vareschi, V. 1980: Vegetationsökologie der Tropen. Ulmer, Stuttgart, 253 S.

Walter, H. 1973: Die Vegetation der Erde, Bd. I: Tropische und subtropische Zonen. 3. Aufl., Fischer, Jena, Stuttgart, 743 S.

Walter, H. 1990: Vegetationszonen und Klima. 6. Aufl., Ulmer/Stuttgart 382 S.

Walter, H. & Medina, E. 1969: Die Bodentemperatur als ausschlaggebender Faktor für die Gliederung der subaplinen Stufe in den Anden Venezuelas. Ber. Dt. Bot. Ges. 82: 275-281

Walter, H. & Medina, E. 1971: Caracterizacion climatica de Venezuela sobre la base de climadiagramas de estaciones particulares. Bol. Socied. Venez. de Cienc. Natur. 29: 211-240

Wattenberg, I. & Breckle, S.-W. 1995: Tree species diversity of a pre-montane rain forest in the Cordillera de Tilaran, Costa Rica. Ecotropica 1: 21-30

Weischet, W. 1980: Die ökologische Benachteiligung der Tropen. 2. Aufl., Teubner, Stuttgart

Wilson, E.O. 1988: Biodiversity. National Academy Press, Washington D.C. ISBN 0-309-03783-2

Zeller, O. 1973 Blührhythmik von Apfel und Birne im tropischen Hochland von Ceylon. Gartenbauwissenschaften 38: 322–342

Part E: ZB II: Zonobiome of Savannahs, Deciduous Forests and Grasslands of the Tropical Summer Rainfall Area

6

Contents

© Springer-Verlag GmbH Germany, part of Springer Nature 2022
S.-W. Breckle, M. D. Rafiqpoor, *Vegetation and Climate*, https://doi.org/10.1007/978-3-662-64036-4_6

Photo 13 Savannah (ZB II) with *Acacia xanthophloia* (in background) and parched grass cover in Serengeti National Park, Tanzania, at the beginning of the summer rainy season (photo: Breckle)

6.1   General

Tropical **zonobiome II**, characterized by a 12-month thermal growing season, is frost-free like ZB I in the lowlands, but already exhibits a noticeable annual variation in temperature. Heavy cenital rains fall during the warm, mostly perhumid season, and the cooler season is arid. The hygric climate of ZB II is characterized by a pronounced hygric seasonality with a rainy and a dry season, whereby the length of the hygric growing season, i.e. the number of humid and arid months, determines the hygric character of the respective savannah landscape. Accordingly, ZB II can be divided into **semihumid** (7–9 humid months) rainfed moist forests and moist savannahs, **semiarid** (2–6 humid months) rainfed dry forests and dry savannahs, and **arid** (1–3 humid months) rainfed thorn forests and thorn savannahs (Fig. 5.53).

In the Americas, this zonobiome climatically occupies a large area south of the Amazon basin, plus smaller areas as far north as above the 20th parallel in Central America and partly extrazonal in Venezuela. In Africa, ZB II covers vast areas on both sides of the equator. South of the equator, on the plateau of the Zambezi, sometimes severe frost damage is observed in cold years, limiting the spread of ZB II to the south. The cold plateau around Johannesburg is already predominantly a grassland. In Asia, the main areas of distribution are India and SE Asia, whereas in Australia it is restricted to the northern part (Figs. 4.22, 4.23, 4.24, 4.25, 4.26, and 4.27). Corresponding to the humido-arid climate of ZB II on the flat areas are the zonal soils. These store so much water during the rainy season that they do not dry out completely during the drought season. This is a prerequisite for the growth of the zonal deciduous forests, which, although they greatly reduce transpiration losses by shedding their leaves during the drought season, must also absorb a certain amount of water from the soil during the drought season. Even the leafless twigs and branches still lose so much water that the quantities of water stored in the trunk are not sufficient for the whole drought period.

6.2   Climate, Soils and Zonal Vegetation

A special feature of ZB II is that the zonal open forest vegetation is absent in many places and replaced by the **savannah** vegetation type. The causes for this are of various kinds. However, a particularly important one is the presence of impermeable layers (laterite crusts and others) in the soil at various depths. Their presence is well known, but their exceptionally wide distribution was first demonstrated by Tinley (1982) on a 200 km profile by very accurate soil profile surveys in East Africa. He established the location of accumulation layers with pits 7 m deep. These impermeable crusts alter the water balance of the soil to such an extent that the formation of zonal forest vegetation is prevented (Fig. 6.1).

> **Box 6.1 The Humido-Arid Tropical Zonobiome**
> Zonobiome II, the humid-arid tropical zonobiome, is characterized by the sharp alternation of rainy and dry seasons. During a short dry season, deciduous forests often cover the entire area, while grasslands and thorn savannahs predominate during a longer dry season.

The savannahs and grasslands are mostly not only climatic, but edaphic, i.e. conditioned by the soil, and can thus largely be regarded as pedobiomes. A detailed description of the conditions is given below.

Lateritization occurs, on the one hand by slow dissolution of the silicic acid, by accumulation and solidification of round pisolite nodules, which can often consist largely of aluminium, iron and manganese oxides and can gradually be cemented into a hard crust; on the other hand, leaching processes and soil erosion play a major role (cf. the individual stages in Fig. 6.2). What remains is an undulating concrete-hard surface on which hardly any plant growth is possible (Fig. 6.3).

The leaching over long periods of time results in another edaphic characteristic: The often very low nutrient content of the soils in the area of ZB II. The land surface in Africa, but also in Australia, in the Near East and especially the Brazilian plate in South America are parts of the Gondwana shield, i.e. the primeval mainland, which split into the corresponding continents many millions of years ago (in the Mesozoic). Since then, the land surface was never covered by the sea; the soils are ancient and their rejuvenation by marine sediments never took place. The rocks were constantly leached and eroded. Therefore, wherever young volcanic rocks are lacking, the weathering products forming the soil are severely depleted in nutrient elements important for plants (phosphorus, trace elements), so that no forest can develop (Campos Cerrados).

On large plateaus, the barely noticeable lower parts of the relief are flooded during the rainy season and the soils are waterlogged. Forest groves grow only on the somewhat higher non-flooded areas, while a tropical grassland develops on the wet areas (Fig. 6.4). A mosaic-like parkland is thus created with forest plots and grasslands that are ecologically not savannahs. This is because savannahs are understood to be an ecologically homogeneous plant community of scattered woody plants in the midst of a relatively dry grassland. However, many geographers take a broader view of savannahs.

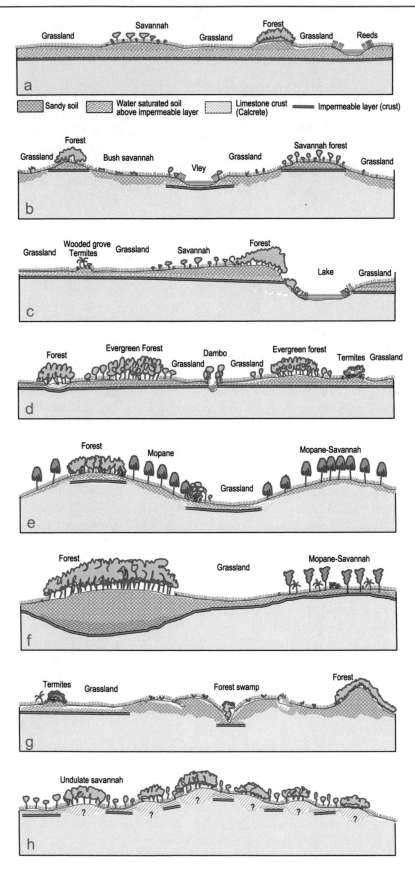

Fig. 6.1 Vegetation as a function of the position of the impermeable layer. Signatures Fig. 6.1a. Further explanations on page 176f. and in the individual legends (modified after Tinley from Walter & Breckle 2004).

(**a**) Sandy area with a high groundwater level above a continuous **impermeable** horizon; the soil is waterlogged. This results in the development of a wet grassland. On sandy, drained flat heights, a tree

Thus, in the ZB II one has to deal with three vegetation types:

1. With zonal deciduous forests
2. With relatively dry savannahs and
3. With the parklands wet in the rainy season.

Many laterite crusts are fossil, i.e. they were formed during the Pleistocene, the geological period characterized by several glaciation phases. These glaciations affected the Saharan desert zone (ZB III), not quite parallel in time, as pluvial periods with ± heavy rain, whereas in the tropical zone (ZB II), as recent pollen-analytical studies prove, they were dry periods extending into ZB I, which led to the formation of laterite crusts and relict savannahs still present today, even in the midst of evergreen rainforests.

> **Box 6.2 The Gondwana Remnants in ZB II**
> In Zonobiome II, peinobiomes are often developed on the old Gondwana shield areas: Biomes characterized by the severe nutrient depletion of the old soils.

Species diversity in savannahs is much lower than in tropical rainforests. Some examples of neotropical areas are given in Table 6.1.

Climatically, ZB II can be divided into two subzonobiomes according to the duration of the thermal and hygric growing season, namely a humid and a dry one. The corresponding climatic diagrams for India have been shown in Fig. 5.73. It is not expedient to give certain climatic limit values for all continents; the conditions in each case are too different for this.

According to the climate, wet and dry zonal tropical deciduous forests are also distinguished. The zonal soils are probably still too little studied or the results often not generally valid to give general distinguishing characteristics for the moist and dry ones. Like those of ZB I, they belong to the group of red-coloured ferrallitic soils, but SiO_2 leaching does not go so far in these soils, which are wet only during the warm rainy season. While the ratio SiO_2/Al_2O_3 is less than 1.3 in ZB I, it is 1.7 to 2 in ZB II. The sorptive power of zonal soils is also somewhat greater, that is, they retain ions important for plant nutrition better by adsorption due to a greater cation exchange capacity (CEC) and are therefore not quite as nutrient poor.

The most striking difference of the zonal vegetation of ZB II compared to ZB I is the leaf shedding, as a seasonal rhythm. It can be seen that in all climatic zones the tree species always develop the type of leaf structure that ensures the greatest production adapted to the respective climatic conditions. The leaf organs are always short-lived structures, because they age very rapidly, which means that they soon lose the ability to assimilate CO_2, which is their main task. The reason for this is probably the accumulation of ballast material, which is supplied to the leaf dissolved in the transpiration stream, as well as of metabolic by-products (tannins, alkaloids, terpenes, etc.), which, however, almost always also have a defensive role against herbivores.

Fig. 6.1 (Continued) savannah develops; on higher plateaus with more root space, forest stands, which can draw water from deeper layers during drought. In depressions, groundwater emerges and a lake or swamp forms with reed vegetation at the edge. (**b**) Hilly sand deposits with interrupted, impermeable crust at different levels, which only comes to the surface on the slope. There water escapes, the wetted ground is covered with grassland. Savannah or forest grows on the heights, depending on the availability of water in the dry season. In depressions, a Vley (periodic water basin) or a lake forms with grassland at the edge and individual gallery-like rows of trees. (**c**) Above a continuous, slightly sloping impermeable layer (crust) the soil is permanently waterlogged, but drains downslope, where in a depression water seeps out, sometimes evaporates and forms calcareous crusts (calcrete). Where the groundwater reaches almost to the surface, grassland grows, only outstanding termite heaps can support woody plants. Where the groundwater is deeper, savannah or even forest may develop. Woody plants grow at springs and seepage horizons and reeds around the lake. (**d**) On the Cheringoma coastal plain, a continuous waterlogged layer also occurs, so that mostly only grassland can develop. Woody plants grow on mounds or termite heaps, and on larger heights the water conditions are still so good, even during the dry season, that an extrazonal edaphic evergreen forest can develop, changing at the edges into a deciduous forest. In the deeply incised valley (Dambo) a gallery forest grows above flowing groundwater. (**e**) In the Urema Valley of the Rift Valley, the crust runs at very different depths and is up to 1.5 m thick. Where it is deep, rainwater is stored in the rooted soil so that a deciduous forest can develop, while on strongly swelling alkaline gleyic soils the special woody formation with mopane *(Colophospermum mopane)* develops. (**f**) In the western Caprivi area as part of the northern Kalahari, deep but coarse-grained sands often occur with low water storage capacity. Then only a very open tree savannah grows; but if there is a impermeable layer at depth, zonal deciduous forest (with *Baikiaea*) can grow. If the crust is shallow, base-saturated black clay-rich soils form with deep drought cracks in the dry season; they support only grasslands with annual grasses in the wet season. On the somewhat elevated areas mopane savannah, on termite heaps also other woody plants are found. (**g**) In the coastal dunes in Mozambique, depending on the water availability of the dune sand, one finds a low forest or grassland with individual shrubs. The crust is not continuous, again favouring grassland; on termite piles individual woody plants and at the base palms. In deep gullies over impermeable clay grows a swamp forest. (**h**) In the beach area of the Tonga coast (S Mozambique), parallel sand ridges have formed due to sea retreat, with hardened layers of fine dust in between. According to the water supply, forest and savannah alternate here again

Fig. 6.2 Scheme of the individual stages and processes of lateritization in an alternating humid Savannah climate by leaching processes and formation of a concrete-hard laterite crust with soil erosion

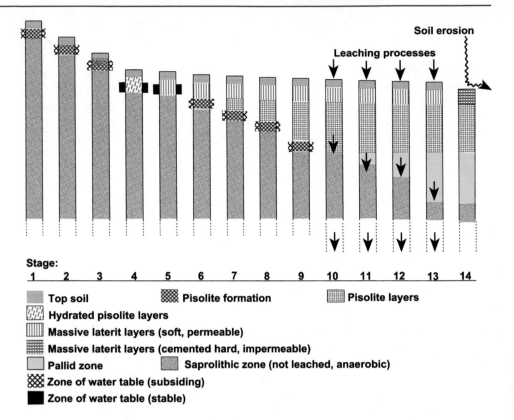

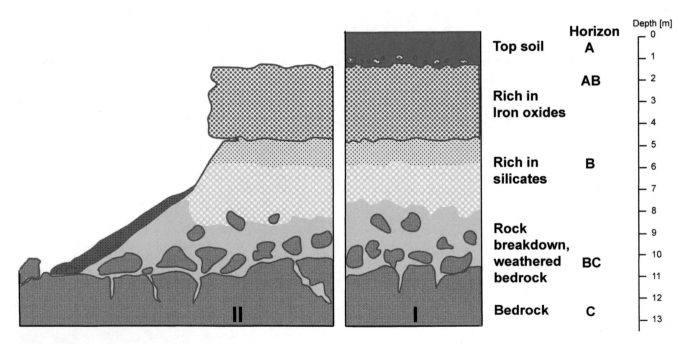

Fig. 6.3 Repeated drying of the **B-horizon** and erosion of the topsoil irreversibly hardens the iron oxide-rich layer (plinthite) to 'ironstone' (laterite) (modified after Schultz 1995, from Thomas 1974). The hard laterite crusts can 'cement' sinkholes and thus contribute to the formation of steps on mountain slopes (mesa formation)

Fig. 6.4 A broad plain in central Australia with imperceptible relief parts, covered with grassland and forest islands. It is flooded during the rainy season; waterlogged soils may then form. In the background the Olga Mountains, about 35 km away (photo: Breckle)

Even the evergreen trees of ZB I soon shed the old leaves when the young ones have become functional. Some species have been observed even in ZB I to be evergreen in good rainy years, but to lose their leaves before the leaf buds sprout when an unusual drought occurs, so the trees are bare for a short time. In ZB II, a long drought season is normal, while the rainy season is mostly very wet. Accordingly, the tree species only develop very drought-sensitive large and thin leaves at the beginning of the rainy season, for which they need less building material per leaf area unit than for the thick leathery leaves of the species of ZB I. Although the thin leaves assimilate CO_2 only during the wet season, the saving

of organic material makes the annual balance of production more favourable. For CO_2 assimilation, i.e. the production of organic matter, the assimilation intensity is decisive in addition to the leaf area. The latter is higher for thin leaves.

The water balance of the trees in ZB II is very balanced during the rainy season. This is because the diurnal variation of the transpiration curve runs parallel to the evaporation curve and hardly shows any midday depressions, which are always indicating a sign of incipient water shortage. The osmotic cell sap potentials of the leaves are also relatively low in the range of -0.7 to -1.9 MPa for all species. At the onset of drought, an increase in the sugar concentration in the

Table 6.1 Floristic richness of some neotropical savannah areas (after Sarmiento 1996)

Region	Area (km^2)	Species numbers			
		Grasses	Semi-shrubs and herbs	Trees and shrubs	Total number of species
Llanos in Colombia	150,000	44	174	88	306
Llanos in Venezuela	250,000	43	312	200	555
Cerrados in Brazil	2,000,000	429	181	108	718

Fig. 6.5 *Colophospermum mopane* forest in northern Namibia greening at the beginning of the rainy season (large image). During this time, the trees also begin to flower (photos: Breckle)

cells of the leaves to six times is noticeable (by 0.2 MPa in absolute terms). Soon thereafter, yellowing or drying of the leaves occurs.

Frost damage has been observed on the southern border of ZB II in Africa in unfavourable years (Ernst & Walker 1973). Sprouting of annual shoots and unfolding of leaves occurs only after the onset of rain (Fig. 6.5). But it is noticeable that the flower buds of many tree species open before the first rains. Since the petals have only cuticular, extremely low transpiration, this is associated with a hardly noticeable greater loss of water; on the other hand, pollination of the flowers by insects is facilitated in the still leafless forest.

The triggering factor for the onset of flowering is likely to be the maximum temperature, which occurs towards the end of the drought period but already before the onset of rain.

The most extensive ZB II forest stands are found in the sparsely populated parts of Africa south of the equator. These are the 'miombo' forests on the watershed between the Indian and Atlantic Oceans and on the Lunda threshold south of the Congo Basin, where there is no drinking water available for settlements during the drought season. At the dry frontier of ZB II, the occurrence of the 'monkey bread' or baobab tree *(Adansonia digitata)* is very striking; its misshapen bizarre trunk, which reaches a circumference of 20 m (Fig. 6.6), can store up to 120,000 litres of water. It can therefore be assumed that in a leafless state it survives the drought without absorbing water from the soil. Bottle trees belonging to the same Bombacaceae family also occur in South America and Australia.

Medina (1968) determined soil respiration in Venezuela in a deciduous forest at 100 m asl (annual temperature 27.1 °C, annual precipitation 1334 mm). It was three times more

Fig. 6.6 Very large baobab *(Adansonia digitata)* east of Tsumeb (Namibia) (photo: Breckle)

Table 6.2 Quantitative comparison of two dry forests

1 light Miombo forest in Zaire (11°37'S, 27°29'E, 1244 m asl)
Tree species: *Brachystegia, Pterocarpus, Marquesia* and others
Soils: Latosols
LAI: 3.5
Phytomass above ground: 144.8 t·ha^{-1} (of which leaves 2.6 t·ha^{-1})
Phytomass below ground: 25.5 t·ha^{-1} (estimated)
Annual net production: Litterfall 4–6 t·ha^{-1}·a^{-1}
Wood production: Not determined
2 dry monsoon forest in India (24° 54'N, 83°E, 140–180 m asl)
Tree species: *Anogeissus, Diospyros, Buchenania, Pterocarpus* and others
Soil: Reddish brown, lessivated sandy loam
Phytomass above ground: 66.3 t·ha^{-1} (of which leaves 4.7 t·ha^{-1})
Phytomass below ground: 20.7 t·ha^{-1} (estimated)   Annual net production: Trunks and branches: 4.4 t·ha^{-1} · a^{-1} Leaves: 4.75 t·ha^{-1}·a^{-1} Undergrowth: 0.35 t·ha^{-1}·a^{-1} Roots: (estimated) 3.4 t·ha^{-1}·a^{-1}

intense during the rainy season than during the drought season. It corresponded to an annually decomposed organic matter amount of on average 11,2 t·ha^{-1}. The annual litterfall was 8.2 t·ha^{-1}. The difference could correspond to root respiration.

Some information on production can be found in Cannel (1982) (Table 6.2).

In Thailand, Ogawa et al. (1961) studied:

1. A sparse Dipterocarpaceous dry forest at an altitude of 300 m with light-standing trees about 20 m tall and a grass layer 20 to 30 cm high.
2. A moist mixed deciduous forest with trees 20 to 25 m tall and sparse grass growth.

The following values for phytomass t·ha^{-1} and primary production.

(t·ha^{-1}·a^{-1}) were obtained:

Forest type	Phytomass	Production	(LAI)
1	65.9	7.8	4.3
2	77.0	8.0	4.2

The deciduous forests are used by the population for shifting cultivation for 3 to 5 years at a time. After 10 to 20 years, a secondary forest grows on the abandoned areas. The trees do not seem to grow older than 100 years.

6.3 Savannahs (Trees and Grasses)

As mentioned, savannahs are tropical ecosystems in which woody species scattered in a tropical grassland compete with grasses (Fig. 6.7).

Grasses and woody species are two ecologically and functionally antagonistic plant types that are usually mutually exclusive (Huntley & Walker 1982). Only in the tropics with summer rains and on deep loamy sands are they in ecological balance with each other. The antagonism is caused by the difference of 1. The root system and 2. The water balance (Walter 1939).

1. The grasses have a very finely branched intensive root system that roots very densely through a small soil volume. It is particularly suitable for fine sandy soils with sufficient water capacity in summer rainfall areas where the soil contains a lot of water during the growing season. The woody species, on the other hand, have an **extensive root system**. The coarse roots reach very far horizontally as well as in depth and root through a large soil volume, but not so densely. This root system is particularly effective in stony soils where water is irregularly distributed, not only in summer rainfall areas but also in winter rainfall areas when water percolates and must be absorbed by the roots from greater soil depths in summer. Therefore, in winter rainfall areas, grasses do not play a role.
2. With regard to the water balance, the typical grasses are characterised by the fact that they transpire very strongly when the water supply is favourable, have an intensive photosynthesis and produce a lot of organic mass in a short time. When water shortage occurs after the end of the rainy season, transpiration is not slowed down, but continues until the leaves and usually the whole above-ground parts dry up. Only the root system and the shoot vegetation tips remain alive, and their meristematic tissue, protected by many sheaths of dry leaf sheaths, is capable of surviving a long dry season. The soil can almost dry out. Only after the first rains does new growth begin.

Fig. 6.7 *Kigelia africana* savannah (Bignoniaceae) in Kenya. The grass layer withers after the rainy season. One has the impression that the tree cover becomes denser on the slopes of the hills, but it is always the same savannah (photo: Breckle)

Woody plants, on the other hand, which have a large shoot system with many leaves, have a balanced water balance. At the first signs of water shortage, the stomata are closed and thus transpiration is greatly reduced. If the water deficiency worsens, leaf shedding takes place. During the dry season, only the axial framework with the buds remains. Although these are well protected against water loss, measurements have shown that even leafless twigs have a very low but measurable water release over the course of hours. The water reserves in the wood are not sufficient to compensate for water losses during a prolonged dry period, which means that the woody plants are dependent on absorbing a certain, albeit very small, amount of water even during the dry period. They dry out and die when the soil no longer contains absorbable water.

If one takes these differences into account, one can understand the ecological balance in the savannah. As an example, consider the conditions in SW Africa with gradually increasing summer precipitation in an area with balanced relief and fine sandy soils that absorb all rainwater and store most of it (Fig. 6.8). This is zonoecotone II/III, that is, the transitional area between ZB II and the deserts with summer rains.

Climatic savannahs occur here with precipitation of 500 mm to 300 mm per year and a drought period of about eight months.

If the annual precipitation is only 100 mm (Fig. 6.8a), the water will not penetrate very deeply into the soil. In the soaked soil layers, the small horst grasses (tussocks) take root, consume all the stored water and then dry up after the rainy season; only the root system with the shoot vegetation tips (apical meristem) remains alive. Woody plants cannot survive because there is no water in the soil that can be absorbed by the plants during the drought (semi-desert). When rainfall is 200 mm, conditions are similar (Fig. 6.8b); the soil is more deeply soaked, and the horst grasses (tussocks) are larger, but they also consume all the water (grassland). Only when rainfall increases to 300 mm (Fig. 6.8c) will the grasses leave some water in the soil at the end of the rainy season; this small amount of water is not enough to keep the grass layer green, but it allows small woody plants to survive the drought period; a **shrub savannah** forms. If the annual precipitation is 400 mm (Fig. 6.8d), then the amounts of water remaining in the soil at the end of

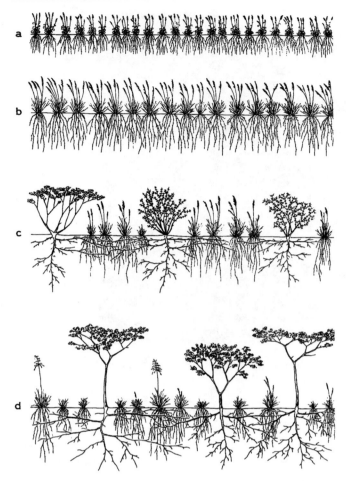

Fig. 6.8 Schematic representation of the transition from grassland (**a** and **b**) to shrub savannah (**c**) and tree savannah (**d**). Explanation in the text

the summer rainy season are greater, so that individual trees establish themselves and a **tree savannah** results.

But even in this, the grasses are still the superior partner. It depends on them how much water is left for the woody plants, and this proportion can vary greatly from year to year.

Only when precipitation is so high that the tree crowns move together and, by shading the grass layer, prevent it from fully developing, does the competitive relationship reverse. In the savannah forests or rain-green tropical dry woodlands, the woody plants then become the determining competitive partner, and the grasses have to adapt to the light conditions on the ground.

However, this unstable competitive balance in the savannah is very easily disturbed when humans intervene through grazing. The grasses are eaten away, thus the water losses through their transpiration cease, more water remains in the soil after the rainy season and this benefits the woody plants (mostly *Acacia* species), which develop luxuriantly and fruit abundantly. Some species additionally form root shoots. The tree seedlings do not suffer from the competition of the grass

roots; the tree seeds are spread with the excrement of the cattle that eat the pods and the mostly thorny shrubs grow so densely that scrub encroachment occurs, i.e. the pasture becomes worthless.

Scrub encroachment is a serious threat in all areas that are not rationally grazed. That is why thorn scrub is now more widespread as a substitute community than climatic savannah (thorn savannah), for example also in the arid parts of India, in N Venezuela and on the offshore islands (Curacao, Dominican Republic and others) (Fig. 6.9). If the area is more densely populated and the woody plants are used as firewood or for thorny enclosure of the cattle crest against predators, an anthropogenic desert with all the signs of desertification usually develops, covered with annual grasses only during the rainy season. During the dry season the cattle starve, having only the straw-like remains of the grasses as poor fodder. Such conditions prevail, for example, in Sudan, but also in N Kenya.

On stony soils the woody plants are absolutely superior to the grasses; grasses are almost entirely absent. With decreasing rainfall, woody plants become smaller and move further apart, because each shrub requires more root space, and the roots run shallow; for only the upper layers of soil are moistened.

Only a few small dwarf shrubs with xerophilous adaptations (dwarf shrub semi-desert) remain at the boundary with ZB III.

Special conditions prevail on two-storey soils (see below), as for example in Namibia, where a bush savannah then still grows with an annual precipitation of only 185 mm (Fig. 6.10).

With this amount of rainfall, pure grassland would be expected on deep sandy soil; however, the soil profile shows sandstone of the Fish River Formation overlain by a 10 to 20 cm layer of sand, which is either finely bedded with small fissures or coarsely bedded with larger cracks. The upper sand layer does not retain all the rainwater, some seeps into the crevices of the sandstone. The grasses make use of the water in the sand layer, while the roots of the shrubs penetrate the deeper sandstone and consume the water contained in the cracks. The water supply in the crevices of the fine layered sandstone is only enough for the small *Rhigozum* bush. Roots in the cracks of the large-banked sandstone allow the larger *Catophractes* shrub to thrive (Fig. 6.11). Thus, the distribution of shrubs reflects the structure of the sandstone and is similarly found where the covering sand layer is absent. Competition exists between the shrubs. In larger crevices both species may germinate, but in time the larger one displaces the smaller, of which only the dead remains. There is no competition between grasses and the woody species in this case.

In ZE II/III, zonal savannahs occur instead of deciduous forests when annual precipitation is too low for the latter. In

Fig. 6.9 Cactus forest with *Opuntia moniliformis* and *Neobootia paniculata* near Jimani, Dominican Republic (photo: Breckle)

ZB II, on the other hand, where despite sufficiently high precipitation the soil contains too little water during the drought period for a forest to survive, zonal savannahs spread. On the other hand, too much water during the rainy season, i.e. waterlogging, also precludes the growth of woody plants. A pure grassland then forms, which can dry out during the drought season and which is typical of parklands. In Fig. 6.1 a-h we have explained this. Some examples are given to show how the crust formed in the soil at different depths affects the vegetation mosaic. The effect of the crust can only be determined and studied in more detail by working in the field during heavy rains to observe runoff and water infiltration into the soil and relate it to the vegetation type in

question (although areas are often difficult to access during the rainy season).

The Fig. 6.1e is a slightly hilly area of the northern Kalahari with deep sand. Adherent water at field capacity is relatively low, so that much of the rainwater percolates and the adherent water is only sufficient for savannah vegetation. But in places laterite crusts prevent water from percolating. Depending on the depth of the crust, dense woody vegetation or dry forest may develop over the moist sand. If the crust is shallow in a depression, the soil above it is waterlogged, and only grassland grows.

On Fig. 6.1f, there is a continuous laterite crust in the sandy soil, which forms a depression on the left, above which seepage water flowing in laterally also collects. The soil is well aerated and moist at depth, so that the zonal deciduous forest finds favourable conditions; in the middle is a depression with grassland that is flooded during the rainy season, on the right on somewhat higher relief parts with clayey, base-saturated soils, the grassland is interspersed with some woody species adapted to heavy soils (mopane, *Balanites*, flute *Acacia*) (Fig. 6.12).

Figure 6.1b shows the vegetation structure again in a low-water-holding sandy area with a savannah in which the woody plants dry down to the ground during the drought and sprout again from the stem base or as root shoots in the rainy season. Where laterite crusts lie at different depths, depending on the water conditions, a grove or a savannah forest develops on the elevations or, in the lowlands, a shallow small lake (Vley) with swamp vegetation that can dry out

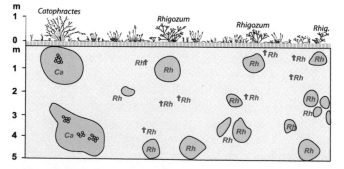

Fig. 6.10 Line profile (1 m wide) through a typical vegetation plot at Voigtsgrund (SW Africa). Grasses droughty during the dry season. Below, plant cover in plain view (without grasses). **Ca** = *Catophractes*, **Rh** = *Rhigozum* († dead)

Fig. 6.11 *Catophractes alexandri* savannah in Namibia. Shrubs grow in the small crevices of the sandstones of the Fish Formation, and even trees in the larger ones (photo: N. Derber)

during the drought. Some trees may also occur over laterally draining excess water at the edge of the crust. One can thus see how, depending on the position of the crusts, forest and savannah alternate or, in the case of waterlogged soils, parklands develop.

In contrast to Fig. 6.1f, the continuous crust in Fig. 6.1a is the same depth everywhere and the overlying soil is waterlogged and covered with grassland in the rainy season; only on small elevations is the soil better drained and supports a tree savannah (left) or, in the case of larger root space, a woody plant. On the far right is a depression with an open water table; a reed bed is developing at the edge of the water basin.

The hardened layer of the **B-horizon** is not infrequently exposed on slopes by erosion. It leads to the formation of mesas. At the edge of these mountains, step formation leads to steep slopes, as shown in the development in Fig. 6.3.

There are other factors that promote savannah vegetation, such as fire (Fig. 6.13), the large game herds (Fig. 6.14), and the various interventions of humans with their livestock. For example, in Pendjari National Park in Benin, West Africa, park management sets small-scale controlled fires annually at the beginning of the dry season to prevent the accumulation of larger dry masses. These fires are considered 'cool' and once out of "topkill height" do not kill trees and shrubs and prevent uncontrollable large-scale 'hot' fires (Fig. 6.13).

Fire has been effective in the climatic region of ZB II as a natural factor long before the appearance of man. Thunderstorms usually usher in the rainy season; since there is a lot of dry grass around this time, lightning can easily start a fire. The frequency of such fires is proved by the many pyrophytes, that is, species of wood which are resistant to the action of fire. The tree or shrub species often have a thick bark which is only charred and protects the cambium, or the shrubs have dormant buds above the root collar in the

Fig. 6.12 *Acacia drepanolobium* scrub (flute *Acacia*) with galls (ant nests) over heavy soils in the grass savannahs of northern Kenya (photo: Breckle)

Fig. 6.13 Fire in the savannah in Pendjari National Park in Benin, West Africa (photo: A. Erpenbach)

Fig. 6.14 Big game herds in the Ngorongoro Crater National Park (Tanzania) in the grass savannah area (photo: Breckle)

ground which sprout when the above-ground shoot parts burn. Many species have underground storage organs that can lignify (Lignotuber, Fig. 6.54) and allow rapid regeneration.

Grass fires were already set up by primitive man in prehistoric times to protect himself and his places of settlement from the danger of surprising fires caused by lightning. For with the high growth of grasses in the wetter zones, fires spread with great speed and violence. Today, burning during the dry season has become a common bad habit to facilitate the hunting of big game, or to destroy vermin (snakes, etc.). After a grass burn, the grasses sprout earlier, which is initially favourable for grazing.

The grass fires can only invade dry forests with grass understory, but they also push back the wet forest at the edge, causing a sharp borderline. Above all, however, they prevent the forest from reclaiming lost terrain as well as the cleared and subsequently grassed areas.

In northern Venezuela, fire also plays an important role in the equilibrium between *Byrsonima crassifolia* (with very low density) and the perennial C4 grasses *Trachypogon vestitus* and *Axonopus canescens*, in addition to water supply and nutrient distributions (Fig. 6.26). However, there is also a very unstable equilibrium between the two grasses, which Inchausti (1995) investigated by transplanting experiments. Only under absolute protection does *Axonopus* gradually replace *Trachypogon*.

A very essential factor for savannahs is grazing by big game (Anderson et al. 1973) (Fig. 6.15). Young tree growth is destroyed by browsing and trampling. Elephants are particularly hostile to forests. They tear out trees or debark the trunks. Elephant tracks clear the forest and allow grass fires to enter the forest. One elephant can destroy an average of four trees per day. Tree losses in miombo woodlands reach up to 12.5% per year. Elephant numbers are increasing rapidly in protected areas. Lake Albert's Murchison Park (in Uganda) is being deforested by elephants over time. In the Serengeti

Fig. 6.15 The volcanic areas in East Africa are blessed with fertile soils. Here, a savannah landscape with a high species richness of big game can be found (photo: E. Fischer)

(in Tanzania), on the other hand, there seems to be a balance between game damage and vegetation regeneration (Fig. 6.16).

It is striking that in Africa, which is rich in game, many woody plants of the savannahs are thorny (Figs. 6.17 and 6.18), whereas this is much less the case in South America and Australia, which are poor in game. This speaks for a selection of species protected from wild browsing.

An indirect impact on vegetation comes from game trails, which easily initiate furrow erosion. This is especially true for hippos, which climb up the riverbanks from the water at night to graze on the grasslands.

Erosion furrows can drain a wet grassy area, which in turn allows woody plants to encroach. A summary of these multiple impacts of big game can be found in Cumming (1982). Even greater is the impact of humans, both livestock producers and arable farmers.

Grazing on the savannahs north of the equator began at least 7000 years ago. Forests are only present in small remnants in this area; a large part of the savannahs is therefore likely to be secondary in nature (Hopkins 1974).

In summary, the following savannah types are distinguished according to their genesis:

1. **Fossil savannahs** formed under formerly different conditions in the area of ZB I
2. **Climatic savannahs** in the area of ZE II/III with annual precipitation below 500 mm
3. **Edaphic savannahs**, that is, ZB II savannahs conditioned by soil properties:
   (a)   On soils whose water balance is less favourable than it should be according to the amount of rainfall due to crusty accumulation horizons (laterite crusts, clay layers, compacted silt or sand layers).

Fig. 6.16 Acacia savannah in Massai land of the Serengeti in Tanzania (photo: Breckle)

    (b) On soils that are primarily so poor in nutrients that forests cannot grow on them.

    (c) Within parklands with wet soils during the rainy season as a special type of palm savannah.

4. **Secondary savannahs** as a result of fires, impact of big game and the various interventions of man.

What type of savannah it is in each individual case cannot be determined by eye, but requires a detailed investigation.

> **Box 6.3 Human interventions against Forests**
> All interventions, such as fire, grazing, clearing in the context of shifting cultivation or firewood extraction are directed against the forest.

Fig. 6.17 Inselberg and thornbush savannah in Olduvai Gorge, Tanzania, west of Ngorongoro, the valley of prehistoric man with numerous Pleistocene fossil sites of hominids (photo: Breckle)

Fig. 6.18 Thorn and succulent bush savannah with arboreal euphorbias on the slope of Ngorongoro Crater (photo: Breckle)

6.4 Park Landscapes

In the case of very flat terrain, park landscapes usually form in the ZB II. This landscape is conditioned by differences in relief that are hardly noticeable in the terrain and which are not perceived during the drought period. During heavy rainfalls in summer, all deeper parts of the relief are flooded, because it takes months for the water to drain away. These biotopes are occupied by grasslands; the soils are grey, while on the higher non-flooded parts, where the woody plants grow, the soils are deep, red sandy loams. The river system here begins on the watershed with barely sunken strips covered with grass, which unite below and gradually change into

Fig. 6.19 Dry bush with young baobab trees, briars and *Sansevieria* in the understory in the Serengeti, Tanzania (photo: Breckle)

incised stream and river beds as the gradient becomes steeper (easily seen from the plane).

A special formation is the termite savannah, which is understood to mean wide grass-covered depressions from which broad, abandoned termite heaps protrude as islands, which are not flooded and which can therefore be covered with tree growth. It is therefore a mosaic of two different communities (Fig. 6.19), i.e. not an actual savannah. However, this depends entirely on the particular termite species and its species-specific burrows. In N Australia, savannahs with numerous small columnar burrows are found. Massive tower-shaped structures in the *Catophractes savannah* in Namibia and Uganda remain for decades (Fig. 6.20).

The deeper depressions with black clays referred to as 'Mbuga' are a special **Amphi-Biome** with alternating wet soils and a hard iron concretion layer at 50 cm depth. Because potential evaporation far exceeds the 1000 mm rainfall, the clay soil dries to a depth of 50 cm in August to December and is divided into polygons by deep crevices (Fig. 6.21). Such biotopes are unsuitable for tree species. Trees only grow where the groundwater table is always below 3 m. The laterite crust is also located at this depth and the roots of the trees reach just as deep.

In contrast to the termite savannah, the **palm savannah** is a homogeneous plant community. As woody monocotyledons, palms have a tufted root system consisting of identical, barely branching roots that spread radially so that the palms stand alone in the grassland. They tolerate intermittent flooding. The soils of palm savannahs are likely to dry out less during drought than those of pure grasslands, but no studies are available on the competitive relationship between palms and grasses ('Palmares').

Fig. 6.20 "Termite savannah", intermittently flooded grassland with tree growth on old termite piles west of Georgetown, Murchison Falls, Uganda (photo: E. Fischer).

Fig. 6.21 Polygonal dry cracks of the clay layer in an Amphi-Biome, in the dry valley of the Rio Chota in Ecuador (photo: Rafiqpoor)

Fig. 6.22 Comparing individual trees in the savannah and the tree-fall gap in the rainforest, there are remarkable parallels in the expression of the 'island biotope' or 'idiotope' (modified after Belsky & Canham 1994)

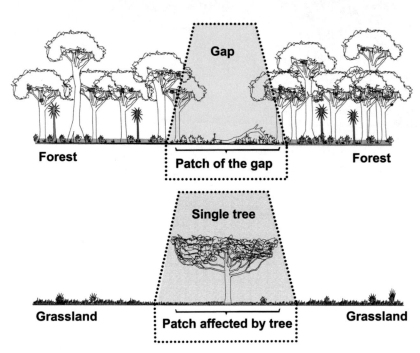

In very open savannahs, the trees stand far apart as isolated individual trees. Belsky & Canham (1994) compared this island biotope situation (idiotope) with that of tree gaps in forests (Fig. 6.22). The area influenced by a tree-fall in the forest and its dynamics up to canopy closure and, on the other hand, the area dominated by the individual tree in the grassland and its development into a group of trees or tree-free grassland are compared as contrasting structural elements in Table 6.3.

Table 6.3 Comparison of some processes in forest gaps of a closed forest and in individual trees in grassland (after Belsky & Canham 1994)

	Forest gap	Isolated tree in the savannah
Origin	Mostly sudden (episodic storm event)	Slow (germination, seedling establishment)
Enlargement	Seldom due to falling branches or trunks of neighbouring trees	Gradually by enlarging the crown
Disappear	Mostly quickly due to overgrowth by neighbouring trees	Possibly very suddenly due to the death of the tree
Duration	Short (5–30 years)	Long (lifetime of the tree, usually well over 50 years)
Resource dynamics	Usually only short, additional release of nutrients	Mostly permanent preference by game and input from outside (detritus)
Secondary succession	Only in large gaps in the forest	Rarely identifiable
Ecological effect on the environment	Short range (5–20 m)	Longer range (50–100 m)

6.5 Examples of Large Savannah Areas

In South America, very extensive savannah-like vegetation types are common along the Orinoco, in central and eastern Brazil, and in the Chaco region.

Along the gradient from the perhumid tropical rainforest to the extremely dry tropical desert, the structure of the vegetation stocks changes fundamentally; the importance of the respective life forms is also very different. Ellenberg (1975) has summarized this in a general scheme (Fig. 6.23) that applies to the lowlands of the Andes but is in principle also transferable to other areas with similar gradients.

A few examples of large-scale savannahs are highlighted below.

6.5.1 Llanos on the Orinoco

The Llanos are located in Venezuela at 100 m asl in a basin landscape that was still a sea in the Tertiary. They occupy a width of 400 km on the left bank of the lower Orinoco and continue for another 1000 km to Colombia. This basin was filled up by the rivers with the weathering products of the Andes. The climate of the central llanos around Calabozo (Fig. 6.44) is very typical of zonobiome II: Annual precipitation over 1300 mm, rainy season 7 months, drought season 5 months. Thus, a moist deciduous forest would be expected in this area. It is also present in typical formation, but only in the form of isolated very small groves—the "Matas". The deep llanos bordering the river and flooded during the rainy season are, as usual in ZB II, a pure grassland (trees only on the embankments as gallery forest: Fig. 6.24). Otherwise, the area is covered by a grassland about 50 cm high with scattered small trees (*Curatella, Byrsonima, Bowdichia*), i.e. a typical savannah. Since this cannot be climatic (the precipitation is too high), only edaphic causes, i.e. soil conditions, come into question.

The often-expressed assumption that it is an anthropogenic savannah created by fire from forest is the simplest, but also the most uncritical. The savannah existed before the arrival of the whites. The Indians had not used it for cropland or pasture. Fires always occur in grasslands by lightning. Certainly, the Indians will have set fire to the dry grass more often than not, but they could only do so because natural grasslands already existed. Fire helped shape the savannah by causing only fire-resistant woody species to grow in the grasslands and at the edge of the Matas, but it was not the primary cause of these vast grasslands. In the central llanos, it has been demonstrated that at a time when groundwater was still very high in the basin landscape, a laterite crust was formed that was cemented by iron hydroxide. It is referred to there as 'Arecife' (Fig. 6.25). It extends at variable but shallow depths (most commonly 30 to 80 cm deep) below the ground surface, rarely sinking below 150 cm, but also emerging at the surface or being eroded out.

The impermeability of the arecife to water is not true in this case; because during the summer rainy season 750 mm of rain falls in 3 months. These quantities cannot be absorbed by the soil above the arecife; there would therefore have to be a flooding of the flat surface, which is not the case. The red colouring of the soil also argues against prolonged waterlogging. On the other hand, a groundwater-rise under the arecife from -575 cm to -385 cm (Fig. 6.25), i.e., by almost 2 m, was observed at the end of the rainy season. Assuming a pore volume of the alluvial deposits of about 50%, this would mean that about 300 mm is retained by the soil above the arecife and 1000 mm percolates through. On an arecife exposed by erosion on the river bank, it could be clearly seen that quite irregular passages lead through the hard crust at individual points.

The grasses take root in the fine-grained soil above the arecife and consume about 300 mm of rainwater for their development. The woody plants, however, stand where their roots growing along the arecife surface, find a passage through the arecife and through this then reach the moist rock layers below. There, water is available to them in

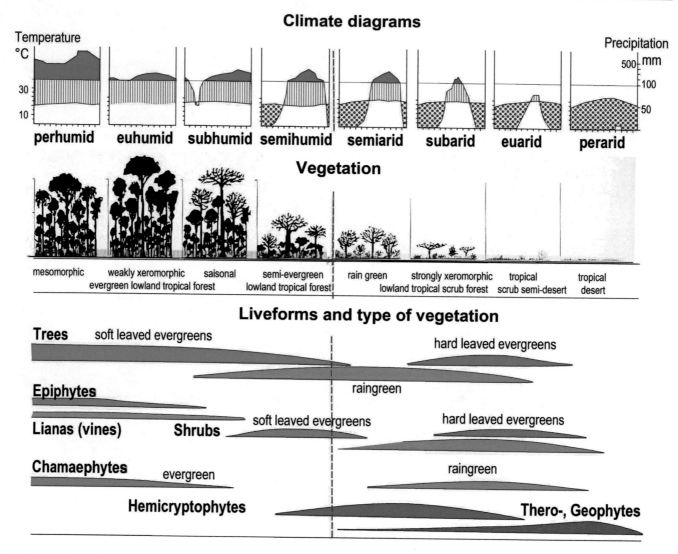

Fig. 6.23 Formation structure and distribution of life forms in the lowlands of the Andean foreland along a climatic gradient (modified after Ellenberg 1975)

sufficient quantity. If the passages are very large or if they lie close together, a group of trees can grow above them; small forest stands, on the other hand, can only be found where the arecife is completely absent in places or lies very deep, so that the vegetation appropriate to the climate develops, i.e. a deciduous forest. One must thus regard this savannah as a stable, natural plant community in which the tree distribution reflects the arecife structure. This is supported by the following facts:

1. Where the arecife is superficial, the grass cover is absent, but scattered small trees at greater distances grow on it; in this case the roots must reach through the arecife into the soil below.
2. *Curatella* remains green during the dry season, in contrast to the other behaviour of woody plants in the typical

savannah, a sign that its water supply is good all year round. Transpiration measurements showed that a small tree transpires about 10 litres per day during the drought season; since the soil above the arecife is dry during this time, the water must come from the soil layers below the arecife. The same is true of the other species of wood.
3. Where the groves ('Matas') grow, arecife is locally absent, allowing tree roots to penetrate deep into the soil unhindered.

The final proof could only be provided by root excavations on larger areas, which, however, are very difficult to carry out. Blasting the arecife with dynamite would have to lead to the spread of the woody plants. Slight depressions are interspersed in the savannahs of the llanos, into which the water drains after heavy downpours (1961: 38 mm in

Fig. 6.24 Example of gallery forests: Here from West Africa along a river in the savannah areas of Comoé National Park (Côte d'Ivoire) (photo: Barthlott)

Fig. 6.25 Scheme for interpreting water conditions in the llanos north of the Orinoco. Below the arecife, the changing groundwater table is accessible only to deep-rooted plants (modified after Walter 1990)

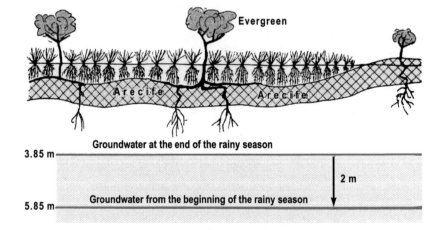

20 minutes) and in which grey clays are deposited, so that the water in the depressions is about 30 cm deep during the rainy season. Towards the end of the drought period the grey soil dries out completely.

This alternating humidity is well tolerated by certain grasses (*Leersia, Oryza, Paspalum* and others), but not by the tree species, with the exception of the palms. The 'Palmares' are then formed, grasslands with the palm *Copernicia tectorum*, i.e. palm savannahs, also widespread in tropical Africa. These areas also often burn down, but palms withstand fire well (as do tree ferns) because they have no cambium to be damaged. The dead leaves of palms that surround the trunk burn, and the outer leading bundles

char; this char layer acts as an insulator in later fires. The cone of vegetation surrounded by young leaves remains. If old leaves are completely missing from the trunk, it is a sign that the palm savannah has only recently burned down; if they enclose the trunk down to the ground, the palm has not yet been exposed to fire; if only the lower part of the trunk is bare, the palm has grown a number of years in height since the last fire.

Part of the water must run off the areas covered with palms; otherwise, the soils would dry out, since a rainfall of 1300 to 1500 mm is countered by a potential evaporation of 2428 mm, i.e. the hydrological water balance is negative. When the soils are permanently wet, *Mauritia minor* palm

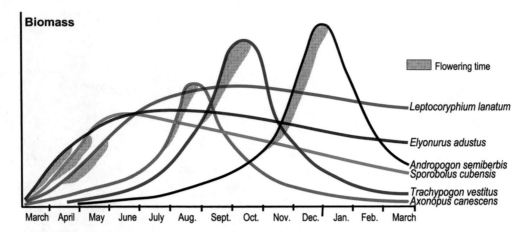

Fig. 6.26 The annual rhythm of green biomass of six dominant grass species of the Venezuelan llanos after the usual fire in March shows a close temporal mixing of the grass species (yellow grid: flowering time) (modified after Sarmiento 1996)

occurs. Black acidic peaty soils with some grasses, *Rhynchospora, Jussieua, Eriocaulon* and the insectivorous *Drosera* species (sundew) and others are formed. These areas, as well as the alternating wet grassland already mentioned, are also a particular form of helobiome and amphibiome.

Today, many of these areas are burned annually to improve pasture. This naturally synchronizes the growth of the different grass species. The periodic niching of biomass production is then particularly clear (Fig. 6.26).

Further east, the llanos merge into a plain with sandy deposits of the Orinoco River, which used to bend north here and flow into the Caribbean Sea through the Unare lowlands.

The quartz sands, which are often quite white, are weathering products of the quarzitic sandstones of the Guyana mesas, which correspond to those of the Brazilian Shield and are likewise almost nutrient-free. Similar leached quartz soils also occur on other ancient Gondwana sites. The savannahs, some of them pure grasslands, are probably due to similar causes as the Campos Cerrados.

6.5.2 Campos Cerrados

These are savannah-like vegetation covering an area of two million square kilometres (= 23% of the total area of Brazil, Ruggiero et al. 2002) in central Brazil (Eiten 1982) (Fig. 6.27). The cover of the 4- to 9-m-high tree stand varies

Fig. 6.27 Campos Cerrados comprise large areas in Minas-Gereis, SE Brazil, with tree heights of 3-4 m. They form a partly impenetrable bush formation (photo: Denis A.C. Conrado, https://pt.m.wikipedia.org/wiki/Ficheiro:Bonfim_047.jpg)

Fig. 6.28 Chaco in Paraguay covers about 60% of the land area in Paraguay and parts of the Bolivian lowlands. In the wet parts, Chaco has the character of a parkland as shown here in the picture. The dry parts, however, are a thorn savannah (photo: Ilosuna, https://t1p.de/av11)

from 3% to 30%. The climate, with annual precipitation of 1100 to 2000 mm, is characterized by a five-month long drought. Rawitscher (1948) was the first to study the water balance of these savannahs and proved that the deep soil remains permanently moist even at a depth of 2 m, so that the deeper-rooted woody species always have sufficient water available, remain evergreen and transpire strongly even during the drought period. Only the grasses and shallow-rooted species dry out or drop their leaves during drought. The soils are weathered products of the granites and sandstones of the Brazilian Shield and are very low in nutrients, especially phosphorus, but also potassium, zinc and boron. Crops of cotton, maize and soya with various applications of fertiliser showed this to be the case. That it is not the water factor but nutrient poverty that prevents the formation of zonal deciduous forests is shown by the fact that near Sao Paulo a zonal semi-evergreen forest grows on basalt soils. The Campos Cerrados were regularly burned. The presence of many pyrophytes shows that fire was also a natural factor here since ancient times. Fires reduce the density of stands, but they are not the real cause of the lack of closed forest vegetation (Coutinho 1982).

6.5.3 The Chaco Area

This is the westernmost part of ZB II in South America, a vast plain between the Brazilian Shield to the east and the pre-Andean mountain ranges to the west. The central part of this plain is only about 100 m above sea level. The plain extends from S Bolivia, most of Paraguay, and well into W Argentina for 1500 km from north to south at an average width of 750 km (Hueck 1966).

During the heavy summer rains, large parts of the plain are flooded, especially in the eastern part (annual precipitation 900 to 1200 mm). The temperature can rise above 40 °C, it is the heat centre of S America. It is a park landscape with forest, wide periodically flooded grasslands, palm savannahs or swamps (Fig. 6.28). In the central part, dry savannahs occur in addition to parkland. The western part in Argentina is heavily scrubby, and salt pans with the halophytes *Allenrolfea* and *Heterostachys* also occur. The southern Chaco leads to the Pampa. The relief is very flat, impermeable layers occur in the soil; the vegetation is mainly a *Prosopis* savannah with a grassy layer of *Elionurus muticus* and *Spartina argentinensis*. The main tree species of the Chaco forests are strongly tannic Quebracho species *Aspidosperma quebracho-blanco* (*Schinopsis quebracho-colorado* and *S. balansae* and others). Of the palms, *Trithrinax campestris* is common, while *Copernicia alba* is typical for moist depressions.

The mammal fauna is not species-rich. Termite eaters are *Myrmecophaga tridactyla* and *Tamandua tetradactyla*. Of predators, the jaguar *(Leo onca)*, puma *(Felis concolor)* and many smaller species are or were represented. Rodents are numerous; on trees are found the sloth *Bradypus boliviensis*, three species of monkeys *(Cebidea)*, the tree porcupine *(Coenda spinosus)* and the mustelid *Eira barbara*, in addition to many insectivores or bats feeding on fruits and flowers, and the blood-sucking vampire *Desmodus rotundus*.

Of the birds, the large ratite *Rhea americana* should be mentioned, the reptiles are represented by two rare species of caiman, three species of turtles, some poisonous snakes (a total of 25 species of snakes) and various lizards; of anurans, 30 species are known so far. In addition, there are countless invertebrates.

Ecosystem research has probably not yet been initiated. The main human interventions occur through deforestation and grazing, which can lead to scrub encroachment. A brief summary with references is available from Bucher (1982).

In the last few decades, the country has been increasingly conquered by large farms and in some cases intensively used for agriculture. The agricultural front is spreading rapidly into the drier areas of the Chaco. The focus is on planting huge monocultures, for example soy, for export, and recently *Jatropha* for biodiesel. The livelihoods of many still nomadic indigenous groups are threatened by the often illegal clearings. In many places the indigenous people therefore see their basic right to food being violated.

6.5.4 Savannahs and Park Landscapes of East Africa

This area, lying at the foot of the great volcanoes, with the giant crater Ngorongoro (Fig. 6.29), the East African Rift

Fig. 6.29 View of Kilimanjaro volcano (5890 m asl, Tanzania) in East Africa. The volcano's snow cap has completely melted except for a small remnant within the last decades (Climate change!) (photo: Breckle)

Valley, and the vast Serengeti range, is widely known, especially for its abundance of game (Figs. 6.14 and 6.15), which may also be related to the nutrient-rich volcanic soils and thus better plant food. But in this equatorial area with a diurnal and monsoon climate, two rainy seasons occur, one small and one large. Usually, these are separated only by a short drought period, which is hydrologically more favourable. Their effect is similar to that of a summer rainy season, so that with annual precipitation of around 800 mm one encounters similar savannahs and park landscapes as in ZB II.

Clearing, annual fires and overgrazing have strongly influenced the plant cover; as a result, different stages of degradation are common. It is often referred to as an 'orchard steppe', but it is a typical tree savannah. When the climate becomes drier, or on dry rocky sites, large candelabra euphorbias (Fig. 6.18) and *Aloe* species occur (Fig. 6.30).

There are today several National Parks and Game reserves. With colonization, big game hunters began to shoot down animals in large numbers and thus to decimate the populations. This arbitrary killing of wild animals ultimately made it necessary to set up nature reserves in order to protect the savannah habitat and the wildlife there. In the nineteenth century the area was still grazing land for the nomadic Massai people. The Massai, who were not to blame for the destruction of nature, were severely restricted in their freedom in their own homeland by the nature reserves (Poole 2006).

The Serengeti was partially declared a game reserve (Serengeti Game Reserve) as early as 1929 to protect the lions that were previously considered pests. In 1940 it was declared a Protected Area. In 1951, the Serengeti National Park was established, which at that time also included the Ngorongoro Crater. In 1959, the wildebeest rainy season pastures in the southeast of the Serengeti at the Ngorongoro Crater were separated from the national park and declared only a conservation area, in which Massai herders are allowed to graze their cattle.

The Serengeti is one of Africa's most complex and least disturbed ecosystems, ranging from dusty summer drought to green winter and lush spring. The focus is on the savannah with scattered acacias. To the south there are wide open short grass plains, to the north there are long grasslands covered by thorn trees, along the rivers gallery forest and in the hilly western corridor extensive forests and black clay pans.

6.5.5   Monsoon Forests in India

India is located in the south of the largest continental landmass—Asia—and is rightly called a subcontinent due to its large expanses, approximately 3220 km in north-south and approximately 2980 km in east-west direction. Connected to

Fig. 6.30 *Aloe* in Serengeti National Park in Tanzania with low summer precipitation (ZE II/III) (photo: Breckle)

Africa, Antarctica and Australia by land bridges until about 180 million years ago, it has been moving northwards ever since, folding up the Himalayas, which also form the natural boundary to the north. This high mountain range is still evolving; the earthquake in May 2015 with its devastating consequences in Nepal is evidence of the steady and current geo-morphodynamics in this region.

The determining factor of the Indian climate is the rain-bringing summer monsoon. It is caused by the summer warming of the inner Asian land mass with the formation of the inner Asian low pressure area. This sucks in air from the summer high pressure area over the Indian Ocean, which is cooler relative to the land mass. The resulting southwesterly air flow (SW monsoon) brings in a lot of moisture from the Indian Ocean and rains down on the mountain flanks of the Western Ghats facing it with an average of 2500 mm to 3000 mm of rain per year (also with maximum values of up to 10,000 mm, e.g. in Cherrapunji: Fig. 2.8) and on the Himalayas.

The climate of India corresponds to the zonobiome II and in the dry NW to the zono-ecotone II-III. It extends marginally to the Afghan border.

Although the Indian flora undoubtedly belongs to the Palaeotropics, it has many peculiarities peculiar to itself.

The reason for this is the certain isolation of the subcontinent due to the shielding effect of the mountain ranges in the north, the long coastline in the south and the arid regions in the northwest. Nevertheless, distinct Holarctic influences are found, completely dominating the higher elevations of the north. Isolated elements of Gondwanaland (e.g. the Podocarpaceae) indicate the former land connection to Africa, Antarctica and Australia. The relative isolation, which allowed immigration of new species only slowly and predominantly from the north-eastern flanks, resulted in numerous endemic genera and species via radiation from existing clades. Again, some species just reach across the Afghan border in the subtropical thorn savannahs of the Khost basin and in Nangarhar.

Champion & Seth (1968) classify the vegetation of the Indian subcontinent on the basis of the dominant woody vegetation types. The respective vegetation type depends primarily on the total amount of local precipitation, its seasonal distribution and the relationship between precipitation and evapotranspiration.

The tripartite division of the forests of the Indian subcontinent is summarized in Fig. 6.31 as follows: Numbers **2** to **7** are the monsoon forests proper, Numbers **8** to **11** indicate forests of the hilly and low mountainous regions, and Numbers **12** to **14** those of the mountainous regions of the Himalayas.

As an example, the particularly extensive alternating green tropical monsoon forest (**4**) is described here (Fig. 6.32c).

This vegetation type covers the largest area of India and is widespread both in the northern part of the Deccan highlands up to the edge of the Ganges basin and in the southern part. The outer canopy reaches 20 m, is built up of only a few tree species, mostly leafless in the dry season, and is not completely closed. The middle canopy layer is almost completely alternate green, allowing enough light to pass through for a patchy shrub layer with a high proportion of grasses and bamboo. In the dry season, this forest is therefore quite bare (Fig. 6.32d).

Representatives are: *Tectona grandis* (= teak; Verbenaceae), *Diospyros tomentosa* (Sapotaceae), *Aele marmelos* (Rutaceae), *Butea monosperma, Anogeissus latifolia, Adina cardifolia, Buchanania langan* etc. Most probably, human use (grazing) has also significantly altered this forest type and probably favoured the occurrence of grasses. Mostly, now this forest type has degraded to open scrub due to intensive grazing (Fig. 6.32e).

The second example is the moist, alternating-green tropical rainforest (3), which is often referred to as the actual monsoon forest (Fig. 6.32b). A forest type, found in large areas of India with 1000–2000 mm of rainfall, it would not have been severely repressed by millennia of continuous use and replaced today by open agricultural landscapes, tea plantations, and planted timber forests. It is found in parts

Fig. 6.31 Spatial differentiation of the vegetation units of the Indian subcontinent (modified after Champion & Seth 1968)

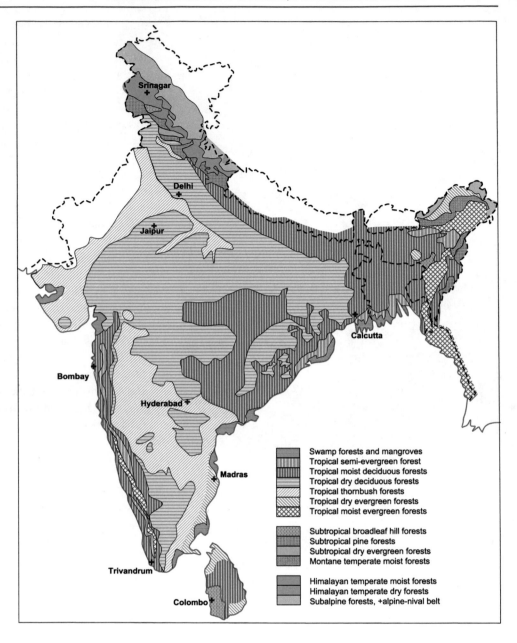

Swamp forests and mangroves
Tropical semi-evergreen forest
Tropical moist deciduous forests
Tropical dry deciduous forests
Tropical thornbush forests
Tropical dry evergreen forests
Tropical moist evergreen forests

Subtropical broadleaf hill forests
Subtropical pine forests
Subtropical dry evergreen forests
Montane temperate moist forests

Himalayan temperate moist forests
Himalayan temperate dry forests
Subalpine forests, +alpine-nival belt

of Assam, in the basin of the Ganges and Godavari rivers, in eastern Sri Lanka and on the western flank of the Western Ghats. In this range alone there are 4000 species with 1500 endemics (Henry et al. 1987, 1989, Nai & Henry 1983, Pascal 1988). With irregular outer alternate green canopy layer, the forest also reaches 40 m or more. The second, lower canopy layer is formed increasingly by evergreen species, and a shrub layer is developed. There are *Pterocarpus dalbergoides, Shorea dalbergoides, S. robusta* = "Salbaum"(Pterocarpaceae), *Terminalia bilata, T. procera* (Combretaceae), *Albizzia lebbeck* (Mimosaceae), *Erythrina indica* (Fabaceae) with its spectacular red flowers, *Bauhinia purpurea* (= "orchid tree"; Caesalpiniaceae), the Bengal fig = *Ficus benghalensis* (Moraceae) which branches out over several hectares, *Lagerstroemia parviflora, Adina*

cardifoia. Bamboo species are widespread as spreading climbers. Massive termite mounds (Fig. 6.32f) are common in these partly dry forests.

On the Indian subcontinent, practically all ecosystems known from the tropical-humid, tropical-arid, from humid-temperate and dry-temperate to alpine areas occur—from lowland rainforest (Fig. 6.2a), swamp forest and the mangroves to the dry forests, thorn-bush and thorn savannahs (Fig. 6.2e) to the semi-deserts (Fig. 6.32f) and deserts, from evergreen deciduous and evergreen hardwood and coniferous forests to high mountain tundras. It is certain that this great diversity of ecosystems has led to the high biodiversity (Wilson 1988, Pullaiah & Ramakrishna 2018) of the subcontinent.

Fig. 6.32 (**a**) Lianas, epiphytes, and hanging aerial roots characterize tropical ever-humid lowland rainforest in northeastern India (e.g., near Siliguri), but it is also still strongly influenced by the monsoon; (**b**) Tropical moist fallow deciduous forest in Kambalkonda Wildlife

Due to the high relief energy, almost all ecosystems of the orobiomes are represented, both from the tropical-subtropical area and from the temperate latitudes. For the tropical orobiomes, especially in the east of the Himalayan mountain flank, the worldwide peculiarity applies that they are characterized by extremely high monsoon rains. There is no Páramo, since—in contrast to the tropical Andes or the tropical high mountains of Africa—seasons are clearly pronounced in the Himalayas.

Human influences on vegetation have a long tradition, especially in India with its early advanced civilizations. Early migrations led to completely different races with completely different languages and religions settling and using the continent in different waves. For this reason, it is often difficult to even address an original, potential vegetation, although India has a high, independent biodiversity with many endemic species.

6.5.6 Vegetation of the Australian ZB II

With the exception of a few small relicts of deciduous forests in NE Australia with Indomalaian flora elements and some deciduous *Eucalyptus* species in N Australia, but almost insignificant, this vegetation type does not exist. But park-landscapes also with palms are common in the ZB II area, but with evergreen *Eucalyptus* species. A little further south, savannahs with a covering grass layer of *Heteropogon contortus* (also with evergreen eucalypts) occur with lower annual precipitation.

In the detailed vegetation monograph by Beadle (1981) the term 'savannah' does not occur. In contrast, the Australian researchers (Walker & Gillison 1982) count all light forests as savannahs if the grasses of the herb layer have a cover of more than 2%, which would have to include most light eucalypt forests.

One can say now that the largest and still relatively unaffected savannah is in Australia. It extends over an area in the north of the continent that is larger than Western Europe, and since the increasing disappearance and cultivation of the savannah areas in Africa, Asia and South America, it now represents more than a quarter of the world's savannah area.

While on other continents the savannah area—the transition zone between the tropical rainforests and the large desert areas characterized by grass and groups of trees—continues to decline due to urban sprawl and animal husbandry, it has remained largely intact in northern Australia. A network of nature reserves and a relatively thin population—mainly in the form of smaller Aboriginal settlements—are responsible for their preservation.

It is to be hoped that gentle partial development will not also lead to exploitation and desertification.

The Australian savannah cannot keep up with the African savannah in terms of fauna. The previously species-rich Australian megafauna has practically completely disappeared already 50,000 years ago—around the time when Australia was first colonized by humans: the large grazing animals such as the hippopotamus-sized giant wombat *Diprotodon*, the largest known marsupial, as well as the large ones predatory mammals. Only a few "larger" mammal species have survived, including the antelope kangaroo, which is slightly smaller than the grey giant kangaroo that lives in southern Australia.

Spectacular herds like those in the African savannah are brought about by an involuntary immigrant from Afghanistan: the descendants of dromedaries that were imported to Australia from the nineteenth century and kept as load carriers by Afghan camel drivers. In the absence of food competition and large predators, they could run wild and fill the free ecological niche. Hundreds of thousands of them live today from the northern grasslands to the arid areas in the middle of the continent and have become a plague that is now being reduced by shooting and control programs.

6.6 Ecosystem Research: Examples

One of them, the Lamto Savannah, is located in West Africa and is a relict savannah in the rainforest area, the other, the Nylsvley Savannah in South Africa. It borders the Kalahari to the west. Grasses and trees are the components in the savannah. The number of species of grasses is relatively low, more significant is their biomass; the reverse is true for legumes (Fig. 6.33). In addition, however, there are numerous other rarer species (about 25%), but their biomass is only about 1.5%.

6.6.1 The Lamto Savannah

This savannah is located in the Guinea forest zone (Ivory Coast area) at 5° W and 6° N, still in ZB I. It is burned every year, so that the rainforest adjacent to it cannot advance, even if the soil conditions would allow it. The mean annual

---

Fig. 6.32 (Continued) Sanctuaryin Visakhapatnam, SE India (photo: Adityamadhav, https://t1p.de/48vg); (**c**) Tropical dry deciduous forest in monsoon area in Mandhya Pradesh (Photo: L.R. Burdak, https://t1p.de/25lf); (**d**) Semi-evergreen in valleys, species-rich dry deciduous forest on slopes at Ajanta and Ellora (with ancient temple caves) and millennia of use; (**e**) Stratified ribs at Nowgong with open tropical deciduous scrub or thorn scrub, valley bottoms cultivated; (**f**) Large termite mounds in dry fallow deciduous forest, the typical monsoon forest, here in Orissa (eastern India) during leaf fall in the dry season in February (photos **a,d, e,f**: Breckle)

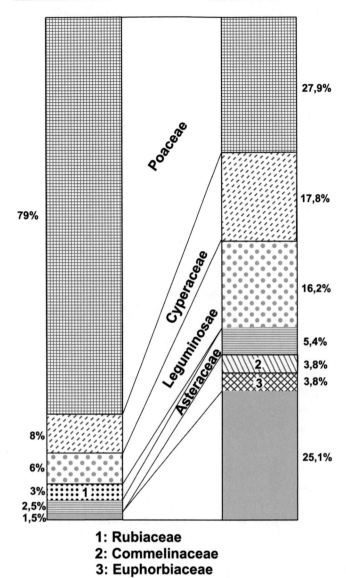

1: Rubiaceae
2: Commelinaceae
3: Euphorbiaceae

Fig. 6.33 The relative biomass (left column) and the percentage species numbers (right column) of the most important plant families in the African savannah (modified after Müller 1991)

precipitation is 1300 mm. There is a drought period of only one month—in August—on the climate diagram, but the weather pattern varies greatly from year to year; rainfall ranges from 900 to 1700 mm per year. On the higher part of the relief, a tree or shrub savannah grows on red savannah soils with laterite concretions. On the other hand, in lower parts of the relief, palm savannah grows on waterlogged soils.

The different plant communities were studied by Menault & Cesar (1982) (Table 6.4).

Lamotte (1975) dealt with the consumers and destroyers of this savannah: Big game occurs only sporadically. The zoomass (per in kg·ha^{-1}) of birds is 0.2 to 0.5; that of twelve rodent species 1.2; that of earthworms 0.4 to 0.6. The mass of termites (grass-, humus- or wood-eating) as well as that of other invertebrates could not be determined.

Soil respiration, which serves as a measure of microbial activity, was determined as 8 t CO$_2$·ha^{-1}·a^{-1}. The attempt to determine the energy flow during decomposition (Lamotte 1975) resulted in the following:

1. Annual fire mineralizes about 1/3 of the primary production. Less than 1% of the primary production is probably eaten by consumers; the decomposition of detritus eaters with the main group of earthworms is also not very effective.
2. 80% of the primary production is degraded by microorganisms, so that the representation of the energy flux as a pyramid seems very questionable. This confirms that the long cycle via consumers is quantitatively almost meaningless.
3. Many faunal data for the various animal groups represented in the savannahs can be found in the volume edited by Bourlière (1983).

6.6.2 The Animal World

The fauna of the *Burkea* tree savannah and that of the *Acacia* thornbush savannah show striking differences for both vertebrates and invertebrates.

In the entire protected area, 18 amphibian species are found (in the Nyl floodplain), while in the experimental area there are 11 species; far from the water, both the toad *Bufo garmani* and the frogs *Breviceps mosambicus* and *Kassina senegalensis* are found. In terms of reptiles, 3 species of turtles, 19 species of lizards and 26 species of snakes were found in the experimental area.

The number of bird species in the entire protected area is 325, of which 197 are permanent. In the experimental area there are 120 species (14 birds of prey, 71 insectivores, 4 berry eaters, 10 grain eaters and 26 omnivores). Of the 62 mammal species in the reserve, 46 were recorded in the experimental area. Rodent species are the most numerous, along with one species each of porcupines, warthogs, and jackals, and two species of monkeys.

Of the particularly important cloven-hoofed animals, mention may be made of: Kudu *(Tragelaphus strepsiceros)*, Impala *(Aepyceros melampus)*, Deuker *(Sylvicapra grimmia)* and Ibex *(Raphicerus campestris)*.

The determination of the number of individuals or the living zoomass is difficult and succeeded only in a few cases approximately. Three animals per hectare are reported for snakes, the most common reptile, the gecko *(Lygodactylus capensis)* is represented with 195 to

Table 6.4 Ecosystem parameters (extreme values) of a low shrub savannah and a dense tree savannah

	Shrub savannah	Tree savannah
Number of woody plants per ha	120	800
Woody plant cover	7%	45%
Leaf area index (LAI)	0.1	1.0
Phytomass above ground (t·ha^{-1})	7.4	54.2
Phytomass below ground (t·ha^{-1})	3.6	26.6
Net timber production (t·ha^{-1}·a^{-1})		
Ditto, above ground	0.12	0.76
Ditto, below ground	0.05	0.37
Net production of leaves and green shoots	0.43	5.53
Net production of the grass layer (t·ha^{-1}·a^{-1})		
Ditto, above ground	14.9	14.5
Ditto, below ground	19.0	12.2

262 animals per hectare, the common lizard *(Ichnotropis capensis)* with 7 to 11 animals per hectare.

The living zoomass of birds is 40 kg per 100 ha in the Burkea savannah, but the number of birds decreases by 25 to 30% in winter, when migratory birds leave the area.

For mammals, catch results were so low and variable that the data mean little. For example, monthly catches of *Dendromus melantois* yielded about 5 (0 to 15) animals per ha, and only 2 animals per ha for other rodents.

For even-toed ungulates, the following mean values (number of animals per 100 ha) are given: Impala 13, Kudu 2, Warthog 1, Deuker 2 and Ibex 1 to 2 (Reedbuck rare).

The former owner of Nylsvley stated that for the last 40 years he only allowed the area to be grazed between January and April because of losses caused by the poisonous species *Dichopetalum cymosum,* a geophyte close to the Euphorbiaceae. The cattle biomass in the 4 months was about 150 kg·ha^{-1} but overgrazing became noticeable in 1975, so that the cattle population was reduced to half in the next few years.

The number of invertebrates is so large that only certain groups of arthropods important for the ecosystem are given: Wood-eating coleopterans, lepidopterans, social insects, root-eaters and spiders.

The zoomass of invertebrates as dry mass was on the woody plants in the average of 135 g ha^{-1} (minimum in August 60 g ha^{-1}, maximum in March 300 g ha^{-1}). The dry mass of insects in the grass layer is greater.

Sporadically, caterpillar masses *(Spodoptera exempla)* or beetle larvae *(Astylus atromaculatus)* occurred on the grass species *Cenchrus ciliaris.* Dung is removed by dung beetles (Coprinae, Aphodiinae) 77% of the time in one day during the warm season by burying it directly under the deposition site, while pill bugs *(Pachilomera* spp.) spread it over a larger area. These coprophages are already leading on to the next group.

> **Box 6.4 *Burkea* Savannah Biomass**
> The following figures represent dry matter in kg·ha^{-1} for *Burkea* savannah and in parentheses for *Acacia* thorn savannah:
>
> Acridoidea 0.76 (2.32), other Orthoptera 0.06 (0.02), Lepidoptera 0.05 (0.03), Hemiptera 0.08 (0.08), other 0.05 (0.05), for a total of 1.0 (2.5) kg·ha^{-1}.

Destructive organisms include the saprophagous small animals in the soil and in the litter layer, which eat dead plant parts and animal remains and at the same time break them down, as well as the protozoa, fungi and bacteria, through which complete mineralisation finally takes place.

The most important saprophages are the termites. Oligochaetes, myriapods and isopods are of minor importance. Acarines and collembolae feed on bacteria and fungi.

Termites are represented by 15 species, the most common species being *Aganotermes oryctes, Microtermes albopartitus, Cubitermes pretorianus* and *Microcerotermes parvum.* Of the 15 species, 4 are humus feeders, the rest feed on dead wood or leaf litter. A mean of 2540 termites was found in the soil under 1 m^2 of area (maximum in November was 8204, minimum in July was 596).

The fauna of the remaining savannah areas in Africa is particularly rich in large mammals compared to the savannahs and grasslands on the other continents. However, it is known that in the other continents mammal fauna was also considerably richer 11–15 thousand years ago.

Among herbivores, the following functional groups can usually be distinguished:

- Grazers (grazing animals; 'grazers').
- Leaf-eaters (shrub and tree foliage; 'browsers').

- Grain eaters (seed eaters; 'granivores').
- Nectivores ('nectivores').
- Fructivores ('frugivores').

6.7 Tropical Hydrobiomes in ZB I and ZB II

With the relatively low potential evaporation, the high precipitation in the humid tropics leads to large water surpluses. As an example, San Carlos de Rio Negro in S Venezuela has a precipitation of 3521 mm and a potential evaporation of only 520 mm. Thus, insofar as runoff is impeded in flat terrain, extensive swamps are created.

In Uganda, such swamps occupy 12,800 km^2, about 6% of the total area. The catchment areas of the river systems there are not separated from each other by watersheds, but are connected in a network-like manner by swamps. On the flight from Livingstone to Nairobi one sees the large Lukango swamps and furthermore those around Lake Kampolombo and Lake Bangweulu. But the largest swamp is formed by the White Nile in S Sudan. With its left tributary, the Bar-el-Ghasal, it fills with water the large basin situated at 400 m asl. It is the marshland known as the Sudd, the greatest extent of which reaches 600 km from north to south and from west to east; the total area is estimated at 150,000 km^2; it varies depending on whether it is high or low tide. Due to evaporation in the Sudd area, the Nile loses half of its water. It is not a free expanse of water with small islands barely protruding above the water, but a green carpet of oscillating grass and floating islands formed by shoots of the grass *Vossia* lying on the water surface as well as *Papyrus*.

Also lawns of floating plants, the *Eichhornia* introduced from South America as well as *Pistia* play a role. In between one recognizes from the airplane individual free watercourses and smaller water surfaces. A part of the land emerges at low water and forms a grassland with the tall *Hyparrhenia rufa* and *Setaria incrassata*. The wettest parts are covered with *Echinochloa species, Vetiveria* and reeds *(Phragmites)*.

The "Great Pantanal"in Mato Grosso, Brazil, on the border of Bolivia and Paraguay, was formerly thought to be a similarly large marsh from which the southern tributaries of the Amazon and the right tributaries of the upper Paraná rise, but this area is flooded only during the rainy season, but during the dry season it is used for grazing, leaving many annular lakes with riparian forests. It is severely threatened by industrial beef-farms.

Swamps and pools of water are also common in the rest of the humid tropics. The aquatic vegetation consists of some cosmopolitan and pantropical species with floristic features peculiar to each area.

6.8 Mangroves as Halo-Helobiomes in ZB I and ZB II

Anyone approaching a tropical coast protected by coral reefs from the sea will notice the mangroves, whose treetops barely rise out of the seawater at high tide. Only at low tide do the lower parts of the trunks with their breathing roots become visible (Fig. 6.34). These forests grow in the intertidal zone in salt water whose concentration is about 35 g·l^{-1}, corresponding to a potential osmotic pressure of 2.5 MPa.

Over 30 species of woody mangroves are known. A distinction is made between the more species-rich eastern mangrove on the coasts of the Indian Ocean and the west coasts of the Pacific Ocean and the less species-rich western mangrove on the coasts of America and the east coast of the Atlantic Ocean. The mangrove reaches its optimum development around the equator in Indonesia, New Guinea and the Philippines. With increasing latitudes, it becomes more and more impoverished, until finally only one species of *Avicennia* remains. The outermost outposts are found at 30°N and 33°S (Sinai, E Africa), at 37 to 38°S (Australia and New Zealand), and at 29°S in Brazil and 32°N in the Bermuda Islands. It can thus be seen that although the mangrove is best developed in the equatorial zone, it extends through the tropical and subtropical zones almost to the winter rainfall area or to the warm temperate zone (Chapman 1976).

The most important genera of mangroves are *Rhizophora* with stilt roots (Fig. 6.35) and viviparous seedlings and *Avicennia* with thin breathing roots growing out of the soil (not viviparous) (Fig. 6.35). *Laguncularia* still belongs to the western mangrove, whereas *Conocarpus* grows only at low salinity. The eastern mangrove also includes species of the genera *Bruguiera* and *Ceriops* (both viviparous and with knee roots), *Sonneratia* (non-viviparous with thick breathing roots), plus *Xylocarpus, Aegiceras, Lumnitzera* species and others. The individual mangrove species usually grow in distinct zones, rarely in mixed stands. The zonation is related to the tides. The closer to the outer edge of the mangrove a species grows, the longer and the deeper it is in salt water (Fig. 6.36).

The tides have a different tidal range (difference in height between low and high water) on the individual coasts; this changes periodically with the position of the moon and the sun. It is greatest at the time of the new and full moon (spring tides) and smallest in between (neap tides). The spring tides are at their highest twice a year at the equinox (equinoxial spring tides).

A distinction is made between **coastal mangroves**, which grow on flat coasts without water supply from the land and are often many kilometres wide, estuarine mangroves, which can be very extensive, especially in the delta region of rivers,

Fig. 6.34 The mangrove zone with *Avicennia* (Acanthaceae) with breathing roots on the coast of New Caledonia (photo: Breckle)

and reef mangroves on dead coral reefs emerging from the water, which play a lesser role. The salinity of coastal mangroves in E Africa has been well studied long ago (Fig. 6.37).

The coast of E Africa at Tanga has a relatively dry monsoon climate. Potential evaporation is likely to be equal to or greater than the annual rainfall. In addition to a small dry season, a pronounced drought season is present. As a result, the shorter the time the soil is inundated, the greater the inland increase in soil salt concentration in the intertidal area.

The conditions are extreme at the inner edge of the mangrove zone, up to which only the equinoctial spring tides reach (Fig. 6.38). Here, the saline water penetrating the soil is strongly concentrated by evaporation during the drought season, whereas during the rainy season the soil can be completely depleted from salt.

No plant species (except cyanobacteria) can cope with these strong fluctuations in concentration, so that these areas are devoid of vegetation. Such areas are found everywhere on the inner edge of the coastal mangroves when the climate is characterized by a period of drought. In N Venezuela, small stands of columnar cacti and opuntia or bromeliads occur in places in the open areas, although these are very salt-sensitive plants. Obviously, the bromeliads take up the water through the leaves and sit here on the ground quite loosely. The cacti, on the other hand, take up the water through shallow roots. They grow here always on small sand

mounds, thus root in these, from which the salt is washed out during the rainy season. The salt soil underneath does not disturb them.

> **Box 6.5 Mangroves as Azonal Vegetation**
> The mangrove is an azonal vegetation tied to the salt water in the tidal area. It always grows on very fine-grained soils, protected from the surf and frost.

Neither the cacti nor the bromeliads contain salts in their tissues; they are therefore not halophytes—another example of the fact that one must examine the ecological characteristics of the plants and the soil conditions in each case very exactly.

The conditions are different in the strongly humid region.

Here, the exposed areas are constantly leached by rainwater, which means that the concentration of soil water must decrease inlandwards, which also applies to the estuarine mangroves upstream. The mangroves thus transition into the freshwater communities via a brackish water zone with the fern *Acrostichum*, *Nipa* (a palm), *Acanthus ilicifolius,* and many other species, without a distinct vegetation-free zone intervening (Figs. 6.35, 6.36, 6.37 and 6.38). Although mangroves are azonal vegetation, their zonation is also determined by climate. It is different in humid ZB I than in a

Fig. 6.35   The mangrove forests of *Rhizophora mangle* (Rhizophoraceae) on the coast of Tulear (SW Madagascar). *Rhizophora* is viviparous and grows in the mud of the intertidal zone (photos: E. Fischer)

climate with a pronounced drought (Fig. 6.39). In this respect, the zonation of mangroves differs fundamentally between ZB I and ZB II or even ZB III.

All plants rooted in saline soils take up a certain amount of salts, which are stored in the cell sap. This is also true for the mangroves with their strongly succulent leaves, in whose cell sap the salt concentration is about the same as in the soil; to this must be added the nonelectrolytes in a concentration common to tropical species. The typical zonation and potential osmotic pressure in the soil as well as in the leaves of the mangroves is shown in Fig. 6.36, while the diagram in

Fig. 6.39 highlights the differences between mangroves in the arid and humid regions.

The zonation is the result of competition between the individual mangrove species, for which the salt factor is decisive in E Africa. *Avicennia*, as the least competitive species, also has the highest resistance to salinity; vestigial specimens of this species therefore form the inner boundary. *Sonneratia is* probably the most competitive species, but is least able to tolerate an increase in salt concentration above that of seawater. As a result, it can only persist on the outer margin. In mangrove permanently humid areas, zonation is

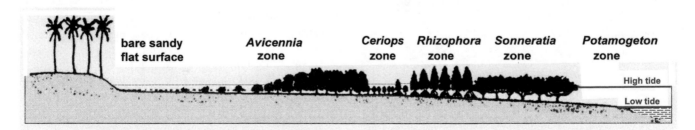

Fig. 6.36   Zonation of the East African coastal mangrove. H.W.L. = high water limit, L.W.L. = low water limit (from Walter & Steiner 1936)

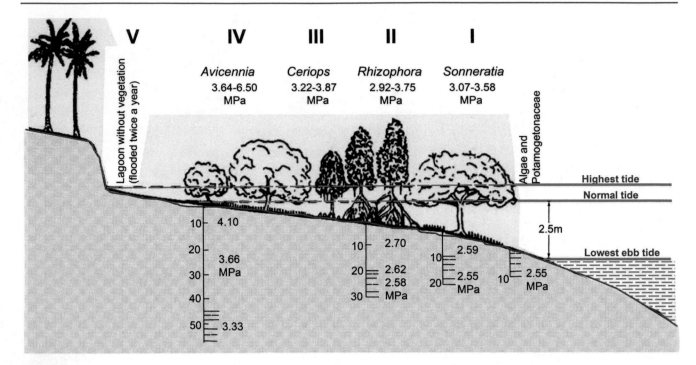

Fig. 6.37 Concentration of cell sap in bar (or MPa) (lowest and highest) of leaves of mangrove species and soil solutions at different depths (in cm). Coastal mangrove of E Africa (arid type) (from Walter & Steiner 1936)

more complicated. *Avicennia* seems to be bound to sandy soil, while *Sonneratia* prefers silty soil. Here, soil type and aeration, inundation duration, water movement, and variations in salt concentration are likely to be more important.

Fig. 6.38 Inner margin of the mangrove zone on the intertidal coast near Maracaibo (Venezuela) with strong fluctuations of the salt concentration in the soil during the dry (high) and rainy (low concentration due to leaching) seasons. Behind the mangroves there are hardly any tides, but constant evaporation in the soil. Partially devoid of vegetation due to high salt concentration. At the right there are stolons of *Sesuvium* (photo: Breckle)

An interesting problem is the salt balance of the mangroves. They cannot simply absorb the seawater as such, because a saturated salt solution would form in the leaves in a very short time, since the plants only release water during transpiration and the salts remain behind. Direct evidence has now been obtained that suction forces of 3.5 to 5.5 MPa are generated in the leaves of mangroves, which are higher than the potential osmotic pressure of the soil solution. These suction forces are transferred to the roots by the cohesive tension in the vessels, which at the same time constitute an ultrafilter, i.e. they allow practically pure water to pass through and feed it to the leaves. Only a very small amount of salt penetrates the plant and is stored in dissolved form in the vacuoles of the leaf cells. It is necessary to generate the suction forces.

How the regulation of the salt concentration takes place is not yet entirely clear. An excess of salts could be eliminated from the plant when the old leaves fall off. This is a general principle in almost all species on saline soils. In *Avicennia*, regulation is also possible by the salt glands located on the underside of the leaves. The concentration of recreted saline reaches 4.1% in *Avicennia*, which is higher than that of sea water. The excreted salts are 90% NaCl and 4% KCl, which corresponds to the ratio in seawater. Recretion occurs in the dark and is most intense at midday. It reaches 0.2 to 0.35 mg per 10 cm^2 of leaf area in 24 hours. In dry periods, the salt accumulates on the underside of the leaf in the form of

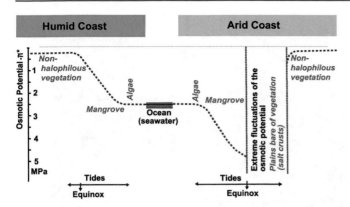

Fig. 6.39 Scheme of soil salt concentration (red curve) and mangrove structure on humid and arid coasts

common salt crystals, which melt and drip off during the night when humidity is high.

It is interesting to note that the viviparous seedlings of *Rhizophora* are almost salt-free and have a potential osmotic pressure of only 1.3 to 1.8 MPa. Thus, water must be supplied to them through a glandular tissue in the cotyledonary body. As soon as the seedlings drop and root in the saline soil, the salinity increases and the potential osmotic pressure rises to the normal level. The radicle initially appears to be permeable to salt.

The function of the respiratory roots (pneumatophores) has also been elucidated. They have lenticels with fine openings that are unwettable and therefore permeable to air but not to water. When the respiratory roots are completely immersed in water, the oxygen in their intercellulars is consumed by respiration and a negative pressure is created because the easily soluble CO_2 escapes into the water. As soon as the breathing roots emerge from the water, pressure equalization occurs and air containing oxygen is drawn in. The O_2 content in the intercellulars of the breathing roots therefore fluctuates periodically between 10 and 20%.

The mangroves, together with their fauna, the many fiddler crabs and with the mangrove fish *(Periophthalmus)* crawling on the trees, are a particularly interesting ecosystem that belongs neither to the sea nor to the mainland. Due to timber exploitation (charcoal burning) and the expansion of crab farming, the mangroves are severely endangered in many places and the coastal regions of the ever-humid tropics have been deserted (Fig. 6.40a). Thus, protection against disastrous tsunamis is lacking. In addition, there is contamination from offshore oil and gas extraction (Fig. 6.40b).

6.9 Shore Formations: Psammobiome

Sandy shore formations of the tropical coasts offer lesser peculiarities. Behind the vegetationless zone exposed to wave action, plants with long runners follow on the sand, of which *Ipomoea pescaprae* and *Canavalia rosea* (Fig. 6.41) are common, as are the halophytes *Sesuvium portulacastrum*, *Batis maritima*, and *Sporobolus virginicus*. Inland, outside the saltwater influence, the sand in the tropics is very rapidly becomes covered by shrubs and trees. These are species whose floating fruits are found in the tidemark of all tropical coasts. *Terminalia catappa* (Fig. 6.42a; its fruits in Fig. 6.42b) is a typical representative; the coconut palm might also be added (Fig. 6.42c), though today palms are almost all planted.

For the eastern oceans *Barringtonia, Calophyllum, Hibiscus tiliaceus* as well as *Pandanus* are typical, for the western *Coccoloba uvifera* (Polygonaceae), *Chrysobalanus icaco* and the poisonous *Hippomane manicinella* (Euphorbiaceae).

Large areas of dunes are absent from the tropics. One of the exceptions is the north coast of Venezuela. Here, in a semi-desert climate near Coro, a lot of sand is blown from the beach by the trade winds blowing constantly from the northeast to the east-northeast, which is collected by *Prosopis*

Fig. 6.40 (**a**) Shrimp breeding pools at Playa near the town of Machala, Pacific coast of Ecuador, after total deforestation of the mangrove (photo: Rafiqpoor); (**b**) Dead mangrove trees due to oil pollution in Venezuela (photo: Breckle)

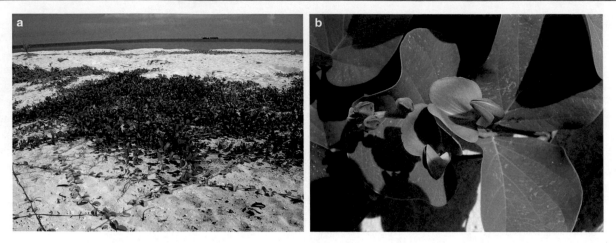

Fig. 6.41 A very salt-tolerant Fabaceae *(Canavalia rosea)* creeping on calcareous sand, with stolons up to 6 m long (**a**), with inflorescence (**b**) on the beach of a small island off Saba, Borneo (photos: Rafiqpoor)

Fig. 6.42 *Terminalia catappa* (**a**, **b**) and the fruit of a coconut palm (**c**) washed by ocean currents onto the beach of an island in Saba, Borneo, where it has begun to germinate (photos: Rafiqpoor)

juliflora. This results in the formation of dunes that continue to grow in the direction of the wind and are repeatedly covered by the *Prosopis bush* (Fig. 6.43). In this way a series of dune ridges is formed, all running side by side parallel to the wind direction and reaching a considerable height. In part of the dune area, probably as a result of logging, shifting sand dunes have developed (barchans), which again merge into dune ridges, but these are oriented perpendicular to the wind direction.

Fig. 6.43 The large sand dunes on the Paraguana Peninsula in northern Venezuela are not covered with *Prosopis*, but mainly with *Conocarpus* scrub (photo with Prof. E. Medina, Caracas: Breckle)

6.10   Orobiome II: Tropical Mountains with an Annual Temperature Cycle

While in the case of orobiome I, a short rainless period in the alpine stage does not yet affect the water supply of the plant, the drought period of ZB II has a significant effect depending on the duration even at high altitudes.

It is true that in the montane belt the amount of precipitation increases and the duration of sunshine decreases as a result of cloud cover to such an extent that an evergreen montane forest occurs, but this has a dry season in the cool season, even though a Fog forest may even develop above it in the trade wind or monsoon area (Fig. 6.45).

In the monsoon region of India, even a smaller mountain range already has a very strong effect on the amount of precipitation, less so on its distribution over the year (Fig. 6.44).

The whole sequence of elevational belts of orobiome II can be traced on the southern slope of the eastern Himalaya, on the very humid Sikkim profile from Darjeeling northward, where it is difficult to distinguish the forest belts. It is further complicated by the fact that in the higher elevational belts the Palaeotropical floral elements are increasingly displaced by Holarctic ones.

At the foot of the mountains a humid deciduous forest with *Shorea robusta* prevails and on wet soils one with

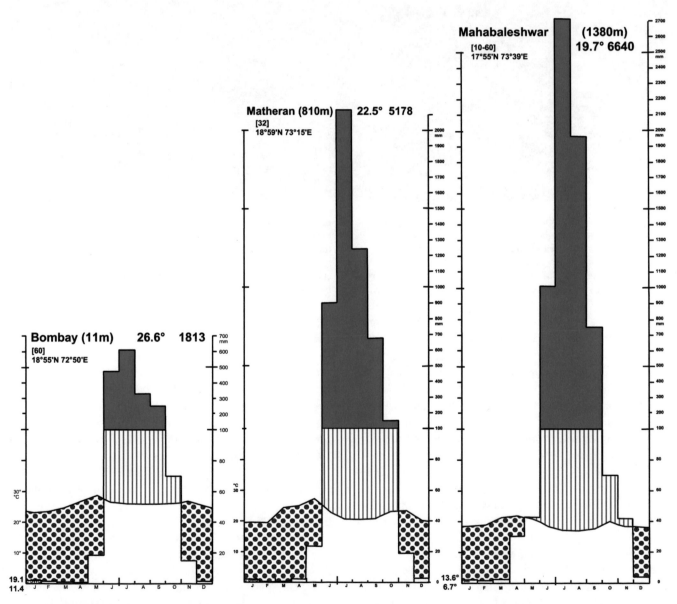

Fig. 6.44 Increase in precipitation with elevation in the monsoon region of India: Climate diagram of Bombay and two stations above it in the mountains. At the upper station at 1380 m asl., nearly 3000 mm of rain falls in July. However, the duration of the rainy season is only extended by 1 month, although the annual rainfall exceeds 6000 mm

Fig. 6.45 (**a**) View from the Bolivian Altiplano (4100 m) to the Cordillera Real with the glaciated volcano Huaina Potosi near La Paz. (**b**) View from the Altiplano to the glaciated volcanoes Parinacota and Pumarape on the Chilean-Bolivian border (West Cordillera, southern Bolivia). The Altiplano in Bolivia is a broad intermontane area fringed on both sides by mountain ranges (Western and Eastern Cordillera).

Large parts of this plateau are covered with a semi-arid grassland. Especially in the vicinity of Lake Titicaca near La Paz, the Altiplano (humid Puna) is intensively used for agriculture; towards the south, as a result of the decrease in the humidity of the climate, the extent of field cultivation also decreases, while grazing remains at almost the same intensity (photos: Rafiqpoor)

bamboo as well as palms. At about 900 m asl an evergreen tropical montane forest *(Schima, Castanopsis)* with tree ferns begins, whereby in the upper part already Holarctic tree genera *(Quercus, Acer, Juglans)*, also *Vaccinium* and others are represented.

Above comes a fog forest with Hymenophyllaceae and mosses. The higher you climb, the more predominant Holarctic genera *(Betula, Alnus, Prunus, Sorbus* and others) are found.

The frost line is reached at 1800 to 2000 m above sea level (potential frost).

In the next higher belt one finds many tall *Rhododendron* and *Arundinaria* species, which are replaced further up by conifers *(Tsuga, Taxus* and others).

At 3000 to 3900 m asl., an *Abies densa* fir forest grows with hardwoods. The forest boundary is formed by *Abies* and *Juniperus*. The subalpine belt is again characterised by tall *Rhododendrons*, which become lower and lower in the alpine

belt with flower-rich mats until *Rhododendron nivale* is only a tiny shrub at 5400 m asl.

Above 5100 m asl., predominantly hemispherical patches occur *(Arenaria, Saussurea, Astragalus, Saxifraga* and others); the snow line is at 5700 m asl.

This orobiome system of the Himalayan mountain ranges is particularly complicated (Troll 1967, Meusel et al. 1971, Miehe in Walter & Breckle 1994, Miehe et al. 2015, Miehe 2021) lying between Zonobiomes.

In the Andes, the altitudinal sequences of the west and east slopes are different, as well as in the inner mountain valleys. A short schematic overview has been given by Ellenberg (1975).

The high plateau of the Altiplano in Bolivia and Peru (Fig. 6.45) is populated and grazed by llama herds, and is thus anthropogenically modified. However, the wild vicuña herds also play a role in the devastation of the landscape. According to the climate, on the western slope the sequence

Fig. 6.46 The forest boundary is located at the high volcanoes of the Western Cordillera in Bolivia (e.g. at Sajama) at about 5300 m asl (**a**, photo Breckle; Fig. 7.53b) and is formed by *Polylepis tarapacana* with *Lepidophyllum quadrangulare* and the tussock grasses from *Festuca orthophylla, Stipa ichu* etc. (**b**, photo Rafiqpoor). Partially frozen Salar surface of Laguna Hediona (3900 m asl) in the Altiplano of Bolivia with flamingos, salt crusts and volcanoes (**c**, photo: Breckle). In the area of the desert Puna in the southern Altiplano large salt lakes like the Salar de Uyuni occur (**d**, photo: Breckle). In this Salar, at 3900 m asl near the village of Uyuni, the salt crust is hewn out, crushed and completely dried in the air for further use (lithium extraction)

of steps becomes more and more xerophytic towards the south. The rain green deciduous forest belts reach higher and higher and the evergreen ones become more hard-leaved and small-leaved.

The presence of a warm season results in an elevation of the timberline up to 4000 m asl.; individual *Polylepis* stands reach up to 4500 (5300) m asl. (Fig. 6.46a). Páramos are replaced by Puna, initially moist Puna with cushion plants such as *Azorella compacta* (Fig. 6.47), more southerly dry Puna with xerophytic tussock grasses such as *Festuca orthophylla* (Fig. 6.48), *Stipa ichu,* and others, until a desert puna with many Salars (salt pans) predominates in the orobiome II area (Lauer 1975, Chong-Diaz 1988) (Fig. 6.46c, d). Correspondingly, soils of the alpine belt change in a southerly direction from peaty soils to chestnut soils and syrosems to solonez and solonchak.

A very detailed ecological and also microclimatic study of the Puna in NW Argentina between 22 to 24 1/2°S is available from Ruthsatz (1977).

Fig. 6.47 *Azorella compacta* (Apiaceae) at a rock site on the Bolivian Altiplano 4500 m asl (photo: Breckle)

Fig. 6.48 *Festuca orthophylla* over weathered volcanic fine material on a gently sloping hillside in southern Bolivia (4750 m asl). Due to the effect of frost heave and the slight slope, the grass tufts are arranged in grass garland structures (photo: Breckle)

6.11 Man in the Savannah

Today, the savannah has been replaced in many places by cattle pastures. Even in earlier times, pastoral nomads were on the move in the vast savannah areas and, with their extensive herds, provided the wild animals with considerable competition for food sources.

Large-scale African grasses have been introduced into neotropical savannahs, drastically reducing the original biodiversity. However, productivity has increased partly to the benefit of extensive cattle grazing (Solbrig et al. 1996).

Due to the frequent grass and bush fires, which are usually deliberately set shortly before the start of the rainy season, the growth of new greenery is to be improved. Over a longer period of time, however, this leads to an ever-increasing nutrient depletion of the soils with greater risk of erosion and thus increasing desertification.

6.12 Zonoecotone II/III

6.12.1 Sahel

This zonoecotone includes the open, climatic savannahs, such as Namibia. Similar conditions are found south of the Sahara in the Sahel zone, which forms the transition to the summer rainfall area of Sudan (ZB II). But the Sahel has been completely degraded by overcrowding and overgrazing as a result of the recurrent drought years typical of this zone (Müller et al. 2006). It can only tolerate very sparse settlement and correspondingly low numbers of livestock, which used to be enforced by the few natural water points in this area. However, as part of development aid, people wanted to develop the land and drilled many wells. This made it possible to water larger herds, and the population increased accordingly as long as the annual rainfall was above the long-term average. Then, however, followed several years of drought, which led to disaster. Water for people and animals was available, but no pasture, for the grasses withered. Starving livestock perished, and people had to flee the land or be supported by outside relief efforts. But the pasture suffered irreparable damage and became a 'man made desert' as the worst effect of desertification. New programs are planned and have started to protect the Southern regions from further desiccation and desertification. The "Green Belt" is one of the ongoing projects, stretching about 8000 km from W to E (Breckle 2021a).

In present-day Namibia, with a similar climate, several drought years in succession also have a devastating effect, but the small number of farmers can survive these years by reducing livestock in time, shifting grazing plans and management, and the economy recovers quickly after a few good rainy years, perhaps because the droughts are also shorter than in the Sahel.

6.12.2 Thar or Sind Desert

Another zonoecotone II/III is located in the border area between India and Pakistan—the Thar or Sind desert. It is a uniform arid area between the Aravalli Mountains in the east and the heights of Baluchistan in the west, which is also called the "Great Indian Desert"(Fig. 6.49). The aridity increases from east to west.

In the area with over 250 mm of rainfall, savannahs are grazed and degraded as a result of overstocking with livestock, with the annual grass species *Aristida adscensionis* becoming rampant as a grazing weed.

If the literature often speaks of a Saharo-Sindian desert zone, this is not correct. For the Sahara belongs floristically for the most part as a rainless area or one with little winter rain to the Holarctic and continues eastward into the Egyptian-Arabian Desert to Mesopotamia. The Sind desert, on the other hand, is the last driest spur of the Indian monsoon area and must be included floristically in the Palaeotropics. The Indian desert, Thar, is climatically already a zonoecotone II/III, which can be compared with the transition area from the Sudan to the southern Sahara, the "Sahel". Both receive light summer rains, but the Indian area is already north of the Tropic of Cancer, the annual temperatures are therefore 2 to 3 K lower than in the Sahel, and frosts may occur in December to February (Fig. 6.49). Only the area in the Indus lowlands receives an average of less than 100 mm of rain a year and would therefore be climatically a desert; but it is a water-rich irrigated area because of the Indus and its tributaries. The 'Great Indian Desert', on the other hand, is largely a 'man made desert'. The area was inhabited four thousand years ago, became more densely populated from the time of Alexander the Great, and is now completely degraded as a result of overgrazing, logging, and partial cultivation (Mann 1977). By nature, the area was a *Prosopis* savannah on deep sandy reddish-brown savannah soils, with 400 to 150 mm of rainfall per year, as evidenced by an area that has been protected for several decades not far from Jodhpur (Rodin et al. 1977).

The thorny shrubs there are: *Prosopis cineraria, Ziziphus nummularia, Capparis decidua (= C. aphylla)* and others. *Prosopis* grows to 8 m tall at annual rainfall of 500 to 600 mm, forming stands of 150 to 200 specimens per hectare; at 300 to 400 mm it grows only 5 to 6 m tall (stands of 50 to 100 expl. Per ha); and at 200 mm it grows only 3 to 4 m tall (stands of 25 to 30 expl. Per ha). Similarly, as rainfall decreases, tall grasses (*Lasiurus, Desmostachya*) are replaced

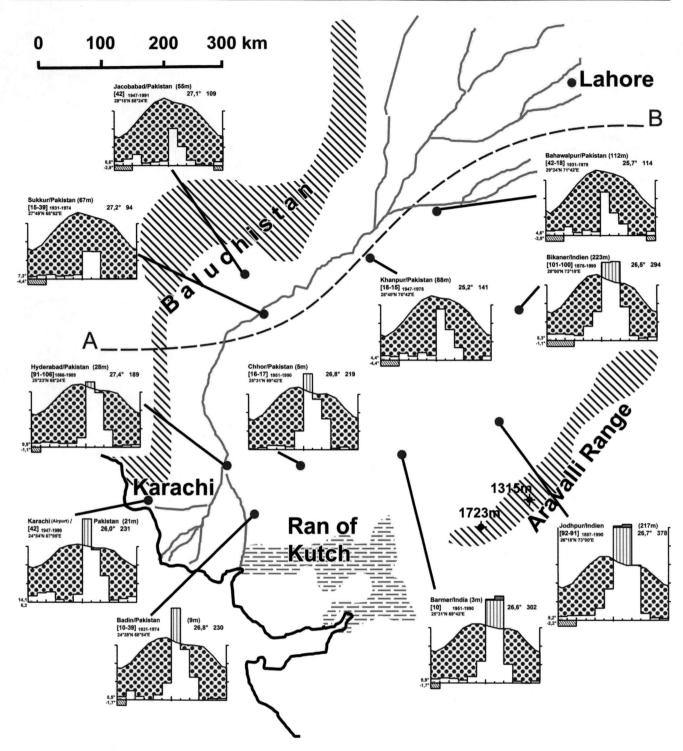

Fig. 6.49 Climate diagram map of the Sind-Thar desert. Northwest of the line A-B the extremely arid area

by low ones *(Aristida)* in the grass layer (Gaussen et al. 1972). Thus, conditions are like those in southwestern Africa (Fig. 6.7).

A very open grassland with *Lasiurus, Desmostachya, Panicum* and *Aristida* species, with some succulents and the widespread *Calotropis procera*, is shown in Fig. 6.50, while in Fig. 6.51 a dense *Prosopis* savannah is shown, with large columnar Euphorbia *(Euphorbia caducifolia)* in the foreground on rock; the stand structure of the thorn savannah with *Prosopis, Acacia, Capparis decidua, Salvadora persica,* with the climbing *Coccinea grandis* is illustrated by Fig. 6.52.

Fig. 6.50 Overgrazed semi-desert near Jodhpur with *Calotropis procera* (photo: Breckle)

Fig. 6.51 Thorn savannah and open rock slabs near Jodhpur, in the transition area of the zonoecotone II/III to the Thar desert (photo: Breckle)

Fig. 6.52 Open species-rich thorn scrub near Jodhpur during the dry season in December (photo: Breckle)

In the Bikaner district (Fig. 6.47), the soils are very sandy. In the vicinity of settlements, mobile barchans, i.e. vegetationless dunes, form as a result of grazing, giving the impression of a real desert (Fig. 6.53). In fact, however, the water content of the sand of such unvegetated dunes is much higher than that of vegetated ones, as shown by data (Table 6.5) from an area with 260 mm of rain per year.

Fig. 6.53 Sand area (Barkhane) in motion in the "Desert National Park"in the vicinity of Huri village about 40 km from Jaisalmer with individual *Acacia* and *Calotropis* shrubs (Photo: http://bit.do/bFP6P)

Table 6.5 Water content (in mm) of the sand of unvegetated (I) and vegetated (II) dunes near Jaisalmer (after Mann 1976)

Depth ranges	Period							
	March		June		Sept.		January	
Depth (in cm)	I	II	I	II	I	II	I	II
0–105	41	10	33	17	45	10	34	7
0–210	106	39	94	48	120	33	105	28

This difference is understandable because a *Prosopis* stand takes about 220 mm of water a year from the soil for transpiration and the often planted grass *Pennisetum typhoides* also takes about 160 to 180 mm.

The population exploits the water content in the sand of the unvegetated dunes by planting watermelons at a distance 2 × 2 m and preventing the sand from blowing away with brushwood. No information can be given about the natural vegetation of the driest part, the Sind Desert in the Indus lowlands. This irrigation area is densely populated; areas of natural vegetation do not exist. Unrationed irrigation has greatly raised the water table, causing secondary salinization of the moist soils. As a result, 40,000 hectares of cultivated land have been lost annually, causing the increase in food production to lag far behind the increase in population. Restoring the brackish soils is very costly in the flat terrain (Breckle 2021b).

Natural saline soils are very common in the south of the Thar Desert on the Gulf of Kutch. Mangroves grow in the intertidal area, followed by salt marshes with *Salicornia, Suaeda, Atriplex* and the salt grass *Urochondra*. In the Rann of Kutch area (Fig. 6.49) with high groundwater levels, nearly sterile clayey saline soils spread with few woody plants in favourable places and with halophytes (*Haloxylon salicornicum, Aeluropus, Sporobolus*) or *Cenchrus* spp., *Cyperus rotundus,* and others (Blasco 1977).

6.12.3 The Caatinga

Ecologically difficult to classify is the Caatinga in NE Brazil, the arid area, "Polygono da Seca". It is characterised by extreme fluctuations in precipitation from year to year. For example, in the driest place, Cabaceiras, the good rainy years 1940 to 1946, with rainfall ranging from 664 to 150 mm, were followed by droughts from 1948 to 1958, with rainfall below 80 mm (only 24 mm in 1952, only 22 mm in 1958), with the exception of 1954, with 170 mm, and 1955, with 187 mm. Such an unreliable climate is best survived by large succulent columnar cacti and large spiny Bromeliaceae growing on the ground, as well as much water-storing bottle trees (*Ceiba* and others) or deciduous shrubs that are leafless for long periods (Fig. 6.54). The area is difficult to exploit and is sparsely populated because periods of drought cannot be predicted, forcing the population to leave the land. Similar conditions are found in the trade wind desert on the north coast of South America in the Venezuela-Colombia border area or in the Galapagos Islands. Years with very high precipitation also occur in this arid region.

6.12.4 Tropical East Africa

Finally, mention should be made of the rather arid areas belonging to the Palaeotropis in the tropical area of E Africa, as well as a small area in the rain shadow between the Pare and W Usambara Mountains with very strange succulents [*Adenia globosa,* rock-block-like *Pyrenacantha, Euphorbia tirucalli* (Fig. 6.17), *Caralluma, Cissus quadrangularis, Sansevieria* (Fig. 6.18), and others]; with an annual temperature of 28 °C and only 100 to 200 mm of rain, it may be the driest area along the equator. Much more extensive are the arid areas of N Kenya, W Ethiopia, Somali, and Socotra with *Adenium socotranum* (Apocynaceae) (Fig. 6.55a), which has misshapen succulent stems 2 m in diameter, and *Dracaena cinnabari* (Fig. 6.55b) with a stem diameter of 1.6 m. The various life forms of the thorny succulent savannah are shown schematically in Fig. 6.56. Most of the life forms can be understood as typical adaptations to long dry periods, but they have not been able to penetrate the actual deserts of zonobiome III after all.

6.12.5 SW Madagascar

Madagascar, with its distinctive flora and fauna, has zonobiome I rainforest on the east coast with up to 2000 mm of rain per year. However, most of the island has a summer rainfall climate and bore deciduous forest. The tree and shrub flora of Madagascar is also very unique overall, with about 94% of the species being endemic. The flora of Madagascar was very rich in species, but today many forests and savannahs have been cleared. Large areas are degraded. Huge areas of grass are burned every year, supposedly to get better pastures for the at least ten million Zebu cattle, in the dry parts goats are kept.

The driest SW corner of Madagascar is characterised not only by Baobab trees (Fig. 6.57) but also by the endemic family of Didiereaceae (four genera with eleven species) (Fig. 6.58), which only occurs here and is reminiscent of columnar cacti. With about 350 mm of rain per year, which moreover usually falls very irregularly, a thorn-bush succulent semi-desert develops here. Numerous succulents of the genera *Euphorbia, Aloe, Kalanchoe, Crassula* occur, plus bottle trees of the genera *Adansonia, Moringa,* and *Pachypodium* (Fig. 6.58). Other species are very small-leaved, thorny, or leafless. Poikilohydric vascular plants and ferns also occur.

Fig. 6.54 Caatinga near Rodelas in Sertão NE Brazil (photo: Glauco Umbelino, http://t1p.de/mb49)

Fig. 6.55 (**a**) *Adenium socotranum* (Apocynaceae) with a trunk diameter of 2 m on western Socotra. (**b**) *Dracaena cinnabari* on the path between Wadi Dirham and the Dicksam Plateau on Socotra. The trunks of the dragon trees have scars left from tapping the resin (dragon's blood). The vegetation along the path shows severe browsing damage (from free-ranging goats). Degraded dragon tree forest is visible in the background. Young trees are largely absent. In the background on the left the Hajhir mountains surrounded by clouds can be seen (photos: Ernst Kluge)

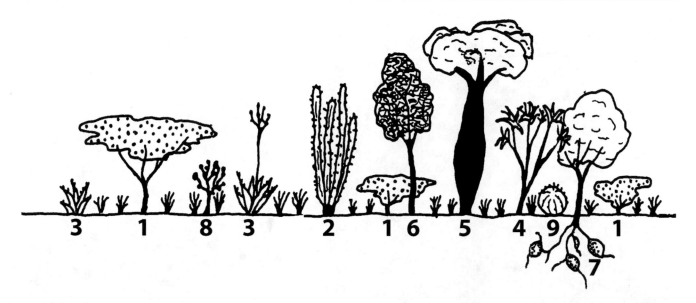

Fig. 6.56 *Characteristic life forms of thorny succulent savannah (after Troll 1960). (1) Thorny fine-leaved umbrella trees (Acacia type); (2) Stem succulent candle or candelabra trees (Cacti type); (3) Succulent and thorny-leaved crested trees (Aloe type); (4) Succulent and thorny-leaved crested trees (Dracaena type); (5) Water-woody barrel-leaved* deciduous trees *(Adansonia* type); (6) Sclerophyllous trees with thorns *(Balanites* type); (7) **Deciduous** trees with xylopodia or lignotuber; (8) Sclerophyllous shrubs and tree-bushes *(Capparis* type); (9) Stem succulents, low plants *(Stapelia* type); (10) Grasses everywhere in between (after Troll 1960)

Fig. 6.57 *Adansonia grandidieri* from Madagascar (photo: E. Fischer)

Fig. 6.58 *Didierea madagascariensis* (Didiereaceae) from Madagascar (**a**, photo: E. Fischer); (**b**) *Pachypodium lamerei* from Tsiman-ampetsotsa, Madagascar (photo: https://t1p.de/cli9)

Photo 14   Rainfed dry savannahs with *Acacia* spec. in Pendjari National Park, Benin, W-Africa (photo: A. Erpenbach)

Photo 15 The big old lion keeps a lookout. Large animal world in the savannah of Tanzania (Zonobiom II) in the Serengeti National Park (photo: Breckle)

Photo 16 Savannah landscape (ZB II) in Serengeti National Park (Tanzania), at the beginning of the summer rainy season (photo: Breckle)

References

Anderson, G.D. & Herlocker, D.J. 1973: Soil factors affecting the distribution of the vegetation types and their utilization by wild animals in Ngorongoro crater, Tanzania. J. Ecol. **61**: 627-651

Beadle, N.C.W. 1981: The Vegetation of Australia, Vegetationsmonographien der einzelnen Großräume Bd. IV. Stuttgart 690 p.

Belsky, A.J. & Canham, C.D. 1994: Forest gaps and isolated savannah trees. BioScience **44**: 77-84

Blasco, F. 1977: Outlines of ecology, botany and forestry of the mangals of Indien subcontinent. Ecosystems of the World, (ed. V. J. Chapman), Vol. **1**: 241-260

BourliÈre, F. (ed.) 1983: Tropical savannahs. Ecosystems of the World **13**: 730 S.

Breckle, S.-W. 2021a: Grüne Mauern ("green walls") gegen die Wüstenausbreitung als Maßnahme gegen Desertifikation. In: Lozán J. L. S.-W. Breckle, H. Grassl, W. Kuttler & A. Matzarakis (Hrsg.). Warnsignal Klima: Böden und Landnutzung. pp. Xxx-yyy. Online: www.klima-warnsignale.uni-hamburg.de.

Breckle, S.-W. 2021b: Bodenversalzung (Meeresspiegelanstieg, Aridität und Fehler bei der Bewässerung. In: Lozán J. L. S.-W. Breckle, H. Grassl, W. Kuttler & A. Matzarakis (Hrsg.). Warnsignal Klima: Böden und Landnutzung. pp. Xxx-yyy. Online: www.klima-warnsignale.uni-hamburg.de

Bucher, E. H. 1982: Chaco and Caatinga -South American arid savannahs, woodlands an thickets, 48-79. In: Huntley, B. J. and Walker, B. H. (eds.), s. there.

Cannel, M. G. R. 1982: World forest biomass and primary production dates. Academic, London, New York, 391 p.

Champion, H.H. & Seth, S.K. 1968: A revised survey of the forest types of India. Governm. of India, New Delhi 404 p.

Chapman, V.J. 1976: Mangrove vegetation. Vaduz, 477 p.

Coutinho, L.M. 1982: Ecological effect of fire in Brazilian Cerrado, 273-291. In: Huntley, B. J. & Walker, B.H. (eds.) s. there

Cumming, D.H.M. 1982: The influence of large herbivores on savannah structure in Africa, 217245. In: Huntley, B.J., & Walker, B.H. (eds.), s. there

Eiten, G. 1982: Brazilian „savannahs": 25-79. In: Huntley, B. J. & Walker, B. h. (eds.), s. there

Ellenberg, H. 1975: Vegetationsstufen in perhumiden bis perariden Bereichen der tropischen Anden. Phytocoenologia **2**: 368-387

Ernst, W. & Walker, G.H. 1973: Studies on hydrature of trees in miombo woodland in South Central Africa. J. Ecol. **61**: 667-686

Gaussen, H., Meher-Homji, V.M., Legris, P. et al. 1972: Notice de la feuille Rajasthan (1:1 Mill.). Travaux Sect. Sc. et Techn., Inst. Français de Pondichéry, Serie No 12

Henry, A.N., Kumari, G.R. & Chitra, V. 1987: Flora of Tamil Nadu, India, ser.1, vol. **2**. Botanical Survey of India, Coimbatore

Hopkins, B. 1974: Forest and Savannah (West Africa). 2. edit., 154 S., Ibadan/London

Hueck, K. 1966: Die Wälder Südamerikas. Vegetationsmonographien der einzelnen Großräume, Bd. II, Fischer, Stuttgart, 296 S.

Huntley, B.J., Walker, B.H. (eds.) 1982: Ecology of tropical savannas. Ecol. Stud. 42, Springer/Berlin

Inchausti, P. 1995: Competition between perennial grasses in a neotropical savannah: the effect of fire and of hydric-nutritional stress. J. Ecol. **83**: 231243

Lamotte, M. 1975: The structure and function of a tropical savannah ecosystem. Ecol. Stud. **11**: 179222

Mann, H.S. (ed.) 1977: The spectre of desertification. Ann. Arid Zone, Jodhpur **16**: 279-394

Medina, E. 1968: Bodenatmung und Stoffproduktion verschiedener tropischer Pflanzengemeinschaften. Ber. Dtsch. Bot. Ges. **81**: 159-168

Menault, J.C. & Cesar, J. 1982: The structure and dynamics of a West African Savannah: 80-100. In: Huntley, B. J., and Walker, B. H. (eds.), s. there

Meusel, H., Schubert, R., et al. 1971: Beiträge zur Pflanzengeographie des Westhimalajas. Flora **160**: 137-194, 370-432, 573-606

Miehe, G., Pendry, C. & Chaudhary, R. (eds.) 2015: Nepal. An introduction to the natural history, ecology and human environment of the Himalayas. A companion to the Flora of Nepal.–Royal Botanic Garden Edinburgh. Edinburgh, VIII + 561 p.

Miehe, G. 2021: Extrem kalt-arides Subzonobiom VII(tIX) der Kälte- und Hochlandwüsten Zentralasiens: Tibet. In: Breckle, S.-W. (Hg.) Ökologie der Erde, Band 3: Spezielle Ökologie der Gemäßigten und Arktischen Zonen Euro-Nordasiens. Schweizerbart/Stuttgart S. 405-454

Müller, H.J. 1991: Ökologie. 2. Aufl. UTB 318 Fischer/Stuttgart 415 S.

Müller, J.V., Veste, M., Wucherer, W. & Breckle, S.-W. 2006: Desertifikation und ihre Bekämpfung–eine Herausforderung an die Wissenschaft. Naturwiss. Rundschau **59**: 585-593

Nai, N.C. & Henry, A.N. 1983: Flora of Tamil Nadu, India, ser.1, vol.1. Botanical Survey of India, Coimbatore

Ogawa, H., Yoda, K. & Kiro, T. 1961: A preliminary survey on vegetation of Thailand. Nature Life SE Asia **1**: 21-157.

Pascal, J.P. 1988: Wet evergreen forests of the Western Ghats of India. Inst Français de Pondicherry, Ashram Press/Pondicherry 345 p.

Poole R.M. 2006: Harte Zeiten für die Savanne. In der Serengeti konkurrieren Tiere, Touristen und Einheimische–Gehen die Massai leer aus? National Geogr. Deutschland, p.40-67

Pullaiah T., Ramakrishna 2018: Biodiversity in India. In: Pullaia, T. (ed): Global biodiversity. Vol. 1. selected countries in Asia. p. xxx-yyy

Rawitscher, F. 1948: The water economy of the "Campos Cerrados" in southern Brazil. J. Ecol. **36**: 237-268

Rodin, L.E., Bazilevich, N.I., Gradusov, B.P. & Yarilova, E.A. 1977: Trockensavanne von Rajputan (Wüste Thar). Aridnyepochvy, ikhgenesis, geokhimia, ispol'novaniye: 195-225, Moskva (Russ.)

Ruggiero, P.G.C., Batalha, M.A., Pivello, V.R. & Meirelle, S.T. 2002: Soil-Vegetation relationships in cerrado (Brazilian savannah) and semideciduous forest, Southeastern Brazil. Plant Ecology **160**: 1–16

Ruthsatz, B. 1977: Pflanzengesellschaften und ihre Lebensbedingungen in den andinen Halbwüsten Nordwest-Argentiniens. 168 S. Diss. Bot., **39**, J. Cramer, Vaduz

Sarmiento, G. 1996 Biodiversity and water relations in tropical savannas. Ecol. Stud. **121**: 61-75

Schultz, J. 1995: Die Ökozonen der Erde. 2. Aufl. Ulmer, Stuttgart

Solbrig, O.T., Medina, E. & Silva, J. F. (eds.) 1996: Biodiversity and savanna ecosystem processes. Ecol. Stud. **121**, 233 p.

Thomas, M.F. 1974: Tropical geomorphology. London.

Tinley, K.L. 1982: The influence of soil moisture balance on ecosystem patterns in Southern Africa. 175-192. In: Huntley, B. J., and Walker, B. H. (eds.), s. there

Troll, C. 1960: Die Physiognomik der Gewächse als Ausdruck der ökologischen Lebensbedingungen. Verhdl. Dt. Geographentag **32**: 97-122

Troll, C. 1967: Die klimatische und vegetationsgeographische Gliederung des Himalaya-Systems. Er-geb. Forsch. Untern. Nepal Himalaya **1**: 353-388

Walker, J. & Gillison, A.N. 1982: Australian Savannahs. 5-24. In: Huntley, B. J., & Walker, B. H., s. there

Walter, H. 1939: Grasland, Savanne und Busch der ariden Teile Afrikas in ihrer ökologischen Bedingtheit. Jb. Wiss. Bot. **87**: 750-860

Walter, H. 1990: Vegetationszonen und Klima. 6. Aufl., Ulmer/Stuttgart 382 S.

Walter, H. & Breckle, S.-W. 1994: Ökologie der Erde, Bd. 3: Spezielle Ökologie der Gemäßigten und Arktischen Zonen Euro-Nordasiens. UTB Große Reihe, 2. Aufl., Fischer, Stuttgart. 726 S.

Walter, H. & Breckle, S.-W. 2004: Ökologie der Erde. Bd. 2: Spezielle Ökologie der tropischen und subtropischen Zonen. 3. Aufl. Spektrum, Heidelberg 764 S.

Walter, H. & Steiner, M. 1936: Die Ökologie der ostafrikanischen Mangroven. Ztschr. f. Bot. **30**: 63-193

Wilson, E.O. 1988: Biodiversity. National Academy Press, Washington D.C. ISBN 0-309-03783-2

Part F: ZB III: Zonobiome of Hot Deserts or Subtropical Arid Climate

7

Contents

© Springer-Verlag GmbH Germany, part of Springer Nature 2022
S.-W. Breckle, M. D. Rafiqpoor, *Vegetation and Climate*, https://doi.org/10.1007/978-3-662-64036-4_7

Photo 17 Sand desert (Erg) with pronounced sand dunes in Merzouga (ZB III), in S Morocco (photo: Rafiqpoor)

Photo 18 Extreme desert (ZB III) at Wadi Allaqui, Southern Egypt, with slaty rocky sculptures and sand-blown depressions, free of any vegetation (photo: Breckle)

Photo 19   Dense dune system with few scattered grasses (*Panicum*) and shrubs (*Tamarix, Calligonum*) and feral camels in the Rub al Khali (ZB III) in S Oman (photo: Breckle)

7.1 Climatic Subzonobiomes

Deserts together account for more than 35% of the Earth's solid surface. In the subtropical desert zone, in **zonobiome III**, there is no cold winter season, which is so characteristic of the arid areas of the temperate zone.

The term desert is relative. For someone who comes from the humid east of North America, the southwest of the country is already a desert, even though Tucson (Arizona) receives more than 300 mm of precipitation per year; but an Egyptian who lives in dry Cairo will no longer consider the Mediterranean coast a desert, even though the annual rainfall there barely reaches 150 mm.

The Earth's desert systems are the result of atmospheric circulation. They generally develop in the transition zones of the large wind and precipitation systems. To illustrate this phenomenon, we would like to consider the dynamics of the atmosphere in the course of desertification using the example of desert formation Africa (Fig. 7.1).

In the **inner tropics** in the Congo Basin (approximately 0°-10° on both sides of the equator), the convergence of the NE and SE trade winds in the area of the Intertropical Convergence Zone (ITCZ) gives rise to high-pressure thunderstorm cells (cumulonimbus clouds) with regular afternoon showers. Due to the large vertical air mass transport into the higher atmosphere, an area of high pressure develops over the equator in the higher troposphere, while an area of low pressure develops at ground level (equatorial low pressure trough). To compensate for this, monsoon-like westerly winds with abundant precipitation at the western sides of the continents (e.g. Chocó >8000 mm, Cameroon Mountain, Southeast Asia) flow in near the ground. Due to the high insolation and the removal of air masses from the areas near the tropics, a mass deficit is created. Accordingly, cold, heavy air masses flow in from the troposphere, but warm up and dry out on their way down. At ground level, these air masses generate high pressure.

In this quasi-stationary high-pressure belt near the tropics between the tropics (with summer rainfall) and the subtropics (with winter rainfall) are the extensive semi-desert and desert areas of the Earth. Central areas of deserts are not reached by either rainfall regime (Fig. 7.2). In "rainier" deserts, on the other hand, the two types of precipitation dovetail (overlap areas). This phenomenon favours the growth of a species-rich flora (e.g. Namaqualand in SW Africa; Sonora Desert in the USA).

In general, therefore, a hot area is called a desert if the annual precipitation is less than 200 mm and the potential evaporation is more than 2000 mm (up to 5000 mm in the central Sahara).

In the arid regions of the Earth, the sparse precipitation falls at different times of the year (Walter & Breckle 2004). Accordingly, the zonobiome III is divided into the following subzonobiomes (sZB):

1. sZB with two rainy seasons (Sonora Desert, Karoo)
2. sZB with a winter rainy season (northern Sahara, Mojave Desert, Near Eastern deserts)

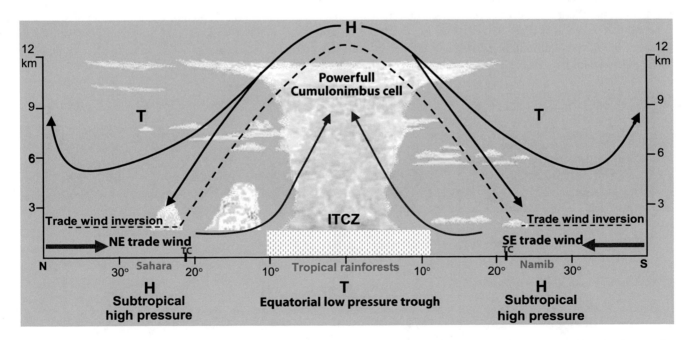

Fig. 7.1 The atmospheric circulation over Africa in July and the formation of wet and dry areas. The red arrows must be thought of as running obliquely backwards. TC = Tropics of Cancer and Capricorn

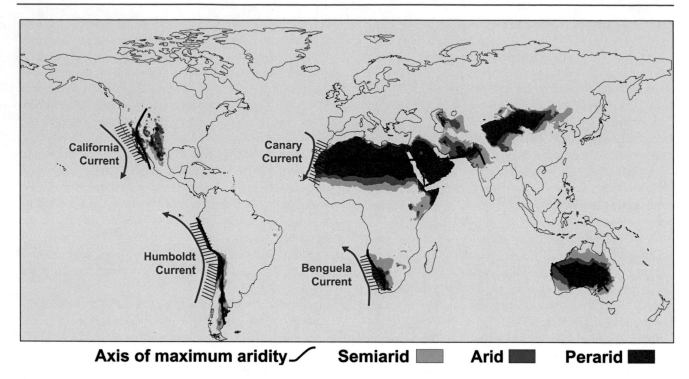

Fig. 7.2 The arid regions of the Earth with axes of maximum aridity (as in the central part of the Sahara) or the intersecting areas of tropical and subtropical rain regimes (e.g. in Namaqualand, the Sonora Desert, or central Australia)

3. sZB with a summer rainy season (southern Sahara, Inner Namib, Atacama)
4. sZB with sparse rain possible at any time of the year (Central Australia, Karoo)
5. sZB of the coastal deserts almost without rain, but with a lot of fog (northern Chilean-Peruvian coastal desert, outer Namib)
6. sZB of the rainless vegetationless deserts (Central Sahara, Central Atacama).

In Fig. 7.3, the climate diagrams of the different sZB are shown with the exception of sZB 5, because the fogs are hardly measurable as precipitation and thus not evident from the diagrams (Fig. 7.36). A very important feature of all arid areas is the large variability of rainfall in individual years. Therefore, the mean values do not mean much. Years with rainfall below the mean are most frequent; however, a few years with very high rainfall occur, which replenish the water reserves in the soil for decades.

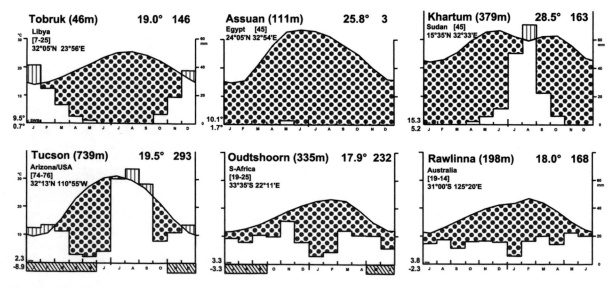

Fig. 7.3 Climate diagrams from desert stations. Top row from North Africa with winter rain, no rain, and summer rain; bottom row with 2 rainy seasons (Sonoran Desert and Karroo) and rain possible at any time of year (Rawlinna, Australia). Fig. 7.33

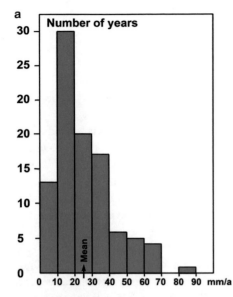

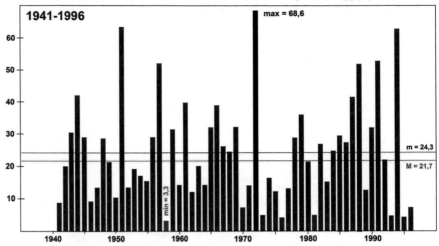

Fig. 7.4 (**a**) Variability of annual precipitation at Cairo from 1906 to 1953 (modified after Walter 1973)—a markedly skewed distribution of decadal steps. (**b**) Annual precipitation from 1941 to 1996 indicating the big variance (driest year 1958 with 3.3 mm, wettest year 1972 with 68.6 mm per year; median M is lower than average m

---

> **Box 7.1 The Deserts in the Arid Regions**
> Deserts are arid areas. In these, the potential evaporation is much higher than the annual precipitation. Semi-arid, arid and extremely arid areas can be distinguished. In zonobiome III the "hot deserts" are summarized, in zonobiome VII the "winter-cold deserts".

The variability of precipitation for Cairo (winter rainfall area) is shown Fig. 7.4a, b. A similar skewed distribution has that of Mulka, the most arid station in central Australia, except that the mean is 100 mm and the extremes are 18 and 344 mm, in Swakopmund (outer Namib): The corresponding mean is 15, extremes zero and 140 mm.

The ecological conditions in the individual years are so different that only long-term observations provide a correct picture of the ecosystems in deserts. Each desert must be considered on its own merits, but let us first discuss the few commonalities.

In all deserts the air is very dry (exception: foggy deserts), the irradiation and radiation and thus also the daily fluctuations of the air temperature are correspondingly strong. Only during the mostly very short rainy season are the extremes moderated.

7.2   Soils and their Water Balance

In the deserts one can hardly speak of soils in the proper sense, because they are raw soils (syrosems), which consist of the weathered debris of the rocks, partly altered by wind or water or by strong diurnal heating and nocturnal cooling as a result of intensive irradiation and radiation (mechanical weathering). Therefore, the properties of the often loose parent rocks are decisive, which means that we cannot speak of **climatic soils.** There is also no definable climatic zonal vegetation in strict sense, but only pedobiomes (lithobiomes, psammobiomes, halobiomes and others).

The water supply of the plants also depends on the substrate. For plants in arid regions, the amount of precipitation is only indirectly important. The decisive factor is rather the amount of adhesive water in the soil that is available to them. It forms only part of the water that falls on the soil as rain, because some runs off and some evaporates (Fig. 7.5). The proportion of adhesive water depends on the structure of the substrate. In humid areas, sandy soils are considered dry because they retain little adhesive water, whereas clay soils are considered moist. In arid regions we have to relearn; there it is just the other way round.

Infiltration to greater depths down to the groundwater does not take place on leveled terrain in the arid region. Only the upper soil layers are moistened. The depth to which the water penetrates depends on the field capacity of the soil. Let us assume that 50 mm of rain falls on a dry desert soil and that it penetrates completely into the soil. In this case, for a sandy soil, the top 50 cm will be moistened to field capacity. In a fine-grained clayey soil with five times the field capacity, the water will penetrate only 10 cm deep, whereas in a rocky soil with only small fissures, the water will penetrate much deeper, perhaps 100 cm to several meters (Fig. 7.6).

After the rain, evaporation begins. If the top 5 cm of clay soil dries out, 50% of the rainwater that has penetrated is lost.

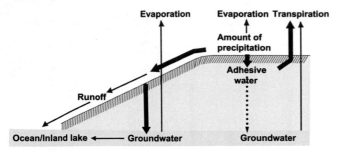

Fig. 7.5 Scheme of the fate of precipitation in arid regions. Adherent water is important for plants. The runoff water percolates into the dry valleys and feeds the groundwater, which is rarely reached by plant roots except in wadis

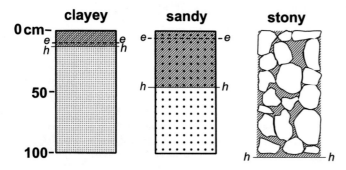

Fig. 7.6 Schematic representation of water storage in different soil types after a rainfall of 50 mm in arid regions. **h--h** = lower limit of soil moisture; **e---e** = lower limit to which the soil dries out again. The clay soil stores 25%, the sandy soil 90% and the stony soil 100%

The sandy soil dries out less. But even if the upper 5 cm lose their water, only 10% of the water would evaporate. With rocky soil there is practically no evaporation at all, i.e. all water is stored. It follows that, in contrast to the conditions in the humid region, clay soils are the driest locations for plants in the arid region, while sandy soils provide a better water supply. Ragged rocky soils are the wettest sites, provided that the rain penetrates them unhindered and there is enough fine soil in the rock crevices to store the water.

These considerations are confirmed by measurements in the Negev Desert. For the same annual precipitation, the amount of water exploitable by plants was found to be 35 mm in loess soil, 50 mm in rocky sites with a relatively considerable runoff, 90 mm in sandy soil, and 250 to 500 mm in dry valleys with a strong inflow. That sandy soils in arid regions are more favourable to plants is evident from the fact that the same type of vegetation occurs on sand with less rainfall than on clayey soils. In Sudan, *Acacia tortilis* semi-desert is found on sandy soils in a zone with 50 to 250 mm of rainfall, but on clay soils only at 400 mm, or *Acacia mellifera* savannah on sandy soils at 250 to 400 mm, but on clay soils

only at 400 to 600 mm of annual rainfall. In the short-grass prairie of the Great Plains, long-grass prairie is found on sandy soils in W Nebraska, which otherwise occurs only farther east at higher precipitation. The more favourable water conditions of rocky soils are often conspicuous in arid areas because of their tree cover amid low vegetation on fine-grained soils.

If in sandy soils or in rock crevices the soil is soaked down to the groundwater, then the roots can grow deep enough to reach the groundwater; the water supply of the plants is then assured. The following example may be mentioned here:

North of Basra in Mesopotamia, groundwater is present at a depth of 15 m, constantly fed by gravel layers from the Euphrates and Tigris rivers. However, since the rainfall is only 120 mm per year, only the upper layers of soil are moistened, and the roots of plants cannot reach the groundwater; the soil covers itself with a scanty ephemeral vegetation after the rains fall in winter. However, the local population have dug wells and use the water to grow vegetables, planting crops in furrows and irrigating several times a day when daily maximum reaches 50 °C. As a result of the greater evaporation, the soil quickly shrivels up, so that the vegetables can only be grown for 1 year.

But between the vegetable plants are inserted salt-tolerant *Tamarix* cuttings, which easily take root. If in the second year the furrow does not receive water, but the soil is soaked to groundwater due to heavy irrigation in the previous year. As a result, the roots of *Tamarix* grow deeper and deeper over the next few years until they reach the groundwater. Trees then develop that are cut for firewood every 25 years, but sprout again from the stump as cane sprouts. All former vegetable land turns into a *Tamarix* forest in this way. It is thus possible to reforest deserts with groundwater at greater depths, if the first few years after planting the trees are so heavily irrigated that the whole soil is soaked down to the groundwater. However, the question of how long the groundwater supply will last remains open.

This example gives us the explanation that phreatophytes, which are bound to groundwater, reach it with their roots, although there are many feet of dry soil above them. They can only do this after very favourable rainy years, when the soil is soaked from the surface to the groundwater, but then they persist until the woody plants reach their age limit. This need not always be groundwater. Often it is only ground moisture, i.e. adhesive water, which is stored in the soil. As soon as it is deeper than 1 m, it is preserved for a very long time, provided that no or very few plants reach and consume it with their roots. Saline soils occur very frequently in the deserts, and especially in the depressions.

7.3 Substrate Dependent Desert Types

The desert biomes can be subdivided into biogeocene complexes according to soil characteristics, which were first studied in the Sahara. Therefore, mostly the local designations there were generally adopted.

> **Box 7.2 Desert soils as Hidden Water Reservoirs**
> The hidden water reserves in desert soils are greater than the superficial observer believes.

7.3.1 The Stone Desert (Hamada)

When the parent rock formed in the course of geological history is at the surface, it is called a rock desert. Such a one is quite seldom to be found, because arid mountains are often almost completely submerged in their own coarse debris by physical weathering. Coarse rock is also found especially on the elevations of mesas, from which all fine weathering products have been blown out, with strong wind erosion on all projecting rocks due to sand blowing (Fig. 7.7a). At the surface, the rock fragments accumulate. They form a stone pavement. The stones are often covered by dark desert varnish (Fig. 7.7b). This gives the landscape a somber appearance. Under the stone pavement there may be a water-repellent accumulation soil, rich in gypsum and salt if marine sediments are present, which prevents plant growth. Hamada surfaces are rugged by deep erosional valleys with steep slopes covered by debris (Fig. 7.8). Some plants can persist in the crevices and rocky clefts; they are not infrequently xerohalophytes.

Fig. 7.7 Hamada types in Egypt (**a**, photo: Breckle) and in Fuerteventura, Canary Islands (**b**, photo: Rafiqpoor). The stones of the Hamada are covered with desert varnish and are mostly gloomy and dark

Fig. 7.8 Fish River canyon in the desert in southern Namibia (Photo: Rudi Bosbouer, https://bit.ly/2zmkONe)

7.3.2 The Gravel Desert (Serir or Reg)

This occurs when the parent rock is heterogeneous, for example a conglomerate. The more easily weathered putty substance decays and is removed by wind. The hard pebbles in turn accumulate at the surface (Fig. 7.9).

These autochthonous gravel deserts are contrasted with the allochthonous ones, which are alluvial deposits of earlier rainy seasons from which the fine material was blown out. Beneath the gravel layer, darkened by desert varnish, there may be a hard crust caked by gypsum and lime. The particularly monotonous gravel desert is only slightly undulating. The shallow, broad valleys are filled with sand and are more conducive to plants gaining a foothold. Among them you can find plants of the sandy soil, but also xerohalophytes.

7.3.3 Sand Desert (Erg or Areg)

They form in the large basin landscapes where the sand blown off the elevations is deposited and contributes to the formation of dunes. If one wind direction prevails, then crescent dunes or barchans form, which slope flat on the windward side and steeply on the leeward side. They move along with the wind direction. If the wind direction changes periodically, only the crest of the dune is rebuilt each time, while the base remains fixed. The sand grains are coated with an iron oxide film on the surface, giving the dunes a bright orange or red colour in dry hot areas (Fig. 7.10), as in the Inner Namib. Near the coast, however, with higher humidity, the colouring is yellow-brownish, e.g. in the Outer Namib.

Movable and therefore vegetation-free dunes are water reservoirs, as rain penetrates easily and evaporates only to the smallest extent. Even with only 100 mm of annual rainfall, a groundwater horizon is formed, so that water extraction from wells is possible or the water escapes in the interdune area.

If the sand cover is not very thick, colonization by plants (non-halophytes, such as dune grasses, *Ziziphus* and others) can occur. Perennial species or shrubs then serve as sand traps. The plants grow back out of the sand deposited around them, so that new sand is always being accumulated. In this way, each plant forms a cluster dune (several meters high) called a **nebkha** (Fig. 7.11) The whole landscape is given a very typical character by these miniature dunes.

7.3.4 The Dry Valleys (Wadis or Oueds)

In S Africa these are called riviers (Fig. 7.12), in America also washes or arroyos. They are an important landscape element of all deserts. Their formation is usually to be sought in the past, when rainfall was higher (pluvial periods). But even today, every few years or decades so much rain may fall in the catchment area that a broad flood further furrows the wadi valley. The dry valleys begin as barely noticeable erosional gullies that merge into deeper ditches or valleys until they often merge into deep canyons.

The water running off after a rain deposits gravel and sand. The salts are partially washed out and the soil is deeply soaked; favourable growing conditions arise, especially for halophytic woody plants (*Tamarix, Nitraria*). In the large dry valleys, the bed is devoid of vegetation because the soil is redeposited by the infrequent water floods. Vegetation is confined to the margins protected from the floods and is more abundant the more water is stored in the alluvial deposits. Often a constant groundwater flow is present, then dense non-halophytic phreatophytic woody plants are found as extrazonal vegetation, often in a row (contracted vegetation).

Fig. 7.9 Serir in Morocco. The fine material has been blown out of the interstices of the boulders and pebbles and carried away. The stones have a desert varnish coating. This desert also looks gloomy and dark like the Hamada (photos: Rafiqpoor)

Fig. 7.10 Erg or Areg (sandy desert) usually occupy larger areas compared to the first two types of desert mentioned. In the Wahiba Desert in Oman, large areas are covered by sandy deserts with characteristic crescent dunes, ripple marks, etc. (photo: Breckle)

Fig. 7.11 The nebkhas are formed in the sandy deserts by the accumulation of sand on the perennial plants. As a result, mounds of sand form around each plant and are widely spaced. Here a *Salsola* nebkha at Cape Cross in the Namib with fog bank in the background of the picture (photo: Breckle)

Fig. 7.12 Wadis are the dry valleys in deserts that only carry water during a sudden rain event (river oases). More details in the text (**a**: Kuiseb-Rivier Southwest Africa, photo: Breckle; (**b**) Morocco, photo: Rafiqpoor)

Box 7.3 Hamada: Serir—Erg—Takyr—Sabkha
This is often a geomorphological sequence and succession of desert types corresponding in their arrangement to a large catena (here landscape links connected by processes of erosion and deposition) with a substrate sorted by grain size.

7.3.5   Pans (Sabkhas, Dayas or Schotts) and Takyr

These are the small hollows and depressions or large depressions in which the silt or clay particles carried by the water of the wadis are deposited. If these pans have an underground drainage (in karstified areas), then no bracking occurs.

The same is true of the **Takyr**, the delta-like formations at the mouth of the valleys, from which part of the water slowly drains away in a broad front after particularly heavy rainfall. Their heavy clay soils, however, are unfavourable sites; usually the water hardly penetrates the soil, which soon dries up again after a flood. Therefore, only algae, cyanobacteria, lichens or few ephemeral species grow predominantly on the takyr soils (Fig. 7.13).

If there is no drainage and all water evaporates from the basin, salt accumulation is the result. In such salt pans, i.e. halobiomes, solid salt layers form on the deepest parts. At the edge, where the salt concentration is lower, hygrohalophytes appear. Often the groundwater is less saline and salt crusts form only on the surface. If a thin layer of sand is deposited locally on the surface of such a salt pan, capillary rise and thus salt accumulation ceases. Plants settle on these sand deposits, which then serve as sand traps, which in turn creates a pile dune or **Nebkha landscape** around the pan.

7.3.6   Oases

The places in the desert endowed with dense plant growth where low-salt water comes to the surface in the form of common or artesian springs are called oases (Fig. 7.14), often also along dry valleys (Fig. 7.12b). Here hygrophilous species can grow. Today, such oases are all densely populated. The natural vegetation has been replaced by cultivated plants or weeds. Oases with strong springs are often adjoined by salt pans (bulkheads, sabkhas) in which the excess water accumulates and evaporates (southern Tunisia, Algeria).

Fig. 7.13 Dry takyr surface with mirage and heat flicker in the Verneukpan clay surface in the Great Karoo, South Africa. (photo: Breckle)

Fig. 7.14 The Rub Al Khali sand desert in Saudi Arabia and Oman is the largest sand desert in the world. Wide dune valleys, which may be loosely vegetated, are fringed by high dunes (photo: Breckle)

> **Box 7.4 What Is the Commonality of all Deserts?**
> As diverse as the individual deserts of the earth are, they all have one thing in common: The low density of plant cover.

7.4 Water Supply for Desert Plants

The great aridity of arid regions tempts researchers who do not know the desert from their own experience to assume that desert plants possess special physiological properties—a physiological drought resistance—which enables them to grow under arid conditions. In particular, the supposedly high cell sap concentrations that enable the plants to absorb water even from nearly dry soil are repeatedly emphasized. However, in-depth ecophysiological studies have shown that these views are incorrect. The water supply of desert plants is not as poor as one is inclined to assume on the basis of low rainfall. This is because rainfall in millimetres means litres of water per square metre of soil surface; one must therefore also calculate the transpiring surface per square metre of soil surface in order to assess the plant's water supply.

The landscape in the deserts is therefore not dominated by the plants, but by the bare rock. If one wants to determine the exact relationship between the amount of rainfall and the density of vegetation, one must compare plants of the same life form (for example, grasses or trees with similar foliage) and select an area in which the rainfall changes over a relatively short distance but the temperature conditions remain approximately the same; these should be flat, large areas with similar soil and the vegetation must not be disturbed by human intervention.

Suitable areas are SW Africa with grass cover at rainfall levels of 100 to 500 mm per year and SW Australia with eucalypt forests at rainfall levels of 500 to 1500 mm. The result of the corresponding studies was a linear function between rainfall depth and plant mass production, or the size of the transpiring area (Fig. 7.15). It also applies to creosote bush (*Larrea divaricata*) stands in SE California as well as to ephemeral vegetation of arid regions with annual precipitation up to 100 mm.

Only at first the grass seedlings consume 16 to 17 mm for the germination process and use the water less well than the perennial grasses, so that the straight line rises more shallowly.

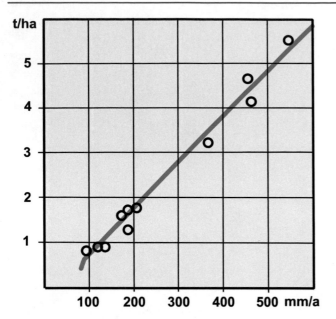

Fig. 7.15 Substance production (aboveground dry matter in t.ha^{-1}) of grasslands in southwestern Africa as a function of annual rainfall (in mm)

Box 7.5 Relationship of Water Supply and Transpiring Surface
It follows that the water supply in terms of the unit of transpiring area remains more or less the same in arid and humid areas (rainfall 100 to 1500 mm per year).

The drier an area is, the further apart the plants move, the more soil space each plant needs to absorb water.

This rule is confirmed in North Africa for olive tree crops: Intuitively, farmers reduce the number of trees per hectare in proportion to the decrease in rainfall, until eventually there are only 25 trees per hectare (Fig. 2.48). Yet the yield per tree remains essentially the same, a sign that its water supply is not changing significantly. It is also true for cereal crops that seed density must be lower as rainfall decreases. To be able to draw water from a larger soil space, the plant must have a larger root system.

The second essential characteristic is that with increasing aridity the plants reduce their transpiring surface more and more, but develop the root system more strongly. Indeed, it is found that when the cell sap concentration is increased, shoot growth is immediately strongly inhibited, while root length growth is even initially promoted. While in humid regions the greater part of the phytomass is above ground, in arid regions this applies to the underground part. In arid regions the roots often do not penetrate deeper into the soil, as is usually represented, but the root system becomes shallower and more widely branched. This is because the more sparse the rain is, the less deeply it soaks the soil. There is no water at all beneath the top layer of waterlogged soil for the plants to absorb. Only plants that are tied to groundwater (phreatophytes) or whose roots penetrate rock crevices have been observed to have very deep taproots. But this should not be generalized.

If we come to extremely arid areas with precipitation below 100 mm, the plant cover changes: The diffuse vegetation with an even distribution of perennial plants over an almost flat surface changes in extremely arid areas into a contracted vegetation, i.e. the perennial plants only grow in often hardly noticeable erosion gullies or depressions, while the higher surfaces remain vegetationless. This is related to the distribution of water in the soil.

In the extreme deserts, with the exception of mobile sand, the soils usually have a biological crust on the surface that is difficult to wet (Fig. 7.16) and is swollen by dew. As a result, rain, although infrequent but usually falling in downpours,

Fig. 7.16 In all deserts, especially in the misty coastal deserts (e.g. Namib or in the Atacama), a film of fine dust can develop on the soil surface which is impermeable to rainwater, mainly because a biological crust always develops which is hydrophobic (of cyanobacteria and unicellular algae), later also with mosses and lichens. Then the rare rainwater can run off superficially as here from sand dunes of the Negev (Photo: Breckle)

hardly penetrates the soil but for the most part runs off superficially even on sand (Breckle et al. 2008). The sandy erosion gullies and the depressions therefore receive much more water than corresponds to the rainfall, and this penetrates deeply into the soil. The plants here root as deeply as the soil is soaked, often several feet deep. Groundwater can even accumulate in places in the valleys.

Even in the desert near Cairo-Heluan with 25 mm/year of rain, vegetation is present in all valleys. Assuming that 40% of the rainwater drains into the low parts of the relief and that the latter account for only 2% of the total area, the same amount of water is available to plants at 25 mm rainfall by inflow at these growth sites as is available on a plain at 500 mm rainfall. In fact, it has been measured that the annual water output of the plant cover at such a site is 400 mm at Heluan by transpiration. The cell sap concentration of the plants increases only slightly even in the rainless summer, which is an indication of good water supply. The sandy depressions in the gravel desert at the Cairo-Suez Strait constantly contain 2.4% water already at a depth of 75 cm (wilting point 0.8%), so they never dry out and support sparse perennial vegetation. In individual erosion gullies, plant roots can reach depths of over 5 m. This depends on the moisture penetration. Regardless of the high aridity, the flora around Cairo still has 200 species.

Thus, the water supply of plants in the extreme deserts is also not as bad as is usually assumed. Where plants grow in the desert, there is at least at certain times always some water available, even if the soil looks superficially still so dry. The plants only have to have the ability to endure long periods of drought. This is made possible mainly by special morphological adaptations. There is no essential plasmatic drought resistance. Cell sap concentration is generally low (the halophytes excepted).

The principle of contracted vegetation has been used by the Berber population in S Tunisia since time immemorial for crops when there is 200 mm or less rainfall per year: Each small gully is provided with a dam impounding the runoff water (Fig. 7.17), and date palms or cereals or field beans are cultivated in the moist soil washed up in front of the dam.

Similar run-off agriculture in pre-Arab times by the Nabataeans has also been found in the Negev Desert. The old dams were rebuilt and the experimental cultivation of various crops was successful (Evenari et al. 1982).

7.5 Ecological Types of Desert Plants

People have called all plants that grow in dry areas xerophytes. This is not appropriate. This is because in every arid region there are sites which guarantee the plants a

Fig. 7.17 Shallow channel with several cross dams to accumulate rainwater and retain fine material as soil. The embankments here are planted with *Opuntia*, and the plots are newly sown with melons and field beans. Tunisia near Sidi Mansour (photo: Breckle)

permanently very good water supply, for example in the oases. In such locations species even of the humid tropics can grow. In the rainless desert near Aswan (Fig. 7.3), for example, coconut palms, mango, mate, papaya, batata, cassava, camphor tree, mahogany tree, coffee, pomegranate, and many species of the Indian monsoon forests are cultivated on an island in the Nile with artificial irrigation. In the dense stand, the microclimate is less extreme than in the open desert. Also, under natural conditions, plants can grow in groundwater-bearing dry valleys that are not exposed to water shortages and therefore have no adaptations to drought. Moreover, in most deserts there is at least temporarily a short wet season. It is lacking only in the Central Sahara, the Namib, and the Peruvian-Chilean deserts. Species that develop in these humid periods (therophytes, ephemerals) and survive the rest of the time as seeds (therophytes) or in the soil (geophytes = ephemeroids) also show no special adaptations to water shortage, apart from the strongly developed root system mentioned above.

A distinction between drought-evading and drought-bearing species is ecologically illogical. All endure drought, some as seeds (ephemerals) or tubers, or bulbs (ephemeroids), others in a latent life state such as the poikilohydric low plants (algae, lichens), but also a number of ferns *(Cheilanthes, Notholaena)* or *Selaginella* species and even flowering plants, of which *Myrothamnus flabellifolia* *(*Rosales) is the best known (Fig. 7.18). The succulents and xerophytes persist in a reduced-active state.

Xerophytes are the ecological groups that require some, albeit minimal, water uptake during drought, as they do not have large water reservoirs. They are three subgroups connected by transitions:

1. Malacophyllous xerophytes more characteristic of semi-arid areas. They have soft leaves that wilt in drought, with the concentration of cell sap increasing greatly; in prolonged drought they shed the leaves, leaving only the youngest leaf systems in the densely hairy buds. Typical examples are many labiates, Compositae and cistroses of arid regions.

2. Sclerophyllous xerophytes with small, hard leaves stiffened by mechanical tissues. They are found especially in areas with a long summer drought. They can reduce their transpiration to a minimum when water is scarce; cell sap concentration increases only under extreme conditions. Examples are the evergreen oaks, the olive tree and others.

3. Stenohydric xerophytes, which immediately close their stomata when water is scarce, thus preventing an increase in cell sap concentration; but this brings gas exchange and thus photosynthesis to a halt, i.e. the plants enter a state of starvation. During prolonged drought, the leaves of these species do not wither but turn yellow and fall off. Some non-succulent spurges can serve as an example, but most extreme desert plants belong precisely to this group.

Survival is achieved with incredible tenacity often only miserable cripples. Plants can become very old in the process, often a hundred years and more. Many branches die, but it is enough if some survive, which grow again after rain.

A group by itself are the succulents, which store water and during the drought consume this water very sparingly; their small sucking roots often die, so that during the drought there is no absorption of water from the soil at all. According to the organs in which the water taken up during the rainy season is stored, a distinction is made:

1. Leaf succulents *(Agave* and *Aloë* or *Cotyledon, Crassula, Sansevieria* and others), which may also become arboreal (Fig. 7.18b).

Fig. 7.18 (**a**) *Myrothamnus flabellifolius* (left) in latent life state (branches and leaves folded) and a *Notholaena* species (right), two poikilohydric species on the escarpment to the Namib Desert (photo: Breckle), (**b**) the arboreal *Aloe dichotoma* in the Namib with low summer rainfall (photo: Rafiqpoor)

Table 7.1 Life forms of succulents in the deserts of southern Africa (modified after von Willert et al. 1990)

Group	Lifestyle
1	Ephemeral (germination possible after every rain event)
	Annuals • Summer annuals (germination only during summer rains) • Winter annuals (germination only during winter rains)
3	Pauciennials (living a few years)
	Perennials (perennial, many years) • Geophytes
	• Flowers and leaves at the same time • Flowers and leaves at different seasons
4	• Root system persistent • Above ground only flowers
	• Above ground persistent plants • No green leaves other than cotyledons • With annual leaf change (rain green) • Periwinkle

2. Stem succulents (cacti, many *Euphorbia* species, *Stapelia*, *Kleinia*, *Aloë dichotoma* and others)
3. Root succulents with non-visible underground stores, such as *Asparagus* species, *Pachypodium* and others, but there are also some legumes with huge tubers in the sandy areas of the Kalahari.

A more precise classification of succulents has been given by v. Willert et al. (1990). They distinguish the types given in Table 7.1 on the basis of seasonal development.

> **Box 7.6 Desert Plants and Dry Season**
> In the deserts, it is less important for the plants to produce a high biomass, but rather to survive the droughts at all. There is no competition between the above-ground parts.

The cell sap concentration of all succulents is very low and does not increase, or increases only slowly, even with large water losses during long dry periods. This is because succulents simultaneously lose organic compounds (sugars, acids and others) as a result of respiration, so that the water content calculated on dry matter can remain unchanged. Many succulents are able to remain alive for more than a year without absorbing water. In many of them the diurnal acid metabolism (CAM—Crassulacean Acid Metabolism) has been proven, i.e. they open their stomata only at night, when transpiration losses are low, take up CO_2, whereby this leads to the formation of organic acids, so that the acidity of the cell sap increases strongly. During the day the stomata are closed and in light the CO_2 bound at night is assimilated, causing the acidity to decrease again. The necessary gas exchange takes place in this way with minimal water loss (Dinger & Patten 1974).

Among the annual **succulents**, summer annuals are predominantly C4 plants (for example *Zygophyllum simplex*), *while* the winter annuals are CAM plants (for example *Opophytum aquosum*).

A very important group in many deserts are the salt plants or halophytes. However, they are more bound to the occurrence of saline soils than to the climate. Their distribution often goes far beyond zonobiome boundaries.

7.6 Productivity of Desert Vegetation

If the individual plants limit their transpiring and at the same time photosynthetically effective surface in times of drought, production decreases. In the case of prolonged drought, it comes to a standstill. On the other hand, in good rainy years the plants develop more luxuriantly, but they cannot use all the water available. The surplus then benefits the ephemerals, which develop particularly strongly and represent, as it were, a vegetation buffer through which the large fluctuations in annual precipitation are compensated.

In bad rain years, the ephemerals almost do not develop or they are represented only by dwarf plants. If the reduction of the surface with the perennial species is not sufficient to achieve a balance of their water balance, large parts of the plants die, because the maximum $\pi*$ is exceeded. For survival it is sufficient if the shoot meristem of a branch remains alive and sprouts again after rain. With all woody plants of the deserts, one sees many dead branches as signs of earlier drought years. Reproduction by seed also occurs only after a good rainy year or when several follow each other, which is seldom the case more than once a century. It needs favourable conditions for germination forming seedlings and successful establishment of saplings. Young growth is therefore usually absent altogether. Under these circumstances it is hardly possible to give average values of production.

The leaf area index of perennial species is very far below 1 even in favourable years, and only very abundant ephemeral vegetation can achieve some production in good years.

In the desert near Cairo, the production of ephemeral vegetation has been determined, following a winter rainfall of 23.4 mm, which soaked the upper 25 cm of the soil. Of this amount of water, 68% was lost unproductively by evaporation; the transpiration of the ephemerals during the winter months corresponded to 7.3 mm, i.e. 32% of the amount of rain, which is 730 kg of water calculated per 100 m^2 of soil area. The ephemerals produced 9.834 kg of fresh matter or 0.518 kg of dry matter on the same area. This gives a transpiration coefficient of 730:0.518 = 1409, which is very high compared to the values of our crops in Central Europe

(400 to 700). This is due to the very low humidity in the desert.

Similar values were obtained by Seely (1978) for annual grasses in the Namib at very low rainfall. The zoomass in the desert and thus the secondary production is vanishingly small; however, the food chains as control loops for the ecosystem are not without importance in the desert.

As an example, we cite the special production studies on agaves and cacti. They were carried out in the westernmost part of the Sonora Desert in California with a summer drought.

(a)   Quantitative data (all mean values) are given by Nobel (1976) for *Agave deserti*, which also occurs in the eastern Sonoran Desert. For plants with a mean of 29 leaves it is stated: Length of leaves 30 cm, area 380 cm^2, weight of a leaf fresh 348 g, dry 47 g Stomata 30 per mm^2. Number of roots per plant 88, their length 46 cm, radially quite flat stroking, so that any rainfall can be used to absorb water.

Stomatal opening occurs during the rainy season (November to May) at a soil water potential of −0.1 bar on 154 to 175 days. If this potential drops to −3 bar at the beginning of the drought, then water uptake no longer takes place, but the stomata open at night on a further eight days. Then they remain closed; a diurnal acid metabolism (CAM) only takes place again after a rainfall.

In 1975, transpiration losses amounted to 20.3 kg per plant, which, converted to the rooted soil area, corresponds to a rainfall of 26.9 mm = 35% of the annual rainfall. The transpiration coefficient, i.e. the ratio of the amount of water transpired to the amount of dry matter produced (both in kilograms), was 25, i.e. very low, which means an extraordinarily economical water use.

Per plant 0.8 kg dry matter was produced in the year. Growth is therefore very slow and only older plants flower once and then die because all the plant's substance and water reserves are used up to produce the large inflorescence.

The following determinations confirm this (Nobel 1977a): The flowering old plant had 68 leaves, which were 4.1 cm thick when the inflorescence just became visible. After formation of the inflorescence they had shrunk, faded and were only 1.4 cm thick.

The entire leaves lose 24.9 kg of fresh weight and 1.84 kg of dry weight during flowering. Water uptake from the soil was not sufficient; 17.8 kg was received by the inflorescence from the leaves. The dry weight of the inflorescence was 1.25 kg and 0.59 kg of dry weight was respired. One flowering plant produced 65,000 seeds, 85% of which were destroyed by animals. Not a single seedling was found in an area of 400 m^2 where there were 300 agave plants.

Propagation by seed takes place only in favourable years, otherwise only vegetatively by runners. These numerical values make it understandable why agaves are hapaxanthic (monocarpic) species, i.e. they flower and fruit only once and then die.

(b)   An equally detailed production analysis was carried out in the same area with the globular cactus *Ferocactus acanthoides* (Nobel 1977b). This is also a species with diurnal acid metabolism (CAM), but in which the effort for the flowering organs is so low that it flowers every year (Fig. 7.25).

The 34 cm high plant with a diameter of 26 cm weighed 10.8 kg with a water content of 8.9 kg. Transpiration losses amounted to 14.8 kg in 1 year; in addition, 0.6 kg was used for transpiration and the build-up of generative organs. CO_2 assimilation produced 1.6 kg in 1 year, of which one third was respired. The measured annual growth was determined as 9%. The transpiration coefficient was 70, higher than in the agave, but still very low. The opening of the stomata was similar to that of the agave.

7.7   Desert Vegetation in the Different Floral Kingdoms

The conquest of the deserts by plants took place in prehistoric times, when the various floral kingdoms had already differentiated. The plant families or generally the taxa of the individual floral kingdoms have a different species population; consequently, the adaptations to the way of life under arid conditions have developed in different directions in the plants of the individual floral kingdoms. Not only are the deserts floristically different, but the life forms need not be the same, although many convergences occur (Walter & Breckle 2004).

7.7.1   Sahara

The Holarctic includes only the northern part of the largest subtropical desert—the northern Saharo-Arabian desert, which in the east passes gradually and directly into the Irano-Turanian and Central Asian deserts with cold winters. The boundary between the two can be drawn by the northern limit of distribution of the productive date culture (e.g., in north-central Iran). In this desert, the Chenopodiaceae are particularly abundant, partly due to the high prevalence of saline soils. Succulent *Euphorbia species* are found only in W Morocco (Fig. 7.19a). Most species are xerophytic dwarf shrubs, some of them broom-like shrubs. Grasses are

Fig. 7.19 The succulent desert with *Euphorbia officinarum* at Cap Rhir north of Agadir, Morocco (**a**: photo Rafiqpoor) and *Zygophyllum simplex* (**b**, photo: http://bit.ly/2fkTT4S) typical of shallow gullies even in the central Sahara near Aswan

represented only by xeromorphic forms with hard leaves: *Stipa tenacissima* and *Lygeum spartum* (transition zone), *Panicum turgidum, Aristida pungens* and others. After good winter rains, many ephemeral species appear.

In the vast Sahara, at least today, the central part is not an overlap zone between the northern winter rainfall regions on the Mediterranean and the southern tropical summer rainfall regions, but this central part is a largely rainless desert, an extreme desert with very rare rain events (Fig. 7.2). Nevertheless, although restricted to gullies and wadis, there is definitely still a flora, albeit species-poor. Small localized showers can suddenly cause some annuals to germinate in a narrowly defined area; *Zygophyllum simplex* (Fig. 7.19b) in particular then occurs sporadically.

The landscape forms are largely determined by the geologically given rock layers with their specific properties in relation to physical weathering (Fig. 7.20), often modelling out large blocks or even small inselbergs that are very dark due to desert varnish overlays.

Shrubs that are bound to moist sites, i.e. occur contracted in small gullies or wadis, would be *Tamarix, Nitraria* and *Ziziphus*. These are already more paleotropical elements, as are the *Acacia* species in the groundwater-bearing dry valleys.

Fig. 7.20 Extreme desert of the southern Egyptian Sahara south of Aswan (Egypt) with long-term mean annual precipitation of only 1–3 mm, dark boulder fields and rocky deserts (Hamada) with individual sand dunes (Erg) (photo: Breckle)

To the Palaeotropis belongs the southern Sahara with the Sahel, as a transition to the Sudanese summer rain area. Here, grasses *(Aristida, Eragrostis,* Paniceae), with less hard leaves play a much greater role. Shrubs and herbs are also more numerous *(Acacia, Commiphora, Maerua, Grewia, Calotropis, Crotalaria, Aerva* and others), which are also found in the Thar or Sind deserts.

7.7.2   Negev and the Sinai

They join the Sahara in the east as a bridge to the Arabian deserts. On the Sinai Peninsula, mountain deserts predominate, in which Irano-Turanian plants already occur at high altitudes. The northern Sinai and the Negev are characterized by extensive sand fields, with mobile sand dunes only when grazed heavily (Breckle et al. 2008).

Precipitation shows a very strong gradient from north to south, as shown in the precipitation map (Fig. 7.21).

The northeastern part of the Sinai Peninsula and the Negev Desert pass over the Arawa Depression rift valley, the Dead Sea and the Jordan Rift Valley to the Jordanian Desert. Ecological research has been very intensive in this area for several decades. The Negev Desert is therefore one of the best researched deserts (cf. Walter & Breckle 1991, Breckle et al. 2008).

As small as the Negev Desert is in terms of area, as great is its importance in floristic terms as a transitional area between different floral regions. Within a short distance, the Mediterranean vegetation from the north, the Iranian-Turanian vegetation from the northeast, the Saharan vegetation from the west and southwest, and the Arabian desert vegetation from the east meet here. In addition, there are even Sudanic enclaves, especially in the low-lying rift valley, for example with *Salvadora persica, Cordia gharaf, Maerua crassifolia. Cyperus papyrus* still occurs in the Huleh swamps on the upper Jordan River, where at the same time *Nymphaea alba* (as a Holarctic plant) reaches here its southernmost point.

7.7.3   Arabian Peninsula

In the same latitude as the Sahara, the Arabian Peninsula continues the desert belt eastward.

The precipitation is almost on the whole peninsula between 15 and 100 mm, on some steep slopes they go partly a little beyond 100 mm and in rainier mountainous areas above 2000 m 250 to 650 mm are measured. In the north of Yemen, a distinctive altitudinal sequence is recognizable with a rich vegetation, with evergreen hard-leaved bush forest, in which already numerous tropical genera occur, likewise in the southeastern Oman.

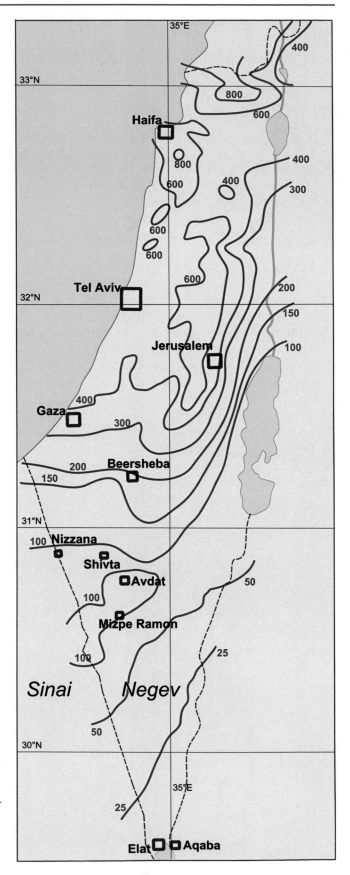

Fig. 7.21 Annual precipitation in the Negev Desert and in Israel. Note the considerable gradient from south to north

The eastern part of the peninsula is occupied by the Rubal Khali, a vast sandy desert area (Fig. 7.13). The same geomorphologically determined vegetation differentiation occurs here as in the Sahara. The vegetation is almost exclusively contracted. The larger wadis are characterized by rows of *Acacia*, among which a whole range of other woody plants, especially *Prosopis,* can also be found. In smaller depressions *Tamarix* species, *Calotropis procera* and *Calligonum comosum* occur. In the southern areas there are already transitions to *Acacia* thorn savannah (ZE III/II). Occasionally precipitation falls already in summer (for example in Sana or in Salala in Oman). Accordingly, succulent euphorbias and many other tropical-subtropical floral elements occur, such as *Adenium* (Fig. 7.22), *Jatropha,* etc. On the rocky slopes of the high mountains above 1500 m asl, woody plants include *Juniperus* spec. (Hall 1984, Fisher 2000) and sporadic *Olea europaea* (Fuellner 1997); on damp sites exposed to moisture, even mosses (Frey & Kürschner 1988; Kürschner & Böer 1999). Floristically, the flora of Yemen is the richest. Prominent floral elements of the mountains are the genera *Euphorbia, Euryops, Dodonaea, Themeda, Lavandula, Solanum, Abutilon* and others (Fig. 7.23).

7.7.4 Sonora

In N America only the deserts in S California and S Arizona can be counted as subtropical deserts, but with Holarctic flora elements. The arid areas in N Arizona, Utah and Nevada already have very cold winters (ZB VII).

Neotropical are several semi-deserts to desert areas: The Sonora Desert (N Mexico and southern Arizona) is located in N America, but floristically it already belongs to the Neotropics. About this desert (perhaps better semi-desert) extensive investigations are available, which were carried out at the Desert Laboratory in Tucson (Arizona). The stands of tall candelabra cacti *(Carnegiea gigantea)* are called "cacti forest"(Fig. 7.24). These succulents can store so much water that they can last for more than a year without taking up water (Fig. 7.25).

The cacti have very shallow roots. As soon as the upper soil layers are moistened, they form fine sucker roots within 24 hours and fill up their water reservoirs. But apart from succulent cacti, the other ecological types are also represented here: Winter and summer ephemerals, poikilohydric ferns, malacophyllous semishrubs *(Encelia),*

Fig. 7.22 *Boswellia sacra* grows in the southern Arabian Peninsula, Oman, Yemen and Socotra. It grows from a tuberous root resembling a lignotuber; sometimes this tree even forms stems that store water (photo: Breckle)

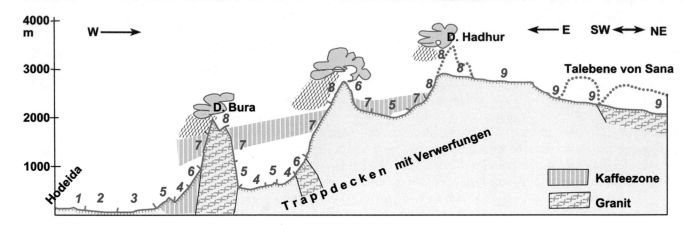

Fig. 7.23 West-east profile of vegetation in central Yemen (after H. von Wissmann 1972): (1) Halophilous desert vegetation; (2) Pile dunes; (3) Umbrella acacia stands; (4) Semi-desert on alluvions; (5) Semi-desert on rocky slopes; (6) Gallery and canyon forests; (7) Tropical evergreen scrub forest with lianas and succulents; (8) Hardwood shrubs; (9) Semi-desert, partly with Mediterranean species

sclerophyllous species, stenohydric and the rain-green *Fouquieria,* which forms new leaves after each heavy rain.

If there is a lack of water, many species turn yellow after a short time. Vast arid areas are covered by the creosote bush *(Larrea divaricata)*, which smells of creosote when the leaves are moistened by rain and is particularly drought-resistant. It is also characteristic of the Mohave Desert, which receives only winter rains and is poor in succulents.

Fig. 7.24 (**a-b**) *Carnegiea gigantea* in the Sonora Desert in Arizona, USA. The gigantic cacti form candelabra-like "trees" that can grow 10–15 m tall. The mechanisms of water uptake are shown in Fig. 7.25 (photos: Barthlott)

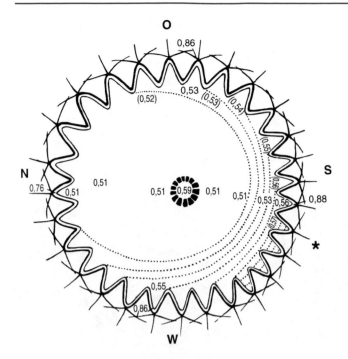

Fig. 7.25 Distribution of osmotic potential or potential osmotic pressure ($-\pi^*$) on the transverse section of *Ferocactus wislizenii*. Isosmoses = lines of equal pressure (cell sap concentration, numbers in atm, highest pressure at * in the southwest)

Larrea is usually joined by *Franseria*, a soft-leaved composite, but also many Opuntias with flat or cylindrical succulent shoots.

A *Larrea* desert also extends in the lee at the eastern foot of the high Andes for 2000 km from N Argentina to cold Patagonia. The main species *Larrea divaricata* is probably identical to that in Arizona (Fig. 7.26).

A monograph on the family Cactaceae as a well-studied model group of neotropical drylands, mapping the range of each cactus species (over 1400 species in total) and including diversity maps of family genera and species, was published in 2014. Quantitative analyses of the centres of diversity of cacti revealed that at the species level they are centred in the Chihuahua-Sonora Desert in North America and in Mexico, while at the genus level they are centred in South America and specifically in the Andes and northeastern Brazil (Barthlott et al. 2015).

The deep tectonic depression of "Death Valley" on California's border with Nevada is characterized by extremely high temperatures in summer. Air temperatures of up to 57 °C have been measured there. The basin is heavily salinized, in some places there are freshwater springs. *Tamarix* species colonize there, but also the very heat-resistant Amaranthaceae *Tidestroemia oblongifolia* (Fig. 7.27) and *Cleomella obtusifolia* (Capparidaceae), both of which have a photosynthetic optimum above 40 °C. The rocky steep slopes at the edge of the basin are a typical hamada, the bizarre rocks are covered with desert varnish (Fig. 7.28), very occasionally there are heat- and drought-resistant dwarf shrubs, such as the completely white-looking *Atriplex hymen-elytra*, in which the typical bladder hairs are glued together to form a radiation shield against overheating.

7.7.5 Australian Deserts

The arid areas in the Australis show very different conditions. The whole of central Australia is arid. The sand dune areas (Gibson desert, Simpson desert) have desert character, but they are not the climatically driest parts of Australia, and the "Gibber plains", bare areas with stone pavements caused by heavy overgrazing. The vegetation of the driest parts, with infrequent rainfall at any time of year, are the 'Saltbush' (Fig. 7.29) *(Atriplex vesicaria)* and the 'Blue bush' *Maireana (Kochia) sedifolia*, both Chenopodiaceae. They occur in pure stands, but also mixed (Fig. 7.30).

The soils under *Atriplex* contain little chloride, about 0.1% by dry weight. However, as the loamy soils dry out considerably, the concentration can be high. This is matched by the high cell sap concentrations of *Atriplex* (usually 4 to 5 MPa), with the proportion of chlorides reaching 60–70%. *Atriplex vesicaria* is thus a euhalophyte; growth is promoted by salt. Some salt recretion is possible due to the short-lived and continually regenerating bladder hairs. This semi-shrub lives about 12 years; like most halophytes it has weakly succulent leaves and a root system that extends far laterally over a calcareous crust at a depth of about 10 to 20 cm. The bushes are therefore quite widely spaced.

In contrast, *Maireana sedifolia is* said to grow very old and to have a deep root system that reaches down 3 to 4 m in the crevices of the calcareous crust, but also about as far laterally. The species grows where rainwater seeps deeper (lighter or stony soils). The cell sap concentration of this species is only half that of *Atriplex* and the chloride content is also much lower (about 20 to 40%). It is therefore possible that it is a facultative halophyte and comes to predominate as the climate becomes more humid.

In the salt bush area, sand dunes or sandy areas occur scattered with more favourable water conditions; the soil here is free of chlorides. Shrubs grow here *(Acacia, Casuarina, Eremophila)*.

The arboreal *Heterodendron* and *Myoporum* species, together with *Eremophila* and *Cassia* species, are attached to silty soils. The most important species of central Australia is *Acacia aneura*, known as "Mulga". It dominates on wide areas, which look like a grey sea from the airplane. The shrub reaches 4 to 6 m in height and has phyllodes covered with resin that are thinly cylindrical or somewhat flattened (Fig. 7.31). The root system is strongly developed and

Fig. 7.26 Semi-desert with
Larrea divaricata
(Zygophyllaceae) in Arizona,
USA (photos: M. Neumann)

penetrates through the hard soil layers about 2 m deep. With the irregularity of rainfall, flowering is not tied to any season, but only to rain. After heavy rainfall, the fruits and seeds ripen. At the same time, a flowering carpet of white, yellow and pink immortelles (everlastings, strawflowers), several

genera belonging to the Compositae (Fig. 7.31), then develops on the ground (Fig. 7.32).

Acacia aneura is sensitive to salt, but can tolerate long periods of drought. In dry sites, the bushes are widely spaced, while in moist depressions they form thickets. This species as

Fig. 7.27 *Tidestroemia oblongifolia* is resistant to overheating in Death Valley, USA (photo: Breckle)

Fig. 7.28 The bizarre rocks on the slopes of Death Valley (USA) are covered in desert varnish (photo: Breckle)

well as *Rhagodia baccata* and *Acacia craspedocarpa* were studied ecophysiologically (Hellmuth 1968, 1969).

Another important group is the hedgehog grasses *(Triodia, Plectrachne)*, which are grouped together as "spinifex grasslands". They have coiled, persistent, and very hard leaves with a resinous coating that terminates in a sharp point, and they form large, rounded cushions, hemispheres up to 2 m high in *Triodia pungens* (Fig. 7.33). We can classify these species among the sclerophylls.

Fig. 7.29 Salt Bush formation with single trees of *Eucalyptus* (with some emus) near Port Augusta in Australia (photo: Breckle)

Fig. 7.30 Blue Bush Formation from *Maireana (Kochia) sedifolia* in South Australia (photo: Breckle)

Triodia basedowii is found on sandy areas in the most arid part of Western Australia. Its dense root system goes 3 m vertically into the depth. Older pads break up into individual festoons. Other characteristic genera, represented by many species, are *Eremophila, Dodonaea, Hakea, Grevillea* and others. The arrangement of the vegetation is conditioned by the nature of the soil and by stratum floods after heavy rains, creating a complicated mosaic of vegetation.

The quaternary history, derived by Crowley (1994) from pollen diagrams of numerous lake sediments, reveals an increase in rainfall and concomitant decreased salinity for the Australian desert areas at the end of the last glacial period, which increased again 5000 years ago and became particularly pronounced after the arrival of European settlers.

7.7.6   Namib and Karoo

Of the South African deserts, the Namib and the Karoo are also palaeotropical. Occasionally, Capensian floral elements already occur. The Namib extends along the coast of SW Africa. This coastal Namib, rich in fog, must be distinguished from the southern Namib in the transition area to the Karoo, which as an actual desert lies between the southern winter rainfall area and the northeastern summer rainfall area and intermittently can have two rainy seasons.

The **Karoo** extends into the Oranje Free State. The two rainy seasons favour the development of innumerable succulents, on rocky sites with the larger *Euphorbia,*

Portulacaria and *Cotyledon* species as well as many small Crassulaceae and *Mesembryanthemum* s.l. species on quartzite veins (Fig. 7.34). The wide areas are covered with dwarf shrubs (mainly Compositae) (Fig. 7.35). In the dry valleys woody plants are found, such as *Acacia, Rhus, Euclea, Olea, Diospyros,* but also *Salix capensis.* In the transition area of the Upper Karoo, the grassland of the summer rainfall area is already growing on deep fine-grained soils, while Karoo succulents are still found on the shallow rocky areas (Fig. 7.36).

As an example of a biome of the zonobiome III and the subzonobiome of the fog deserts, the **Namib** on the coast of Southwest Africa will be discussed in more detail, because it differs strongly from the other deserts. Although it is a subtropical and extremely rainless desert, the coastal strip is characterized by high humidity with about 200 foggy days per year and low temperature fluctuations as in oceanic climate areas.

The mean annual temperature (16 °C) on the Namibian coast is about 5 K too cool for the latitude because of the cold Benguela Current. The quasi-stationary high-pressure area and the cold Benguela Current here, like the Humboldt Current on the Chilean coasts, cause a pronounced temperature inversion at about 600 to 1000 m asl with the formation of a blanket of high fog at the inversion lower boundary (Figs. 7.11 and 7.37). The thickness of the high fog is about 300 m and depends on the cooling from below. Inland, the temperature increases so that the fog cover dissipates about 50–70 km inland. The vegetation has adapted well to these

Fig. 7.31 Over 500 *Acacia* species occur in Australia. We present here some examples. (**a**) *Acacia cf. lasiocalyx*; (**b**) *Acacia cf. cuneata*, (**c**) *Acacia* sp., (**d**) *Acacia aneura*, (**e**) *Acacia maidlandii*; (**f**) *Acacia dyctiophylla*, (**g**) *Acacia tetragonophylla*; (**h**) *Acacia pyrifolia* (photos: Breckle)

Fig. 7.32 Mulga vegetation in the interior of Australia near Wiluna after rain. Large shrubs—*Acacia aneura*, small bush—*Eremophila* spec., ground densely covered with short-lived immortelles, such as *Waitzia aurea* and white *Helipterum* species (photo: Breckle)

ecological conditions. In the fog desert area (Fig. 7.38), a rich small-leaved succulent flora occurs. Outside the fog zone, where periodic summer precipitation intervenes, a stem succulent flora tends to dominate (Fig. 7.38).

The temperatures are always cool, hot days are few in the year. These strange conditions are caused by the cold Benguela current (water temperature 12 to 16 °C). Above it lies a 600 m high layer of cold air with a fog bank, so that due

Fig. 7.33 *Triodia pungens* grassland near Ayers Rock Australia is broken up into individual bushes (photo: Breckle)

Fig. 7.34 In the Karoo, an exceedingly exotic-looking leaf succulent formation of the Knersvlakte (a with *Oophytum nanum*) has developed on the weathered white quartzite veins from various *Mesembryanthemum* species and other genera of the Aizoaceae ("Living Stones"), etc. Some examples: b: *Mesembryanthemum crystallinum;* c: *Malephora purpureo-crocea* (Aizoaceae); d: *Drosanthemum diversifolium;* e: *Argyroderma delaetii;* f: *Mesembryanthemum nodiflorum* (Aizoaceae) (photos: Rafiqpoor)

Fig. 7.35 Great Karoo near Laingsburg (South Africa) with succulent euphorbias, *Rhigozum obovatum, Rhus burchellii* and dwarf shrubs (photo: Breckle)

to the inversion the warm easterly current does not reach the ground. Rather, a sea breeze sets in daily from the southwest, bringing the fog and cool air into the desert (Logan 1960, Besler 1972).

When the inversion layer is broken, thunderstorms form and rain falls, which is the case in very few years. Exceptions are rare, heavy rains only once or twice in a century, such as in 1934/35 with 140 mm of rain and in 1975/76 with over

100 mm and in 2006 and 2011 there was also abundant rainfall. The long-term annual mean of 16 mm for Swakopmund (other sources: 28 mm) therefore means little (Fig. 7.38).

The humidification of the soil by dew or fog is minimal, on average 0.2 mm, maximum 0.7 mm per day; the annual sum of fog precipitation of about 40 mm remains ineffective because the individual fog precipitations evaporate again

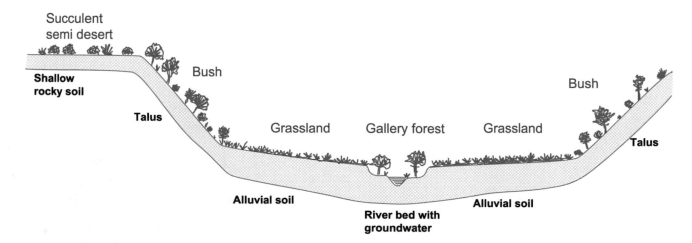

Fig. 7.36 Vegetation profile through a valley of the Upper Karoo near Fauresmith (South Africa). Plant cover structure conditioned by differences in soil. Scrubland with *Olea, Rhus* and *Euclea*

Fig. 7.37 A dense blanket of fog develops on the Peruvian-Chilean coast due to the effect of the cold Humboldt Current. It is so saturated with moisture that water is exuded from it when it comes into contact with objects. This fog humidity leads to the development of fog-adapted flora on the fog coasts of the earth (Southwest Africa, Northwest Africa) (photo: Rafiqpoor)

without being stored by the soil. They benefit only the poikilohydric lichens and soil algae (Fig. 7.39b), which cover all the rocks in the fog zone with variegated colours when the humidity is high, as well as the **window algae** found on the underside of transparent quartz pebbles (Fig. 7.39a), where the fog moisture is retained longer. True mist plants, like the tillandsias in the Peruvian desert, which do not draw water from the soil, do not exist in the Namib.

Only where the drifting mist collides with a rock face does water condense and penetrate deep into the crevices. There plants (mostly succulents, Fig. 7.40) can establish themselves. This is the case with the inselbergs that rise above the almost flat hull platform of the Namib.

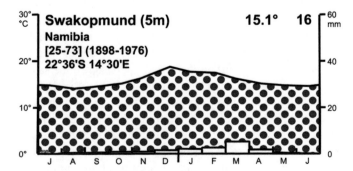

Fig. 7.38 Climate diagram of Swakopmund in the Namib. Almost rainless area, but with 200 foggy days per year (not measurable precipitation)

This hull plain rises with a gradient of 1:100 from the coast to the east and has a width of 100 km to the foot of the escarpment from the African highlands (Escarpment). The fogs are noticeable up to 50 km inland. They also contain droplets of seawater sprayed by the surf, which come to be deposited so that the soils of the outer Namib are brackish.

Perennial plants are only found in the Namib where the soil contains water at a depth of less than 1 m. These water supplies come from good rain years. After the 140 mm of 1934, the desert was green and covered with flowers (Fig. 7.41a). These were mainly ephemerals, including especially many succulent *Mesembryanthemum*-species and other Aizoaceae, Portulacaceae etc. (Fig. 7.41b). These stored so much water in the shoot that they still flowered and fruited the next year, although the root and the shoot base had already dried out (Fig. 7.42). In them, almost the entire supply of assimilates and of water is used to form the fruits and seeds (V. Willert et al. 1990). Many seedlings also grow from the perennial species in such years, their roots penetrating rapidly into the depths and reaching the lower soil layers that remain moist longer. However, they can only survive for the next decades where larger water reserves are stored in the soil.

After rare heavy rains, the water flows in wide, sand-filled channels, the Riviers (wadis) to the sea, without reaching it. Rather, it seeps into depressions filled with alluvial soil and penetrates deep into the ground. Only the upper soil layers dry out to a depth of 1 m (less deep in sandy soil). Below that, the water persists for decades and can be

Fig. 7.39 In the Outer Namib in SW Africa, window algae occur here and there on the underside of quartz pebbles in the area of fog influence (**a**). Migratory lichens *Omphalodium convolutum* (**b**) are concentrated in depressions and gullies (photos: Breckle)

exploited by deep-rooted plants. In the gullies, the sand is desalinated by the rainwater run-off; in the depressions, on the other hand, the salt is washed in. This results in two distinct sites—in the small and large erosion gullies with non-halophilic biogeocoenoses *(Citrullus, Commiphora, Adenolobus and*, where there is more groundwater, the shrubs *Euclea, Parkinsonia* and *Acacia* spp.), while on the wide flat depressions halophilic species settle. These are mainly *Arthraerua* (Amaranthaceae), *Zygophyllum stapffii* and *Salsola* species (Chenopodiaceae), with sand blown onto each plant from which it grows above. Low mounded dunes form, creating a typical Nebkha landscape (Fig. 7.43; Fig. 7.11). It can be assumed that all the plants germinated in the same rainy year; they are also fairly equal in size and can last as long as the water supplies in the soil last; if no new rainy year comes for a long time, they slowly die and the

dune sand is blown away. If, on the other hand, they receive good rain again in time, they continue to grow.

For the survival of these plants, fog (caused by cold ocean currents) plays a major role; because in water-saturated air, the plants can assimilate CO_2 without significant transpiration losses. Their water consumption is thus low. *Arthraerua* is now thought to be able to absorb fog moisture from the air with specially constructed, low-lying stomata at the end of capillary passages (Loris 2004).

Apart from the three biogeocoene complexes with salt-free sandy soil and the brackish depressions near the coast, the oases of the large Rivier valleys (dry valleys) should be mentioned: Omaruru, Swakop and Kuiseb in the Central Namib (Fig. 7.44), in the northern Namib: Ugab, Huab, Uniab, Hoanib, Hoarusib and Khumib up to the Angolan border at the Kunene. In the sand Namib, only the Kuiseb (border river between rock and sand Namib) reaches the Atlantic Ocean, Tsondab and Tsauchab seep into the sand masses beforehand.

The Riviers all originate on the highlands with summer rains (average 300 mm) and are partly cut deep into the Namib platform. The riverbed is filled with sand, in which the water seeps away after rains on the highlands and only after very good rains it flows down to the sea. But the rest of the time there is a constant flow of groundwater in the sand, so that water can be obtained from wells. Partly it is slightly brackish due to the inflows from the Namib. This groundwater creates the possibility for the development of gallery forests of *Acacia albida, A. erioloba, Euclea pseudebenus, Salvadora persica* or in somewhat brackish places *Tamarix* and *Lycium* species. Where the woody plants are protected from the high tides, the forests can reach a great age. On the often shifted sand grow *Ricinus, Nicotiana glauca, Argemone, Datura* and others, on sand dunes the thorny

Fig. 7.40 *Hoodia currorii* flowering in front between white marble rocks (Witport Mountains, Namibia) (photo: Rafiqpoor)

Fig. 7.41 The flowering Namib Desert in Namaqualand in October 2008. The gently sloping slopes and the entire area is transformed into a sea of flowers of annual Asteraceae, Mesembryanthemaceae etc. after the first episodic rains (see also Cowling et al. 1999) (photos: Rafiqpoor)

and leafless *Acanthosicyos* (Naras gourd) and *Eragrostis spinosa*—a woody thorny grass; where the ground water forms pools, *Phragmites, Diplachne, Sporobolus* and *Juncellus* stand.

All these plants are abundantly supplied with water and have a high productive power. These oases are also rich in animal life: Birds, rodents, reptiles, arthropods and others. Even today, elephants (Fig. 7.45) and giraffes roam the Rivier valleys. Elephant and rhinoceros used to be abundant. They have been almost exterminated by man. Only the baboons have persisted in the rocky gorges.

The fauna of the Nebkha landscape is poor. There are: Some rodents, reptiles and scorpions, as saprophages and beetle species. More species are found in the inselbergs, especially if they are further inland and already receive frequent summer rains, so that there are waterholes between the

rocks and shrubs can grow in crevices. In the Sand Namib, too, the fauna is much richer in species.

The given description referred to the outer Namib. As soon as one moves further than 50 km from the sea, the inner Namib begins with sparse summer rains and changing grass growth. The desert conditions are not so extreme and give the mobile game the opportunity to find food and visit individual waterholes. This part is rich in game. The most common animals are: Zebra, oryx antelope, springbok, hyena, jackal, occasional lion as well as ostrich and other birds. This is because this uninhabited area has been declared a nature reserve in the central Namib; it is explored from the Namib Desert Station Gobabeb.

In the central Namib, on the border between the outer and inner Namib, the famous *Welwitschia mirabilis* occurs in numerous specimens. It grows in wide and very shallow erosion gullies with hardly noticeable slopes (Fig. 7.46), where the sparse summer rains converge and penetrate deeper into the soil. *Welwitschia* absorbs this water with its roots, which reach well over 1.5 m deep. Underneath is a hard crust of lime. If this species is absent in the deeper erosion gullies, it is probably because *Welwitschia* seedlings are very sensitive to flushing water and to being filled in with sand. At present it only rejuvenates in the northern Namib.

Welwitschia has only two ribbon-shaped leaves, which constantly grow from a meristem on the turnip-shaped stem and dry up at the tip. In good rainy years the living part is quite long, in bad ones the leaves dry up almost to the meristem, so that the transpiring surface is greatly reduced, whereby the transpiration falls almost to zero. The leaves are very xeromorphic anatomically and have sunken stomata. There is no evidence of dew uptake. An age determination with the C14 method gave an age of about 2000 years for the oldest measured specimen. The wood shows annual rings and tracheae.

Fig. 7.42 *Mesembryanthemum cryptanthum* in the Skeleton Coast near Möve Bay, Namib even after many months of drought with thick-fleshy leaves and fruits (photo: Breckle)

Fig. 7.43 *Arthraerua leubnitzia*-Nebkha (Amaranthaceae) in the Namib near Swakopmund (photo: M. Loris)

Transpiration and photosynthesis were studied by v. Willert et al. (1982): *Welwitschia* is a C3 plant; the water consumption of a medium-sized plant is about one litre per day. Calculated on the rooted area, this would correspond to a rainfall of 2 mm per year. Thus, water supply is guaranteed even in this arid area. They are pollinated both by the wind and by a species of bug *(Probergrothius sexpunctatis)* that feeds on the nectar of the female flowers (Fig. 7.47).

Fig. 7.44 The dry riverbed of the Kuiseb (Wadi, Rivier) near Gobabeb with tree population of *Acacia albida, A. erioloba, Tamarix usneoides* and *Salvadora persica*. In the background the mighty dunes of the Sand Namib (photo: Breckle)

Fig. 7.45 Big game at the waterhole in the Etosha Pan in northern Namibia (**a**); elephant family in the Huanib Rivier, Namibia (**b**) (photos: Breckle)

Unique are special ecosystems of the Namib:

1. The almost vegetationless dunes south of Kuiseb (Fig. 7.43)
2. The Guano Islands
3. The mating grounds of the seals
4. The lagoons behind the beach.

In the dune valleys, organic detritus is found from blown-in grass remains, protein-rich animal remains and perished insects (butterflies). The detritus is eaten by psammophilous wingless tenebrionids (black or darkling beetles), which in turn are eaten by small predators (spiders, solifuges) or by larger lizards, snakes living in the sand and goldmouths (Chrysochloridae) (Kühnelt 1975).

Since the sand heats up to over 60 °C during the day, almost all animals hide in the cool sand and only come out at night. The water source is the mist, which they ingest in a special way (Seely & Hamilton 1976, Hamilton & Seely 1976). Some species have comb-like appendages on the hind legs with which to comb out the mist droplets, while others stand perpendicular to the wind and suck up the mist droplets, which condense on the hind legs and abdomen and then drip towards the head (Fig. 7.48). The fauna is rich in endemics.

The guano islands are the nesting sites of, among others, the cormorants, which find their food in the cold seawater rich in fish. In the rainless climate, the birds' excrement accumulates and prevents any plant growth, but it is decomposed as guano (phosphate fertilizer). Similar

Fig. 7.46 *Welwitschia mirabilis* on the Welwitschia Vlakte between Khan and Swakop Rivier (photos: Breckle)

Fig. 7.47 Male (**a**) and female inflorescence (**b**) of *Welwitschia mirabilis* and its pollinator the bug *Probergrothius sexpunctatis* (**c**) (photos: Breckle)

conditions prevail on the mating grounds of the seals (Cape Cross).

The lagoons are cut off from the sea by sand bars, with only occasional waves breaking over during storms. The evaporated water is replaced by seawater that seeps from the sea through the sand. They are therefore aquatic ecosystems with very high concentrations of salt, which we will not go into details. Like the Namib, each desert has its ecological specificity and must be treated monographically (cf. Walter 1973 and Walter & Breckle 2004).

7.7.7   Atacama

The Peruvian-Chilean coastal desert is very strongly divided into subregions (Fig. 7.49). In its extreme part it is as rainless as the Namib, but the fog here has a greater effect only in the coastal region, because the coast rises steeply in places. Here the only known true fog plants among the flowering plants are tillandsias (Bromeliaceae), which cannot absorb water from moist air like lichens, but nevertheless suck in the condensation droplets during fog directly with special scales

Fig. 7.48 A mist-catching tenebrionid on Namib sand dunes in the early morning (photo: M. Seely)

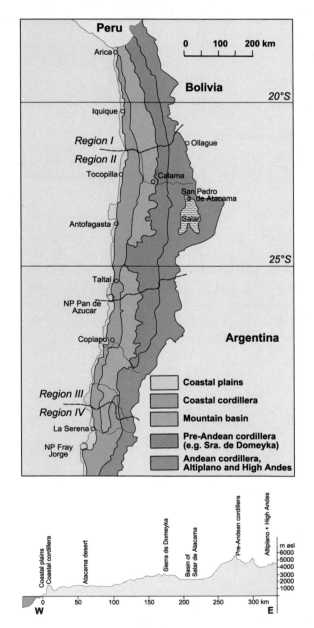

Fig. 7.49 General map (top) and transect (bottom) of northern Chile and region of the Atacama Desert proper between the Pacific Ocean and the Andes (modified after Wickens 1993)

Fig. 7.50 *Tillandsia straminea* (**a**), *Tillandsia purpurea* (**b**) in the Atacama Desert of southern Peru forms festoon structures (**c**) but without forming roots in the soil. They comb out the water they need from the fog with special suction scales (photos: Rafiqpoor)

on their leaves. They sit as epiphytes on columnar cacti or lie loosely as rosettes on the sandy soil (Fig. 7.50).

At an altitude of 600 m, the blanket of fog known as "Garua"in Peru lingers for months during the cooler season. The soil of the slopes is so heavily wetted that a carpet of herbs—the "Loma vegetation"—develops, which is grazed. Woody plants are absent, but were formerly present. Under planted eucalypts, dripping of condensed fog could collect amounts of water equivalent to 600 mm of precipitation. In the coastal cordillera itself in N Chile, sporadic, mighty columnar cacti (*Echinopsis atacamensis*) up to 8 m high are found, densely covered with lichens (Fig. 7.51), but only on slopes exposed to the fog. Further south at Fray Jorge (now a national park in central Chile), even a true cloud forest occurs.

Fig. 7.51 *Echinopsis atacamensis* (**a**: on an island in the Salar de Uyuni, Bolivia) grows about 8 m high in the Atacama Desert. When exposed to fog, they are used as a base by epiphytic tillandsias and lichens (**b**: Chile). Salt lakes have formed in the closed basin landscapes (photos: Breckle)

In N Chile in the area of the large saltpetre deposits, shielded from the coastal fog by the coastal cordillera, the desert is devoid of vegetation. Plant populations and cultures can only be found along the river courses, which are fed by the snowfields of the high Andes.

The inner basins lie at higher altitudes. However, up to the high altitudes of the Andes and into southern Bolivia, they are characterized by huge salt pans (Fig. 7.51a): Salars in which not only NaCl but a number of other minerals have accumulated (probably due to the extremely active volcanism and the arid climate). The extreme conditions allow only a few species to make a meagre living (Fig. 7.52). Only above 3500 m, where even occasional summer rains occur, is there a puny dwarf shrub semi-desert (with *Baccharis, Fabiana, Parastrephia,* etc.), which from 4100 m changes into the tussock grass mountain desert (Ichu grass: *Festuca chrysophylla, F. orthophylla, Stipa venusta*), in which lama and guanaco, but also nandu, graze.

For the western slope of the Andes in N Chile Ellenberg (1975) gives a perarid full desert up to the montane belt, then a subalpine dwarf shrub semi-desert and above 4500 m asl in the alpine altitudinal belt a tropical andine grass semi-desert or "desert puna". But even between 5200 and 5500 m there are still dwarf shrubs, for example at the volcano Ollagué (approx. 5900 m) and in the lava debris even occasional shrubs or small trees of *Polylepis tarapacana* up to 4 m high (Wickens 1993). A snow line is hardly detectable (Fig. 7.53).

Fig. 7.52 Salt-encrusted extremely halophilous cushion-shaped *Salicornia pulvinata* in the Salar regions of Bolivia (photo: Breckle)

Fig. 7.53 In the highlands of the Andes on the eastern edge of the Atacama: Ollague volcano (5900 m) (**a**). Mountain flank with high altitude desert. Even at 5800 m there are hardly any remnants of snow: the area is so dry that a climatic snow line cannot be defined. (**b**) At Sajama Volcano (6542 m), which lies somewhat further north in the Bolivian altiplano, the upper forest line rises to 5300 m (**b**), a lower forest line is at about 4400 m (Fig. 5.46a) (photos: Breckle)

7.8 Orobiome III: The Desert Mountains of the Subtropics

In extreme deserts, the air contains so little water vapor that there is no uphill rain even at high altitudes. We have already met orobiomes in the previous sections as well.

In the Tibesti Mountains (3415 m asl) in the Central Sahara, only annual precipitation of 9 to 190 mm has been measured at 2450 m altitude (4 years) with frequent cloud cover in the winter months. Accordingly, arid conditions persist to high altitudes; but the occurrence of a number of Mediterranean elements suggests somewhat humid conditions. *Erica arborea* was found in gorges at 2500 to 3000 m asl, and *Olea cuspidata, a* wild form closely related to the olive tree, was found as a relict in the Hoggar Mountains at 2700 m.

In the sequence of stages in the much less arid Sonoran Desert in S Arizona, one finds a belt with *Prosopis* grass savannahs and many leaf succulents (*Agave, Dasylirion, Nolina*) above the *Larrea* or giant cacti desert, then several elevational belts with evergreen *Quercus* species and *Arctostaphylos, Arbutus*, and a *Juniperus* shrub layer, followed by coniferous forest belts: *Pinus ponderosa* ssp. *scopulorum* (higher with *Pinus strobiformis*), *Pseudotsuga menziesii* with *Abies concolor,* and only on San Francisco Peak in N Arizona on north-facing slopes to nearly 3700 m asl *Picea engelmannii*. Here, annual precipitation increases very rapidly with elevation. This varies greatly among

deserts. However, this does not apply to the altitudinal zones in the Andes on the Atacama side (p. 259).

7.9 Man in the Desert

The inhospitable conditions make it seem astonishing that people have lived in all deserts, in some cases for a very long time. They have adapted their way of life, almost always travelling as nomads in order to maintain a livelihood in a larger area (Fig. 7.54). Settlement was in each case confined

Fig. 7.54 Bedouin tents in the southern Egyptian Sahara, at Wadi Allaqui, today near the eastern bank of the Nasser reservoir of the Nile. Supplies are stored on stilts (photo: Breckle)

to the oases; these therefore served as base stations for the long migrations. Livestock served as a food reserve (pastoral nomads with sheep and goats) and the camel as a versatile transport and livestock animal.

There are very different ethnic groups living in the desert. They are groups of people that have to keep moving in caravans in search of places with water and food, defying the greatest risks: sandstorms, silted up wells and loss of bearings due to the lack of points of references. Some of these peoples among others are the Berbers of North Africa, that include the Kabilis and the Tuaregs, the Bedouins of the Arabic deserts, the Bejas in Namibia, the Sans in the Kalahari desert and the Australian Aborigines.

In the peripheral areas of the deserts, as well as in the mountains, simple agriculture was possible as rainfed agriculture (run-off, Lalmi). Irrigation cultures became the basis of developing early cultures only in the area of the large foreign rivers (Egypt: Nile; Mesopotamia: Euphrates and Tigris).

7.10　The Zonoecotone III/IV: The Semi-Deserts

Where at the edge of the deserts, as a result of the increasing winter rains, the contracted vegetation changes into a diffuse one, the boundary between the desert proper and the semi-desert can be drawn. It is not, however, always sharply marked. The ground cover in the semi-desert is up to about 25% of the total area. The floristic composition of this vegetation is as different in the individual floral kingdoms as that of the deserts. North of the Sahara the most important species are the malacophyllous *Artemisia herba-alba* and the sclerophyllous grasses *Stipa tenacissima* (half-grass) and *Lygeum spartum* (esparto grass). *Artemisia* grows mostly on heavy loess soils or clayey soils. In Tunisia, calcareous precipitates were found at a depth of 10 cm. Dense rooting was present at 5 to 10 cm depth, with individual roots going as deep as 60 cm. *Stipa* tends to grow on elevations covered with stone pavement. A soil profile shows the following: 2 to 5 cm of stone pavement, with loamy soil well rooted to 30 cm below, followed by a firmly crusted gravel that appears to be an obstacle to roots, but probably also provides a water reservoir (lots of capillary water that can be absorbed by roots through close contact). The tufted roots originating from the base of the tussock sweep widely horizontally at a depth of 10 to 20 cm, so that the 0.5 to 1 m (2 m) apart tussock touch each other with their root systems. In both cases, scattered *Arthrophytum* plants are found between them. The soils are not brackish, *Lygeum spartum*, on the other hand, is characteristic of gypsum soils and also tolerates some salt.

Stands of half-grass are cut and provide material for wickerwork, for making coarse rope or for papermaking. *Stipa tenacissima* is distributed from SE Spain and E Morocco only as far as Al-Khums in Libya; the natural habitat is sparse Aleppo pine forests. *Artemisia herba-alba* s.l. also occurs in

Fig. 7.55 Mallee formation in Australia with shrubby eucalypts (photo: Breckle)

the Near East; it has spread in many cases at the expense of former grassland as a result of overgrazing.

With further increases in precipitation, solitary trees appear in northern Africa, such as *Pistacia atlantica* in the west and *P. mutica* in the east, or *Juniperus phoenicea*. The sparse tree stands finally lead over to the hardwoods (ZB IV).

In California, *Artemisia californica* occurs in the transition zone along with semi-shrubby *Salvia* and *Eriogonum* species (Polygonaceae).

In N Chile, in the transition zone, one finds a dwarf shrub semi-desert with Compositae *(Haplopappus)* as well as columnar cacti and *Puya* (large Bromeliaceae), after which a savannah with *Acacia caven* begins: the grass layer is now formed by annual European grasses.

In S Africa, the Renoster formation (Renosterbos with *Elytropappus rhinocerotis,* Asteraceae) can be considered typical of the low rainfall winter rainfall area.

In Australia, the Mallee formation forms the transition (Fig. 7.55), consisting of shrubby *Eucalyptus* species whose branches arise from an underground, tuberous stem (Fig. 7.56). However, sparse stands of *Eucalyptus* with *Maireana sedoides* understory may also occur.

Fig. 7.56 In Australia, *Eucalyptus* species in many places form a mighty lignotuber from which new shoots emerge after fire (photo: Breckle)

Photo 20   Sonora-Desert (ZB III) in SW USA and Mexico, with some few summer-rains; it is a floristic rich desert with many cacti (*Carnegiea gigantea*, *Ferocactus* spec.), at the left with Ocotillo (*Fouquierea splendens*), and chamaephytic various small shrubs (photo: Barthlott)

Photo 21 Atacama-desert (ZBIII) in S Peru and N Chile is almost rainless, but exhibits rather often cool foggy days, here with Werner Rauh in between (*Copiapoa cinerea*)-cacti (photo: Rauh)

References

Barthlott, W., Burstedde, K., Geffert, J.L., Ibisch P.L., et al. 2015: Biogeography and Biodiversity of Cacti. Schumannia 7: 205 S.

Besler, H. 1972: Klimaverhältnisse und klimamorphologische Zonierung der zentralen Namib. Stuttgarter Geogr. Stud. 83

Breckle, S.-W, Veste, M., Yair, A. (eds.): 2008: Arid dune ecosystems–The Nizzana Sands in the Negev desert. Ecol. Stud., vol. 200, 475p.

Cowling, R., Esler, K., Rundel, P. 1999: Namaqualand, South Africa–an overview of a unique winter-rainfall desert ecosystem. Pant Ecology 142: 3-21

Crowley, G. M. 1994: Quaternary soil salinity events and Australian vegetation history. Quarternary Science Reviews 13: 15-22

Dinger, B.E. & Patten, D.T. 1974: Carbon dioxide exchange and transpiration in species of Echinocereus (Cactaceae) as related to their distribution within the Pialeno Mountains, Arizona. Oecologia 14: 389-411.

Ellenberg, H. 1975: Vegetationsstufen in perhumiden bis perariden Bereichen der tropischen Anden. Phytocoenologia 2: 368-387

Evenari, M., Shanan, L. & Tadmor, N. 1982: The Negev. The challenge of a desert. 2nd Cambridge, MA, 437 p.

Fisher, M. 2000: Dieback in the montane woodlands of Arabia: a conservation matter of gravest concern. In: Abuzinada, A.H. & Joubert, E. (eds.): Proceed of the workshop on the conservation of the flora of the Arabian Peninsula. NCWCD, Riyadh; IUCN, Gland: 86-92

Frey, W. & Kürschner, H. 1988: Bryophytes of the Arabian Peninsula and Sokotra. Floristics, phytogeography and definition of the xerothermic Pangean element. Stud in Arabian bryophytes 12. Nova Hedwigia 46: 37-120

Fuellner, G. 1997: First observation of Olea cf. europaea (the Wild Olive) and Ehretia obtusifolia in the Arab Emirates. In: Tribulus 7 (1): 12-14

Hall, J.B. 1984: Juniperus excelsa in Africa; a biogeo-graphical study of an afromontane tree. J. Biogeo-graphy 11: 47-61

Hamilton III, W.J. & Seely, M.K. 1976: Fog basking by the Namib Desert beetle, Onymacris unguicularis. Nature 262: 284-285

Hellmuth, E. 1968, 1969: Eco-physiological studies on plants in arid and semiarid regions in W Australia. I., II., J. Ecol. 56: 319-344; 57: 613-634

Kühnelt, W. 1975: Beiträge zur Kenntnis der Nahrungsketten in der Namib (Südwestafrika). Verh. Ges. f. Ökologie/Wien 4: 197-210

Kürschner, H. & Böer, B. 1999: New records of bryophytes from the southern Musandam Peninsula and Jebel Hafit (United Arab Emirates). Studies in Arabian bryophytes 23. Nova Hedwigia 68: 409-419

Logan, R.F. 1960: The Central Namib Desert, South West Africa. Publication 758, 162 S. Nat. Ac. Sc., Washington D.C.

Loris K. 2004: Nebel als Wasserressource für den Strauch Arthraerua leubnitziae. In: Walter,H., Breckle. S.-W: Ökologie der Erde, Vol. 2, p. 485-489

Nobel, P.S. 1976: Water relations and photosynthesis of a desert CAM plant Agave deserti. Plant Physiol. 58: 576-582

Nobel, P.S. 1977a: Water relations of flowering Agave deserti. Bot. Gaz. 138: 1-6

Nobel, P.S. 1977b: Water relations and photosynthesis of a Barrel Cactus, Ferocactus acanthoides in Colorado Desert. Oecologia 27: 117-133

Seely, M.K. 1978: Grassland productivity. S. Afric. J. of Sci. 74: 295-297

Seely, M.K. & Hamilton III, W.J. 1976: Fog catchment sand trenches by Tenebrionid beetles, Lepidochora, from the Namib Desert. Science 193 (4252): 484-486

Walter, H. 1973: Die Vegetation der Erde, Bd. I: Tropische und subtropische Zonen. 3. Aufl., Fischer, Jena, Stuttgart, 743 S.

Walter, H. & Breckle, S.-W. 1991: Ökologie der Erde, Bd. 4: Spezielle Ökologie der Gemäßigten und Arktischen Zonen außerhalb Euro-Nordasiens. UTB Große Reihe, Fischer, Stuttgart. 586 S.

Walter, H. & Breckle, S.-W. 2004: Ökologie der Erde, Bd. 2: Spezielle Ökologie der Tropischen und Subtropischen Zonen. 3. Aufl., UTB Große Reihe, Fischer, Stuttgart. 764 S.

Wickens, G. E. 1993: Vegetation and ethnobotany of the Atacama desert and adjacent Andes in northern Chile. Opera Botanica 121: 291-307

Willert, J. V, Eller, B.M., Brinckmann, E. & Baasch, R. 1982: CO₂ gas exchange and transpiration of Welwitschia mirabilis Hook fil. in the Central Namib Desert. Oecologia 55: 1 21-29

Willert, J. V, Eller, B.M., Werger, M.J.A. & Brinckmann, E. 1990: Desert succulents and their life strategies. Vegetatio 90: 133-143

Part G: ZB IV—Zonobiome of Sclerophyllic Woodlands Mediterranean Winter Rain Areas

8

Contents

© Springer-Verlag GmbH Germany, part of Springer Nature 2022
S.-W. Breckle, M. D. Rafiqpoor, *Vegetation and Climate*, https://doi.org/10.1007/978-3-662-64036-4_8

Photo 22 Typical maquis on the Cape Corse of the Island of Corse (ZB IV) with white-flowered *Cistus salvifolius, C. monspeliensis* and the sclerophyllous evergreen oaks *Quercus coccifera* and *Qu. ilex* in the background (photo: Rafiqpoor)

Photo 23 Fruiting dragon tree (*Dracaena draco* var. *ajgal*) in S Morocco (ZE III/IV). It was described in 1997 as a new taxon from the Anti-Atlas (east of Agadir) (photo: Breckle)

Photo 24 Mediterranean rock garrigue (ZB IV) on steep limestone slopes with high biodiversity of herbs, geophytes (several orchids, *Muscari*) and annuals in southern Albania (photo: Breckle)

8.1 General, Climate, Soils

It is convenient to divide **zonobiome IV** into five floristic biome groups according to the floristic kingdoms that condition strong floristic differences, each of which forms typical, often similar-looking vegetation units (Fig. 8.1). Of these, the Mediterranean is the largest, as winter rains extend from the Atlantic Ocean into Afghanistan. However, severe winter frosts already occur in Anatolia and further east, so that these areas must better be placed in ZB VII.

The Mediterranean climatic regions of the ZB IV are mostly followed by arid zonoecotones (ZE), in which the winter rain regime still prevails, but the drought or the winter frosts have a stronger effect. However, this type of climate is also generally described as Mediterranean. In southern Australia, the southwest, but also the south, has Mediterranean features; there are two separate sub-areas (Fig. 8.1).

The climate diagrams for the individual biome groups are very similar, except that the summer drought is sometimes more pronounced, sometimes less. However, the range of different manifestations of this climate type is also very wide in the western Mediterranean region (Fig. 8.2).

The sclerophyllic vegetation of ZB IV, typical of winter rainfall areas with only sporadic frosts, does not tolerate prolonged cold. The most favourable growing season is spring, when the soil is moist and temperatures rise, and autumn after the first rains. The winter period is already too cool for good growth at temperatures around 10 °C or below.

Box 8.1 The Five Mediterranean Regions of the World
The five winter rainfall areas: (1) the Mediterranean (with evergreen forest, maquis, garigue, asphodel, etc.) (2) the Californian (with sclerophyllic forest, chaparral, partly encinal, etc.) (3) the Chilean (with matorral, espinal, etc.) (4) the Capensian (with fynbos, renosterbos, etc.) (5) the Australian (with jarrah forest, sclerophyllic bush = mallee, etc.).

Box 8.2 Similarities and Differences of Mediterranean Woody Floras
The Mediterranean vegetation is dominated by sclerophyllic woods, which are similar in appearance but mostly belong to completely different genera in the different areas.

The individual Mediterranean areas are geographically far apart from each other. Superficially, the vegetation units and the biotopes sometimes look strikingly similar. This similarity is particularly great between the Mediterranean area, California and Chile (Fig. 8.3), as well as between the Cape region and Australia. This certain dichotomy is not least related to geological history. Although the climate as a formative primary factor is similar in all five areas, the

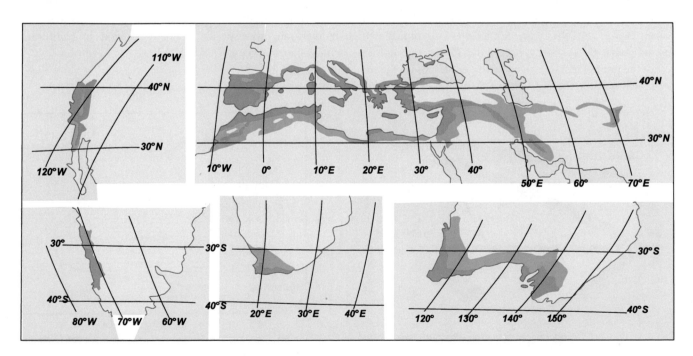

Fig. 8.1 Areas with a Mediterranean climate, arranged at comparable latitudes. They are preferentially located on the western side of the continents. **Green**: Mediterranean climate type with Mediterranean zonobiome IV; **Orange**: arid areas with predominantly winter rainfall, various ZE of ZB IV, especially ZE III/IV; ZE III/VII (modified after Walter and Breckle 1991)

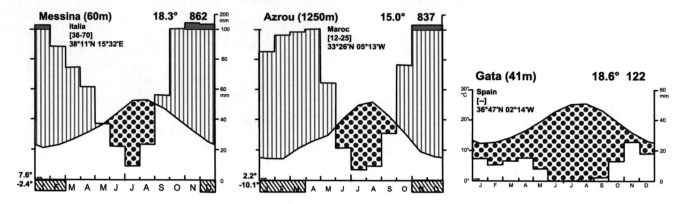

Fig. 8.2 Climate diagrams of Messina in Sicily, Azrou in the montane belt of the Middle Atlas (Morocco) and Cabo de Gata (SE Spain) = the driest place in Europe (desert, ZEIII/IV)

geological history of the areas is very different. Australia and the Cape region are parts of the ancient Gondwana mass, they have been depleted for millions of years, and the soils are very poor in nutrients (Fig. 8.4). Much younger and strongly influenced by Tertiary mountain formation are the other three areas. The nutrient status of their soils is up to a factor of 10 better in terms of nitrogen and over 100 times better in terms of phosphate.

In discussing the climatic subzonobiomes of each biome group, the zonoecotones (ZE) will also be discussed subsequently. The transition can take place from ZB IV to ZB V, VI or VII.

The present climate of the ZB IV was not always so. Both the wide distribution of fossil soils and the development rhythm of the main representatives and other facts (fossils) indicate that in the Tertiary the climate was still tropical with summer rains. Only shortly before the Pleistocene did the shift of the rainfall maximum to the winter months take place. The plants had to adapt: A sharp selection took place, and

only those species with small xeromorphic leaves that grew on dry sites in the previous climatic epoch survived. The present reduction of activity in summer is imposed by drought. It is absent when plants have sufficient water available. The ephemerals and ephemeroids, which serve as vegetation buffers, are restricted in their development to the favourable spring or the again wet autumn.

Consideration of these historical facts facilitates understanding of the ecological behaviour of vegetation (Specht 1973; Axelrod 1973; Castri 1981; Arroyo et al. 1995). Close relationships exist between many taxa of ZB IV and ZB V or ZB II (for example, species of the genus *Olea, Eucalyptus,* and others). Thus, *Quercus baloot* (closely related to *Q. ilex* s.l.) grows in Afghanistan with additional summer rainfall. The Encinal vegetation of the mountains in Arizona with summer rain corresponds to the Chaparral in California with winter rain only.

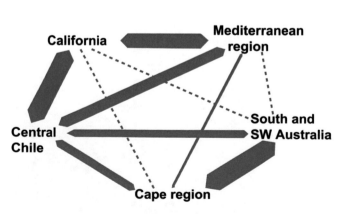

Fig. 8.3 The five Mediterranean regions. The thickness of the connecting lines schematically indicates the similarity of the five regions in terms of the phylogeny of flora, phenology, morphology and vegetation types, as well as climate and land-use patterns (modified after Castri 1981)

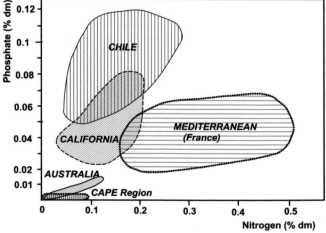

Fig. 8.4 The phosphate and nitrogen contents in soils (total contents in %) of the five Mediterranean regions (modified after Rundel 1982; di Castri 1981)

8.2 Origin of Zonobiome IV and Their Relations to Zonobiome V

In their monograph Castri & Mooney (1973), in addition to various aspects of ZB IV, historical questions of the origin of this ZB IV are discussed, which is closely related to those of ZB V. Both go back to a common root, the tropical vegetation of the Tertiary reaching the higher latitudes.

Axelrod (1973) has summarized the further development of vegetation up to the present for California and comparatively applied it to the Mediterranean area.

Fossil finds show that at the beginning of the Tertiary in the Eocene, tropical evergreen but also deciduous species grew in the Northern Hemisphere in the area of today's temperate climate, indicating a tropical climate with a pronounced summer rainy season at that time. Studies of fossil marine molluscs allow the conclusion that in California the minimum sea surface temperature was about 25 °C around 50 million years ago. During the Oligocene and Miocene, a steady cooling of the sea occurred, and by the end of the Tertiary in the Pliocene, the minimum was only 15 °C. Correspondingly, the climate on the mainland also became cooler and cooler and the flora poorer in species with high heat requirements. At the same time, however, the rainfall distribution in California changed. The summer maximum became less pronounced and towards the end of the Miocene it disappeared; in the Pliocene a shallow minimum was already noticeable in summer. During the Pleistocene with the ice ages, cold ocean currents developed on the western sides of the continents and at the same time a climate with pronounced summer drought and rain only in the winter months, i.e. the type of ZB IV.

During the Tertiary the ever-rising mountains also bulged fully in western North America, and in Europe the alpine ranges. The consequence of this was that in the Tertiary tropical zone of today's higher latitudes arid climates and arid local sites in unfavourable exposure developed, so that among the evergreen species a selection took place into species with the typical leathery leaf of the humid tropics (often called laurel-leaved—lauriphylls; leaf mesophyll thickened usually by thick cellulose cell walls) and into more drought-resistant sclerophyllous species (hard-leaved species; leaf mesophyll cell walls usually thickened by lignification). Then, during the Pleistocene, as the summer-dry climate (termed Mediterranean) developed on the western side of the continents, the sclerophyllous species gained dominance in this climatic region and the woody species flora became impoverished, while on the eastern side of the continents, favoured by warm ocean currents, the humid climate with summer rains at somewhat lower annual temperatures was maintained as zonobiome V. On the humid eastern coasts of the continents of N and S America,

as well as SE Africa, SE Asia, and E Australia, the transition from tropical humid to subtropical humid and warm-temperate species-rich flora with evergreen leather-leaves is still quite gradual.

The sclerophyllous vegetation of ZB IV did not evolve by adaptation to summer drought, but the tertiary species were already preadapted to dry sites. Only a limited number of new species evolved, in California for example in the genus *Ceanothus* with 40 species, *Arctostaphylos* with 45 species, others, as mentioned, spread strongly, for example *Adenostoma*. *Arbutus* has a more leathery leaf (Fig. 8.5).

Of the 113 woody genera (with 169 species) of the hardwood range in Chile, only 13 genera are the same as the 109 genera (with 272 species) in California. Australia, with 66 genera (with 140 species), has even only 2 genera in common with California and 3 with Chile. However, the total number of species is much higher. Especially the partly very small areas of ZB IV represent a certain exception to the rule that species richness increases from the poles to the equator (Table 8.1).

The corresponding but much smaller area in the S African Cape region is thought to contain about 8000 species, SW Australia the same 8000 species, while the much more extensive and richly divided Mediterranean area is estimated to contain about 24,000 species of vascular plants.

> **Box 8.3 Relationship of ZB IV with Neighbouring ZBs**
> The formation of zonobiome IV is closely related to zonobiome V, it occurred only during the late Tertiary.

This evolutionary history also makes it understandable that between ZB IV and ZB V the same genera but represented by vicariant species are often present, for example sclerophyllous *Quercus* species in California and the leather-leaved evergreen *Quercus virginiana* in south-eastern North America (ZB V). In Australia, the leather-leaved *Eucalyptus* species of ZB IV in SW and S Australia differ little from those in the summer rainfall area of ZB V of the east coast. There, as in the west, a rich Proteaceae vegetation is also found on dry calcareous soils, only the species are different. Also, the occurrence of the fossil "terra rossa"soils in the Mediterranean area becomes understandable. In these one finds relics of the tropical soil fauna, which at greater depths does not feel the summer drought. The rest of the fauna of ZB IV confirms the remarks made with regard to the vegetation (contributions in Castri & Mooney 1973).

As Axelrod emphasises, the fossil record in N Africa also points to a similar history of Mediterranean vegetation. However, the conditions in Europe are more complicated. Since the postglacial period, the climate of Western Europe has been determined by the warm Gulf Stream.

Fig. 8.5 The strawberry tree (*Arbutus*) is a typical Mediterranean element. This tree occurs with two species in the Mediterranean area: *Arbutus unedo* (**a** and **b**) in the western Mediterranean region, *Arbutus andrachne* (**c–e**) with red trunk in the eastern Mediterranean region

(photos **a–b**: Rafiqpoor; **c:** http://bit.ly/2mYGaZd; **d–e**: Breckle). A third species (*Arbutus canariensis*) is only known from ZB V (Macaronesia)

The cold Canary Current (Fig. 7.2) only makes itself felt south of this archipelago as far as Senegal (fog coast). The ZB IV extends from the west along the coasts of the Mediterranean Sea, due to the extensive coastlines, far to the east.

The last ice ages had a particularly negative effect in Europe and practically destroyed the flora. Remnants did not migrate again from the few refugia until the postglacial period. The flora remained poor, so that continuous fossil records from the Tertiary to the present, as in California, are lacking. But the prevailing view is that the history of ZB IV was essentially similar everywhere, and that a climate corresponding to ZB IV with zonal sclerophyllic vegetation

did not yet exist in Tertiary times, although the hardleaf species did on dry local habitats.

8.3   The Mediterranean Area

The climatic conditions in this zone are shown in the diagrams (Fig. 8.2). In winter the cyclones bring rain, while in summer the Azores high-pressure causes hot and dry summers. Since the Mediterranean area is one of the oldest cultural regions, the zonal vegetation had to give way to cultivated crops on most sites.

Table 8.1 Number of genera and species in zonobiome IV of California and Chile, winter rainfall area (after Arroyo et al. 1995)

Parameter	Chile	California
Area in km^2	294,600	278,000
Number of genera	681	806
Number of species	3385	4240

Nevertheless, there can be no doubt that the zonal vegetation was an evergreen sclerophyllous forest with *Quercus ilex* (Fig. 8.8).

Based on small remnants, the following information can be given about the original forests:

Holm oak forest (Quercetum ilicis):

Tree layer: 15 to 18 m high, closed, largely formed by *Quercus ilex* alone.

Shrub layer: 3 to 5 (up to 12) m high,

- *Buxus sempervirens,*
- *Viburnum tinus,*
- *Phillyrea media,*
- *Phillyrea angustifolia,*
- *Pistacia lentiscus,*
- *P. terebinthus,*
- *Rhamnus alaternus,*
- *Rosa sempervirens* and others

As lianas:

- *Smilax,*
- *Lonicera*
- *Clematis.*

Herb layer: about 50 cm tall, sparse but rich in species

- *Ruscus aculeatus,*
- *Rubia peregrina,*
- *Asparagus acutifolius,*
- *Asplenium adiantum-nigrum,*
- *Carex distachya* and others.

Moss layer: Very sparse.

Under these low forests, in calcareous areas, one usually finds a terra rossa soil profile with a litter layer, a blackish humus horizon and below it a 1 to 2 m thick, clayey, plastic, red terra rossa horizon. In cultivated soils the upper horizons are missing (erosion), so that the colour is already visible at the soil surface. They are mostly fossil soils of a more tropical climatic period. Today brown loams are developing (Zinke 1973).

The aspect sequence begins in March with the flowering of many shrubs. The main flowering time, also for *Quercus ilex,* is May; in June *Rosa, Clematis* and *Lonicera* are still flowering. The coincidence of the highest temperatures with the greatest drought causes a relative dormancy. Only with the autumn rains recommences new growth and sometimes a renewed flowering of the sclerophyllous woods. The holm oak (*Quercus ilex*) is widespread in the western Mediterranean area as far as the Peloponnese and Euboea (Greece), and the cork oak (*Quercus suber*) is also found (not on calcareous

Fig. 8.6 Cork oak forest (*Quercus suber*) in southern Spain. The oaks are freshly debarked, the cork sheets are collected and then transported away for processing. Cork can be obtained from older cork oaks about every 10 years (photo: Barthlott)

soils) in the far west (Fig. 8.6). Its growth is promoted by cultivation, especially by repeatedly clearing these forests of competing species and regular debarking of older tree trunks (Breckle 1966).

In the eastern Mediterranean, the Kermes oak (*Quercus coccifera*) replaces the previously mentioned tree species. In Palestine it occurs as an arboreal race (*Qu. calliprinos*) (Fig. 8.7).

In the hot lower level of Spain and N Africa, the wild olive tree (*Olea oleaster*) and the carob tree (*Ceratonia siliqua*) (Fig. 8.8a) grow in the tree layer with *Pistacia lentiscus* (Fig. 8.8b); in addition, there are various *Cistus* species (Fig. 8.8c) as well as the European palm (*Chamaerops humilis*) (Fig. 8.8d). Of particular interest on Crete are Tertiary relict sites of a wild form of the date palm (*Phoenix theophrasti*) already mentioned by Theophrastus. A large stand grows in front of a small lagoon near Vai (at Cape Sideron, NE corner of Crete) above groundwater and in a few other locations and also on the SW coast of Turkey.

Quercus ilex shows a montane distribution in N Africa from Morocco to Tunisia (Fig. 8.9) above an intercalated coniferous forest belt with *Tetraclinis* (*Callitris*) and *Pinus halepensis* (Aleppo pine). The SE corner of Spain, with 130 to 200 mm rainfall, has almost desert-like conditions (Fig. 8.2, Gata).

Today, real *Quercus ilex* forests can only be seen in a few places in the mountains of N Africa.

Box 8.4 Human impact on plant cover
The slopes were deforested and grazed, so that severe soil erosion set in and today only various stages of degradation remain.

Fig. 8.7 Tall maquis and scrubland with *Quercus calliprinos* in Galilee (Keziv Park) with species-rich herb layer (**a**, photo: Breckle). (**b**) The typical maquis formation on Cap Corse in Corsica. It consists mainly of *Quercus ilex, Erica arborea, Arbutus unedo, Pistacia lentiscus, Cistus monspeliensis, Cistus albidus, Cistus incanus* etc. with a species-rich herb layer (photo: Rafiqpoor)

Otherwise, they are felled as coppice every 20 years and regenerate by means of shoots from old stumps. The result is a man-high shrubbery with sparse areas in between, which is called **Maquis or Macchia**. Macchia can also be found on slopes where the shallow soil does not allow a tall forest to grow. The sclerophylls, which are usually only known as shrubs, can form real trees in favourable locations when they reach a higher age; mighty trees of *Quercus ilex* can be seen in old gardens or parks. If the felling takes place every 6 to 8 years and the areas are regularly burned and grazed, then higher woody species are absent and we get open communities called **Garigue** (Fig. 8.10) (or Garrigue, in Greece 'Phrygana', in Spain 'Tomillares', in Palestine 'Batha').

Fig. 8.8 Floral elements of lower altitudinal Mediterranean vegetation: *Ceratonia siliqua* (**a**), *Pistacia terebinthus* (**b**, photo: Breckle), *Cistus incanus* and *Cistus albidus* as understory (**c**), *Chamaerops humilis* (**d**) (photos: Rafiqpoor)

Fig. 8.9 *Quercus ilex* forest on limestone above Azrou in the Middle Atlas (Morocco) (photos: Rafiqpoor)

In the garigue, single species often predominate, such as low bushes of *Quercus coccifera, Juniperus oxycedrus* (in the east also *Sarcopoterium spinosum* bushes) or *Cistus, Rosmarinus, Lavandula, Thymus,* etc. The most favourable conditions for grazing are provided by the *Brachypodium ramosum-Phlomis lychnites* community in southern France on limestone. In spring, many therophytes (ephemerals) occur on bare patches. Geophytes (ephemeroids) such as *Iris,* Orchids *(Serapias, Ophrys)* and *Asphodelus* species are also not absent. Finally, on sites very degraded by fire and grazing, an almost pure **Affodill** vegetation with many bare soil (Fig. 8.11) remains. The garigue and open affodill sites are a sea of flowers in spring, while they scorch out severely in late summer. If the cultivation of crops or grazing are abandoned, successions become noticeable, tending towards the direction of zonal vegetation, as shown in the scheme (Fig. 8.12) for S France.

Fig. 8.11 Affodil site (*Asphodelus aestivus*) as a degradation form of the Mediterranean cultivated landscape in Cap Corse on Corsica (photo: Rafiqpoor)

Fig. 8.10 Garigue is an open community mostly of thorny cushion plants, in which also here and there some small trees of *Juniperus communis* or *Pistacia lentiscus* etc. from the macchie may occur. The picture shows the rock garigue on Corsica, in the background on the slopes the macchie stands as a somewhat higher formation (photo: Rafiqpoor)

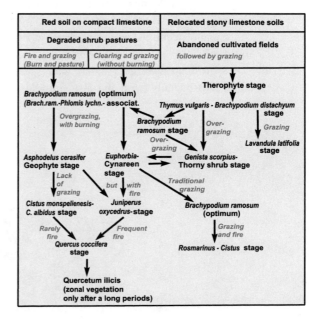

Fig. 8.12 Scheme of the regeneration stages of degraded pastures or cultivated soils on calcareous soils in the Languedoc (southern France) to holm oak forest (*Quercus ilex*) or, in the case of persistent grazing (and fire), to *Rosmarinus-Cistus* garigue. The dependence of the changes on the type and intensity of use is indicated (modified after Walter and Breckle 1991)

Fig. 8.13 Regression and progression (regeneration) stages in the *Quercus calliprinos* zone at Jebel Ansariye in Syria on limestone rock (modified after Nahal 1962). **G** = ongoing grazing; **GD** = interrupted grazing; **F** = deforestation; **U** = vegetation turnover

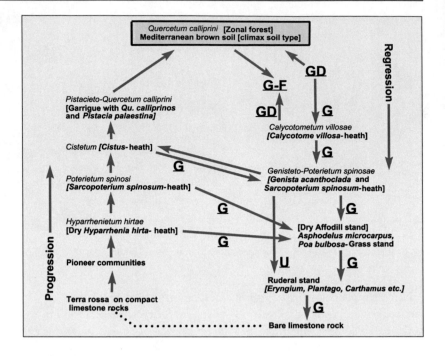

On sandstone or acid gravel soils, succession proceeds similarly, except that the individual stages have a different floristic composition; characteristic species are, for example, the strawberry tree *(Arbutus)* and the tree heath *(Erica arborea)*.

In the eastern Mediterranean, the arboreal *Quercus calliprinos* (closely related to the western Mediterranean, mostly shrubby *Qu. coccifera*) takes over the role of *Qu. ilex* and represents the zonal forest type (Fig. 8.13 and Fig. 8.7). The stages of progression and regression are similar to those in the western Mediterranean region, although other species usually dominate in the genera represented in each case. The manifold influence of man often leads to a barely usable thorny, low-growing garigue (Batha) with sparse, thorny dwarf shrubs (especially with the also fire-tolerant

Sarcopoterium) or even to completely open rock heaths (Fig. 8.12), where the soil is largely washed away, leaving bare rock in many places. Progression (regeneration) seems almost impossible here without appropriate measures. In the continental Mediterranean area of S Anatolia, the pine *Pinus brutia* (close to *P. halepensis*) plays a greater role (Fig. 8.14). It often forms the tree layer, while the sclerophyllic species occur as macchie in the shrub layer. Since pine does not regenerate in the macchie due to lack of light, these stands can only rejuvenate after forest fires, which explains the evenness of the tree layer being all much the same age. The umbrella pine *(Pinus pinea)*, which is frequently planted in the Mediterranean region, probably had its natural habitats on poor sandy areas on the coast.

Fig. 8.14 *Pinus brutia* forests at the Anatolian Plateau mountain fringe on the western slope of the Taurus Mountains in Turkey (photos: Rafiqpoor)

8.4 Importance of Sclerophylly in Competition

If one is interested in the ecophysiological conditions in the Mediterranean area, the question immediately arises to what extent the plants are affected by the long summer drought. One must first distinguish between the sclerophylls and the malacophylls, which are strongly represented by *Cistus, Rosmarinus, Lavandula, Thymus,* and others. Furthermore, one must take into account that the favourable euclimatopes are nowadays occupied by crops, for example vineyards, and that the Mediterranean species are relegated to the shallow sites, i.e. grow under relatively unfavourable conditions.

If the rock is deeply fissured, the abundant winter rains penetrate deeply and are stored in the soil. In the rock crevices, the roots of the woody species can be traced 5 to 10 m deep, down to layers that still contain sufficient exploitable water even in summer (Breckle 1966).

Cell sap studies throughout the growing season revealed in sclerophylls that the water balance is not significantly disturbed during the drought period. However, this can only be achieved by restricting gas exchange through partial stomatal closure when water supply is impeded. Transpiration measurements showed that at dry sites water release in summer is about three to six times lower than at moist sites. At extremely dry sites with only stunted growing specimens the cell sap concentration increases strongly ($-3.0 - -5.0$ MPa). It must be remembered, however, that on the good soils, where the vineyards yield is heavy in the autumn, the water conditions are much more favourable. Thus, summer dormancy caused by drought was hardly an option in the original sclerophyllic forests.

In contrast to the hydrostable sclerophylls, the malacophylls are hydrolabile. *Cistus, Thymus* and *Viburnum tinus* show increases in cell sap concentration up to 4.0 MPa in summer. At the same time, a strong reduction of the transpiring area occurs in them, as a large part of the leaves is shed. Often only the buds remain. These species do not root so deeply. The laurel *(Laurus nobilis),* which does not belong to the sclerophylls, has its natural habitat in the Mediterranean region always in the shade, in valleys or on northern slopes. Today it forms forest stands only in the fog belt of the Canaries or in a macchie in the winter rain area without pronounced summer drought (N Anatolia), *Prunus laurocerasus* behaves in the same way.

The ecological importance of sclerophylly is probably to be seen in the fact that the sclerophyllic species are active in gas exchange when water supply is good (number of stomata 400 to 500 per mm^2), but in the case of water shortage they are able to strongly throttle water losses by closing the stomata. This gives them the ability to survive months of drought while maintaining plasma hydration and without loss of leaf area until the next rainy season, when they can immediately resume substance production in autumn.

But the conditions change immediately if the summers in the humid winter rainfall areas are not decidedly dry, or if in a typically Mediterranean climate the site is permanently damp, for example on north-facing slopes or in floodplain forests. On the former, the sclerophylls are first displaced by laurel-like evergreen species and then by deciduous trees. *Quercus ilex* is replaced by the deciduous oak *(Qu. pubescens)* with greater growth.

In the floodplain forests of the Mediterranean area, deciduous tree species grow, such as *Populus* and *Alnus* species, *Ulmus campestris, Platanus orientalis* and in SW Anatolia the Tertiary relict species *Liquidambar orientalis.* However, as soon as the rivers dry up in summer, we do not find deciduous woody species, but the evergreen sclerophyllous oleander *(Nerium oleander).*

Box 8.5 Dependence of Substance Production on the Assimilation Budget

Substance production depends mainly on the assimilate balance of the plants; it is the greater:

1. the greater the proportion of assimilates used to increase the productive leaf area,
2. the greater the leaf area/leaf dry weight ratio, i.e. the less substance is used to build up the leaf area,
3. the higher the intensity of photosynthesis,
4. the longer the time during which the leaf surface can assimilate CO_2.

More precise data for 1 (Box 8) are not available, but it can be assumed that the proportion of leaf mass to total phytomass is more favourable in deciduous species than in sclerophylls. Regarding 2, the ratio is two times greater in the thin deciduous leaves than in the evergreen ones, and for 3, the measurements show that the intensity of photosynthesis in deciduous and evergreen leaves, calculated per unit leaf area, does not show great differences. As for 4, the evergreen leaf is obviously more favourable. Two points are thus in favour of the deciduous species and one point for the evergreen species.

Box 8.6 Competitiveness of Sclerophyllous Species

Sclerophylls outcompete both non-sclerophyll, more lauriphyllous evergreen species sensitive to drought, and deciduous trees only in winter rainfall areas.

More precise calculations showed for the humid, mild climate of Lake Garda, where both *Quercus ilex* and *Qu.*

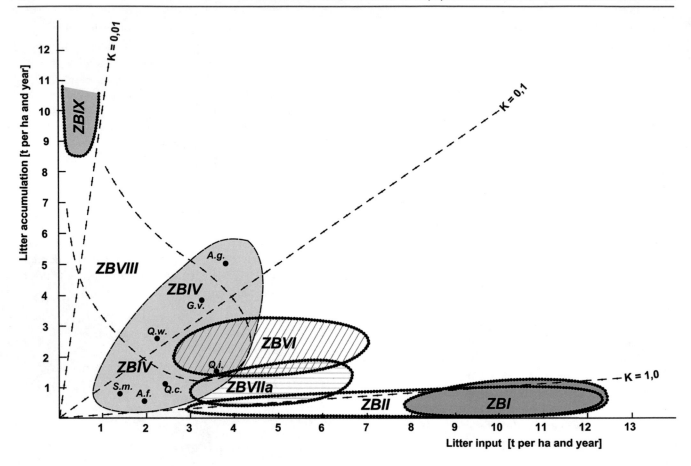

Fig. 8.15 Litter input and litter accumulation in different zonobiomes. ZB IV is highlighted, here individual species are indicated (A.f. = *Adenostoma fasciculatum*; A.g. = *Arctostaphylos glauca*; G.v. = *Garrya veatchii*; Q.c. = *Quercus coccifera*; Q.i. = *Qu. ilex*; Q.w. = *Qu. wislizenii*; S.m. = *Salvia mellifera*). The range of some other zonobiomes is circumscribed. k is the rate of decomposition, assuming uniform negative exponential decomposition (modified after Read and Mitchell 1983)

pubescens grow, a substance yield in grams per gram of branch weight of 22.9 for *Qu. pubescens* compared with only 17.9 for *Qu. ilex*, confirming the observation of the greater competitive power of deciduous species under these climatic and site conditions (Larcher 1961). In the same climate, but on steep cliff faces from which much of the rainwater runs off so that the site is dry in summer, we find evergreen *Qu. ilex* shrubs. At such growing sites *Qu. pubescens* is not competitive (Freitag 1975). In addition, on steep rocky slopes *Qu. ilex* is protected from cold air stagnation in winter. This is because its northern limit is mainly due to winter cold, affecting seedlings and saplings.

Of course, sclerophylly also has an effect on soil formation, because the decomposition of leaves with high lignin content and high crude fibre content is slower than that of malacophyllous leaves. The decomposition rates of leaves depend on their mechanical strength on the one hand, and on their mineral content on the other. Ashy leaves rich in minerals are decomposed faster by the decomposers in the soil.

In comparison with other zonobiomes, the Mediterranean region with the hardwoods is about average, so to speak, with regard to litter production and the accumulation of litter (due to reduced decomposition rates by decomposers) (Fig. 8.15). The needle litter of conifers in ZB VIII is mineralized even more slowly, of course also due to the unfavourable climate with very long winters, as is that of the tundra (ZB IX), so that raw humus accumulation occurs there. In the Mediterranean area, litter input and accumulation are more or less balanced, whereas in ZB I the permanent litter input is very high, but the accumulation is insignificant; there, everything that accumulates is continuously degraded (k = 1, Fig. 8.15).

8.5   Arid Mediterranean Subzonobiome, N Africa, Anatolia, Iran

Small arid areas are found in the Ebro Basin in NE Spain (Walter 1973), where winter cold already plays a role, and even more pronounced in SE Spain (Freitag 1971a), the only corner of Europe that can almost be counted as having

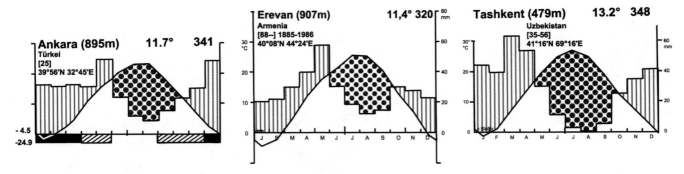

Fig. 8.16 Climate diagram of Ankara, arid Mediterranean. Homoclimates are Erevan (High Armenia) and Tashkent (Central Asia, slightly lower and warmer)

conditions like ZB III (Fig. 8.2), it is the European ZE IV/III—corner.

As an example of a larger area, however, Central Anatolia should be mentioned, which still belongs entirely to the winter rainfall region and represents a central basin landscape enclosed by high peripheral mountains at over 900 m asl. The mountains hold off a large part of the winter precipitation. In May, the already heated but still humid rising air leads to thunderstorm formations and a rainfall maximum (Nisançi 1973) (Fig. 8.16).

Total precipitation is less than 350 mm, summer drought is very marked, but the months of December to March are cold (minima to −25 °C), although interrupted by thaws (ZE IV/VII). No forest can grow under these conditions. The *Pinus* forests of the encircling mountains (montane—Mediterranean belt) pass over a scrub zone with *Juniperus, Quercus pubescens, Cistus laurifolius, Pyrus elaeagrifolia, Colutea, Crataegus* and *Amygdalus* (dwarf almond) species into a steppe. It is therefore a ZE IV/VII. The steppe has now become mostly arable land (winter wheat cultivation as "dry farming = Lalmi"), or it is heavily grazed. This results in a degradation to an *Artemisia fragrans-Poa bulbosa* semi-desert with very many early therophytes and geophytes.

At higher altitudes, many species of *Astragalus (Tragacantha)* and *Acantholimon* (Plumbaginaceae) occur as spherical thorny cushions, which are especially characteristic of the cold Armenian-Iranian highlands.

Originally, a herb-rich grass steppe *(Stipa-Bromus tomentellus-Festuca vallesiaca* community) prevailed in Central Anatolia, already reminiscent of the Eastern European steppe, except that the species are Mediterranean elements. The soil has a typical chernozem profile, but with an A-horizon not very rich in humus. The vegetation period in this steppe is shortened to four months by the winter cold and summer drought. Very important is the rain maximum in May.

The most favourable season is the spring. Already from February to March the first geophytes (*Crocus, Ornithogalum, Gagea* and others) bloom. They are followed, especially in case of overgrazing, by the numerous small therophytes, which root only in the upper 20 cm and therefore disappear already by June. The actual perennial steppe species reach their development maximum in May and dry up only in July. As the soil contains enough water in spring, the cell sap concentration of these species is low (1.0 to 1.5 MPa) and increases just before withering. A number of species, which include also the spherical thorny cushions, flower only during the main drought. These species are characterized by a deep taproot, allowing them to draw water from deep soil horizons that are still moist in summer. In the case of camelthorn *(Alhagi)*, a root depth of 7.65 m has already been measured in a 30-month-old plant. The cell sap concentration is also below 1.5 MPa.

The periphery of the Mediterranean steppes are among the areas settled by man particularly early and are the cradle of human culture and civilisation. This is true not only for the Hittites in Anatolia, but also for the area of the "Fertile Crescent", that is, for the mountain slopes that surround Mesopotamia from the west, north and east. Here (around Jericho, Beidha, Jarmo) have been found the oldest traces of the cultivation of grain, for which the steppe is especially favourable. At the same time, livestock farming was possible in this area. The neighbouring forest served hunting purposes and provided wood. In these primeval settlement areas, man has destroyed the natural vegetation thoroughly and almost completely over the past millennia and in some cases transformed formerly fertile areas into deserts. Today, these processes are referred to as desertification. The onset of soil erosion has created many "bad lands" in which all plant growth is absent.

We will come back to the very different zonoecotones in the north of the Mediterranean area, which extends very far in west-east direction.

8.6 California and Neighbouring Regions

This area is restricted in western N America by the mountain ranges (Cascades, Sierra Nevada) to a narrow strip on the Pacific coast. The winter rainfall area extends up the west

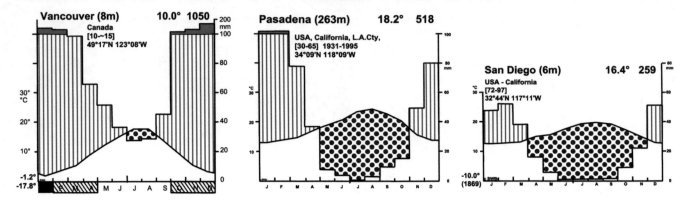

Fig. 8.17 Climate diagrams of stations on the Pacific coast of N America (from N to S) in the coniferous forest, hardwood vegetation, and desert transition area

coast from British Columbia to lower (Baja) California, but in the north the rainfall is so high and the summer drought so short that these are species-rich hygrophilous to mesophilous coniferous forests, which should already be considered zonoecotone IV/V (Barbour & Major 1977). Only central and southern California are a sclerophyllous region, while Lower California is already too arid (Fig. 8.17 and Fig. 8.18).

California ZB IV corresponds to the proper Mediterranean Californian floral province, which is very species-rich (Table 8.1). Since the present-day western American flora still largely resembles the Pliocene flora, i.e., it did not undergo Pleistocene depletion, all plant communities are very species-rich; genera such as *Quercus, Arbutus,* and others are represented by a large number of species, plus a

great amount of genera that are entirely absent in Europe, for example, the important genus *Ceanothus* (Rhamnaceae) with 40 species; of *Arctostaphylos,* 45 shrubby species are present. A leading species is the Rosaceae *Adenostoma fasciculatum* ('Chamise') with needle-like leaves. The distribution of this shrub fairly reflects the extent of the sclerophyllic zone.

Ecologically, a site of an *Adenostoma* chaparral near San Diego in the mountains (458 to 1678 m asl) south of the Mojave Desert, protected for 40 years, was studied in more detail by Mooney & Parsons (1973, in Castri & Mooney 1973). Climate data for the station at 815 m asl are as follows: Mean annual temperature 14.3 °C, abs. Maximum 42.5 °C, abs. Minimum −7.8 °C, frost may occur from October to May; mean annual rainfall 670 mm mainly in December to March; evaporation 1625 mm a year, mainly in the four hot summer months. The soil can dry out to a depth of 1.2 m in low rain years, below which it is always moist.

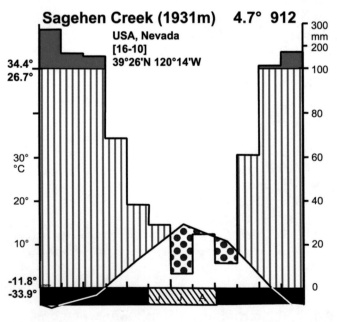

Fig. 8.18 Climate diagram of Sagehen Creek at the pass summit (1931 m asl) of the Sierra Nevada before Reno. The small rain maximum in August is due to summer thunderstorms. Absolute temperature maximum 34.4 °C, −minimum −33.9 °C (from Walter 1990)

> **Box 8.7 The Arrangement of the Plant Cover Is Climate-Dependent**
> The north-south gradient means that evergreen sclerophyllous oak forests occur only in the northern part of the California hardwood range, sometimes even mixed with deciduous species, while in the southern part a scrub formation known as chaparral predominates. It corresponds to the more Mediterranean macchie.

Fires after lightning are common, with flame temperatures reaching 1100 °C, 650 °C at the soil surface, and 180 to 290 °C at a depth of 5 cm. *Adenostoma* sprouts more than 50% even during drought, often in 10 days after fire, forming 25 cm shoots in 30 days. All plants of *Quercus agrifolia* and *Rhus laurina* sprout. *Adenostoma* reaches greatest cover 22 to 40 years after a fire, and growth almost ceases after 60 years. Stand regeneration occurs after a new fire. About 50% of shrub species rejuvenate by sprouting, the others by seed. About 20 years after a fire, the stand is closed again. In the first few years after fire, severe soil erosion occurs on steep slopes. Aboveground phytomass reaches 50 t·ha^{-1}, and belowground is probably twice that. Net aboveground production in a year is about 1 t·ha^{-1} in young stands and decreases with age. The shrubs are normally photosynthetically active all year round.

In the spring develops a very rich ephemeral vegetation. Some of these species germinate only after fire. *Adenostoma* dominates on southern slopes, whereas *Quercus dumosa* grows in the denser stands on northern slopes.

The coastal strip of California immediately bordering the sea north of 36 degrees latitude is not part of the sclerophyllous zone because the fog caused by the cold ocean current (California current) makes the summer season cool and moist, allowing the hygrophilous northern tree species to grow.

The chaparral, unlike the macchie, is a natural zonal vegetation that corresponds to the relatively low winter precipitation of 500 mm. It is true that fires are also very frequent here; but these fires were a natural factor before human intervention. Accurate statistics from the "National Forest Administration" have shown that fires caused by lightning are exceptionally common in the chaparral area, so that a constant fire watch is necessary during thunderstorms. It has been found that fires that recur about every twelve years do not change the chaparral because the shrubs keep striking out. If the fires stay out for a very long time, then species such as *Prunus ilicifolia* and *Rhamnus crocea* invade. If one fire is followed by another in two years, the seedlings of the shrub species that do not sprout after fire are killed, thus pushing back these woody plants.

The root systems of sclerophyllic species reach far down into the soil, because its uppermost soil layers usually dry out completely in summer. The maximum depths of the roots far into the crevices are 4 to 8.5 m (more detailed information with root system profiles can be found in Kummerow 1981). A certain water absorption is therefore possible in summer. This can be seen from the fact that after a fire in midsummer, the shrubs sprout very soon; after the loss of the transpiring surface, even a small amount of water absorption is sufficient to make the buds grow. The autumn rains do not have a direct effect. It takes over a month for the water to reach a depth of 1 m. Meanwhile, the temperature drops so much that the shoots stop growing. The peak of development is April, when the temperature rises with a good water supply. The old evergreen leaves assimilate into the spring. They do not fall off until June, when the young ones become fully functional. Almost all species of the chaparral have a mycorrhiza. The *Ceanothus* species, however, form nodules that assimilate nitrogen.

A very detailed vegetation monograph with much ecological information was published by Barbour & Major (1977).

Evergreen oak sclerophyllic forests are also found in N America as a montane belt in the mountains of S and C Arizona above the cactus desert at 1200 to 1900 m elevation. It is the **Encinal belt**, which is divided into a lower and an upper belt on the basis of the various *Quercus* species. The latter is replaced by the *Pinus ponderosa* belt. The chaparral species *(Arbutus, Arctostaphylos, Ceanothus)* occur as shrub layer below the tree layer. Although there are two rainy seasons in Arizona, the vegetation is very reminiscent of that in California, but the sclerophyllic forests in the mountains are much better developed and still native. The occasional summer thunderstorms supplement the scant winter precipitation. Nevertheless, the summer drought is very pronounced. East of the Sierra Nevada, in the state of Nevada, winter precipitation decreases to about 150–250 mm.

The cold season lasts 6–7 months at 1300 m altitude. This is shown by the climate diagram of Sagehen Creek (Fig. 8.18) at the top of the pass with still relatively high precipitation and a forest as well as bog vegetation. Reno (Fig. 8.19) is already in the lee. Only an *Artemisia tridentata* semi-desert called "Sagebrush" persists there (ZB VIIa). The extent to which precipitation levels depend on relief in this area can be seen in Fig. 8.20. The *Artemisia* semi-desert occupies vast areas in Nevada and Utah and adjacent states. It replaces the

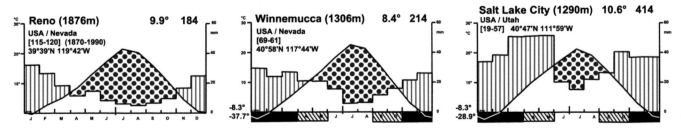

Fig. 8.19 Climate diagrams from the sagebrush area (*Artemisia tridentata* semi-desert): Reno, Winnemuca, and Salt Lake City (already transitioning to grasslands)

Fig. 8.20 The dependence of precipitation amounts (top) on relief (bottom), shown on a W-E profile through western North America at about 38°N (modified from Walter 1960)

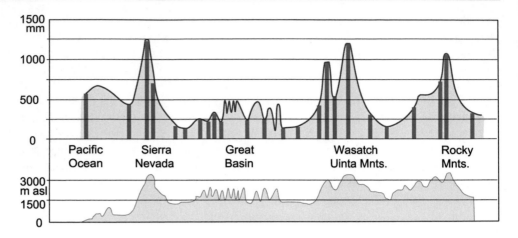

southern *Coleogyne* and *Larrea* semi-deserts in the cold climate. *Artemisia* prefers the heavy soils of the basin landscapes and is replaced on the elevations by the "Pinyon". These are low scattered *Pinus monophylla* or *P. edulis-Juniperus* tree communities, which include some cold-resistant chaparral species. In the mountains at about 2000 m elevation the true coniferous forests begin with *Pinus flexilis* and *P. albicaulis*, while further east *Pinus ponderosa* occurs, replaced higher by *Pseudotsuga* and *Abies concolor*, whereas *Picea engelmannii* and *Abies lasiocarpa* form the tree line at above 3000 m. The dry southern slopes often remain devoid of trees, so that *Artemisia* reaches up to the alpine belt; however, the sequence of elevational belts can change spatially very strongly. Aspen *(Populus tremuloides)* (Fig. 11.12) also plays a major role with extensive clonal suckers (shoots from the roots) on water-rich soils.

8.7 Central Chilean Winter Rain Region with the Zonoecotones

The state of Chile forms a strip about 200 km wide, extending 4300 km at the western foot of the High Andes from 18 to 57 °S, showing all transitions, from rainless subtropical desert in the north to a sclerophyllic wood area to very humid temperate and subarctic forests in the south. Winter rains predominate in central Chile (Fig. 8.21). The cold Humboldt Current, which is flowing along the entire coast, moderates the summer drought so that temperatures are lower compared with California; the annual temperature of Pasadena at 34° N, for example, is 16.8 °C, whereas that of Santiago at 33°S is only 13.9 °C. A comparison of the climate of the two areas was made by Castri (1973).

Since Chile belongs to the Neotropics, the floristic conditions are completely different from those in the Mediterranean region and in California. Only the cultural landscape is very similar. The same species are grown and cultivated in the gardens.

The sclerophyllous region occupies the central part of Chile and connects to the arid areas in the north. It is also present only in remnants.

We may mention *Lithraea caustica* (Anacardiaceae), which causes skin rash and fever on contact, the soap bark tree *Quillaja saponaria* (Rosaceae), *Peumus boldus* (Monimiaceae) or the Lauraceae *Cryptocarya* and *Beilschmiedia,* which prefer damp ravines. In addition, there are a number of shrubby species. In a narrowly defined area northeast of Valparaiso grows the endemic palm *Jubaea chilensis.* On dry rocky sites, columnar cacti *(Neoraimondia arequipensis)* (Fig. 8.22) and the large *Puya* species (Bromeliaceae) are found, along with the thorny Rhamnaceae *Colletia* and *Trevoa.*

Externally, the sclerophyllic species of California, Chile and Australia look similar, but there are considerable differences, for example, in the fruit shapes, as Table 8.2 shows. According to this table, there are particularly many species with fruit appendages, with spines or hooks in Australia and almost half of the species have small dry fruits, whereas in Chile and California many species also have large and fleshy fruits. The colour of the fleshy fruits also differs significantly, which allows conclusions to be drawn about the fruit-dispersing animals.

The actual Matorral area is very small in terms of surface area, because the Andes drop very steeply on the Chilean side. The almost 7000 m high Aconcagua is only about 100 km away from the sea coast.

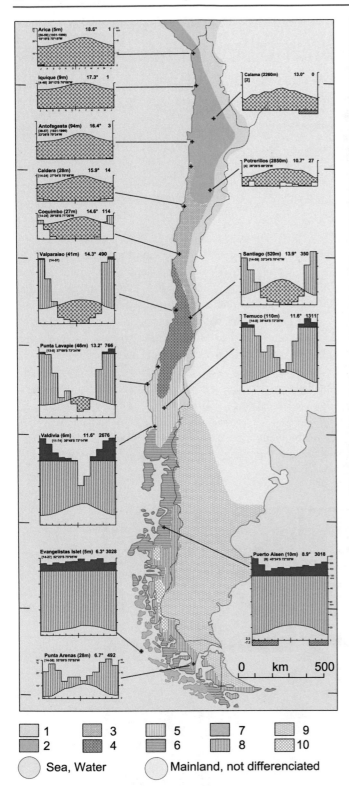

Fig. 8.22 Landscape view of the cactus rock desert with *Neoraimondia arequipensis* at the foot of the Andes on the Chile-Peruvian border (photo: Barthlott)

Fig. 8.21 Climatic diagram map of Chile with vegetation zones (modified from Schmithüsen 1956): **1** Northern High Andes, **2** Desert area, **3** Dwarf shrub and xerophytic shrub area, **4** Sclerophyllous wood region, **5** Deciduous forest, **6** Temperate evergreen rainforests, **7** Tundra-like cold zone vegetation, **8** Sub-Antarctic deciduous forest, **9** Patagonian steppe, **10** Southern Andes

In the mountains, debris communities predominate, and the elevational belts are difficult to discern. The sclerophyllous vegetation only goes up to about 1500 m (Fig. 8.23). Shrub communities transition to the alpine belt, with the coniferous species *Austrocedrus (Libocedrus) chilensis* occurring in places. Widespread are alpine debris shrubs, such as *Tropaeolum* species, *Schizanthus* (a Solanaceae with zygomorphic flowers), as well as Amaryllidaceae *(Alstroemeria, Hippeastrum)* and *Calceolaria* species.

For the upper alpine belt flat cushion plants (*Azorella* and other Apiaceae) are characteristic. The species at these elevation of the orobiome, but also south of the sclerophyllous, are already Antarctic elements, which include the arboreal *Nothofagus* species. South of Concepcion, the forest with *Nothofagus obliqua* (zonoecotone IV/V), which sheds its foliage in the cool winter months, begins with decreasing summer drought (Fig. 8.24), and still farther south, with

Table 8.2 Fruit shapes of the Mediterranean flora in C Chile, California and Australia and percentage distribution of fruit colours of fleshy fruits (after Hoffmann & Armesto 1995)

Condition of the fruit	Chile	California	Australia
Small, fleshy fruits	34.2%	29.1%	12.1%
Small, dry fruits	19.8%	43.7%	45.0%
Large fruits (> 15 mm)	14.4%	6.3%	0
Anemochore (for example winged fruits)	29.7%	19.4%	23.6%
Other (with aril, hooks, spines, etc.)	1.8%	1.5%	19.3%
Colouring of fleshy fruits:			
Black/Purple	48%	27%	–
Red	16%	43%	–
Green	12%	2%	–
Other	24%	28%	–

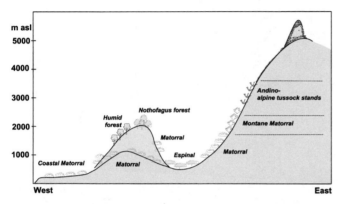

Fig. 8.23 West-east transect through central Chile indicating the main vegetation formations up to the Andes (modified after Rundel 1982)

precipitation above 2000 to 3000 mm, transitions into ZB V of the evergreen Valdivian temperate rain forest (Quintanilla 1974) (Fig. 8.25). It is hardly inferior to the tropical one in luxuriance, and the standing wood mass may be even greater. The woody species are partly neotropical elements, and bamboos *(Chusquea)* also play a major role; some are already Antarctic elements such as the evergreen *Nothofagus dombeyi*. Very old conifers are also represented, especially in montane locations. Besides *Austrocedrus* and *Podocarpus* species, *Saxegothea, Fitzroya, Araucaria araucana* (= *A. imbricata*) and *Pilgerodendron uviferum* should be mentioned. In the very moist and cool, but frost-free climate, this evergreen forest merges into the Magellanic

Fig. 8.24 The *Nothofagus* forests (here an example from S Chile), belong to the Antarctic elements and occur in all continents of the S hemisphere as evidence of their former connection (Gondwana continent). Leaves of *Nothofagus dombeyi* from Argentina (photo: https://t1p.de/gg7j) (photo small: https://t1p.de/n714)

Fig. 8.25 The *Nothofagus* forests in Cangillo National Park, southern Chile consisting of Arctic elements are in the humid regions of the southern hemisphere as evidence of the connection of the Gondwana continent in the geological past (photo: M. Neumann)

Fig. 8.26 The ever-humid cool Magellanic forests are common at the windswept southern tip of the South American continent in Chile (photo: https://t1p.de/im68)

forest, which extends almost to the southern tip of the continent (Fig. 8.26); in the process it becomes increasingly sparse in species and lower, finally only 6 to 8 m high. All the offshore islands to the west are covered by cushion bogs (*Sphagnum* occurs but plays no role). This vegetation is floristically close to that on the Antarctic islands. Similar Antarctic elements can be found on New Zealand as well as on the mountains of Tasmania - a sign that these areas used to be in direct contact with each other via the Antarctic continent. The bogs can be described as Antarctic tundra (ZB IX).

> **Box 8.8 Vegetation of Central Chile**
> The typical vegetation of C Chile is a 10–15 m high woody shrub, the matorral, with xerophytic hardwood species.

8.8 The Cape Province in South Africa

The South African winter rainfall area is confined to the extreme south-western tip of Africa, but nevertheless encompasses an entire floral kingdom—the **Capensis**. The species richness in this small area is quite extraordinary. In the Jonkershoek Conservation Area alone, some 2000 species have been recorded on 2000 ha, as well as on the 50 km stretch from Table Mountain to the Cape of Good Hope. The genus *Erica* comprises 963 species (Fig. 8.27), *Restio* (Restionaceae) 108 species, *Muraltia* (Polygalaceae) 115 species, *Cliffortia* (Rosaceae) 117 species, *Protea* about 100 species. The Proteaceae (Fig. 8.28) play a particularly important role among the sclerophyllic plants. This family is otherwise only strongly represented in Australia, but by another subfamily; a few genera also occur in S America.

The species richness is certainly partly due to the deep valleys and steep mountains with a strong dissection (Knapp

Fig. 8.27 The genus *Erica* in the Cape region, as a small but independent floral kingdom, achieves the greatest species differentiation worldwide with 936 species. We bring some examples collected only on Table Mountain. (Above **a**: *Erica* spec.; **b**: *E. versicolor*; **c**: *E. galdulosa*; **d**: *Erica sessiliflora*; **e**: *Erica* spec.; **f**: *E. formosa*; **g**: *E. subdivaricata*; **h**: *Erica* spec.; **j**: *E. cerinthoides*) (photos: Rafiqpoor)

Fig. 8.28 The genus *Protea* also reached its greatest differentiation and speciation in the Fynbos Formation in the Cape region. Here are some examples from the Cape of Good Hope Fynbos Formation: (**a**: *Leucospermum cordifolium*; **b**: *Protea cynaroides*; **c**: *Mimetes fimbriifolius*; **d**: *Leucospermum cordifolium*; **e**: *Leucospermum cordifolium*; **f**: *Protea aurea* subsp. *aurea*; **g**: *P. eximia* (photos **b**: Breckle; rest: Rafiqpoor)

Fig. 8.29 The famous Rooibos tea of South Africa is obtained from *Aspalathus linearis* in the Cape Province. This species is an element of the Renosterbos in the northern Cape. However, the shrub is also cultivated on a large scale (photos: Rafiqpoor)

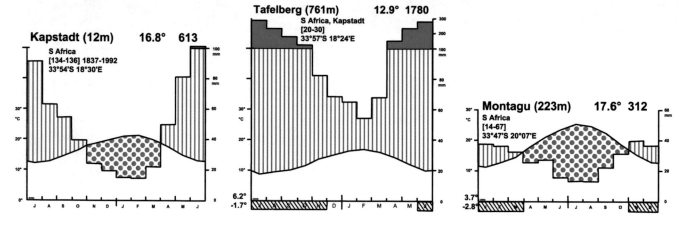

Fig. 8.30 Climate diagrams from South Africa: typical sclerophyllic area, humid montane climate (rich in fog), transitional area and typical Karroo

1973), but also due to the isolation to the north, which has existed for a long time due to the climatic aridity of the Karroo semi-desert.

The endemic genus *Aspalathus* (Fabaceae) comprises 43 species. Among other things, it also provides the rooibos tea (*Aspalathus linearis,* Fig. 8.29). Its centre of distribution is concentrated in the Cape area. Among our houseplants many originate from the Cape (*Pelargonium, Zantedeschia = Calla, Amaryllis, Clivia* and others). The climate diagram of Cape Town corresponds to that of Tanger; only the annual precipitation is 260 mm lower; the summer, however, is somewhat less dry (Fig. 8.30).

The Fynbos, like the Matorral, also has only a very small area (Fig. 8.31). The only tree species *Leucadendron argenteum* (silver tree: Fig. 8.32) has a very small range on

Fig. 8.31 The Fynbos formation occupies large areas in the winter rainfall region of the southern Cape. *Restiona* and *Protea* are widespread in the Fynbos (photo: Rafiqpoor)

Fig. 8.32 The silver tree (*Leucadendron argenteum*) is a typical element of the Cape Fynbos (photos: Rafiqpoor)

the moist slopes of Table Mountain below 500 m asl. Woodland-like stands occur in moist ravines; however, these are the last outcrops of moist temperate forests on the SE coast of Africa (ZB V). Some of the leaves of *Protea* are very large (Fig. 8.28); they have little mechanical tissue but a thick cuticle and are therefore hard. The water balance of proteaceous shrubs, as in all hardy shrubs, is balanced, that is, the cell sap concentration shows little variation during the year. The soil is always likely to contain exploitable water in the rooted deeper layers, even in summer. The soils in the Cape region are acidic and very low in mineral nutrients, which particularly suits the Proteaceae and Ericaceae (with obligate mycorrhiza).

The most important ecological factor is the fire. After a fire, countless geophytes (*Gladiolus, Watsonia,* etc.), in which the Cape flora is very rich (about 350 species), appear in the first year, followed by herbaceous species, together with dwarf shrubs (Fig. 8.33).

Fig. 8.33 The Cape region is rich in geophytes. We bring here some examples from the Renosterbos area as examples. **a**: *Dietes grandiflora*; **b**: *Moraea sisyrinchium*; **c**: *Moraea flaccida* **d**: *Ferraria crispa*; **e**: *Gladiolus alatus*; (photos: Rafiqpoor)

Fig. 8.34 The "Tablecloth"on Table Mountain near Cape Town. It develops not only on Table Mountain but also on other higher ridges on the south coast of Africa. The northern flank is sunny and the southern flank is shrouded in dense clouds that spill over to the north (photo: Breckle)

After about seven years the proteaceous shrubs have grown up again, either as stick shoots or as seedlings. They can reach a great age, but then become woody and flower weakly, thus seeming adapted to periodic burning. Again, fire from lightning is likely to be a natural factor.

Today, burning is caused deliberately or due to human negligence. It is interesting that the bulbous plants only come to flower after a fire, but otherwise grow vegetatively. Fertilization by the ashes does not play a role, rather the suddenly for some time reduced root competition of the burnt shrubs seems to be the triggering cause. With increasing elevation in the mountains, the rainfall increases namely on the SE slopes where the moist warm air from the Indian Ocean is forced to rise. Table Mountain station, which is 750 m above Cape Town, records three times the amount of precipitation (Fig. 8.30). The Cape province is a mountainous region with isolated basins between the mountain ranges. On top of these very often lies the "Tablecloth", i.e. a cloud cover produced by warm moist winds from the Indian Ocean, which creeps up the SE slope to dissipate again on the NW slope (Fig. 8.34).

It forms a weeping mist on the plateaus of the mesas, so that they are moist and tend to become scrubby *(Restio, Erica)* or even boggy (moss mats with *Drosera* and

Utricularia species, Fig. 8.35). Succulents *(Rochea coccinea* and others) grow in dry niches between boulders. Inland, winter precipitation decreases (Fig. 8.30), especially in the rain shadow of the individual mountain ranges. In the rain shadow, the dry formation form of the cape vegetation, the Renosterbos, occurs first, with *Elytropappus rhinocerotis* (Asteraceae) (Fig. 8.36) as the dominant, rod-shaped shrub species. This is the transitional area, the ZE IV/III. This is then replaced by the semi-desert vegetation types of Karroo.

> **Box 8.9 Fynbos in South Africa**
> The sclerophyllous vegetation of S Africa is called **Fynbos**. It is a macchia-like proteaceous scrub 1 to 4 m high.

The sclerophyllous vegetation of the Fynbos has spread widely since the settlement of the Cape province, i.e. after 1400 AD. In former times, the evergreen temperate forest with palaeotropical elements extended along the entire SE coast of Africa to beyond the S tip of Africa (Cape Agulhas) (ZB V).

Fig. 8.35 Boggy areas covered with Restionaceae develop on the wet parts of the Table Mountains in South Africa. In these boggy sites, the soil surface is in many places covered with moss mats in which carnivorous *Drosera* species (e.g. in the picture *Drosera trinervia* as an endemic on Table Mountain in Cape Town) grow (photos: Rafiqpoor)

Fig. 8.36 The Renosterbos Formation with *Elytropappus rhinocerotis* (Asteraceae) is still widespread in the winter rainfall area in northern Cape province (photo: Rafiqpoor)

8.9   SW and S Australia

Perth in SW Australia occupies almost the same latitude as Cape Town. The climate is also very similar (Fig. 8.37). But not only the SW corner of this continent has winter rains, but also the area around Adelaide in S Australia.

As a result of the special floristic conditions, the sclerophyllic vegetation is distinguished by a different character than in the other winter rainfall areas of the world. The tree form (*Eucalyptus* species) is dominant, the Proteaceae form the shrub layer among these or predominate on the sandy heaths. The eucalypts do not have hard, but leathery leaves. Many shrubby or low *Eucalyptus* species grow in the Mallee, forming a lignotuber (Fig. 8.38). Lignotuber formation is a genetically fixed trait, although it may be highly modified by environmental factors. Lignotubers are interpreted as an adaptation to survive adverse events (fire, drought, cold). They occur in all Mediterranean areas. Many species in Australia in particular have lignotuberous formations.

The ecological significance of lignotuber formations is not always clear. In California, the lignotuber-forming *Arctostaphylos glandulosa* grows alongside the lignotuber-free *Arctostaphylos glauca* in the same habitat. In addition, *Adenostoma fasciculatum*, *A. sparsifolium*, *Ceanothus*, *Quercus dumosa*, *Rhus laurina*, and others possess a lignotuber in California. *Eucalyptus camaldulenesis* does not have a lignotuber in S Australia, those growing more northerly are ecotypes with lignotubers. Numerous species of the W Australian Eucalypts, but also *Banksia*, etc., form lignotubers; in Chile, for example, *Colliguaja odorifera*, *Quillaja saponaria*, *Lithraea caustica*, *Cryptocarya alba*; in the Mediterranean region lignotuber formation is regularly known only from *Quercus suber*. In all cases, the possibility of rapid sprouting after fires certainly plays an important role.

A special feature of SW Australia are the grass trees (*Xanthorrhoea*, *Kingia*, Fig. 8.39), the Cycadeae *Macrozamia* and the *Casuarina* species. The Ericaceae are replaced by Epacridaceae. The soils are as poor and acid as in the Cape region. They are quartz-rich with iron concretions representing laterite crusts of an earlier period of tropical climate. The parent rocks are among the oldest geological formations on earth due to their long Gondwana history. An indication of soil poverty is the fact that the herb layer of the forest around Perth contains 47 carnivorous species of *Drosera* (sundew), especially climbing ones, which thrive on nutrient-poor sites. The bracken fern is also widespread when there is sufficient moisture.

South of Perth, rainfall increases (to over 1500 mm), but decreases to the N and inland. With each change in climate, other *Eucalyptus* species come to dominate. The wetter the climate, the taller the trees grow and the greater the leaf area per hectare. The vertical position of the leaves allows a lot of light to penetrate the trunk area, so that the shrub layer is usually well developed if it is not reduced by frequent fires.

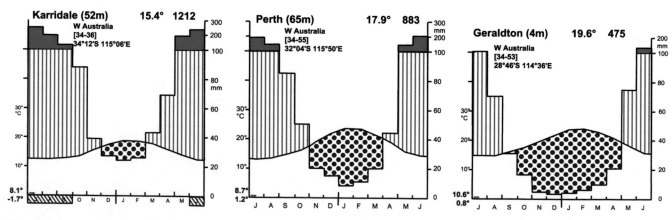

Fig. 8.37 Climate diagrams from SW Australia. Stations in the Karri forest, the Jarrah forest and the Shrub Heath (Fig. 9.10, Adelaide)

Fig. 8.38 In all Mediterranean regions of the world, a large number of plant species form lignotubers. This is true for many plant families also in Australia, especially for Myrtaceae, Araliaceae, etc. We bring here two examples from Australia: **a**: *Eucalyptus botryoides* (photo: P. Woodard, https://t1p.de/ezkh); **b**: *Cussonia paniculata* (photo: Gent, https://t1p.de/ucwx)

For the climate comparable to the Mediterranean, with 625 to 1250 mm of rain and a summer drought, the **'Jarrah' forest** is characteristic, in which *Eucalyptus marginata* absolutely predominates. This species can become 200 years old and reaches a height of 15 to 20 m (maximum 40 m). In the more humid southern part, the **'Karri' forest** is found with *Eucalyptus diversicolor,* which reaches 60 to 75 m (maximum 85 m) in height (ZE IV/V). With a canopy closure of 65%, a shrub layer and a dense herb layer, often with fronds of bracken up to 1.5 m high, are developed (Fig. 8.40).

The drier **'Wandoo' zone** with *Eucalyptus redunca* receives only 500 to 625 mm of rain. The woodlands are sparser. It is slightly more inland, but is now almost entirely converted to sheep pasture. In the absence of suitable native grasses, *Lolium rigidum* is sown with the Mediterranean clover *Trifolium subterraneum,* which is annual but buries its fruits in the soil, as a nitrogen source; prior superphosphate fertilization is essential given the poverty of the soils. Fertilization and seeding are done from the airplane given the large extent of the land. The species-rich Mallee with numerous shrubs, also many Proteaceae and an enormous species richness of small shrubs, herbs and geophytes is almost today only preserved in protected areas (Fig. 8.41).

In the zone with 300 to 500 mm of rainfall, many loosely standing *Eucalyptus* species occur (ZE IV/III), but this area is now the winter wheat zone with farms of several 100 ha in size, managed by two to three men when the farms are fully motorised. Growing wheat in the wetter zones is unprofitable due to the incidence of rust fungus damage.

Fig. 8.39 Grass trees (*Xanthorrhoea*) are a special feature of Australia (photo: Breckle)

Soil temperatures at 15 and 30 cm depth ranged from 4.1 to 36.0 and 5.8 to 29 °C, respectively. Root systems of 91 species were excavated. The dominant sclerophylls are the shrubby *Eucalyptus bacteri*, 9 Proteaceae, 2 *Casuarina* species, *Xanthorrhoea*, Leguminosae and others.

The main growing season is the dry summer, as the soil remains moist at greater depths. The smaller perennial species (42%) root only in the upper 30 to 60 cm; they develop in spring. *Drosera* and orchids are ephemeral species because they root only 5 to 7 cm deep. Water is found to be very unevenly distributed in the sandy soil with a wilting point of 0.7 to 1%; this is because the large species channel rainwater to the stem. The composition of the heath is determined by the fires. After a fire the grass-tree *Xanthorrhoea* sprouts first; it flowers only after a fire. The Proteaceae *Banksia* rejuvenates by seedlings after fire. Its share of the aboveground phytomass increases to 50% by the 15th year. The main mass of dry matter in 25-year-old specimens is accounted for by the large inflorescences, which open only after a fire.

Banksia thus belongs to the **pyrophytes** that are very common in Australia, i.e. species that can only rejuvenate after fires because the woody fruits do not open otherwise (Fig. 2.33). This fact suggests that fires caused by lightning were also a natural factor in Australia.

Today, forest and heath are very often burned because the woody plants seem to have no monetary value and they interfere with grazing. "One blade of grass is worth more than two trees" says the farmer - but for how long?!... crazy!

The pyrophytes include a great amount of Proteaceae and Myrtaceae, the conifers *Actinostrobus,* etc. *Eucalyptus* spp. also seed themselves particularly abundantly after a fire. In a heath that has not been burnt for a long time, all the nutrients are bound in the fruits of *Banksia,* in the old leaves of *Xanthorrhoea*, and in the accumulating litter. A 50-year-old stand therefore degenerates. It is only by fire that mineralization of the nutrients occurs, and new succession is initiated.

The ecophysiological conditions of *Eucalyptus marginata* correspond fairly closely to those of sclerophyllic vegetation. The roots go partly through the hard laterite crust up to more than 2 m deep. There is no summer dormancy, transpiration is restricted only at noon from 10 h to 15 h by partial stomatal closure, so that the water balance can be maintained. The cell sap concentration was 1.6 MPa in winter and probably only slightly higher in summer.

Not only the **flora** and therefore the vegetation of Australia differs strongly from that of other continents, but also the **fauna**.

Only in Australia do the primitive mammals—the cloacal animals (Monotremata)—occur, to which the platypus *(Ornithorhynchus anatinus)* belongs, which still lays one to three eggs that are incubated by the mother. In contrast, the echidna *(Tachyglossus = Echidna)* hatches only one egg in the brood pouch. Altogether Australia is thus the home of

If the mean annual precipitation falls below 300 mm, the eucalypts disappear and the very extensively grazed shrub semi-desert begins (Fig. 8.42). S Australia lacks the wet winter rainfall areas. Conditions are otherwise similar to SW Australia, but more complicated because various mixed stands of several *Eucalyptus* species each are found. The area is also mountainous, which in turn causes a strong differentiation of the vegetation.

In addition to the forests described, proteaceous heaths ½ to 1 m high are common over wide areas. They grow on sands so poor that even the undemanding *Eucalyptus* species are not competitive on them. They are peinobiomes. They are also uncultivated and rarely grazed. The strange thing, however, is that the species richness on these poor sands is particularly high; on 100 m^2 we could count 90 species, including 63 small woody species, mostly Proteaceae or Myrtaceae; *Drosera* species and an *Utricularia* with tubers were not missing.

Results from an ecophysiological study are available for such a heath with 450 mm rainfall and seven months of drought in summer in S Australia.

Fig. 8.40 *Eucalyptus diversicolor* forest in SW Australia. Undergrowth *Acacia pulchella* and bracken fern (*Pteridium esculentum,* picture foreground left). Fire tracks are visible on the tree trunks (photo: S. Porembski)

5 remaining species of egg-laying monotremes. It passes on to the marsupials *(Marsupialia)*. With few exceptions, these are also restricted to Australia. Among them are herbivorous and carnivorous representatives. The best-known group are the kangaroos *(Macropodidae)* with the large kangaroo

(Macropus), which as a grazing game certainly influences the vegetation.

In total the fauna of Australia includes a large number of different animal species that are only found on this continent. 83% of mammals, 89% of reptiles, 90% of freshwater fish

Fig. 8.41 Species-rich Mallee with several *Eucalyptus* species and shrubby Proteaceae (*Banksia*), herbs and geophytes west of Raventhorpe, SW Australia (photo: Breckle)

Fig. 8.42 The dry savannah in the vicinity of Devils Marbles, Australia, was formerly *Eucalyptus* open forest land. Today, this region has been converted to pasture (photo: Breckle)

and insects and 93% of amphibians are endemic species. This high proportion, as with the flora, is due to Australia's long geographical isolation and the geological stability of the continent.

The Australian continent is also home to the five remaining species of egg-laying monotremes. The high number of poisonous spiders, scorpions, octopuses, jellyfish, mussels and stingrays is also striking. It is also unusual that Australia is home to more venomous than non-venomous snakes.

In the Pleistocene, Australia was home to a rich fauna of large animals, including the giant *Diprotodon*, the Tasmanian rhinoceros *Zygomaturus*, the Marsupial Lion, the Marsupial Tapir *Palorchestes*, large short-snouted kangaroos (*Procoptodon, Simosthenurus*), the giant rat kangaroo *Propleopus* and the giant bird *Megalyania*.

Most of these large animal species are thought to have died out around 50,000 years ago, which correlates strongly with the first appearance of humans on the continent (Roberts et al. 2001) It is, however, still an open dispute to what extent the Aborigines are involved in the extinction of the Australian megafauna.

8.10   Mediterranean Orobiome

In the mountains of the Mediterranean region, we have to distinguish the humid elevational sequence and the arid elevational sequence (Walter 1975):

(a)   The arrangement of the humid elevational belts of mountains occurs at the northern edge of the western, maritime Mediterranean zone, where with increasing elevation not only the temperature decreases, but at the same time the drought disappears. In both cases, several vegetation units corresponding to the zonobiome (hypsozonal or orozonal vegetation) form the arrangement of the elevational belts.

Here the evergreen sclerophyllous belt is followed by a deciduous sub-Mediterranean deciduous forest belt with deciduous oaks (*Quercus pubescens*) or sweet chestnut (*Castanea*) and above it, at the height of the summer cloud cover as a fog forest, a beech *(Fagus)* and fir *(Abies)* belt. The beech forms the tree line in the Apennines as well as in Catalonia (Montseny Mountains); it still occurs on Etna and in N Greece. In the Maritime Alps, above the beech belt, we have a subalpine spruce *(Picea)* belt, in the Pyrenees one with *Pinus sylvestris* and *P. uncinata.*

(b)   The arrangement of the arid elevational belts occurs in the continental climatic region with a summer drought that is noticeable up to the alpine belt. Here a deciduous forest belt is absent; the Mediterranean sclerophylls belt is immediately followed by a series of different coniferous forest belts, for example, on the southern slope of the Taurus in Anatolia an upper Mediterranean with *Pinus brutia,* a weakly developed montane with *Pinus nigra* ssp. *pallasiana,* a high montane with *Cedrus liban* and

Fig. 8.43 *Cedrus atlantica* forests at the Zeida-Midelt Pass, form an open forest belt in the upper montane belt (2200 m NN) of the Atlas Mountains in Morocco. Underneath, *Quercus ilex* stands occur as montane forests (photo: Rafiqpoor)

Abies cilicica (wetter) or *Juniperus* species (drier), and a subalpine with *Juniperus excelsa* and *J. foetidissima*. But in the rainy northeast corner of the Mediterranean with the Amanus Mountains (= Kohe Nur in the Turkish-Syrian border area) a cloud belt with *Fagus orientalis* is present. *Cedrus libani* also occurs on Cyprus and as a small remnant in Lebanon at 1400 to 1800 m asl. In Cyprus and Crete, as well as in Cyrenaica, cypress *(Cupressus sempervirens)* always occurs in its natural form with horizontal branches in the upper Mediterranean belt. The frequently planted columnar variety is a mutation. Cedars *(Cedrus atlantica)* also form the high-montane belt (>2300 m asl) in the Atlas Mountains from the eastern High Atlas to the Tunisian border; (Fig. 8.43) but the elevational belts change greatly depending on the course of the range and the slope exposure. The likewise complicated arrangement of the elevational belts of the Spanish mountains are shown on Fig. 8.44.

The difference between the arid and the humid sequence of the elevational belts can be seen even above the tree line in the alpine belt. Whereas in the humid sequence conditions like in the Alps can be found, in the arid sequence a hemispherical cushion belt (Fig. 8.45) occurs, sometimes with many convergent species of different families, which can be easily distinguished only in the flowering state; this is followed by a dry grassland belt, and only in places kept moist by thawing snow in summer can be found hygrophilous, mostly endemic species of arctic-alpine affinities.

The interlocking of the Mediterranean vegetation is particularly complicated in the mountains of SE Europe, where transitions to ZB VI and in part expressions of ZB VII

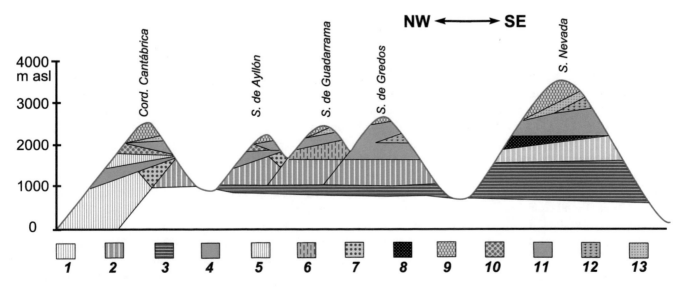

Fig. 8.44 Elevational belts of the crystalline high mountains of the Iberian Peninsula on a NW-SE profile (modified after Ern 1966). **1** Deciduous oak forest *(Quercus robur, Qu. petraea)*, **2** Pyrenaean oak forest *(Qu. pyrenaica)*, **3** Holm oak forest *(Qu. ilex)*, **4** Beech forest *(Fagus sylvatica)*, **5** Birch forest *(Betula verrucosa)*, **6** Pine forest *(Pinus sylvestris)*, **7** Mixed deciduous forest *(Quercus, Tilia, Acer)*, **8** High altitude forest of Serra Nevada *(Sorbus, Prunus,* etc.), **9** High alpine grassland and herbaceous meadow, **10** Dwarf shrub heath *(Calluna, Vaccinium, Juniperus)*, **11** Broom heath *(Cytisus, Genista, Erica)*, **12** Thorn cushion belt, 13 *Festuca indigesta* dry grassland

Fig. 8.45 Thorn cushion belt with *Erinacea pungens* (Fabaceae) in the mountainous region of Teruel (Spain) at the Linares Pass (2000 m NN) (photo **a**: Breckle) and in the subalpine elevational belt of the Atlas Mountains at the Tubkal Massif in Morocco (photo **b**: Rafiqpoor)

become effective and frosts occur more frequently. The submediterranean deciduous forests there are almost always degraded to a deciduous scrub, the **Shibljak**, by logging, slash-and-burn and forest grazing. Towards the east, more and more deciduous shrub species appear in the macchia-like formations, representatives of the East European Shibljak from the ZE IV/VI of Bulgaria and Yugoslavia, for example *Ostrya carpinifolia, Cotinus coggygria, Fraxinus ornus, Pyrus spinosa* and others. Such a mixed vegetation formation with evergreen species of the maquis and deciduous species of the Shibljak is called **Pseudo-macchie**.

An overview of the elevational zones of the Mediterranean region is given by Ozenda (1975). Particularly interesting conditions are found in the orobiomes of Macaronesia, especially the Canary Islands, which are exposed to the NE trade winds.

8.11   Climate and Vegetation of the Canary Islands

Macaronesia includes the archipelagos of the Azores, Madeira, the Canary Islands and Cape Verde. The first three are characterised by a climate with winter rain and summer drought and thus belong to Zonobiome IV, partly with hints of Zonobiome V, while the climate on Cape Verde is so dry and uniformly warm that this group of islands south of the Tropic of Cancer must already be counted as Zonoecotone II/III. Of these island groups, the Canaries and especially the islands of Tenerife and Gran Canaria are the most interesting. They are also the most studied botanically. Since Alexander Von Humboldt interrupted his journey to Venezuela on Tenerife in 1799 and distinguished five altitudinal zones on the basis of a brief survey, numerous botanists have subsequently studied the flora of this island.

The corresponding bibliography lists 1030 titles (Sunding 1973). Works on the sociology of plants include those by Oberdorfer (1965) and Sunding (1972), and ecological studies (Voggenreiter 1974; Kunkel 1976, 1987; Kull 1982; Peinado and Rivas-Martines 1987; Lösch 1988; Höllermann 1991; Pott et al. 2003).

The origin of the volcanic islands goes back to the Cretaceous period. Gran Canaria rises to almost 2000 m above sea level, Tenerife even to just over 3700 m. They are very steep orobiomes, which differ from those of the others of ZB IV in that they rise directly from the ocean and are located around the 28th latitude (north), thus exposed to the trade winds. As a result, their northern slopes, which are exposed to the wind, have different climatic conditions than the southern slopes, which are in the lee.

On the northern slope, the trade wind clouds accumulate, causing upslope rains with additional fog precipitation, so that a summer drought is absent. From the fog cover, *Pinus canariensis* trees comb out the fog droplets (Fig. 8.46). The warm, humid climate of the middle slopes is more like that of ZB V with evergreen laurel forests (Fig. 8.47). In contrast, the southern slope, especially at lower elevations, is particularly dry and more frequently exposed to the hot Saharan winds. As a result, site conditions corresponding to ZB III-IV are found on these islands, and at higher altitudes even those under increasing frost exposure. On Tenerife, the Pico de Teide above 3000 m asl is covered with alpine debris deserts, which are actually typical for the tropical mountains.

The volcanic islands were colonized with plants from neighbouring Africa at various times, especially in the Tertiary, when evergreen Tertiary forests grew there; these tree species have been preserved on the moist and warm northern slopes of the islands to the present day as in a living museum, while they became extinct on the neighbouring mainland.

This results in floristic relationships with now distant elements at the humid southern tip of Africa *(Ocotea foetens),*

Fig. 8.46 On the northern slope of the Canary Islands orobiome, *Pinus canariensis* trees comb out moisture from the saturated fog cover (photo: http://is.gd/MDOFZ1)

with India (*Apollonias*), with other tropics *(Persea, Visnea— a Theaceae, Dracaena draco)* or with the humid Mediterranean area, such as *Laurus azorica, Laurocerasus (Prunus) lusitanica, Phoenix canariensis.* On the other hand, elements of the arid regions have also immigrated, finding suitable niches at low altitudes and rocky sites (*Launaea, Zygophyllum,* succulent *Euphorbia*—and *Kleinia* species). Many species are endemics, for example the numerous succulent Crassulaceae, which used to be placed with *Sempervivum* s.l., but are now considered endemic genera (*Aeonium* with 33 species, *Aichryson* with 10, *Greenovia* with 4, *Monanthes* with 15 species).

In the *case* of quite a few species, it is evident that more primitive, namely woody, representatives of genera that are herbaceous on the Euro-African mainland occur on the islands in Macaronesia. Examples are: *Plantago*

arborescens, P. webbii; Centaurea webbiana; Carlina salicifolia; Sonchus congestus; Echium giganteum; Isoplexis canariensis. The reason for this is probably the low competitive pressure and similar environmental conditions over long periods of time. The different islands show closely related species, so-called vicariant species.

Lösch (1988) has shown in extensive studies how succulent Crassulaceae have adapted to small-scale sites and how the evolutionary history with the typical adaptive radiation of certain genera can be deduced from this. The adapted cold and heat resistance and the interplay of temperature and water availability have differentially expressed the existing ability of CAM photosynthesis, but each suitable to the microclimatic site conditions. Based on traits compiled according to morphological criteria and ecophysiological findings, Lösch (1988) summarized the kinship circles and stages of radiation of species in a phylogenetic tree-like manner, in a scheme that is still widely accepted today. On islands isolated from the mainland, competition from immigrating species is low or non-existent, so ancient forms tend to be preserved. The geological history of such island groups thus plays an important role in developmental processes (Kull 1982).

The age of the dragon tree individuals (*Draceana draco*) was discussed for a long time. These monocotyledonous trees have no annual rings, only by indirect methods Mägdefrau (1975) could show that none of them is more than a thousand years old, rather all trees are younger than 300 years, only the oldest tree in Icod de los Viños is about 365 years old.

Dracaena draco is frequently planted, it occurs only in a few wild sites (Fig. 8.48). The genus has a very disjunct distribution, there are relict species with isolated occurrences in the Anti-Atlas, in Socotra (Fig. 6.54b), on islands of the

Fig. 8.47 Laurel forest of *Laurus azorica* on the fog-exposed northern slope of the Anaga Mountains in Tenerife, Canary Islands. The dry leeward side of the mountains is generally free of forest (photo: Rafiqpoor)

Fig. 8.48 This stately dragon tree *Draceana draco* (**a**, photo: K. Rafiqpoor) has stood here on the island of Tenerife since Alexander von Humboldt's voyage to South America. *Draceana draco* occurs in a 1996 newly discovered variety *"Draceana draco* var. *ajgal"*(**b**, photo: Breckle) at Jebel Imai/Djebel Imzi, north of Etnine in the Anti-Atlas about 80 km east of Tiznit at an elevation of 600–700 m asl before in Morocco (see also Foto on page 265; Photo 23)

Indian Ocean, in West and South Africa, in Central America, in Cuba, in Hawaii, etc.

In addition, real Mediterranean elements were probably added in the Pleistocene at the earliest.

Since the islands were settled by Spain 500 years ago, the immigrants brought other Mediterranean species as well as the goats. The settlements with the cultivated areas spread more and more. As a result, the original vegetation became severely endangered. This is especially true for the unique moist evergreen laurel forest. This forest is felled for its valuable woods, its litter layer and humus soil are removed to improve the cultivated soils, which makes it impossible for the forest to regenerate on the felled areas. More undemanding species (*Erica arborea, Myrica faya*) are spreading, or the forest is being replanted with *Pinus* and even *Eucalyptus*. On Gran Canaria, the laurel forest remnants can only be found on 2% of the original area (Fig. 8.49), and on Tenerife the forests are also shrinking more and more.

Fig. 8.49 Comparison of the original natural vegetation structure on the island of Gran Canaria (**a**) with the present-day one altered by humans (**b**). Vegetation belts: **1** Succulent semi-desert (today mostly cultivated land in lower, flat areas), **2** Laurel forest or Myrico-Ericetum, **3** Pine forest (today partly *Cistus* heaths), **4** Broom heaths, **5** *Cistus*-broom mixed stands (modified after Sunding 1972)

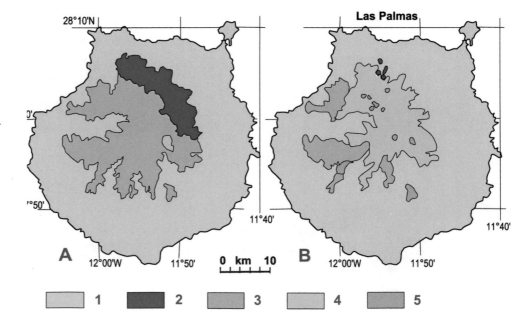

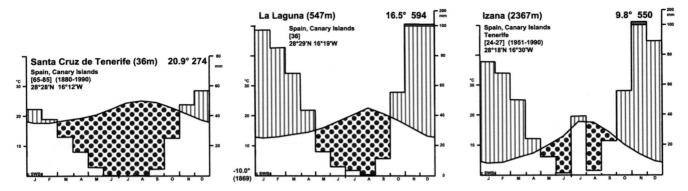

Fig. 8.50 Climate diagrams: Santa Cruz de Tenerife at sea level, La Laguna at the lower cloud level boundary, Izaña at the upper forest line

---

Box 8.12 Human Impact on Islands
"One is shocked when one visits them [the Canaries] again after 40 years and finds only concreted-over fairgrounds with motor roads. Nature conservation usually only becomes effective when there is hardly anything left to protect. Today's youth can no longer get to know the quiet and yet so sublime untouched nature" (after Walter).

Just like everywhere else in the world, the most impressive landscapes of these beautiful islands have the last decades been threatened by mass tourism, which is only aimed at profit.

Kämmer (1974) and Höllermann (1991) have dealt with the climatic conditions on Tenerife in great detail, especially with regard to the significance of the fog precipitation combed out by the trees in the cloud belt (Fig. 8.47) (Kämmer) and the question of land use and forest fire (Höllermann). On the basis of his measurements, which were extended over several years, Kämmer comes to the conclusion that the steeply increased gradient rainfall in the laurel forest belt is of greater importance than the relatively small additional fog precipitation. The information given by Sunding (1972) that a rain gauge set up in an open place in the laurel forest showed an annual precipitation of 956 mm, while another one, which collected the dripping water under trees, showed 3038 mm, may probably not be generalized. Kämmer estimates fog precipitation at only about 300 mm per year. For epiphytes, as we know from the tropics, it is not so much the amount of precipitation that matters, but the frequency of wetting and, in the case of epiphytic mosses, the low evaporation. The short duration of sunshine and consequently high humidity in the cloud belt, namely in summer, is also an important factor for the laurel forest.

The climate diagrams on Fig. 8.50 provide information about the general climate character on Tenerife. The climate of Santa Cruz on the seashore corresponds to a semi-desert climate. On the south coast, the amount of rainfall per year is

likely to exceed 100 mm only slightly, so that one can speak of a desert climate. The climate of La Laguna still under the cloud level, on the other hand, is typically Mediterranean and frost-free (exception 1869). Izana at 2367 m asl at the upper cloud level again receives somewhat lower precipitation, which decreases further at even higher altitudes. The upper timberline is a dry line, as it is in Mexico. Izana does not yet have a cold season, but frosts can occur from October to April (see Kämmer 1982 for more details).

The climate diagrams of Gran Canaria (Sunding 1972) show the same climatic character, the most arid station on the southeast coast receives only 91 mm of rain a year, Las Palmas 174 mm, the stations above 1500 m asl more than 900 mm of rain. The clouds here often envelop the lower peak.

The vegetation structure of Tenerife can be seen from the vegetation map and the profile (A to B) on Fig. 8.51 and Fig. 8.52. On the S shore, in the shade of the trade winds,

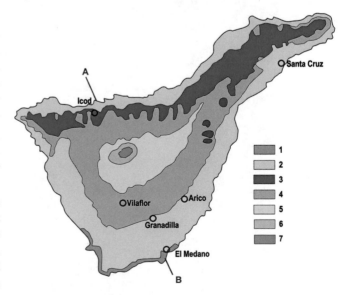

Fig. 8.51 Vegetation map of Tenerife: **1** *Zygophyllum-Launea* desert, **2** *Kleinia-Euphorbia* belt of succulent semi-desert, **3** Laurel forest and *Erica* belt in the north (trade wind side), **4** Pine forest-broom heath belt, **5** *Spartocytisus* mountain semi-desert (temperate), **6** Rocky debris belt with *Viola* and *Silene*, **7** Mountain desert with cryptogams (cold). **A-B** course of the profile on Fig. 8.52 (modified after Walter 1968)

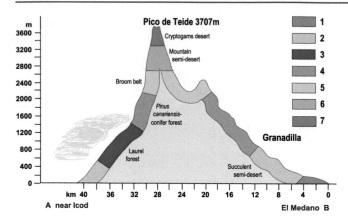

Fig. 8.52 NNW-SSE profile through the island of Tenerife (Fig. 8.51) with indication of elevational belts. Colour legend Fig. 8.51 (modified after Walter 1968)

there is a narrow desert-like lower belt with Saharo-Arabian elements, such as *Launaea (Zollikoferia) arborescens, Zygophyllum fontanesii* (on Gran Canaria also *Suaeda vermiculata*), etc.; above this, on the steep slopes, there is the semi-desert with succulents, which is especially pronounced on the southern slope. The montane forest belt consists of the laurel forest remnants in the cloud level, and above them *Pinus canariensis* forests (Fig. 8.53, Fig. 8.46),

which form the whole forest belt on the dry southern slopes. This three-needled pine species is related to *Pinus longifolia* in the Himalayas.

The summit of Mount Teide (3718 m asl) usually rises completely above the cloud cover (Fig. 8.53). Above the timberline, it is covered with shrubby broom species (*Adenocarpus, Cytisus* spp.); above this, the alpine belt begins. In its lower part, closed stands of white-flowered broom *(Spartocytisus supranubium)* still grow, while the plant cover becomes more and more open with increasing elevation and the endemics *Sisymbrium bourgaeanum,* the purple-flowered *Cheiranthus scoparius* as well as the several-meter-high viper's bugloss (*Echium bourgaeanum*) with reddish inflorescences appear (Fig. 8.54).

Above 2600 m asl, the alpine debris belt begins, which is constantly in motion due to **solifluction** (frost-induced slope sliding) on frost-change days. Here only single debris creepers like *Nepeta teydea, Viola cheiranthifolia* and *Silene nocteolens* persist. Above 3300 m asl only cryptogams occur: some cyanobacteria *(Scytonema)*, mosses *(Weissia verticillata* and *Frullania nervosa)* and lichens (*Cladonia* spp. and others).

The plant communities on Gran Canaria have been studied in detail by Sunding (1972). The elevational zonation is the same as on Tenerife. However, it only reaches up to 2000 m

Fig. 8.53 Trade wind cloud in the area of fog forests of *Pinus canariensis* on the northern slope of the island of Tenerife (ca. 1800 m asl). The leeward side of the mountain (photo location) is free of forest (photo: Rafiqpoor)

Fig. 8.54 *Echium wildpretii* (Boraginaceae) in the alpine belt of the Teide massif, Canary Islands (photo: Rafiqpoor)

asl, i.e. hardly above the timberline. Human encroachment has resulted in some irreversible changes in the sites, for example, severe soil erosion on deforested areas, which consequently do not reforest (Fig. 8.49b). The map of potential vegetation (Fig. 8.49a) does not show the very narrow desert-like zone along the seashore predominantly on the south and east coasts. Above this, occupying over half of the total area, follows the succulent semi-desert belt on the north side below 400 m asl, and on the drier south side below 800 m asl. The rest is occupied by the forest belt, namely by the *Pinus canariensis* coniferous forest; only in the lower part of this belt, but only in NE exposure, the evergreen laurel forest in a broader sense (the drier form with *Myrica faya* and *Erica arborea* included) might have prevailed in former times. The

natural range of the broom belt above the timberline was, in Sunding's opinion, confined to the small summit area.

If one compares this map with today's vegetation (Fig. 8.49b), leaving aside the settlements with cultivated areas on the lower flat slopes, one can see the enormous change: the desert-like vegetation on the flat seashores will soon be completely displaced by hotels or holiday homes with bathing beaches. The succulent semi-desert has expanded enormously at the expense of the forest belt and now covers 78% of the total area. In the upper part of the forest belt, mainly broom heaths replace the former forest, and the remaining forest area has shrunk very much, with almost only pine forests left today. Of the formerly extensive evergreen laurel forest, only in some ravines on the northern side such small remnants remain that they could only be entered as dots on the reduced map.

Natural vegetation is therefore only found today on the steep rocky slopes of the succulent semi-desert belt, which are often difficult to access. From an ecological point of view, this is a highly heterogeneous unit almost with a micro-mosaic structure ranging from dry rocky surfaces and shallow soils, to crevice-rich rocks and scree slopes on which deep-rooted species are relatively well supplied with water, to groundwater-bearing valleys and gorges or dripping wet rock faces. Therefore, a wide variety of ecological types find suitable niches here and often occur side by side, but under quite different conditions. At one extreme are the stem-succulent euphorbias, which can tolerate long periods of drought, and at the other the delicate Venus fern *(Adiantum capillus-veneris)*, which is constantly found on wet rock faces in the shade. Beneath it are found moss cushions encrusted with lime, which remains after evaporation of the water. The small amounts of NaCl in the water can also accumulate, so that even a halophilic species, *Samolus valerandi,* occurs next to the fern. Even small-scale sociological inventories yield random lists of quite heterogeneous ecological types, shallow-rooted and deep-rooted, succulent and non-succulent, tied to quite different niches. Annual therophytes are of no informative value; for they develop during the short rainy season, when all soils are moist, where they are protected from competition in an open place, but very variable from year to year.

Only a careful ecological analysis, taking into account the rooting and watering of the soil in different seasons, can clarify the presence of certain ecological types. Such an analysis is very lengthy. It requires very careful observations with targeted experiments in the field during a long period in all seasons.

In this elevational belt of the succulent semi-deserts, the palms *(Phoenix canariensis)* probably also grew in former times, of which wild specimens no longer exist. It is the palm that is found in the parks in the area of ZB IV, partly also ZB V. It is more ornamental than the related date palm *(Phoenix dactylifera),* but has inedible fruit. It was certainly bound to sunny locations with easily accessible groundwater, i.e. in the water-draining ravines.

Also the famous dragon tree of the Canary Islands *(Dracaena draco)* probably occurred on similar biotopes. Today, however, it is almost only found planted in gardens and parks.

8.12 Afghanistan at the Eastern Edge of the Winter Rain Zone

The treatment of the ZB IV remains incomplete if one is content with the description of the winter rain areas in California, the Mediterranean area, Chile, Cape region and SW Australia. The Eurasian winter rain zone extends from the Atlantic coasts of the Iberian Peninsula over the entire Mediterranean rim towards Asia Minor, the Near East, Iran and Afghanistan. The winter rains are caused by the cyclones of the planetary westerly wind belt, which shifts its position slightly to the south in winter due to the southward shift of the ITCZ and thus of the entire general circulation system.

The precipitation amounts resulting from this cyclogenesis show a strong W-E gradient in the Eurasian Mediterranean zone; i.e. the further the cyclones move eastwards from the Atlantic, apart from the convective intensification effects over the warm Mediterranean, the lower their precipitation intensity also becomes, so that they arrive in Afghanistan only in a weakened form and can cause only sparse rainfall there (Fig. 8.55). An important difference, however, is the greater continentality, which leads to regular frosts in winter. We therefore place Afghanistan largely at ZE IV/VII (Breckle et al. 2004; Breckle 2021).

The climate of Afghanistan and the spatial distribution of precipitation are strongly influenced by its high mountain character. The distribution of precipitation in Afghanistan is shown in Fig. 8.56. They originate to a considerable extent from the cyclones of the westerly wind drift. But in summer a narrow strip in the east of the country additionally comes under the influence of the Indian summer monsoon. An analysis of the percentages of summer precipitation to total precipitation shows (Fig. 8.57) that as the distance from the Afghanistan-Pakistan border to the west increases, the percentages of summer precipitation successively decrease (Rafiqpoor 1979; Frankenberg et al. 1983). In the Khost basin in E Afghanistan, of the total annual precipitation, about 30% (>600 mm) falls in summer; whereas in Kabul, less than 3% (0–20 mm) of the annual precipitation is recorded in summer. C and E Hindu Kush record considerable summer precipitation (Breckle & Frey 1976a, 1976b), the exact amounts of which are unfortunately not yet known due to lack of measurements. In the central mountainous areas in the deserts and semi-deserts of N and SW Afghanistan there is no rain at all in summer.

Thermally, Afghanistan is characterized by a high degree of continentality with very low absolute temperatures in winter (Panjao: −52 °C) and very high temperatures in summer (Zaranj: +52 °C) and thus a temperature fluctuation of >100 K (Breckle 2004). A map of the climate diagrams of Afghanistan (Fig. 8.58) provides a quick orientation on the ecological conditions of the country, showing the annual cycles of precipitation and temperature as well as the months with water deficit in different areas of Afghanistan.

The enormous differentiation of abiotic factors (climate, topography, geology, soils) in Afghanistan leads to a strong diversification of sites for a rich flora (Breckle et al. 2013). The latter found that, due to the geodiversity of the country, according to current knowledge, Afghanistan has about 5000 different plant species, of which 25% are endemic. Thus, Breckle et al. (2013) characterized central and eastern Afghanistan as a biodiversity hotspot in the Near East. The flora of Afghanistan, like its climate, varies greatly from west to east, from north to south, and from the plains to the high mountains, and is influenced by different floral elements (Fig. 8.59).

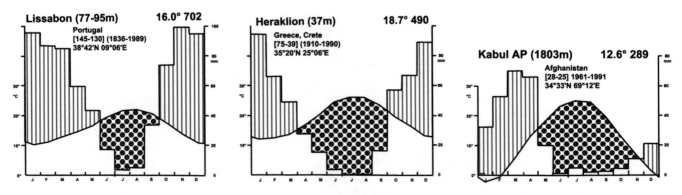

Fig. 8.55 In the climate diagrams of Lisbon, Heraklion and Kabul, the decrease in precipitation and the increase in continentality and winter frosts from west to east in the winter rain areas is clear

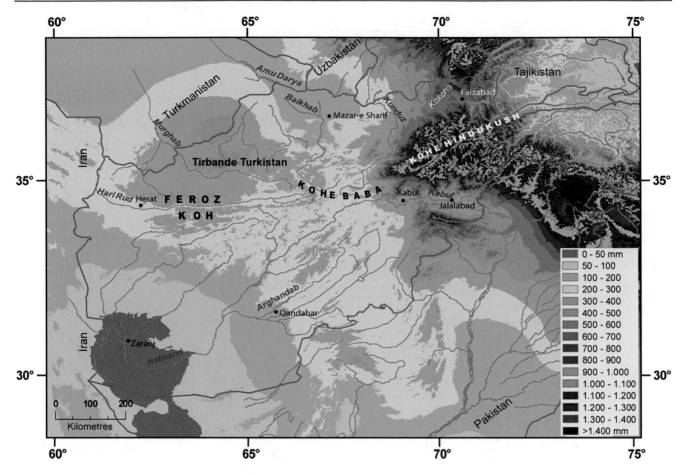

Fig. 8.56 Mean annual precipitation in Afghanistan

Figure 8.60 (Barthlott et al. 2014) gives the biodiversity of Eurasian winter rainfall areas. Table 8.3 shows the diversity numbers and their degree of endemism for the taxa occurring in Afghanistan. Figure 8.61 gives the number of species in the families with more than 40 species.

This shows that more than half of the species in Afghanistan are contained in seven families. Figure 8.62 shows the number of genera and Fig. 8.63 the number of species in the extended families. Again, it is evident that the seven extended families harbour more than half of the genera and species, although the ranking of the families now shifts a little here. The floristic diversity of Afghanistan is largely based on the spatial configuration of the plant geographic regions: About 92% of the country's area comprises the Irano-Turanian floral region, and about 7% is part of the Sino-Japanese floral region. Although temperature conditions in both regions are very similarly continental, precipitation has a different seasonal distribution. The latter floral region is favoured by a second summer rainy season (see above). A very small region (predominantly the Jalalabad basin in east Afghanistan) is Saharo-Sindian. Saharo-Sindic elements are also mixed with Irano-Turanian in S and SW Afghanistan. In the mountains, Central Asian floral elements occur in the

upper levels, Himalayan in the east, but also Euro-Siberian, boreal and even Arctic floral elements.

The plant geographic classification is not seen consistently by different authors (Hedge & Wendelbo 1970; Leonard 1988; Browicz 1997; Breckle 2004). The differentiation of the floristic regions of Afghanistan presented here is mainly based on Freitag et al. (2010).

8.12.1 Irano-Turanian Floral Elements

In accordance with the very different site conditions from the various lowland semi-deserts to the montane tree-slopes and alpine meadows, the occurrence of the individual species varies considerably. Many genera are distributed between C Anatolia and the E Hindu Kush at similar sites, usually with very different species. Especially the large genera, such as *Astragalus, Cousinia, Acanthophyllum, Acantholimon, Allium,* and *Eremurus* (Fig. 8.64) occur at all altitudes, but each with different species.

The lowland species usually have a very wide distribution. Most species occur in S and N Afghanistan, but also in neighbouring Iran, Turkmenistan, Uzbekistan and Tajikistan.

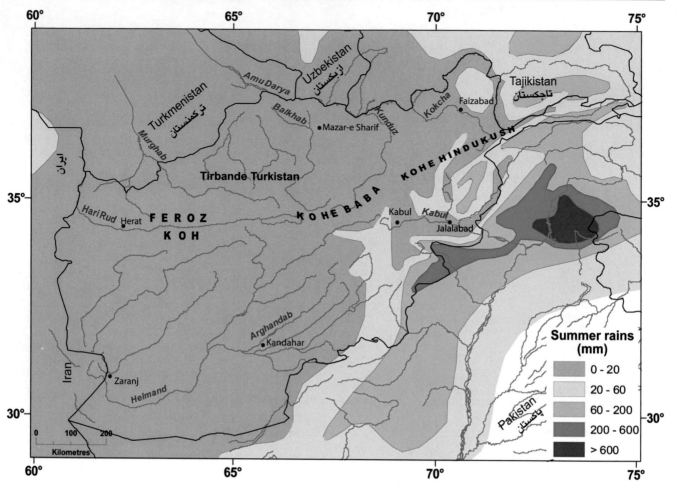

Fig. 8.57 The proportion of summer rainfall in total annual precipitation in Afghanistan

Others range from C Asia just to N Afghanistan, while a third group occurs mainly in the southern deserts of Iranian Baluchistan, S Afghanistan to Pakistan. In the mountain belts, because of the great habitat differentiation, the floristic diversity is much greater, the distribution patterns of the species much more variable, a few extending even to Pakistan, others having a much smaller range or being endemic. Endemic here means that they occur only in the Afghan mountains, although they may also spread to neighbouring mountain ranges in Tajikistan and Pakistan. However, many also occur only in a smaller part of the mountain range. Thus, due to the strong geographical isolation, other species have evolved in each chain; this explains the large number of species in the genera *Astragalus, Cousinia, Acantholimon, Allium* and some others.

8.12.2   Sino-Japanese Floral Elements

With the Himalayan subregion, the Sino-Japanese floral region extends into E Afghanistan (Fig. 8.65). Here it

occupies mainly the lower elevational regions. Biodiversity is particularly high considering the inherently small area. Different forest types characterise the E and SE slopes of the Hindu Kush. Thus, Himalayan cedar *(Cedrus deodara)*, pines *(Pinus gerardiana* and *P. wallichiana)*, oaks *(Quercus baloot, Qu. dilatata* and *Qu. semecarpifolia)*, fir *(Abies spectabilis)* and spruce *(Picea smitheana)* each form small-scale forests with numerous species in the understory, which also have a mostly eastern Afghan-Himalayan distribution. The rare *Rhododendron afghanicum* as understory in the upper coniferous forest belt and *Rh. collettianum* in the subalpine *Juniperus* belt are examples of narrow endemic species of the region (Fig. 8.66).

8.12.3   Saharo-Sindian and Other Floral Elements in Afghanistan

The Saharo-Sindian floral region is less homogeneous, although the climatic conditions with high aridity are very similar from the W Sahara to Pakistan with very hot summers

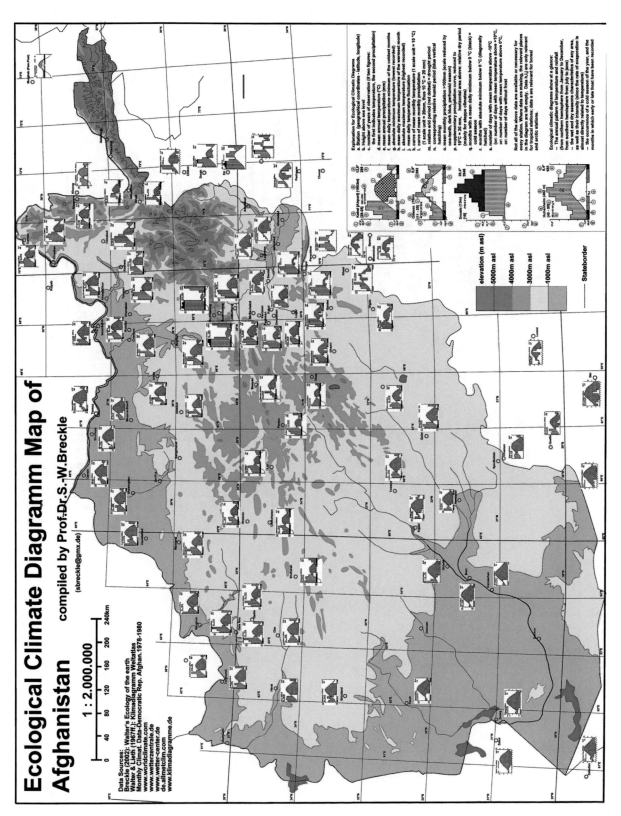

Fig. 8.58 Climate diagram map for identifying regions of the same climate (homoclimates) based on the ecological climate diagrams in Afghanistan

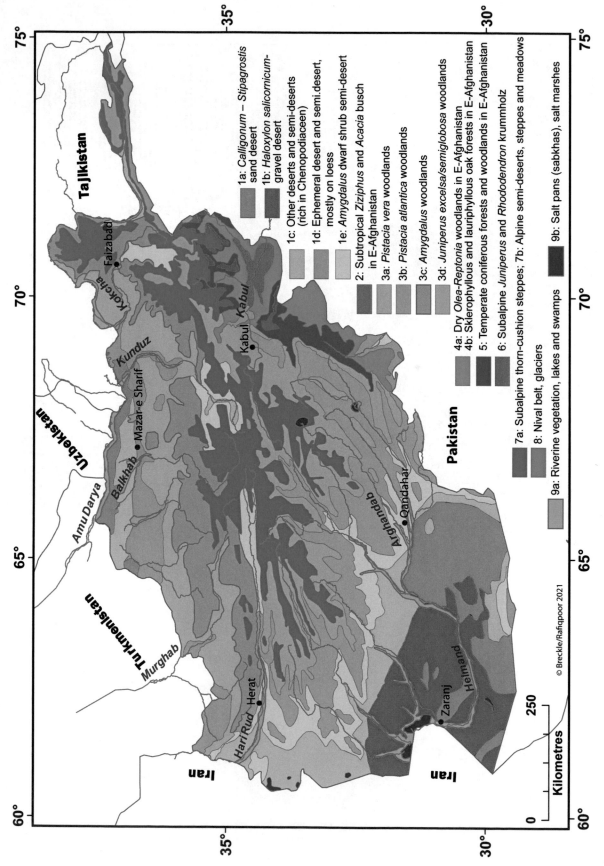

Fig. 8.59 The distribution of natural vegetation in Afghanistan (after Freitag et al. 2010)

Fig. 8.60 Spatial diversity patterns of vascular plants in SW Asia (from Barthlott et al. 2014)

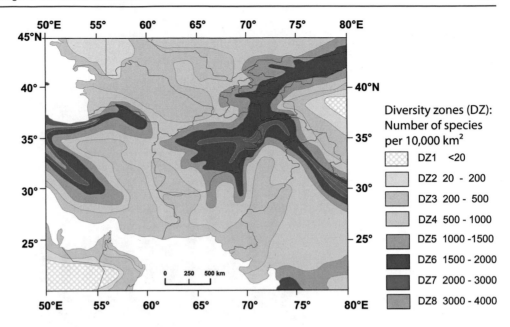

Diversity zones (DZ): Number of species per 10,000 km²

DZ1   <20
DZ2   20 - 200
DZ3   200 - 500
DZ4   500 - 1000
DZ5   1000 -1500
DZ6   1500 - 2000
DZ7   2000 - 3000
DZ8   3000 - 4000

Table 8.3 Number of families, genera, species, taxa and endemics in the Afghan flora, according to the checklist by Breckle et al. (2013, 2019)

Taxon group	Number				Endemics and sub-endemics	Immigrant species	Endemics (%)	Sub-endemics (%)	Endemics Total (%)
	Families	Genera	Species	Taxa					
Pteridophytes (56)	11	23	50	56	0	0	0 (0%)	0 (0%)	0 (0%)
Gymnosperms (24)	4	8	24	24	2	3	0 (0%)	2 (8.0%)	2 (8.0%)
Monocotyledons (840)	28	195	817	840	75	40	57 (6.8%)	18 (2.1%)	75 (8.9%)
Dicotyledons (4115)	106	860	3935	4115	1138	148	898 (21.9%)	243 (5.9%)	1138 (27.8%)
Total (5035)	149	1086	4826	5035	1215	191	955 (19.0%)	263 (5.2%)	1215 (24.2%)

Fig. 8.61 Number of species in the major vascular plant families of the Afghan flora (from Breckle et al. 2013, 2019)

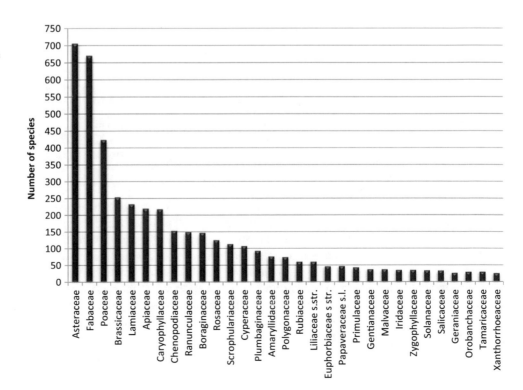

Fig. 8.62 Number of genera in the major vascular plant families of the Afghan flora

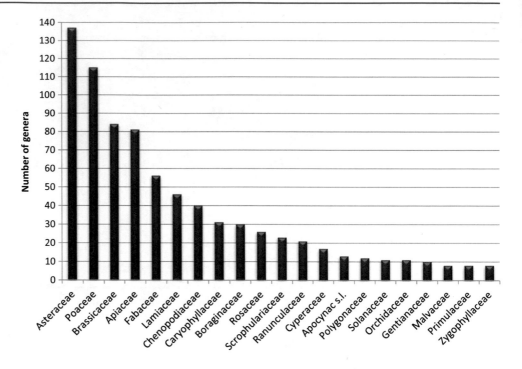

Fig. 8.63 Number of species in the major plant genera of the Afghan flora

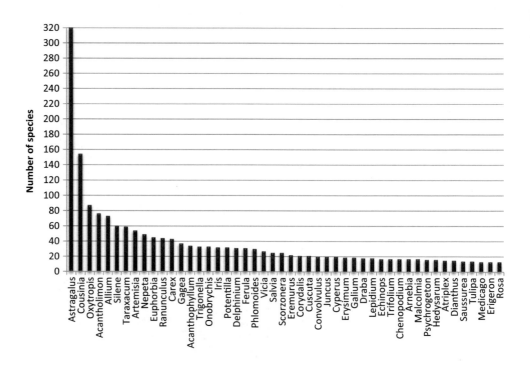

and mild but never quite frost-free winters. There is considerable variation in the distribution patterns of the individual species, perhaps explained by their large range but also by the different climatic histories in historical times. *Haloxylon salicornicum* (Fig. 8.67), *Cornulaca monacantha* (both Chenopodiaceae), and *Gymnocarpus decander* (Caryophyllaceae) are examples of particularly widespread lowland species. They overlap in range with Irano-Turanian species, or they are species that have penetrated far into the

Saharo-Arabian deserts from Irano-Turanian floral regions, such as the chenopodiaceous shrubs *Haloxylon persicum* (white Saxaul) and *Seidlitzia rosmarinus*.

Only in the basin of Jalalabad and less clearly around Khost (Paktia) one finds quite a few Saharo-Sindian species due to the mild winters and summer rains. The thorn bushes or small trees of *Acacia modesta* (Mimosaceae), *Zizyphus nummularia* and *Z. oxyphylla* (Rhamnaceae) and the evergreen shrubs *Calotropis procera, Periploca aphylla* and

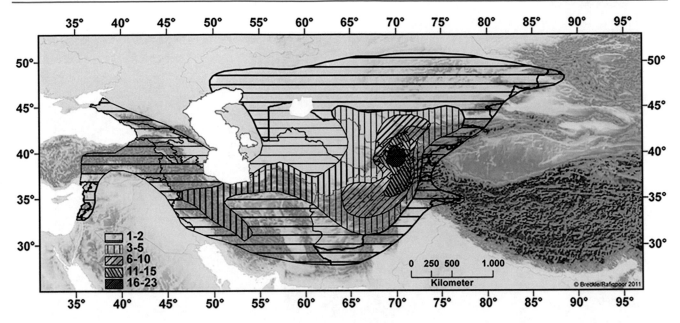

Fig. 8.64 Species density and distribution of *Eremurus* species as an example of Irano-Turanian distribution (Hedge and Wendelbo 1970)

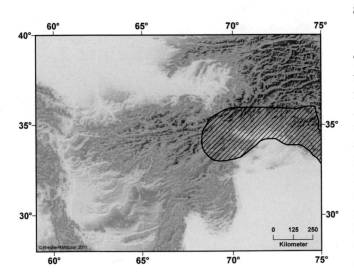

Fig. 8.65 The distribution of *Quercus baloot* as a Sino-Japanese element from the montane elevational range of the Himalaya and E Hindu Kush (as cited by Browicz 1978)

Rhazya stricta (Apocynaceae) can be cited. Another southern floral element is characterised by sclerophyllous trees and shrubs such as *Olea ferruginea*, *Reptonia buxifolia* (Sapotaceaae), the dwarf palm *Nannorrhops ritchieana* (Fig. 8.68), *Maytenus royleanus* (Celastraceae), *Ebenus stellatus* (Fabaceae), and *Dodonaea viscosa* (Anacardiaceae). They occur from the driest parts of W Himalaya, the Suleiman ranges and Baluchistan to the E and S of the Arabian Peninsula (Fig. 8.68).

8.12.4 Floristic Elements of the Afghan High Mountains

The alpine and nival elevational belts passes continuously through the C and E Hindu Kush, but become more isolated in the W, with outriggers in the Kohe Baba. At this altitude, C Asian floral elements are common, not infrequently mixed with Tibetan (e.g. *Delphinium brunonianum*, *Sibbaldia cuneata*, *Chorispora macropoda* and *Primula algida*), with Himalayan (e.g. *Anaphalis nubigena*, *Juncus membranaceus*, *Lamium rhomboideum*, *Primula macrophylla*, *Rheum tibeticum*) and with Irano-Turan mountain species. The uppermost belts are characterized mainly by Euro-Siberian, boreal (e.g., *Androsace villosa*, *Cerastium cerastioides*, *Cystopteris fragilis* incl. *C. dickieana*, *Lloydia serotina*), and arctic species (e.g., *Epilobium latifolium* (Fig. 8.69), *Smelowskia calycina*, *Koenigia islandica*) (Breckle 1974, 1988). However, in addition to cosmopolitan high altitude species such as *Luzula spadicea*, *Oxyria digyna*, *Polygonum viviparum*, *Phleum alpinum*, there are also narrowly restricted endemics in the alpine belt such as *Didymophysa fedtschenkoana*, *Papaver involucratum*, *Polygonum myrtillifolium*, *Polygonum chitralicum*, *Aconitum rotundifolium*, *Corydalis metallica*, *Gentiana longicarpa*, *Gynophorea weileri*, *Potentilla coelestis* or *Potentilla collettiana*.

To illustrate the elevational belts of the vegetation, two schematic profile diagrams are shown, each of which also reveals the typical asymmetrical character of the elevational belts (Fig. 8.70 and Fig. 8.71). This is particularly striking in the formation of the elevational gradients on the SE slope of

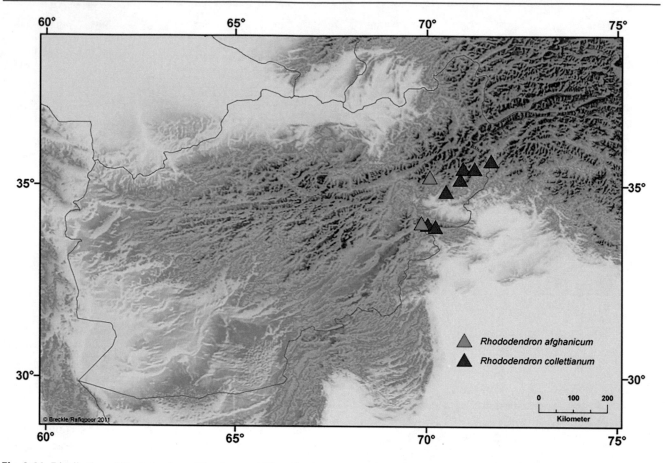

Fig. 8.66 Distribution of *Rhododendron afghanicum* and *Rh. collettianum* as two Himalayan elements in monsoon-influenced eastern Afghanistan (based on Flora Iranica data and new records)

Fig. 8.67 Distribution pattern of *Haloxylon salicornicum* as a lowland element of the Saharo-Sindian floral region with a wide spatial range of variation (as indicated by Browicz 1978)

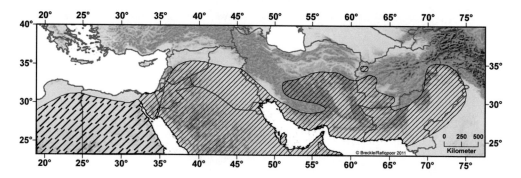

the Hindu Kush and Safed Koh, where the influence of the Indian summer monsoon becomes clear (Breckle & Frey 1976a, 1976b).

In general, the vegetation mosaic is determined on the one hand by the total amount of precipitation, which can vary greatly depending on the windward or leeward position, and locally by the accentuated topography, but on the other hand, especially under monsoon influence, by the additional occurrence of summer rains. Only then do dense forest belts form

(Neubauer 1954a, 1954b; Volk 1954; Freitag 1971a, 1971b). In the remaining parts of the Hindu Kush, one finds (today almost completely disappeared) at most very open tree vegetation, woods with wild almonds and pistachios. Most parts of the mountains are accordingly characterised by open scrubland, mountain steppes and semi-deserts. The elevational levels of the individual mountains are strongly stratified by plant geography (Breckle & Frey 1974; Breckle 1974, 1983, 1988; Agakhanjanz & Breckle 2002). The

Fig. 8.68 Distribution pattern of *Nannorrhops ritchieana* as a representative of the subtropical lowlands just extending into E Afghanistan (as cited by Browicz 1978)

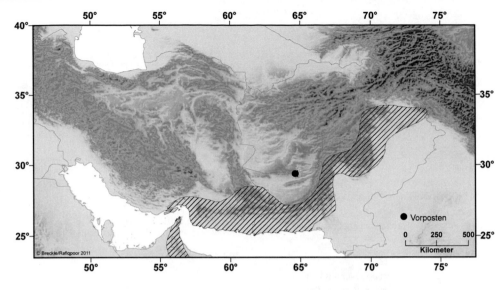

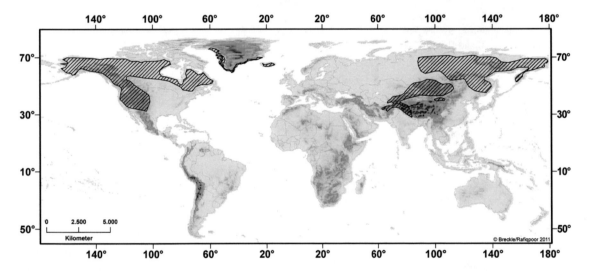

Fig. 8.69 Distribution pattern of *Epilobium latifolium* as a boreal element absent only from Europe and W Asia (according to Flora Iranica)

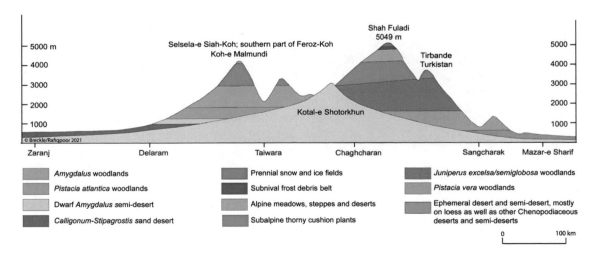

Fig. 8.70 Schematic arrangement of the elevational belts of vegetation in the mountains of C Afghanistan between Zaranj and Amu-Darya (adapted from Freitag et al. 2010 and Breckle 2004, 2007)

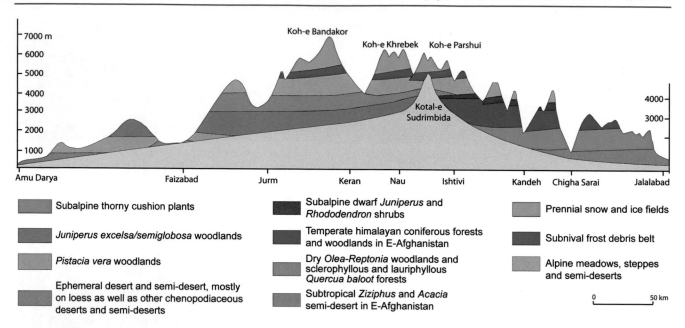

Fig. 8.71 Schematic arrangement of the elevational belts of vegetation in the mountains of E Afghanistan between Jalalabad and Amu-Darya (based on data from Freitag 2010 and Breckle 2004, 2007)

respective climatic influence also has an effect on the alpine belt. On the one hand, mountain semi-deserts or on the other hand, mats with a krummholz belt are the opposite expressions. Figure 8.70 gives an overview of the elevational zonation in the W compared to the SE Hindu Kush in Fig. 8.71. The well-developed forest formations with several altitudinal belts in the SE mountain slops also show a plant-geographical stratification (Breckle 2021).

Fig. 8.72 Remnants of the woody floras of *Cercis griffithii* near Topdarah in Charikar, Parwan Province, Afghanistan, show that this formation must once have been very widespread (photo: M. Keusgen)

The difference is particularly marked in the middle belts with the (originally) dense forest areas on the monsoon side and the dry steppe- to semi-desert-like mountain parts of the NE, N and W slopes (Fig. 8.71). But also in the W and C Hindu Kush, in the Kohe-Baba, in the Paghman Mountains, etc., there are sometimes great differences between the N and S slops, which are expressed in different occurrences of various open scrub vegetation. In the N one still finds *Pistacia vera* woodlands with the wild form of the edible pistachio. In C Afghanistan and further to W, woodlands with wild almonds and thorny dwarf almond shrubs (*Amygdalus* species) occasionally occur, and more rarely with *Cercis griffithii* (Fig. 8.72), a tree species that must still have been very common in the mountains around Kabul in the last century. On other mountain slopes one encounters isolated remnants of *Pistacia cabulica,* a species also endangered by fuelwood scarcity and by overgrazing. Single trees at sacred places ("Ziarat vegetation") indicate the once wider distribution.

Diamond (1997) notes that of the 56 total wild grasses with heavy seeds, 32 occur in the Mediterranean zone of western Eurasia alone, only 1 in England, 4 North America, 2 South America, and 2 in Australia (Table 8.4). Seed heaviness ranges from 10 mg to 40 mg. Diamond considers this advantage as one of the reasons why this area developed into an early centre of civilization.

Table 8.4 Regional occurrences of the 65 species of the family Poaceae with heavy seeds in their importance for cultivation

Region	Number of species
W-Asia, Europe, N-Africa	33
of which in Mediterranean area	32
England	1
E-Asia	6
Africa south of the Sahara	4
America	11
of which in N America	4
Central America	5
S America	2
N-Australia	2
Total	56

It can be seen that the majority of occurrences are concentrated in the Mediterranean regions (according to Diamond 1997)

8.13 Man in the Mediterranean

The Eurasian winter rainfall zone shows major differences compared to the winter rainfall areas in California, Chile, Cape Land and SW Australia. It occupies a considerably larger space and has a rich landscape differentiation both horizontally and vertically because of the W-E orientation of the Tertiary mountains of the former Thethys geosyncline (Pyrenees, Alps, Balkans, Carpathians, the Pontides, Caucasus, Taurus, Zagros, Suliman Mountains to Hindu Kush-Karakoram-Himalayan system). This favours an enormous floristic diversity, which Barthlott et al. (2014) summarize into a "Caucasus-SW Asia Centre of Diversity". Culturally and historically, this area is also the cradle of domestication of domestic animals and crops, especially starchy bread cereals (wheat, barley), compared to all other winter rain areas of the world, which was domesticated in the area of the Fertile Crescent and radiated from here to the other regions of the world.

The influence of humans in the Eurasian winter rain zone is therefore very old, starting with the early advanced civilizations in the Near East, and has also been very great for thousands of years (Table 8.5). Deforestation as early as several thousand years ago (for example by the Phoenicians in Dalmatia) led to the large-scale loss of the original sclerophyllous forests. Regeneration is no longer possible due to the completely eroded soil down to the bedrock. Soil formation on the exposed bare rock requires millennia.

Grazing and early agriculture in the Orient have led to a strong selection of species. Thorny and poisonous plants have spread.

The diversity of species was probably not significantly changed by man at first, some species were introduced, additionally promoted, so the olive tree probably came from northern E Africa and/or S Arabia several millennia ago; but today it is considered a character tree of the actual Mediterraneis. From the New World came agaves and opuntias, from S Africa aloes and Crassulaceae, from Australia acacias and eucalypts. The "eucalyptisation" of Portugal has dangerously increased the risk of forest fires there.

For large parts of the Mediterranean region (Mediterraneis), forms of use had mostly developed over the centuries that were quite well adapted to the ecological conditions. The widespread cork and holm oak stands in the W Mediterranean (Fig. 8.73), where firewood was cut, grazing animals were herded into them, and cork was also peeled (Fig. 8.6), were quite fire-resistant. They were interspersed with other small-scale crops, which provided even more fire protection. Today, many of these cultivated areas are abandoned, they become overgrown with bushes, and other areas are reforested with fast-growing *Pinus pinea* or *Pinus maritima,* which drastically increases the fire hazard.

With increasing human influence, biodiversity and ecosystem dynamics (the number of functional groups, interspecific interactions, etc.) clearly decrease, as shown in the diagram in Fig. 8.74. However, the rich mosaic of diverse

Table 8.5 Time scale for the influence of humans in Mediterranean ecosystems; the figures given are years before today (after Groves et al. 1983)

	Mediterraneis	Australia	South Africa	Chile	California
First appearance of humans: Hunter/gatherer, fire use	400,000	40–70.000	500,000	11,000	14,000
First appearance of pets	10–6000	150	20,000	400	400
First appearance of agriculture	10–6000	150	300	1000?	150
Intensive farming	2000-1000	50	300–200	400	50

Fig. 8.73 *Quercus ilex* forests in their uppermost distributional range on the eastern slope of the High Atlas Mountains in Morocco (photo Rafiqpoor)

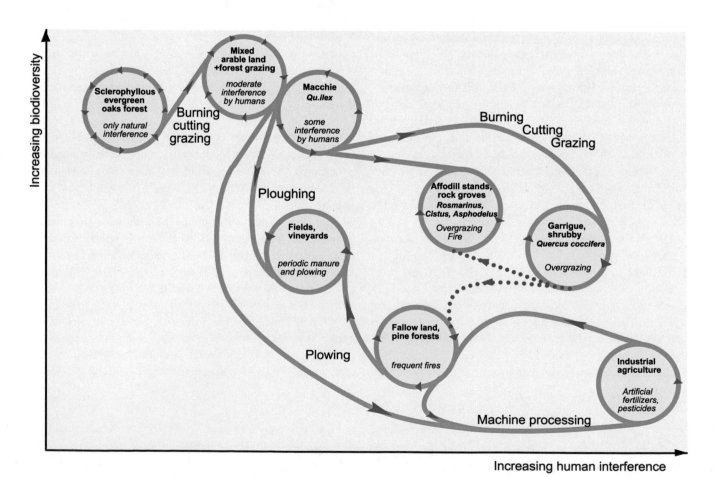

Fig. 8.74 Biodiversity, ecosystem dynamics and human influence in Mediterranean formations in the western Mediterranean region (modified after Blondel and Aronson 1995)

Fig. 8.75 Water is a valuable source, especially in areas with dry periods during the year as in mediteranean arid areas. The use of large water tanks in many countries is essential for water supply during dry summer as here in Kairouan / Tunisia near the famous mosque (photo: Breckle)

land uses, the mixture of small-scale agriculture, pastoralism, coppicing, transhumance, etc. of the late Middle Ages may have had the highest biodiversity until the beginning of this century (Blondel & Aronson 1995). Macchia, garigue and asphodel rock heaths are also often still very rich in species.

However, increased degradation and overuse, and the industrialisation of agriculture, have recently led to considerable impoverishment, at least for many groups of organisms and in many landscapes despite the long tradition of water storage and historical irrigation management (Fig. 8.75).

Photo 25 Matorral Scrub (ZB IV) at Cajón Del Maipo, Región Metropolitana, Chile (panoramio-Nelson Pérez)

Photo 26 Low-growing montane chaparral (ZB IV) with *Ceanothus* in bloom, Cleveland National Forest, Desconso Ranger District; Anza-Borrego desert in background (23 April 2017, Photo: Chaparralian)

Photo 27 West-Australian Acacia-woods (ZBIV) with grasstrees (*Dasypogon hookeri*) and many other endemic Australian plant species (photo: Breckle)

References

Agakhanjanz, O.E. & Breckle, S.-W. 2002: Plant diversity and endemism in High Mountains of Central Asia, the Caucasus and Siberia. In: Körner, C. & Spehn, E. (eds.): Mountain Biodiversity - A global assessment. Parthenon Publ. Group, Boca Raton, New York etc., Chapter **9**: 117-127

Arroyo, M.T.K., Zedler, P.H. & Fox, M.D. 1995: Ecology and biogeography of Mediterranean ecosystems in Chile, California, and Australia. Ecol. Stud. **108**: 455 S.

Axelrod, D.I. 1973: History of the Mediterranean ecosystems in California. Ecol. Stud. **7**: 225-277

Barbour, M.G. & Major, J. 1977: Terrestrial vegetation of California. Wiley-Intersci. Publ. 1002 S.

Barthlott, W., Erdelen, W. & Rafiqpoor, M.D. 2014: Biodiversity and Technical Innovations: Biomimicry from the Macro-to the Nanoscale. In: Lanzerath, D. & M. Friele (eds.): Concept and Value in Biodiversity. Routledge Studies in Biodiversity Politics and Management, 2014: 300-315. ISBN 978-1-415-66057-0

Blondel, J. & Aronson, J. 1995: Biodiversity and ecosystem function in the Mediterranean basin: human and non-human determinates. Ecol. Stud. **109**: 43-120

Breckle, S.-W. 1966: Ökologische Untersuchungen im Korkeichenwald Kataloniens. Diss. Univ. Hohenheim 190 pp.

Breckle, S.-W. 1974: Notes on alpine and nival flora of the Hindu Kush, East Afghanistan. Bot. Notiser (Lund) **127**: 278-284

Breckle, S.-W. 1983: Temperate Deserts and Semideserts of Afghanistan and Iran. In West, N.E. (ed.): Temperate Deserts and Semideserts. Ecosystems of the World (ed.: Goodall, D. W.), Elsevier, Amsterdam **5**: 271-319

Breckle, S.-W. 1988: Vegetation und Flora der nivalen Stufe im Hindukusch. In: Grötzbach, E. (Hrsg.): Neue Beiträge zur Afghanistanforschung. Schriftenreihe der Stiftung Bibliotheca Afghanica **6**: 133-148

Breckle, S.-W. 2004: Flora, Vegetation und Ökologie der alpin-nivalen Stufe des Hindukusch (Afghanistan). In: Breckle, S.-W., Schweizer B., Fangmeier, A. (eds.): Proceed. 2nd Symposium AFW Schimper-Foundation: Results of world-wide ecological studies. Stuttgart-Hohenhei 97117

Breckle, S.-W. 2007: Flora and Vegetation of Afghanistan. Basic and applied Dryland Research (BADR online) **1** (2): 155-194

Breckle, S.-W. (Hg.) 2021: Ökologie der Erde. Band 3. Spezielle Ökologie der Gemäßigten und Arktischen Zonen Euro-Nordasiens, Zonobiom VI - IX. 3. Aufl. Schweizerbart/Stuttgart 804 S.

Breckle S-W, Frey W 1974: Die Vegetationsstufen im Zentralen Hindukusch. Afghanistan J. (Graz) **1**: 75–80

Breckle, S.-W., Frey W 1976a: Beobachtungen zur heutigen Vergletscherung der Hauptkette des Zentralen Hindukusch (Afghanistan). Afghan. J. (Graz) **3**: 95-100

Breckle, S.-W., Frey W 1976b: Die höchsten Berge im Zentralen Hindukusch. Afghan. J. **3**: 91-95

Breckle, S.-W., Hedge, I.C., Rafiqpoor, M.D. 2013: Vascular Plants of Afghanistan - an augmented Checklist. Scientia Bonnensis, Bonn, Manama, New York, Florianópolis, 598 S.

Breckle, S.-W., Hedge, I.C., Rafiqpoor, M.D. 2019: Biodiversity in Afghanistan. In: Pullaiah, T. (ed.): Global biodiversity. Vol. 1. Selected countries in Asia. 33-92

Breckle, S.-W., Schweizer, B., Fangmeier, A. (Hrsg.) 2004: Ergebnisse weltweiter ökologischer Forschungen (Results of worldwide ecological studies, Proceed. of the 2nd Symposium of the A.F.W. Schimper-Foundation, establ. by H. & E. Walter, Hohenheim) Verlag Günter Heimbach, Stuttgart. 397 S.

Browicz K 1978ff. Chorology of trees and shrubs in SW Asia and adjacent regions. Vol. 1-10. Warszawa, Poznan

Browicz K 1982-1997: Chorology of trees and shrubs in SW Asia and adjacent regions. Phytogeographical analysis 1-10 + supplement. Polish Scientific publishers

Castri, F. di 1973: Climatographical comparison between Chile and the western coast of North America. Ecol. Stud. **7**: 21-36

Castri F di 1981: Mediterranean-type shrublands in the world. In: Castri F di et al (eds): Mediterranean-type shrublands. 1-52. Ecosystems of the World. Vol **11**, 643pp.

Castri, F. di & Mooney, H.A. (eds.) 1973: Mediterranean type ecosystems. Ecol. Stud. **7**, 405 pp.

Castri, F. di1981: Mediterranean-type shrub-lands of the world. In: Castri, F. di, Goodall, D.W. & Specht, R.L. (eds.): Mediterranean-type shrub-lands. Ecosystems of the World, Amsterdam, vol. **11**: 1-52

Diamond, J. 1997: Arm und Reich - Die Schicksale der menschlichen Gesellschaft. Fischer Verlag

Ern, H. 1966: Die dreidimensionale Anordnung der Gebirgsvegetation auf der Iberischen Halbinsel. Bonner Geogr. Abh. **37**: 136 S.

Frankenberg, P., Lauer, W. & Rafiqpoor, M.D. 1983: Zur geoökologischen Differenzierung Afghanistans. In: Lauer, W. (Hrsg.): Beiträge zur Geoökologie von Gebirgsräumen in Südamerika und Eurasien. Abh. d. Math.-nat. Kl. d. Akad. d. Wiss. u. d. Lit. Mainz. Franz Steiner Verlag, Wiesbaden: 52-71.

Freitag, H. 1971a: Die natürliche Vegetation des südostspanischen Trockengebiets. Bot. Jahrb. **91**: 147-208

Freitag, H. 1971b: Die natürliche Vegetation Afghanistans. Vegetatio **22**: 285-344

Freitag, H. 1975: Zum Konkurrenzverhalten von *Quercus ilex* und *Quercus pubescens* unter mediterranhumidem Klima. Bot. Jb. **96**: 53-70

Freitag, H., Hedge, I.C., Rafiqpoor, M.D. & Breckle S.-W. 2010: Flora and Vegetation Geography Afghanistan. In: Breckle, S.-W. & Rafiqpoor, M.D.: Field Guide Afghanistan - Flora and Vegetation. Scientia Bonnensis, Bonn, Manama, New York, Florianópolis: 79-111

Groves, R.H., Beard, J.S., Deacon, H.J. et al. 1983: The origins and characteristics of mediterranean ecosystems. In Day, J.A. (ed.): Mineral nutrients in Mediterranean ecosystems. S. Afr. Nat. Sci. Progr. Rep. No. **71**: 1-18

Hedge, I.C. & Wendelbo, P. 1970: Some remarks on endemism in Afghanistan. Israel J. Bot. **19**: 401-417

Hoffmann AJ, Armesto JJ 1995: Modes of seed dispersal in the Mediterranean regions of Chile, California and Australia. Ecol Stud **108**: 289-310

Höllermann, P. 1991: Studien zur Physischen Geographie und zum Landnutzungs-Potential der östlichen Kanarischen Inseln. Erdwissenschaftliche Forschung **25**. F. Steiner Verlag, Stuttgart. 276 S.

Kämmer, F. 1974: Klima und Vegetation von Teneriffa besonders im Hinblick auf den Nebelniederschlag. Scripta Geobot., Göttingen **7**, 78 S.

Kämmer, F. 1982: Flora und Fauna von Makaronesien. 179 S., Selbstverlag, Freiburg i. Br.

Knapp R 1973: Die Vegetation von Afrika. Veget-Monogr der einzelnen Großräume. Band **3**, 626pp. Fischer/Stuttgart

Kull, U. 1982: Artbildung durch geographische Isolation bei Pflanzen - die Gattung *Aeonium* auf Teneriffa. Natur und Museum **112**: 33-40

Kummerow, J. 1981: Structure of roots and root systems. Ecosystems of the World (Amsterdam), vol. **11**: 269-288

Kunkel, G. (ed.) 1976: Biogeography and ecology in the Canary Islands. Junk, The Hague

Kunkel G 1987 Die Kanarischen Inseln und ihre Pflanzenwelt. Fischer/ Stuttgart 205pp.

Larcher, W. 1961: Zur Assimilationsökologie der immergrünen *Olea europaea* und *Quercus ilex* und der sommergrünen *Quercus pubescens* im nördlichen Gardaseegebiet. Planta 56: 607-617

Leonard, J 1988: Contribution à l'étude de la flore et de la végétation des deserts d'Iran. Fasc 8, Meise. Jardin Bot Nat Belgique, 190 p.

Lösch, R. 1988: Funktionelle Voraussetzungen der adaptiven Nischenbesetzung in der Evolution makaronesischer Semperviven. Habil.-Schrift Kiel 491 S.

Mooney, H.A. & Parsons, D.J. 1973: Structure and function of the Californian chaparral - an example from San Dimas. Ecol. Stud. **11**: 83-112

Nahal I 1962: Contribution á l'étude de la végétation dans la Baer-Bassite et le Djebel Alaquite de Syrie. Webbia **16**: 477-641

Neubauer, H.F. 1954a: Die Wälder Afghanistans. Angewandte Pflanzensoziol. Festschr. Aichinger **I**: 494-503

Neubauer, H.F. 1954b: Versuch einer Kennzeichnung der Vegetations-verhältnisse Afghanistans. Ann. Naturhist. Mus. Wien **60**: 77-113

Nisançi, A. 1973: Studien zu den Niederschlagsverhältnissen in der Türkei unter besonderer Berücksichtigung ihrer Häufigkeitsver-teilungen und ihrer Wetterlagenabhängigkeit. Dissertation, Bonn

Oberdorfer, E. 1965: Pflanzensoziologische Studien auf Teneriffa und Gomera. Beitr. Naturk. Forsch. SW-Deutschl. **24**: 47-104

Ozenda, P. 1975: Sur les étages de végétation dans les montagnes du bassin méditerranéen. Doc. Cartogr. Ecol., Grenoble **16**: 1-32

Pott, R., Hüppe, J., Wildpert de la Torre, W. 2003: Die Kanarischen Inseln – Natur und Kulturlandschaften. Ulmer, Stuttgart, 320 S.

Peinado, M. & Rivas-Martines, S. 1987 (eds.): La vegetación des España. Ser. Publ. Univ. Alcalá de Henres, Madrid, 554 P.

Quintanilla V 1974: Les formations végétales du Chili temperé. Doc Cartogr Ecol Grenoble **14**: 33-80

Rafiqpoor, M.D. 1979: Niederschlagsanalysen in Afghanistan – Der Versuch einer regionalen klimageographischen Gliederung des Landes. Unveröf. Diplomarbeit am Geogr. Inst. d. Rheinischen Friedrich-Wilhelms-Univ. Bonn.

Read, D.J. & Mitchell, D.T. 1983: Decomposition and mineralization processes in mediterranean-type ecosystems and in heath-lands of similar structure. Ecol. Studies **43**: 208-232

Roberts, R. G., Flannery, T. F., Ayliffe, L. A. et al. 2001: New ages for the last Australian megafauna: continent-wide extinction about 46,000 years ago. Science **292**: 1888–1892

Rundel, P.W. 1982: The matorral zone of central Chile. Ecosystems of the world **11**: 175-201

Schmithüsen J 1956: Die räumliche Ordnung der chilenischen Vegeta-tion. Bonner Geogr Abh **17**: 86pp.

Specht, R.L. 1973: Structure and functional response of ecosystems in the mediterranean climate of Australia. Ecol. Stud. **7**: 113-120

Sunding, P. 1972: The vegetation of Gran Canaria. Norske Vid.-Akad. Oslo, I Math.-Nat. Kl. Ny Serie No. **29**, 186 p.

Sunding, P. 1973: A botanical bibliography of the Canary Islands. 2. ed., Bot. Garden, Univ. of Oslo

Voggenreiter, F. 1974: Geobotanische Untersuchungen an der natürlichen Vegetation der Kanareninsel Tenerife. Diss. Bot. Cramer/Lehre, **26**, 718 S.,

Volk, O.H. 1954: Klima und Pflanzenverbreitung in Afghanistan. Vegetatio **5** (6): 422-433

Walter, H. 1960: Standortslehre. 2. Aufl., Ulmer, Stuttgart. 566 S.

Walter, H. 1968: Die Vegetation der Erde, Bd. II: Gemäßigte und arktische Zonen. 1001 S., Fischer, Jena-Stuttgart

Walter, H. 1973: Die Vegetation der Erde, Bd. I: Tropische und subtropische Zonen. 3. Aufl., Fischer, Jena, Stuttgart, 743 S.

Walter H 1975: Betrachtungen zur Höhenstufenfolge im Mediterrangebiet (insbesondere in Griechenland) in Verbindung mit dem Wettbewerbsfaktor. Veröff Geobot Inst Zürich **55**: 72-83

Walter H 1990: Vegetationszonen und Klima. 6. Aufl, Fischer/Stuttgart, 382pp.

Walter, H. & Breckle, S.-W. 1991: Ökologie der Erde, Bd. 4: Spezielle Ökologie der Gemäßigten und Arktischen Zonen außerhalb Euro-Nordasiens. UTB Große Reihe, Fischer, Stuttgart. 586 S.

Zinke K, 1973: Analogy between the soil and vegetation types of Italy, Greece and California. Ecol. Stud **7**: 61-82

Part H: ZB V—Zonobiome of the Laurel Forests or of the Warm Temperate Humid Climate

9

Contents

© Springer-Verlag GmbH Germany, part of Springer Nature 2022
S.-W. Breckle, M. D. Rafiqpoor, *Vegetation and Climate*, https://doi.org/10.1007/978-3-662-64036-4_9

Photo 28 Laurel forests (ZB V) in Japan with a temple near Kyoto, Japan (photo: Breckle)

Photo 29 Laurel forest (ZB V) on the slopes of El Bailladero mountain on the island of La Gomera, Canary Islands with festoons of epiphytic mosses (photo: E. Fischer)

Photo 30 Tertiary relict forest (ZB V) in the Surami Mountains in Colchis (Georgia) with endemic ivy (*Hedera colchica*) on *Fagus orientalis* (photo: Rafiqpoor)

Photo 31 Most of the lauriphyll relict forests (ZB V) on the Canary Islands have now given way to agricultural use and settlements. View from the foothills east of La Laguna towards Pico de Teide (3700 m, top right in the background; photo: Rafiqpoor)

9.1   General, Climate, Soils

Zonobiome V cannot be sharply delimited; it is a transition zone between the tropical-subtropical (ZB I–III) and the typically temperate areas (ZB IV, VI, VII). However, it covers too large an area to be treated solely as an ecotone.

The climate is a typical transitional climate, which can be very different, but is generally quite mild. The ecological climate diagrams (Figs. 9.1 and 9.2) show the wide range of variation. Two main subzonobiomes can be distinguished:

The very humid **sZB V(s)** with rain throughout the year, with heavier summer rains, so with some minimum in the cool season. The main vegetation season is always humid and muggy because of the high temperature. These areas are mostly on the eastern sides of the continents about between the 30th and 35th degrees in the southern and northern hemispheres, and are under the influence of trade or monsoon winds. During the cool season, temperatures already drop quite low, light frosts may occur, but a cold season with temperatures below 0 °C is absent (Fig. 9.1); however, winter is already a dormant season for vegetation.

To **sZB V(s)** can be placed most of the E coast of Australia, most of New Zealand, the SW states of the United States, the SE corner of Brazil and Uruguay, the S coastal strip in South Africa, much of C China, the S half of Japan, and SE Korea.

The other **sZB V(w)** is predominantly bound to the western sides of the continents, somewhat further poleward to 40° latitude, i.e. somewhat shifted with respect to the first sZB V (s); it mostly adjoins the humid subzonobiome of ZB IV, where, indeed, winter rains predominate, but the summer drought is largely absent (Fig. 9.2). Both subzonobiomes are characterised by lauriphyllous tree species and/or large-growing conifers, each often rich in relict forms from the Tertiary.

The **sZB V(w)** may include small parts of the NW USA and SW Canada, the W coast areas of Chile including and N of Chiloe Island, small parts of N Portugal and NW Spain, the SE coast of the Black Sea (Colchis) and the S coast of the Caspian Sea (Hyrcania). The isolated laurel forests of Macaronesia can also be included.

On the E sides of the continents, as a result of the trade or monsoon winds, we are dealing with an almost continuous series from zonobiome II via a humid subtropical ZE II/V to ZB V and via a ZE V/VI to ZB VI. This is realised at several regions, and they are often species-rich transition areas, but also densely populated.

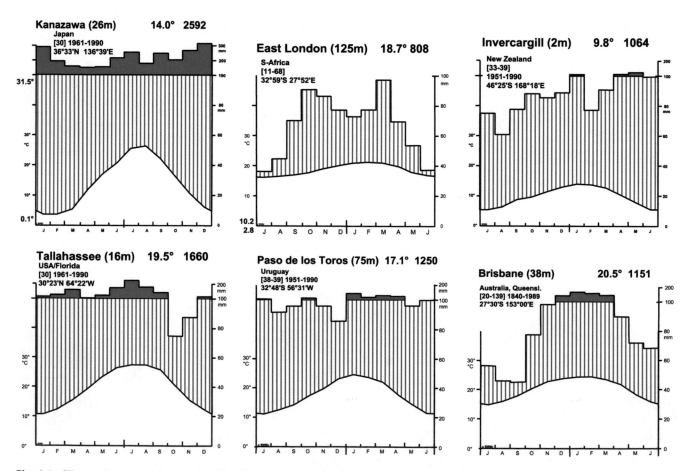

Fig. 9.1 Climate diagrams of the very humid sZB V(s) with rain in all seasons or especially in summer (Kanazawa in Japan; East London in South Africa; Invercargill in New Zealand; and Tallahassee in Florida, USA; Paso de los Toros in Uruguay and Brisbane in eastern Australia)

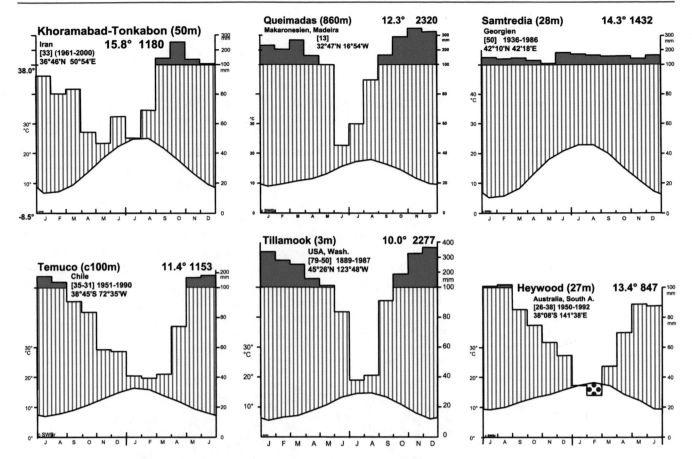

Fig. 9.2 Climate diagrams of sZB V(w) with rain at all seasons or a small maximum in winter (Khoramabad on the Caspian Sea, Iran; Queimadas in Madeira; Samtredia in the Colchis region in Georgia; and Temuco in central Chile; Tillamook in NW USA, Washington and Cap Leeuwin SW Australia—the latter is ZE IV/V)

Box 9.1 The zonobiome V is a transition zonobiome

sZB V(s): On the eastern sides of the continents transition from ZB I and ZB II rather with summer rains to temperate regions (ZB VI) with light frost.

sZB V(w): On the western side of the continent's transition from ZB IV rather with winter rain to ZB VIII with oceanic character.

Both the climatic and the vegetation-ecological demarcation of the above-mentioned sections is difficult. The tropics cease to exist where frosts become noticeable or the mean annual temperature falls below 18 °C when there is no frost, so that tropical crops such as coffee, cocoa, cocos, pineapple, etc. are no longer profitable and only tea, citrus and individual palms remain. In the area of zonobiome V, slight frosts already occur, but the mean daily minima of the coldest month are still above 0 °C, which means that a cold season does not occur. The annual means are slightly above or below 15 °C, the tree species of the forests are at least partly evergreen, while in ZE V/VI this is only true for some shrub species. For the ZB VI, a cold season of two or more months is already typical; the woody species almost all shed their leaves in autumn. Ecologically, the different regions of ZB V have probably not yet been studied intensively, nor can details be given about the ecosystems. It is also, particularly, difficult because most of the forests are rich in species and the growing conditions are favourable, so that the settlement density is usually high and the natural vegetation has almost completely disappeared. One must assume that the decisive factor is certainly the competition between the evergreen and the other species, but this is elusive.

9.2   Tertiary Forests, Lauriphylly and Sclerophylly

If the sclerophyllous vegetation of ZB IV evolved from a lauriphyllic vegetation form in geologically recent times, then lauriphyllic relict species should still be found in the Mediterranean region. *Laurus nobilis* occurs in the areas with higher precipitation and in diverse protected sites, especially also in the W Mediterranean. *Arbutus* (Fig. 8.5) is also lauriphyllic rather than sclerophyllic. Lauriphyllous species occur azonally in canyon forests or orozonally as cloud forest plots. Sclerophylls with woody cellular elements in the leaf (sclereids) have failed to displace lauriphylls only in very temperate sites and in nearly permanently humid areas.

As zonal vegetation, laurel forests occurs over large areas only in East Asia (China, Japan) and in the SE USA. But many of the evergreen temperate forests are now only present in remnants, their species richness (for example in China, Korea or Japan, but also in the SE USA) is remarkably high, especially in S Brazil. Many species are endemic to their respective regions.

The impoverished remnants of ZB V are the stands in N Portugal, which are often degraded to heaths.

The Euxinian (Colchis) and Hyrcanian relict forests (Fig. 9.3) are characterised by their Tertiary relict species (e.g. on the S coast of the Caspian Sea in Iran), where numerous genera are closely related to the Tertiary representatives attested by fossils. This is equally true for the other ZB V regions.

The laurel forests, which today still occur to a considerable extent on the E coasts of the continents, are classified by Klötzli (1987) as thermophilic (20 to 25 °C monthly mean in the vegetation period) and frost-sensitive (minima hardly below −10 °C) as well as drought-sensitive (hardly any arid months in the annual cycle). The distinction of ZB V from subtropical/tropical rainforests is given by their more and more regular precipitation and more balanced temperatures, from sclerophyllous forests by their lower and more sporadic rainfall (winter) and regular fires, from deciduous broadleaf forests by their colder winters with late frosts and often drier summers.

Fig. 9.3 The hyrcanian forests in northern Iran along the Caspian Sea show Tertiary relict species. The forests are small-scale and under considerable anthropogenic pressure (photo Breckle)

9.3 Subzonobiome on the Western Sides of the Continents

9.3.1 North America, Forests with Giant Conifers

The sZB V(w) with winter rain extends in N America from N California to S Canada in the coastal zone (Fig. 8.17, Vancouver). It is the zone of *Sequoia sempervirens* forests, followed further N by forests of *Tsuga heterophylla, Thuja plicata,* and *Pseudotsuga menziesii* (Fig. 9.4). *Prunus laurocerasus* and *Rhododendron ponticum,* but also *Araucaria excelsa,* thrive luxuriantly in the gardens here - a sign of the mild winters. Further N, temperatures slowly drop. The climate becomes increasingly humid, with little diurnal or annual variation in temperature. The maritime-toned and frost-sensitive Sitka spruce (*Picea sitchensis*) comes to dominate. In this meridional zone, which extends into the subarctic on Alaska, sections corresponding to ZB VI or ZB VIII can hardly be identified. It is an extremely humid oceanic ecotone in which no agriculture can be practiced, which is therefore sparsely populated.

Within the framework of the International Biological Programme (IBP), probably the most productive coniferous forests in the world, especially Douglas-fir *(Pseudotsuga)* ecosystems, were studied here. Edmonds (1982) contains the results of the work from 1971 to 1978 in eleven articles. Klötzli (1987) has given an overview of the distribution of evergreen forests.

In the **western USA**, in Oregon and Washington, these are humid, less frost-resistant coniferous forests, which have reached stand heights of over 100 m. The relict species

Fig. 9.4 Oceanic coniferous forest of the humid and mild western side with *Pseudotsuga menziesii* (Olympic National Park, USA) (photos: Barthlott)

Fig. 9.5 The *Sequoiadendron giganteum* forests, along with the redwood forests (*Sequoia sempervirens)* as Tertiary relicts, are famous and enormous forests of the temperate zones of the earth. The height of the *sequoia* trees exceeds with >100 m the highest rainforest trees on earth (photo: M. Neumann)

Sequoia sempervirens (Fig. 9.5), which occurs partly mixed with *Abies grandis, Pseudotsuga menziesii,* or is replaced to the N by *Tsuga heterophylla* and *Thuja plicata*, forms an upper canopy. Many deciduous tree species are represented in the lower tree layer *(Acer macrophyllum, Alnus rubra,* etc.). Many of the trees are richly covered with epiphytic ferns, mosses and lichens. These conifer forests of the warm-temperate zone, which are photosynthetically active almost all year round, must be regarded as relict forests of Tertiary origin. They were apparently little affected by the ice ages due to the N-S running mountains. Further S, larger refugial areas for vegetation were preserved during the ice age, so that, in contrast to the W-E mountain barriers in Europe, northward dispersal could take place rapidly.

9.3.2 Valdivian Rainforest in Southern Chile

In southern Chile, quite analogous conditions prevail. The sZB with winter rains, but without summer drought, corresponds to the likewise very lush Valdivian evergreen rainforest. The climate is permanently humid with high annual precipitation (Fig. 9.6).

Here in southern Chile, the Valdivia rainforest corresponds to ZB V(w). It is species-rich and its lushness reminds us of tropical rainforests. The climate is cool, without frost and permanently humid. Several relict conifers occur (among others *Fitzroya cupressoides, Austrocedrus chilensis, Podocarpus nubigenus, Dacrydium foncki, Araucaria araucana),* but never dominantly. Forest-forming are *Nothofagus* species (Fig. 9.7, Fig. 9.26); deciduous *N. obliqua* can grow over 40 m tall, and evergreen *N. dombeyi, Eucryphia cordifolia,* and others reach 35 to 40 m. The adjoining Magellanic forest to the S, with evergreen but also deciduous *Nothofagus* species and the strong formation of bogs, forms the perhumid transition zone to the subantarctic of Tierra del Fuego and the islands.

Fig. 9.7 *Nothofagus* forest with *N. dombeyi* and other *Nothofagus* species in Nahuelbuta National Park with large araucaria (*Araucaria araucana)* (photo: J. Renz)

9.3.3 Western Australia

In Australia, the SW tip belongs to this sZB with winter rainfall without summer drought (karri forest). This is indicated by the climate diagram of Dwellingup (Fig. 9.1). The Karri tree (*Eucalyptus diversicolor*) (Fig. 9.8) reaches heights of growth not infrequently exceeding 60 m, in some cases up to 90 m; they are giant trees. The famous Gloucester Tree (Fig. 9.8b) near Pemberton, for example, has a viewing platform at 61 m, which used to serve as a monitoring hut to spot forest fires in the area in time. But not only this species is endemic for SW Australia. Also the *Eu. calophylla* (up to 60 m) and *Eu. jacksonii* (up to 70 m), which also form mighty trunks, are the western tree species of this tall mixed forest on partly very poor and boggy soils. The streams and rivers carry brown humus-rich water. Carnivorous species, such as several *Drosera* species (including the climbing *Drosera macrantha)* or on rocks the red-flowered *Urticularia menziesii* (Fig. 9.9), occur in the understory of the forest

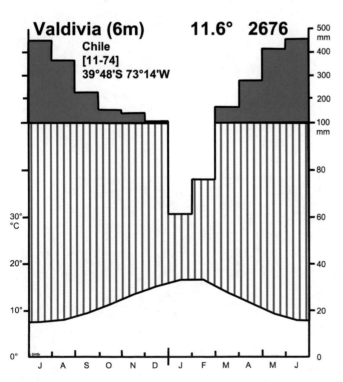

Fig. 9.6 The climate diagram of Valdivia demonstrates the ecological conditions of the Valdivian rainforests in southern Chile

Fig. 9.8 Karri forests of the Australian winter rains with **Eucalyptus jacksonii, Eu, calophylla** (**a**) (Photo: Breckle) and *Eu. diversicolor* (**b**) here the 72 m high famous Gloucester tree; Photo: Sean Mack, http://t1p.de(gxpo/) at the SW tip of the continent near Nornalup consist partly of giant trees, their growth heights are not infrequently 60 m, sometimes even surpassing 90 m

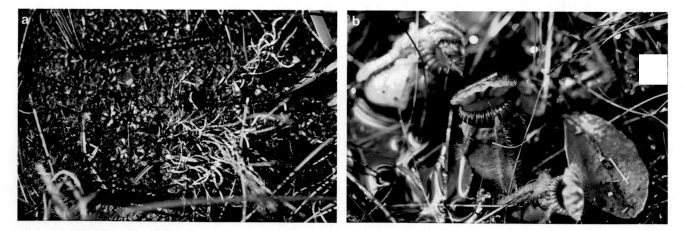

Fig. 9.9 In SW Australia, carnivorous plants reach their highest diversity on the old nutrient-poor soils. Here are two examples: *Utricularia menziesii* (**a**) and the Australian pitcher plant *Cephalotus follicularis* (**b**) (Photo: Breckle)

and on small interspersed bog areas, and on larger bog areas the ground-growing Australian pitcher plant *Cephalotus follicularis*. In the mostly very open shrub layer grows the myrtaceous *Calytris* and grass trees such as *Dasypogon hookeri* and *Kingia australis*.

9.3.4 Western Europe

In Western Europe, the frost-sensitive large conifers of the Pacific coast of N America are completely absent. They became extinct during the Pleistocene ice ages (fossils in the Rhineland lignite coal district in Germany). The closest equivalent to the sZB is the N Spanish and SW French coast with heath formations (Les Landes). The perhumid transition zone is as fragmented as the W European coastal zone. It is spread over Wales, W Scotland, the island groups with Ireland and the wettest parts of the Norwegian west coast with the Lofoten Islands, and extends into the subarctic. Heath moors with birch and willow species are the predominant vegetation today.

In Western Europe, therefore, there is a lack of vegetation corresponding to ZB V, although the climate today would allow such vegetation. Remains of some relict species can be found in the mountains near Algeciras (Campo de Gibraltar), where the evergreen *Rhododendron ponticum* ssp. *baeticum*, *Quercus lusitanica* and *Prunus lusitanica* still occur. In addition, the partly epiphytic fern *Davallia canariensis* and the primitive fern *Psilotum nudum* occur. The carnivorous plant *Drosophyllum lusitanicum* also indicates that these soils are poor in nutrients.

9.3.5 The Colchis and Hyrcania

In the Euxinian forest area of N Anatolia and W Georgia, only deciduous trees are found, although a whole range of evergreen species occur in the understory (*Prunus laurocerasus, Ilex, Buxus, Daphne pontica, Vaccinium arctostaphylos, Ruscus,* etc.). Similarly, the **Colchis** on the E shore of the Black Sea and the Hyrcanian forests on the S shore of the Caspian Sea. This area of **N Anatolia** and **W Georgia**, belonging to sZB V(w), with Colchian forests inhabited by *Rhododendron ponticum* and *Prunus laurocerasus* (Fig. 9.10), is an offshoot of the lush forests

Fig. 9.10 The *Fagus orientalis* forests (**a**) with *Rhododendron ponticum* (**b, d**), *Rhododendron luteum* (**e**) *Prunus laurocerasus* (**c, f**) in the isolated sZB V in the Colchis, Georgia (photos: Rafiqpoor)

Fig. 9.11 In Colchis, tea *(Camellia sinensis)*, a plant of cool-humid tropical regions, is also grown as a sign of lack of prolonged and severe frosts in winter (photos: Rafiqpoor)

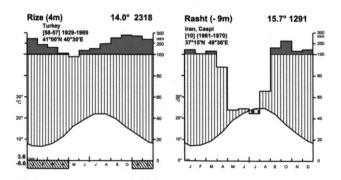

Fig. 9.12 Climate diagrams of Rize as an example of the mild climate of the Colchis and the Hyrcanian forests in the S of the Caspian Sea (Rasht)

in the Colchian Triangle between the Caucasian mountains and the Black Sea with evenly distributed precipitation up to 4000 mm. In this Tertiary relict forest, the evergreen understory has been preserved, but the tree layer with the relict species *Zelkova* and *Pterocarya* as well as *Dolichos* and the lianas *(Vitis, Periploca),* shed their leaves. Isolated cold spells occur, but *Citrus and* tea *(Camellia sinensis)* crops are present (Fig. 9.11), showing that the climate is mild, as indicated by the climate diagram of Rize (Fig. 9.12). The occurrence of the delicate skin ferns (*Hymenophyllum*, Fig. 9.13) also documents this. Skin ferns also occur in most other regions of ZB V.

The Hyrcanian relict forest (Fig. 9.3) on the S coast of the Caspian Sea (Fig. 9.12) is similarly formed with the relict species *Parrotia* (Hamamelidaceae) and *Albizzia julibrissin* (Mimosaceae), among others.

Fig. 9.13 Hymenophyllaceae skin ferns occur in the ever-humid forests of Batumi on the eastern Black Sea in Colchis (Georgia) (photo: Rafiqpoor)

9.4   Humid Subzonobiome on the Eastern Sides of the Continents

9.4.1   East Asia, China, Japan

In East Asia, which is exposed to the E Asian monsoon and therefore has a ZB II, this humid sZB of ZB V occupies a particularly large area. The N boundary at about 35°N still reaches the S tip of the Korean Peninsula with its many islands, bends northward in the Sea of Japan, and passes through the S part of the main Japanese island of Honschu (Fig. 9.1, Kanazawa). The island of Cheju-Do and Ullung-Do (Fig. 9.14) in SW Korea (Figs. 9.15, 9.16) have corresponding laurophyllous forest vegetation, to the extent that these still survive. Here, in addition to the evergreen Fagaceae *Cyclobalanopsis, Quercus* and *Castanopsis*, the Myrsinaceae *Ardisia* and the Lauraceae *Machilus*, among others, occur as forest-forming tree species. But also the

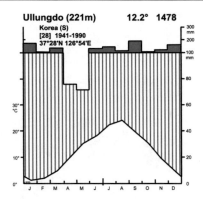

Fig. 9.14 Climate diagram of Ullung-Do as an example of the climatic region of laurophyllous forest vegetation in South Korea

ornamental shrubs *Aucuba japonica, Euonymus japonica, Ligustrum japonicum* and the frost-sensitive *Camellia*, which are commonly cultivated in N Italy (Insubria), originate from there. Further north, deciduous tree species gain the upper hand (Numata et al. 1972), as do higher elevations (Fig. 9.16), even with beech species, on Ullung-Do the relict species *Fagus multinervis,* on Japan with *Fagus japonica* and *F. crenata.* Surprisingly, an elevational belt (>2800 m) with the relict species *Fagus hayatae* is also developed on Taiwan (above the tropical lowland forests).

In China, the N border recedes somewhat inland to the S, to the extent that the cold spells from the Siberian high make themselves felt in winter. Much less sharp is the S boundary towards the evergreen tropical-subtropical forests of S China. Canton still belongs to ZB II. We bring here the division according to Ahti & Konen (1974) (Fig. 9.17). The orobiome in Japan is also discussed there.

In Japan, as in China—both densely populated countries—every easily cultivable spot is used for agriculture. It is therefore understandable that all deep, zonal soils are now cultivated land. The original vegetation has been completely repressed; it is found in remnants on lower slopes that are not suitable for either cropping or grazing. These give some idea of the original vegetation and show that it was extremely species-rich with numerous laurophyllous species. The climate is permanently humid (Fig. 9.1, Kanazawa). The area of evergreen forest extends over nearly 13 latitudes in Japan. The floristic composition also changes accordingly. In Kyushu, the evergreen forest reaches up to 800 m and is then replaced by a conifer-rich forest, but also with deciduous broadleaf forest species, which then dominate from 1500 m and correspond to a ZB VI deciduous forest. The evergreen forests there consist mainly of *Distilium racemosum, Castanopsis sieboldii, Cyclobalanopsis acuta* and *C. salicina* in the canopy layer and the same species as well as *Camellia japonica, C. sasanque, Machilus japonicus* and *Cleyera japonica* in the lower tree layer.

Fig. 9.15 Laurel forest remnants on Ullung-Do Island (South Korea) in individual valleys, transitioning on the upper mountain slopes to a beech forest with *Fagus multinervis*, as a relict forest (photo: Breckle)

Fig. 9.16 Open, young beech forest on upper mountain slopes of Ullung-Do Island (South Korea) with numerous *Fagus multinervis* stems. In the understory, individual evergreen laurel forest species still occur (e.g. *Euonymus japonicus*), but also many herbaceous species (cf. also Albert et al. 1996) from the genera *Helleborus, Hepatica, Maianthemum* and *Sasa* (photo: Breckle)

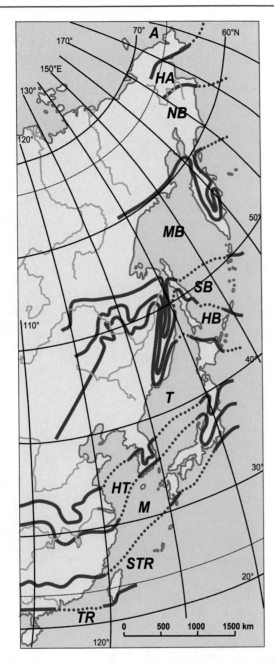

Fig. 9.17 Bioclimatic classification of East Asia (modified after Ahti & Konen 1974). TR = humid tropics, STR = humid subtropics. M = maritime warm temperate ZB V. HT = ZE V/VI and T = temperate ZB VI. HB = hemi-boreal mixed forest zone. SB, MB and NB = southern, middle and northern boreal zone (= ZB VIII). HA and A = hemi-arctic and arctic zone (= ZB IX)

9.4.2 Southeastern North America

In southeastern North America, the S tip of Florida is still tropical, but even Miami and Palm Beach experience light frosts. The evergreen oak forests with *Quercus virginiana* extend along the coast into N Carolina. The total area of ZB V is not very large because inland cold snaps extend to the Gulf of Mexico (Klaus 1975; Lauer 1999). It can be seen on the climate diagram (Fig. 9.18) that the temperature gradient is already clearly continental. In addition, psammobiomes are present on widespread

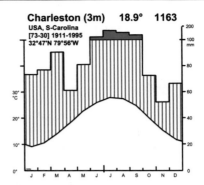

Fig. 9.18 Climate diagram of Charleston as an example of the marginal tropical climate region of SE North America with an already pronounced annual variation in temperature

sandy areas, namely pine forests of *Pinus clausa, P. taeda, P. australis*, and others, partly with evergreen understory. In addition, there are the extensive *Taxodium-Nyssa* swamp forests (hydrobiomes) and the evergreen *Persea magnoia* swamp forests as well as heather moors with the Venus flytrap (*Dionaea muscipula*) as helobiomes. Directly on the coast, salt marshes (halobiomes) occupy large areas.

9.4.3 Araucaria Forests of Southeast Brazil

In South America, the evergreen forests in E Brazil extend from tropical to subtropical and warm-temperate far to the south. The tropics stop at the coast between Porto Alegre and Rio Grande. Even in N Argentina, in Misiones and Corrientes, one speaks of subtropical forests. Along the great courses of the rivers Paraná and Uruguay they penetrate the pampas as gallery forests. On the coast the ZB V stops at La Plata, the ZB VI is missing. The climate diagram from Uruguay (Fig. 9.1, Paso de los Toros) illustrates the typical year-round humid climate.

On the plateau above 500 m, especially in S Brazil, the area of conifer forests of *Araucaria angustifolia* can be found (Fig. 9.19). In any case, these would have to be counted as

Fig. 9.19 *Araucaria angustifolia* forests in SE Brazil occupy larger areas but are now increasingly subject to deforestation (photo: Barthlott)

part of ZB V. In general, it is precisely in this part that the forest area has been greatly reduced by clearing.

9.4.4   South Africa

In Africa, whose SE coast is also exposed to the SE trade wind and receives very heavy rainfall from the wind jam off the Drakensberg, evergreen tropical-subtropical forests are common near the coast as far as East London. The section along the S coast can be described as warm-temperate. In former times the forests extended without interruption to the E slope of Table Mountain near Cape Town. However, most of it has been cleared or secondarily occupied by the fynbos of ZB IV. A larger forest reserve with old tall *Podocarpus* trees and a large number of broadleaf evergreens, among which the "stinkboom" *(Ocotea foetens)* provides valuable timber, is preserved only at Knyshna (Figs. 9.20 and 9.21).

Fig. 9.20 Remnants of the ever-humid Knyshna forests are preserved at the S tip of Africa in the vicinity of Knyshna. These forests owe their existence to the combination of summer and winter rains, which guarantee a year-round water supply. The forests get part of their moisture from frequent mist condensation (photos: Breckle)

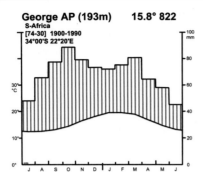

Fig. 9.21 Climate diagram of George as an example of the ever-humid climate area of the Knyshna forest in S Africa

9.4.5 Biomes of *Eucalyptus-Nothofagus forests* in South-eastern Australia and Tasmania

The moist tropical-subtropical evergreen forests of the E coast of Australia, which extend on nutrient-rich mostly volcanic soils into S New South Wales, consist predominantly of Indomalayan elements alien to Australis. Only in S New South Wales, Victoria and Tasmania does the Australis element predominate with the genus *Eucalyptus*. At the same time, however, some significant Antarctic elements are already mixed in. Here in the humid climate without cold season (Fig. 9.22), *Eucalyptus regnans* reached a proven height of up to 110 m (older statements of 145 m cannot be verified with certainty).

Today, tree heights of between 75 and 95 m can be found (Fig. 9.23). They were probably once the tallest giant trees on earth equal to the redwoods of California.

Eu. gigantea and *Eu. obliqua* grow almost as tall. The most important Antarctic species are the evergreen *Nothofagus cunninghamii* and the tree fern *Dicksonia antarctica,* in Tasmania also a number of other species as *Blechnum procera* and many mosses (Beadle 1981).

The particularly perhumid transition zone, on the other hand, comprises only W Tasmania with small *Eucalyptus*

species and bogs as well as the SW of New Zealand's South Island with the offshore Stewart Island. Thus, the transition to the subantarctic islands is given.

The composition of forests depends on the frequency of forest fires.

1. In the moist parts of W Tasmania, where forest fires do not occur, a tree layer of *Nothofagus* with *Atherosperma moschata* (Monimiaceae) 40 m high develops, and below this a 3 m high layer with the tree fern *Dicksonia,* which can still grow in illumination of 1% of daylight. In these humid forests, Hymenophyllaceae (Fig. 9.14) and mosses are very common as epiphytes.

2. If forest fires recur about every 200 to 350 years, then mixed forests are formed, which are three-layered. In addition to the above two layers, there is a 75 m (to 90 m) high layer of the three largest *Eucalyptus* species. This tree layer is of equal age, a sign that the germination of the trees occurred on larger areas after a forest fire. After such a forest fire, the tree layer of *Eucalyptus* and *Nothofagus* is destroyed, but the fruits open, and the intact seeds fall out and germinate. As *Eucalyptus* grows more rapidly, it overtakes *Nothofagus,* so that two layers of trees are formed. Tree ferns lose their leaves to fire, but regrow new ones at the top of the trunk. Regeneration of *Eucalyptus* under *Nothofagus* is not possible due to lack of light. It occurs again only after another fire.

3. If forest fires occur once or twice a century, *Nothofagus* is replaced by other fast-growing low tree species *(Pomaderris, Olearia, Acacia).*

4. After forest fires every 10 to 20 years, pure low *stands of Eucalyptus* develop.

5. Even more frequent fires cause degradation of the forests; an open moorland with the "button grass" *Mesomelaena sphaerocephala* (Cyperaceae) develops, in which Myrtaceae bushes are interspersed and *Drosera* and *Utricularia* occur alongside Restionaceae.

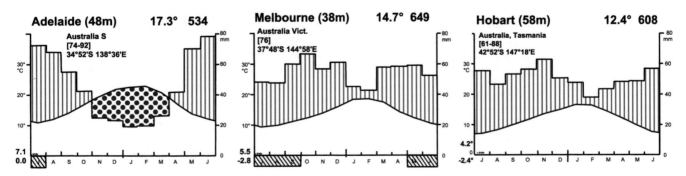

Fig. 9.22 Climate diagrams from the hardleaf region of S Australia and the warm-temperate region of Victoria and Tasmania

Fig. 9.23 *Eucalyptus regnans* high forest in Russel Falls National Park 60 km NW of Hobart on Tasmania (photo: Meraj Amiri)

9.4.6   Warm Temperate Biomes of New Zealand

The forests of New Zealand deserve a special mention. Although the two islands are relatively close to the Australian continent and there was probably a direct connection in the geological past, this must have been interrupted before the Australis flora had developed. There is not a single native species of *Eucalyptus* or *Acacia* on New Zealand. The Proteaceae are also represented by only two species.

In the N of the North Island, one still finds subtropical forests with the conifers *Agathis australis* as well as *Agathis microstachya* (Fig. 9.24) and palms; even mangroves of low *Avicennia* bushes grow along the coast. The species of the forests are Melanesian elements of the Palaeotropis. *Agathis* is also a giant conifer, with stem diameters measured up to 8.54 m and girth (BHD) of 26.6 m! but with not very high crown growth up to about 50 m.

Forests of this character extend even to the South Island, although the climate there is decidedly temperate, but at low altitudes without a cold winter season. The conifer genera *Podocarpus* and *Dacrydium,* which are widespread throughout the southern hemisphere, play a major role.

At the same time, however, the Antarctic element is important, with five evergreen *Nothofagus* species in the forests not only on the South Island, but also on the North Island. These mutually exclusive forest types are arranged in a mosaic fashion, with no clear climatic or ecological explanation for their distribution. One gets the impression that the plant cover is not in equilibrium with the present environment, but that historical factors play a very large role.

Large parts of the North Island were covered with a thick layer of hot volcanic ash by the massive eruption of Taupo 1830 years ago. The first pioneers to appear were the Podocarpaceae, which were spread by birds. They are slowly displaced by forests with tropical elements, and in the mountains partly by *Nothofagus* forests. The South Island was covered by large glaciers in the Pleistocene, so that recolonisation is still in progress there as well, especially since *Nothofagus is* slow to spread (Fig. 9.25).

In the extremely humid SW Fjordland with more than 6000 mm rainfall, the *Nothofagus* forests are already quite similar to those of S Chile. A special feature here are the **forest fall strips** reminiscent of avalanches, which, however, begin in the middle of the forest on steep slopes and are 2 to 6 m wide (Fig. 9.26). When the weight of the trees with heaps of epiphytic mosses growing on rock faces becomes too great, gravity causes the entire layer of vegetation, including the root system and the soil layer, to be removed. The bare rock left behind is recolonised with lichens, mosses and ferns until shrubbery and finally a stand of trees develops, whereupon another fall occurs.

A grave danger to the forests of New Zealand, where originally there was no mammalian species except bats, is represented by the abandoned European red deer, whose reproduction is beyond all control and which prevents regeneration of the often inaccessible *Nothofagus* forests, causing very great damage in the mountains by soil erosion and flooding. Equally dangerous is the Australian opossum (Kuzu: Fig. 9.27), introduced as a fur-bearing animal, which specialises on a tree species that forms the tree line, completely defoliates it and causes it to die, which also increases the risk of soil erosion on steep slopes.

New Zealand is an example of how dangerous it is when humans interfere with the natural balance by introducing new animals or plants. The damage often cannot be repaired.

Fig. 9.24 (**a**) Kauri forests with giant specimens of *Agathis microstachya* in N New Zealand; (**b**) Fruit stand of *Agathis* (photos: Breckle)

Fig. 9.25 The *Nothofagus* forests in Tianau, South Island of New Zealand (**a**), belong to the Antarctic elements and occur in all continents of the Southern Hemisphere as evidence of their former connection (Gondwana continent). (**b**) Leaves of *Nothofagus* from left to right *Loranthus micranthus* (as parasite), *Nothofagus fusca, Nothofagus menziesii* and the small-leaved *Nothofagus solandri*. In these forests all three species occur with a slight dominance of *Nothofagus fusca* (photos: Breckle)

Fig. 9.26 Forest fall tracks in the *Nothofagus* forests of the South Island (Milford) in New Zealand (photo: Breckle)

Fig. 9.27 Opossum, an invasive imported species that has become an ecological threat to New Zealand (photo: http://bit.do/bjTw6)

Photo 32 *Eucalyptus regnans* (Myrtaceae) is the main species forming very high trees with huge stems, formerly up to 130m, in the humid, temperate South-Eastern mountain slopes of Australia, but mixed with many other tree species and tree-ferns, mainly *Dicksonia antarctica* (photo: Breckle)

Photo 33 *Leucogenes leontopodium* (Asteraceae), the New Zealand edelweiss in the alpine altitudinal zone of the New Zealand Alps (Orobiom V) between rocks densely covered with lichens (photo: Breckle)

Photo 34 Thick carpets of *Sphagnum* mosses cover slopes and hills on the Acores, forming blanket bogs, which function also as huge water reservoirs; they develop under cool, perhumid climate (photo: Breckle)

References

Albert R, Breckle S-W, Choo Y-S 1996: Südkorea – Ökologische Exkursion der Universitäten Wien und Bielefeld. 1996, Wien, 276pp.

Ahti, L. T. & Konen, T. 1974: A scheme of vegetation zones for Japan and adjacent regions. Ann. Bot. Fenn. **11**: 59-88

Beadle, N.C.W. 1981: The vegetation of Australia. Veget.-Monographien der einzelnen Großräume, Band IV. Fischer/Stuttgart 690 pp.

Edmonds, R.L. (ed.) 1982: Analysis of coniferous forest ecosystems in the Western United States. US/IBP Synthesis Series **14**: 419 p.

Klaus, D. 1975: Niederschlagsgenese und Niederschlagsverteilung im Hochbecken von Puebla-Tlaxcala. Ein Beitrag zur Klimatologie der rand-tropischen Gebirgsregionen. Bonner Geogr. Abhandlungen **53**

Klötzli, F. 1987: On the global position of the evergreen broadleaved (non ombrophilous) forest in the subtropical and temperate zones. Veröff. Geobot. Inst. ETH, Stift. Rübel, Zürich **98**

Lauer, W. 1999: Klimatologie. Das Geographische Seminar. Westermann Verlag Braunschweig

Numata, M., Miyawaki, A. & Itow, D. 1972: Natural and semi-natural vegetation in Japan. Blumea **20**: 435-481

Part I: ZB VI—Zonobiome of Winter Bare Deciduous Forests or Temperate Nemoral Climate

10

Contents

© Springer-Verlag GmbH Germany, part of Springer Nature 2022
S.-W. Breckle, M. D. Rafiqpoor, *Vegetation and Climate*, https://doi.org/10.1007/978-3-662-64036-4_10

Photo 35   Temperate nemoral deciduous forest (ZB VI) with beech (*Fagus sylvatica*), oaks (*Quercus robur, Qu. petraea*) and hornbeam (*Carpinus betulus*) in the Eifel, Germany (photo: E. Fischer)

Photo 36 Old beech trees (*Fagus sylvatica*) near Lemgo (Teutoburg Forest) in a nemoral mixed oak forest (ZB VI) with tree fungi on the left trunk in winter; the tree roots are partly exposed by soil erosion over the decades (photo: Breckle)

Photo 37   Mixed deciduous forest (ZB VI) with dense shrub layer in early fall, N of Beijing, China (photo: Breckle)

10.1   Leaf Shedding as an Adaptation to the Winter Cold

A temperate climate zone, **zonobiome VI**, with a distinct but not too long cold season (Fig. 10.1) is clearly developed only in the northern hemisphere. It is absent from the southern hemisphere except for certain mountainous areas of the southernmost Andes and New Zealand. We had already become acquainted with facultatively deciduous tree species in the tropics, whose leaves fall when the water balance is disturbed during a prolonged drought, thus reducing the water losses of the trees.

The triggering factor that causes yellowing of the foliage in autumn before the first frosts is usually not known more precisely. It is probably partly the shortening of the day length. Strikingly, the foliage discoloration of the various tree species takes place in a relatively short period of time. According to the phenological calendar in the twentieth century, yellowing in Central Europe occurs between October 10 and 20, with no sharp difference between places in the west and in the east, nor between low and high mountainous areas. Trees near street lamps remain green longer. In recent years one observes an extension of the vegetation period in Central Europe from the 20th to the 30th of October.

The evergreen foliage is neither resistant to cold nor to frost-dryness, i.e. prolonged temperatures below 0 °C. In Central Europe, the evergreen cherry laurel *(Prunus laurocerasus)* always freezes back in gardens and parks during severe frost (Fig. 9.10c,f). Several subspecies of these have perhaps also become widely established in central European front gardens, forest edges, etc. in the course of Climate Change, which is now noticeable after all.

Even with moderate frost, the leaves show CO_2 release in the light, which means that respiration continues, but photosynthesis is blocked. *Ilex aquifolium* (holly) has an Atlantic distribution. *Hedera helix* (ivy) is a sub-Atlantic evergreen

species that avoids eastern continental areas with cold winters. The same is true of the broom species *Ulex* and *Sarothamnus*. The evergreen alpine roses *(e.g. Rhododendron ferrugineum)* and the cranberry *(Vaccinium vitis-idaea)* can only withstand the winter cold in Central Europe under snow protection.

The shedding of the thin, deciduous leaves in winter and the protection of the buds from water loss mean a saving of substance compared to the freezing of thick evergreen leaves. The prerequisite, however, is that the newly formed leaves in spring have a sufficiently long and warm summer period of at least four months to ensure the growth and maturation of the lignifying axial organs and the accumulation of substance reserves for fruiting and for budding in the following year. But even in the leafless state, the branches lose water in winter, and to varying degrees in the different hardwood species. The Central European beech therefore avoids the zone of cold Eastern European winters. The oak, on the other hand, even reaches the Urals. In extremely continental Siberia, deciduous trees are absent except for the small-leaved tree species birch *(Betula)* and trembling poplar *(Populus tremula) as* well as mountain ash *(Sorbus aucuparia)* with its small leaflets.

> **Box 10.1 Forms of Adaptation of the ZB of Deciduous Forests**
> In ZB VI, the zonobiome of temperate deciduous forests, leaf drop is an adaptation to a cold season. However, it is not facultative but obligate, so it occurs even if tree plants are protected from the winter cold in a greenhouse.

If the summers are too short and too cool, the evergreen conifers take the place of the deciduous trees. Their

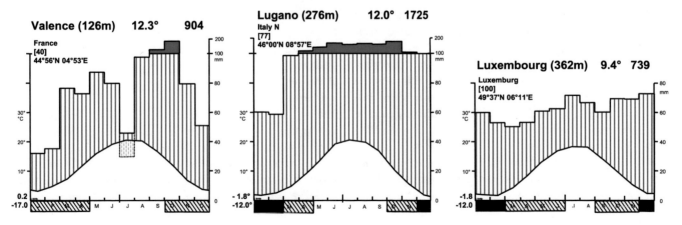

Fig. 10.1 Climate diagrams from the sub-Mediterranean zone (without cool winters: Valence), from the warm and humid deciduous forest zone (Lugano), and from the Central European beech forest zone (Luxembourg)

xeromorphic needles acquire a higher resistance to cold in winter and are ready to produce again when the warm weather arrives in spring. This makes better use of the short growing season. While deciduous trees require a growing season with daily averages above 10 °C of at least 120 days, conifers can manage with as little as 30 days. However, the resistance of the individual species also varies. The yew *(Taxus baccata)* does not go further east in Europe than the ivy. *Pinus sylvestris* (pine) and *Picea abies* (spruce) are very resistant. *Abies sibirica* and *Pinus sibirica (P. cembra)* hold out in Siberia, but farthest into the continental Arctic (to 72° 40' N) advances the deciduous conifer, the larch *(Larix dahurica)*, which has a very high productive power in the short summer. We thus see that, depending on the external conditions and the ecophysiological properties of the species, those with evergreen assimilation organs soon outperform those with short-lived deciduous ones in competition and achieve dominance (compare ZB II, 2).

10.2   Importance of Winter Cold for Species of the Nemoral Zone

In Zonobiome VI, as we have seen, winter frost, even if usually brief, plays an important role in species adaptation. The damage that occurs in cold winters can have two causes:

1. These are direct cold damages, which are related to the freezing of water in the tissues; they are then called **frost damages**.
2. There is a drying out of the above-ground organs, which have a certain transpiration even at low temperatures and are not able to cover the water losses from the frozen soil due to blockage of the conductive vessels by ice. In this case, therefore, we are dealing with frost drought or **frost desiccation**.

Plants have no protection against the effects of low temperatures. Their temperature adapts to the prevailing air temperature. The only adaptation to prevent the damage caused by low temperatures is to become hardened, the **hardening process**. If one tests the cold resistance of plant parts in summer by exposing them to various temperatures below 0 °C in the freezer, for example for two hours, it is found that even low freezing temperatures are sufficient to cause irreversible damage. The same plant parts, on the other hand, can withstand exposure to much lower temperatures in winter without damage because they are hardened off. Hardening is a physiological process that takes place in autumn when the first cold nights begin. It is replaced in the warm spring by the opposite process of "**softening", dehardening**.

Hardening is associated with certain physicochemical changes in the protoplasm. The stability of the membranes (for example due to additional sulphur bridges -S-S-) increases, as does the viscosity of the plasma. It can be recognised by the fact that during plasmolysis (shrinkage of the protoplast of a plant cell) a concave plasmolysis occurs instead of a convex one. This change is accompanied by a sudden increase in cell sap concentration due to an increase in sugar concentration and other osmotica. In the hardened state, the protoplasm is largely inactivated. Cold resistance can increase in overwintering buds of our deciduous trees from −5 °C in autumn to over −25 °C, even −35 °C in January to February. The increase in cold resistance is greater in cold winters than in mild ones, and in related species of a genus the greater the further a species advances into the continental region.

Hardening is a very complicated process that occurs in several stages. The first, which leads to a certain state of dormancy, is initiated in autumn by the shorter length of the day. Further hardening takes place when the temperature drops to a few degrees above 0 °C. The strongest hardening off is observed in species that have already been exposed to very low temperatures, i.e. after the first heavy frosts have occurred. If hardened plant parts are suddenly cooled extremely strongly so that **vitrification** of the protoplasm occurs (without ice crystal formation), freezing in liquid nitrogen (at −190 °C), even at −238 °C, is possible. However, it is necessary to carry out the heating slowly in several steps until thawing, so that subsequently no plasma-damaging ice crystal formation occurs. Then the hardy species of the cold climate zones remain alive. In eastern Siberia around the cold pole, forest vegetation is normally exposed to winter temperatures of −60 °C or lower. Tropical species and even those of ZB IV or ZB V cannot be hardened off.

Hardening generally prevents frost damage to native trees even in severe winters, whereas planted exotics from warmer home areas without hardening- ability often suffer such damage. Frost damage, on the other hand, often occurs when early frosts set in before plants have been hardened, or a late frost sets in after dehardening has already occurred. It is especially common to see late frost damage to young foliage that has sprouted and is very sensitive to frosts. Cambium damage due to late frost also occurs when the trees are already "in sap", i.e. the plasma is already in an active state.

The eastern limit of the beech area is probably due to frequent late frost damage, which reduces competitiveness. For the herbs of the forest, an increase in cold resistance can also be observed in winter due to hardening. However, they are not exposed to such low temperatures under a litter and snow cover. In line with this, cold resistance (for example of *Hepatica triloba*) increases even in the evergreen leaves only down to −15 °C, in the better protected flower buds down to −10 °C and in the rhizomes only down to −7.5 °C.

It is more difficult to detect damage caused by frost-drought. The shedding of the strongly transpiring leaves, the protection of the buds by hard bud scales, and of the twigs by layers of cork, prevent greater losses of water in deciduous trees in winter. Nevertheless, some transpiration of the leafless branches in winter can be detected; it is higher than in the evergreen conifers, and higher in the hardwood species with a southern distribution than in those occurring farther north. These transpiration losses become critical when the intensity of insolation increases in spring and the air temperature rises, but the ground is still frozen solid. Buds and twigs may then dry out. Particularly sensitive in this respect are evergreen species, such as the holly *(Ilex)* or rod shrubs such as the broom species.

In general, frost damage occurs during the coldest part of the year, while frost drought damage occurs in the transitional period to spring and on warm southern slopes (in the northern hemisphere). They should not be confused with late frost damage.

10.3 Distribution of the Zonobiome VI

The climate of ZB VI, with a warm growing season of four to six months with sufficient rainfall and a winter season of three to four months that is not too long or extremely cold, is particularly suitable for the deciduous tree species of the temperate climate zone (Ellenberg & Leuschner 2010). These trees avoid the extreme maritime, as well as the extreme continental areas. We speak of the nemoral zone. Such a climate with a precipitation maximum in summer is found in the Northern Hemisphere in eastern N America and in E Asia between the warm temperate and the cold or arid temperate zones. In Western and Central Europe, it is the region north of the Mediterranean zone where, under the influence of the Gulf Stream, winter rains are replaced by evenly distributed precipitation or those with a summer maximum, and the cold season is relatively short.

The Mediterranean winter rainfall area with sclerophyllous vegetation extends south of it very far from west to east and passes northwards into very different vegetation zones. In the very maritime area in SW and S Spain (for example near Gibraltar) one finds, as already mentioned, elements of the evergreen warm temperate laurel forests. In Portugal, however, this vegetation merges into the Atlantic heaths, which extend into the coastal area as far as Scandinavia (Fig. 10.2). They are replaced by birch forests in the far north. True laurel forests are found only on the humid windward side of the Canary Islands (e.g., Tenerife) as fog forests (Fig. 8.48) or in the very similar ZE IV/V in northern Anatolia (Colchis). A sub-Mediterranean zone is interposed between the Mediterranean and nemoral zones. In it, winter rains still prevail, but summer drought is no

Fig. 10.2 Distribution of Atlantic heaths in W Europe (modified after Hüppe 1993)

longer pronounced and mild frosts occur regularly in all winter months (Fig. 10.1, Valence).

Northeast of the sub-Mediterranean zone in SE Europe is the steppe zone, which is replaced further N by forests of various types. Finally, in the Near East, the Mediterranean sclerophyllous zone gradually leads to the Mediterranean steppes and semi-deserts.

10.4 Atlantic Heaths

The Atlantic heaths (Fig. 10.1, Fig. 10.2) are almost always degradation stages of deciduous forests. The destruction of woodland in this area dates back to prehistoric times; it is now so complete that heathland was long thought to be zonal vegetation. The historical development can be easily traced in pollen diagrams (Fig. 10.3). Along with the clearing, the bogging initially increased, and very soon *Calluna* appeared over large areas, in parallel with charcoal remnants.

The soils in this region are usually extremely poor and acid, and it was supposed that, in consequence of the humid climate, they were naturally exhausted, and could support only a poor heath vegetation. But the same is true in this case as we have stated with regard to the tropical rainforest, which is also very humid. As long as the natural forest vegetation is not touched, there is no leaching of nutrients from the biogeocoen; the nutrient stock remains for the most part stored in the above-ground phytomass. However, once

Fig. 10.3 Atlantic heaths in the fog coasts of NW Norway near Bergen (photo: D. Killmann)

the woodland is cleared and burnt, most of the now mineralised nutrients are lost and only the poor soil remains. If the heath vegetation is subsequently exploited or repeatedly burned, the reforestation, already difficult here, cannot take place. One knows of uninhabited very extreme oceanic climates with similar temperature conditions and even double to four times the amount of precipitation on the Pacific coast of NW North America, in the SW of S America, on Tasmania and on New Zealand, where the virgin forests grow in great luxuriance and show no signs of degradation by a leaching of the nutrients.

> **Box 10.2 Nearly All Plants of the Sub-Mediterranean Zone Are Not Sclerophyllous Species**
> In the sub-Mediterranean zone, the evergreen woody species are absent except for *Buxus*. Tree species such as downy oak *(Quercus pubescens)*, manna ash *(Fraxinus ornus)*, French maple *(Acer monspessulanum)*, hop hornbeam *(Ostrya carpinifolia)* or the frequently cultivated chestnut *(Castanea sativa)* are all deciduous, which is why this zone is classified as a temperate deciduous forest zone and not as a Mediterranean zone. It can be called ZE IV/VI.

How the original forests in W Europe were composed is not easy to say. Oaks *(Quercus petraea* and *Qu. robur)* probably played the main role, in the N also birches *(Betula)*; in addition, *Ilex aquifolium* was an evergreen species. The heath *(Calluna)* was formerly present as an undergrowth in these forests and formed independent communities only in sparse places on shallow or peaty soils. After the destruction of the forests, it then took possession of the entire area (Fig. 10.4). The regeneration of forest formations after the cessation of grazing and ploughing (in which the upper 10 cm of the raw humus layer is cut off in square pieces, used as litter in the stable and then brought to the field as stable manure for fertilisation) proceeds differently depending on the land endowment. This has been pointed out by Leuschner (1993) (Fig. 10.5) and he has reconstructed the corresponding regeneration stages with those originally decisive before heath formation (Fig. 10.6). Also indicated are the factors responsible for the regeneration of woodland on heathland and, conversely, for the maintenance of heathland.

In the S part of this coastal zone (Fig. 10.2), broom species *(Ulex, Sarothamnus,* and *Genista* species) play the main role, along with various *Erica* species. In the central part, the broom species recede more; *Ulex europaeus, Sarothamnus scoparius* and *Genista anglica* remain as the most important representatives; in return, the Ericaceae become more prominent in terms of quantity, besides *Erica cinerea* and *E. tetralix* especially the heather *(Calluna vulgaris)*. In

Fig. 10.4 Development of heathland in the postglacial period as shown by a simplified pollen diagram from the raised bog (modified after Hüppe 1993)

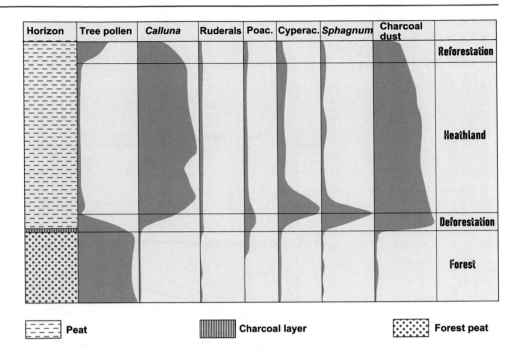

Peat Charcoal layer Forest peat

the N, *Empetrum, Vaccinium, Phyllodoce* and *Cassiope* dominate.

Calluna heaths account for ¼ to ½ of the total area in Scotland (Fig. 10.3); the soil type is iron podsole with a hard cemented B horizon often formed as a hardpan (**"Ortstein"**, a subsoil cemented **horizon = eluviated soil).** The heath is periodically burnt. *Calluna vulgaris* is the absolutely dominant species. It is a dwarf shrub, growing to about 50 cm high, forming a dense web of roots in the upper 10 cm of the soil, with individual roots going down 75 to 80 cm to the cementation **horizon.** *Calluna*'s very small leaves are densely attached to short shoots, a large proportion of

Fig. 10.5 Hypothetical scheme of present-day regeneration stages and forest dynamics of the Lüneburg Heath on nutrient-poor sands under different human interventions. Processes inhibited by high game densities are drawn with dashed lines, those with fragmented forest cover are dotted dashed lines (modified after Leuschner 1993)

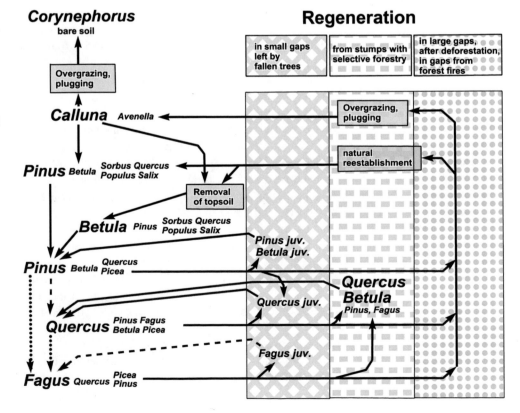

Fig. 10.6 Hypothetical scheme of regeneration stages and forest dynamics of the Lüneburg Heath under natural conditions before forest destruction (ca. 800 BC) (modified after Leuschner 1993)

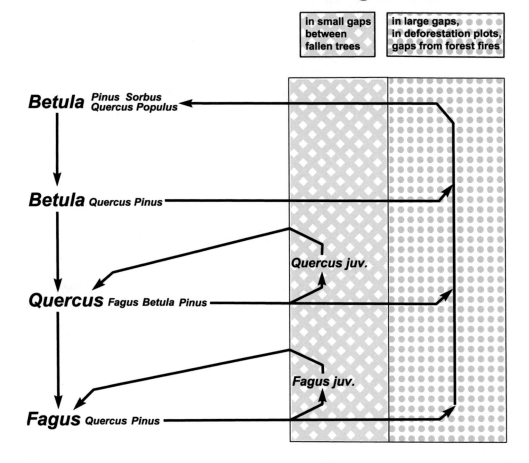

which are shed in autumn, reducing the risk of frost-dryness during cold spells. The annual litter production in a dense stand is 421 kg per hectare.

If burning occurs every 30 years, three phases (each 10-year) of stock development can be distinguished:

1. the **build-up phase** (reconstructive phase) of the dwarf shrub layer after the fire; some of the nutrients are fixed in the litter,
2. the **ripening phase** (phase of maturity) with increasing litter production but a decreasing increase in phytomass,
3. the **degeneration phase**, in which litter production remains constant and litter decomposition increases until a state of equilibrium is reached. After 35 years, the standing phytomass is 24,000 kg/ha and the litter quantity is 17,000 kg/ha.

Most often, people do not wait for the degeneration phase, but burn down the heath already after 8–15 years. In this humid area the fires are caused only by man. Natural fires caused by lightning hardly ever occurred in the original forests, so without human intervention there is no degradation of the forest. From heath to moor there are all transitions. We list four stages of increasing waterlogging, in each of which the species are named according to the decreasing abundance:

1. *Erica cinerea—Calluna vulgaris—Deschampsia flexuosa—Vaccinium myrtillus,*
2. *Calluna vulgaris—Erica tetralix—Juncus squarrosus,*
3. *Erica tetralix—Molinia coerulea—Nardus stricta—Calluna vulgaris—Narthecium ossifragum,*
4. *Erica tetralix—Trichophorum caespitosum—Eriophorum vaginatum—Myrica gale—Carex echinata.*

In Scotland, heathland is used for hunting and as extensive sheep grazing, with 1.2 to 2.8 ha of grazing area calculated for one sheep. Due to grazing and the anthropogenic nitrogen input coming from the atmosphere today, the growth of the grasses is more promoted. But even in earlier times, there was a cycle in the heathlands between a *Calluna* and an *Avenella* phase (Fig. 10.7).

In the Lüneburg Heath (Fig. 10.8a), which is also of purely anthropogenic origin, arable farming was formerly

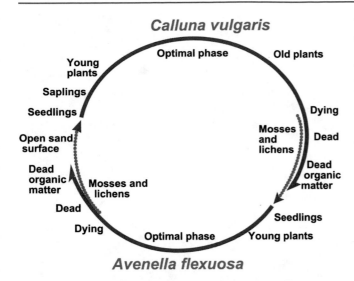

Calluna vulgaris

Avenella flexuosa

Fig. 10.7 Possible cyclic change in dominance between _Calluna_ and _Avenella_ on Dutch heathland (modified after Kaagman & Fanta 1993)

practised on sandy podsol soils (Fig. 10.8b) often with buckwheat (Fig. 10.9); in the process, the heath was ploughed off for peat. This "plaggenhieb" prevented the reforestation. Today, after the heath is no longer used, it becomes wooded through the growth of birch and pine seeds, or it is systematically reforested.

In the extremely maritime area, besides the heath, **blanket bogs** play a major role. The climate is very balanced; on Ireland, for example, the temperature in January is 3.5 to 3.7 °C, whereas in July it is 14 to 16 °C. Frost may occur, but snow is only from 3 to 10 days a year. Precipitation ranges from 350 to 1000 mm per year and is very regular throughout the year. It also varies from year to year by no more than 25%. With the heavy cloud cover, the sunshine duration is only 31% of the maximum possible. Under these circumstances, the risk of swamp formation (paludifikation) after forest destruction is very high. The forest releases more water than low herbaceous vegetation due to the transpiration of the tree layer. Therefore, after clear-cutting in the humid area, one can notice a rise in the water table, which favours the growth of _Sphagnum_ mosses. In addition to _Sphagnum species_, _Racomitrium lanuginosum_ plays a major role. In areas with more than 235 rainy days, the bogs can cover the entire area even in an undulating terrain. Such **blanket bogs** are found in W Ireland, Wales and Scotland, where the largest bog covers 2500 km² and on Acores islands (Photo 33).

Fig. 10.8 Landscape of the Lüneburg Heath in Lower Saxony (**a**) on the nutrient-poor sandy podsol soils (**b**) (photos: Breckle)

Fig. 10.9 Buckwheat *(Fagopyrum esculentum,* Polygonaceae) was once one of the most important foods in Central Europe. Today, it has been largely replaced by wheat, but above all by potatoes (photo **a**: Breckle; **b**: http://bit.do/6nfR)

In areas farther away from the Atlantic coast, the heath is not a danger because all heath species have a low resistance to frost-drought, although the leaves of *Calluna* are very small and have a thick cuticle; the stomata lie in a groove lined with hairs. *Calluna* is distinguished from the true xeromorphic leaves by the very loose structure of the mesophyll. Transpiration may be relatively brisk in summer when water supply is good, and in shady locations it equals that of wood sorrel *(Oxalis acetosella)* when calculated on fresh weight; it may be greatly reduced when water is scarce. But these characteristics are not sufficient to prevent water loss during prolonged frosts. Even in the mild winters of the Upper Rhine Valley, *Calluna* very often dries up without snow protection. Even in the north it is only found where there is a snow cover every year.

Heath occurs inland on the western slopes of the low mountain ranges with oceanic climate, also island-like (Ardennes, Hohes Venn, Eifel, Vosges and even in the Black Forest at the Feldberg). It also extends as a narrow strip to the Baltic Sea.

10.5   The Deciduous Forest as an Ecosystem

10.5.1   General

The deciduous forest is a multi-layered plant community. It often consists of one or two tree layers, a shrub and a herb layer. In the latter, one finds hemicryptophytes, but also many geophytes that develop only in spring. For therophytes, i.e. annual plants, the development conditions are too unfavourable in the poor light conditions on the forest floor. A missing ground layer of mosses would be covered by the falling leaves. Mosses therefore only grow on boulders, tree stumps etc. that protrude above the ground surface. (Fig. 10.10) These plant groups each form synusia.

In the European deciduous forest area, we do not know of any primeval forest stands on euclimatopes, on flat areas with normal soils (perhaps apart from the area near Bialowieca in eastern Poland), where, however, beech already no longer occurs (Fig. 10.10, Fig. 10.11, Fig. 10.12).

Fig. 10.10 Primeval forest of Bialowiez (Poland). The mosses grow on tree bark and hardly on the forest floor. In the mixed forest area, numerous ash, hornbeam and oak trees grow densely high (photo: E. Fischer)

The structure of forests in Central Europe is entirely determined by the type of management. For forestry, the wood species are of importance; the herb layer is only indirectly influenced by it. In the case of forest grazing, on the other hand, it is precisely the herb layer that is exposed to selective cattle browsing, from which the young tree growth also suffers (Fig. 10.12). Rationally managed high forests come close to virgin forests, but differ essentially in the low number of species in the tree layer, their uniformity, the absence of dead wood rotting on the ground, and the homogeneous structure. Primeval forests, on the other hand, usually show a mosaic-like structure.

Managed beech forests are pure stands with only a herb layer. Oak forests are often mixed stands of different hardwood species and have a shrub layer (Fig. 10.10, Fig. 10.11). Of the various deciduous forest biogeocoenoses, a western mixed forest in Belgium, beech forests and spruce forests in the Solling, and oak forests in the east on the forest-steppe

Fig. 10.11 Primeval forest of Bialowiez (Poland). An open forest area with very old oaks *(Quercus robur)* and younger hornbeams *(Carpinus betulus)*. In the background a forest gap with numerous young trees (photo: Breckle)

border were studied in more detail, among others, during the IBC-programm (see below).

In deciduous forests, the canopy is the active layer in which most of the direct solar radiation (including diffuse radiation) is converted into heat. Only a small fraction of the daylight penetrates the forest stand.

In ZB VI of Central Europe, beech is the dominant forest tree (Fig. 10.13). In ZB VI of East Asia and North America,

Fig. 10.12 Primeval forest of Beloweschkaja Pushta (Belorussia). In late winter, the bison "kept semi-wild" there are together in large herds. In gathering places, the shrub and herb layer of the tall, old oak forest is considerably degraded (photo: Breckle)

Fig. 10.13 The autumn aspect of a deciduous forest of the Middle Moselle (Germany), in which *Fagus is* the dominant element and is distinguished by yellow-red foliage colouration (photo: Rafiqpoor)

beech species also occur, but there the number of tree species is several times higher, (again due to the glacial refugial history), so that the number of forest types is much greater (Peters 1997). There, *Fagus* species often occur only as relicts in very small areas (Fig. 10.7). *Fagus sylvatica* has the largest range of all twelve beech species, extending from Scandinavia to N Spain and from England to Turkey (Fig. 10.14), forming a number of different beech forests.

In addition to beech forests, several other forest types occur in Central Europe, although they are often very

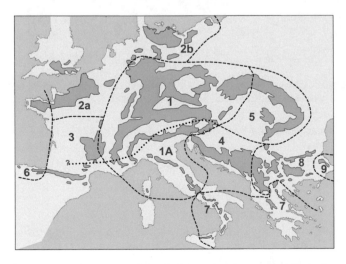

Fig. 10.14 The ranges of beech forests in Europe (modified after Ozenda 1994)

strongly influenced by humans. The very soil-acidic oak-birch forests (Quercus-Betula forests) or the continental oak-hornbeam forests (Quercus-Carpinus forests) (Fig. 10.15) can be cited as examples. However, conifers in particular have been very strongly promoted by forestry in the last century (Fig. 10.16 a, b), so that today pine forests (Fig. 10.17), into which, however, beech and oak are gradually migrating again, are the rule on poor sandy soils.

While in the years after 1980 considerable dieback of spruce forests was observed in particular, in the Ore Mountains and the Sudetes entire slopes also died earlier, in the Harz Mountains later (Fig. 10.18), significant damage to deciduous trees has also been detected, especially since 1990. However, the causes of this forest damage are very complex and can certainly not be attributed to air pollution and altered inputs of pollutants (SO_2, O_3) or excessive nitrogen loads (NO_x) alone. The additional acceleration of the leaching of nutrients from the leaves and the upper soil layers (cryptopodsolisation) also amplified the effects, as did soil fatigue caused by silviculturally one-sided crops (monocultures). One can only speculate about the effects of Climate change (Breckle 2005).

Recently, forest dieback in Germany became a new dimension. In 2020, about 80.4 million m^3 of wood were felled in German forests. This means that logging reached a new record: never before since German reunification has more wood been felled in Germany than in 2020. As the Federal Statistical Office (Destatis) also reports, logging rose

Fig. 10.15 Tall oak-hornbeam forest on basic soils of shell limestone in the Wölmisse National Park in the Thuringian Forest near Jena (photo: Rafiqpoor)

Fig. 10.16 (**a**) Forest management in a *Picea abies forest* near Ulm, southern Germany (photo: Barthlott). (**b**) Large pieces of stem wood (here often *Picea abies*) are stored for wood industry, especially after drought and bark beetle damages like 2020 (photo: Breckle)

again by 16.8% compared to the previous high of 68.9 million m³ in 2019. This development is due to increased forest damage as a result of insect infestation, which is also favoured by drought and heat of preceding years: For example, damaged wood felling due to insect damage (Fig. 10.16b) accounted for more than half (53.8%) of the total logging in 2020. The increased amount of damaged wood in coniferous wood caused by insects underscores a problematic development that has often been discussed in

recent years: the bark beetle spreads rapidly in local forests and primarily attacks spruce. Common bark breeders in Central Europe and their preferred tree species are:

Letterpress (*Ips typographus*) (spruce)
Engraver (*Pityogenes chalcographus*) (spruce)
Large and Small Forest Gardener (*Tomicus piniperda, T. minor*) (Pine)
Oak bark beetle (*Scolytus intricatus*) (oak)

Fig. 10.17 Open pine forest with invading birch and oak in the Senne, south of Bielefeld, in the Augustdorf dune field with late glacial fossil dunes. The seedlings initially have a hard time against the competition of the dense *Avenella flexuosa* carpet (photo: Breckle)

Fig. 10.18 Montane, almost dead spruce forest in the western Hochharz (highest forest damage class). In the background the Brocken (photo: Breckle)

Frequent wood breeders (xylomycetophagous, or ambrosia beetles) in Central Europe and preferred tree species:

Striped timber bark beetle (*Trypodendron lineatum*) (on lying softwood)

Black timber bark beetle (*Xyleborus germanus*) (native to East Asia, introduced to North America and Central Europe. On a large number of deciduous and coniferous trees).

10.5.2 The Beech Forest in the Solling as an Ecosystem

Within the framework of the IBP (International Biological Programme and subsequently), three beech forest and three spruce forest plots with differently fertilised meadows and one arable field were studied and compared in great detail in the **Solling** from 1966 to 1986. Characteristic vegetation types were also studied in other countries over many years. As an example of the main structures and processes in a nemoral forest of ZB VI, we choose results from the Solling and occasionally give comparative figures from studies of a continental oak forest on the Worskla, the left tributary of the middle Dnieper (IBP project of the then Leningrad University). The mixed oak forest is a 1000 ha forest area for planed experiments, of which 160 ha is a protected primeval forest-like 300-year-old stand.

The beech forest in the Solling (Ellenberg et al. 1986; Breckle 2021) is a soil acidic beech forest (Luzulo-Fagetum) at about 500 m asl. The soils are predominantly slightly podsolic brown earths, formed from loess layers overlying the red sandstone.

The climate in the Solling beech forest is characterised by a high degree of oceanity, although there are significant differences between the individual years. The mean precipitation (1967 to 1981, IBP-period) amounts to 1045 mm per year, resp. 1013 (1960–2018) which corresponds to the typical German rainy low mountain range climate (Fig. 10.19a, b). However, the driest year 1976 brought only 686 mm, the wettest more than double, namely 1647 (2007). To illustrate the long-term trend it is necessary to have long observation periods. From the Solling there is a sequence of data from 1960 until 2018 (Climatogram Fig. 10.19a, and Climate-diagram Fig. 10.19b), which we take as an example to demonstrate the changes from year to year. Rather seldomly there is a month which is dry or even arid (indicated by red, Fig. 10.19a), more often there are months, which are hyperhumid with more than 100 m rain per month.

Such long-term data enabling us also to see general trends. The increase of the mean annual temperature is considerable and reaches distinctly more than 1.5 K from 1960 to 2018 (Fig. 10.19c). The trend with the mean annual precipitation is less clear (Fig. 10.19d), but it seems that it is also increasing from about 1070 mm to 1150 mm per year.

As indicated above, drier periods occur again and again, but they cannot be assigned to any particular season in Central Europe. Accordingly, days with relatively dense cloud cover and correspondingly low irradiation occur at all times of the year. Some typical diurnal variations of global radiation (Fig. 10.20) illustrate this. When the sky is overcast (Fig. 10.20, 3.7.1972), the global radiation can often hardly reach □ of the radiation of a clear day (Fig. 10.20, 13.7.1972).

This then also acts as a limiting factor for the photosynthesis of the beech trees.

The weather conditions are predominantly determined by W and SW winds, as illustrated by the wind roses (Fig. 10.21). E winds occur occasionally during high-pressure weather in winter, N winds are practically non-existent.

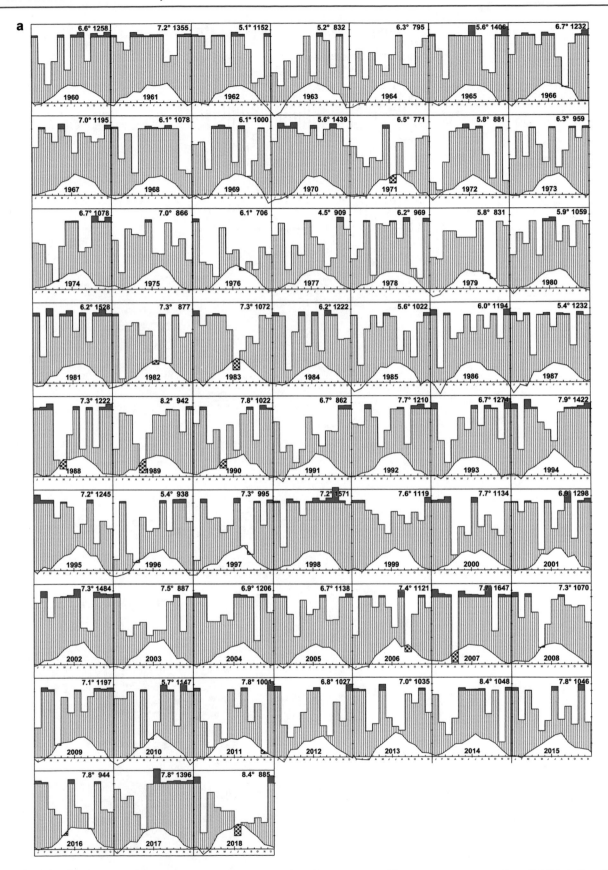

Fig. 10.19 (**a**) Climatogram (1960–2018) from the Solling. The number of days per year with temperature averages above 10 °C is indicated in the middle (partly from Ellenberg et al. 1986, with courtesy from Markus Wagner, Nordwestdeutsche Forstliche Versuchsanstalt/Göttingen). (**b**) Climate diagram of Solling. (**c**) Annual temperature (°C) trend from 1960 to 2018. (**d**) Precipitation per year (mm/year), trend from 1960 to 2018

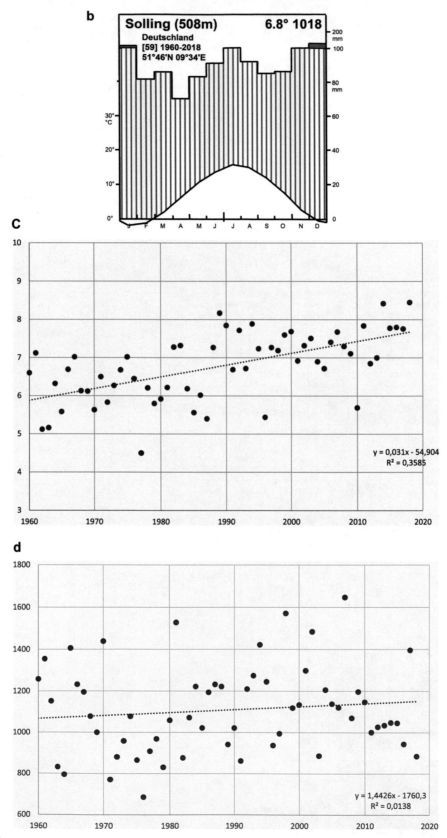

Fig. 10.19 (continued)

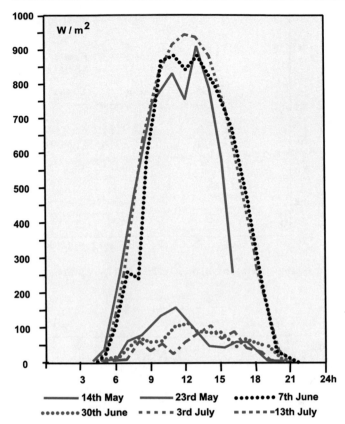

Fig. 10.20 Diurnal variations of insolation (global radiation) in early summer and summer 1972 for three clear and three cloudy days each (modified after Ellenberg et al. 1986)

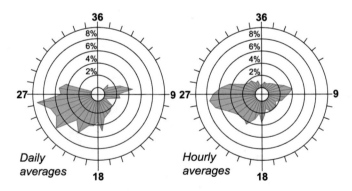

Fig. 10.21 Daily and hourly mean values of the percentage frequency of prevailing wind directions in the Solling (modified after Ellenberg et al. 1986)

10.5.3 Ecophysiology of the Tree Layer

A tree is not a favourable object for experimentation because of its size. Its shape depends very much on the stand. A free-standing tree has a dome-shaped to spherical crown, while this is very small in a dense stand. However, since the leaves are arranged in several layers, the outer ones are exposed to full daylight, while the inner ones develop in the shade. A distinction is therefore made between **sun leaves** and **shade leaves**, which are connected by transitions. The anatomical-morphological and ecophysiological characteristics are different for both.

Sun leaves are smaller, thicker, have a denser venation and more stomata on the lower side of the leaf per square millimetre, that is, they are more xeromorphic than the large and thin shade leaves.

The structural differences are controlled by the less favourable water balance in the establishment of buds sprouting next spring as a result of the greater transpiration of the sun branches, indicated by the increased cell sap concentration; the latter is, for example, 1.6 MPa for sun leaves and 1.2 MPa for shade leaves in beech. Differences are also noticeable with respect to CO_2 assimilation. In laboratory experiments, it was found that in the dark, shade leaves respire less intensively per square decimetre of surface area than sun leaves; for example, shade leaves of beech excrete only 0.2 mg of CO_2 per dm^2/h compared with 1.0 mg of sun leaves. Therefore, in spring, the light compensation point (where respiration = gross photosynthesis) of shade leaves is already at 350 lux, whereas that of sun leaves is at 1000 lux. With increasing illumination, photosynthesis increases proportionally with light intensity until a maximum is reached (Fig. 10.22). In shade leaves and shade plants, this is already less than 20% of the maximum daylight, whereas in sun leaves it is only about 40%. Shade leaves thus make better use of low light intensities, whereas sun leaves make better use of higher light intensities.

In the case of the sun leaves, it strangely appears that they do not make sufficient use of the full daylight. But these figures only apply to leaves oriented perpendicular to the light, while the sun leaves at the crown of the tree are always quite steeply erect. This prevents them from overheating too much in the sun, i.e. suffering too much water loss, but at the same time allows more light to pass through the outer crown, which benefits the lower leaves. In addition, thus the morning (from the east) and evening (from the west) sunlight is better utilised.

> **Box 10.3 Sun and Shade Leaves**
> Sun leaves usually are oriented vertically, parallel to the incoming rays of light, whilst shade leaves often stand horizontal, perpendicular to the incoming rays.

The leaves standing in deep shade are oriented perpendicular to the incident light, allowing an average positive mass balance even with a LAI = 5 or more.

A detailed production analysis by direct measurements of CO_2 assimilation of beech at the site was carried out by Schulze (1970). A single day measurements are shown in Fig. 10.23. It could be deduced that the production of sun and shade leaves per dry weight is the same during the entire

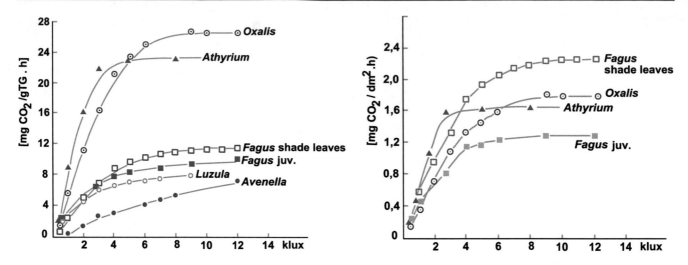

Fig. 10.22 Light saturation curve of photosynthesis (left: leaf area-related; right: weight-related) of some species of the beech forest in Solling (juv. = young growth) (from Ellenberg et al. 1986)

vegetation period because the shade leaves remain active longer in autumn (Fig. 10.24).

If the illumination falls permanently below a certain light minimum, the respiration of the leaf is no longer compensated by photosynthesis, substance losses occur, the leaf turns yellow and is shed. This light minimum, expressed in % of full daylight, varies from one tree species to the other. A distinction can be made between shade woods with a very dense crown and low light minimum (1.2% for beech) and light woods with a light crown and high light minimum [birch (*Betula*) and aspen (*Populus tremula*) 11%]. In between are maple (*Acer*) and oak (*Quercus*). This light minimum in the canopy need not coincide exactly with the light minimum that must be exceeded for tree seedlings to grow on the forest floor, but the values do go in parallel. Beech seedlings manage with little light, birch seedlings need at least 12 to 15% of daylight.

An oak forest reflects 17% of the incident radiation when it is leafy, and only about 11% when it is leafless, i.e. significantly less than meadows and crops (25%). Halfway up the stand or on the ground, only 1.2% and 0.6% of the daylight is measured in the fully leafy state in young stands, and about 20% and 2% in very old stands.

Light conditions are crucial for the competitiveness of tree species. In a clearing, the light woods can grow in a few years. Under their canopy, the shade woods germinate and only very slowly grow higher and higher. Their canopy is so dense that the lightwoods underneath do not gain any material and do not rejuvenate. In time, the most shade-tolerant species becomes dominant, if the other site conditions suit it.

In Central Europe this is the beech (*Fagus sylvatica*), it is the species of the zonal forests. Only on very poor soils or at high groundwater levels, or in the driest basin landscapes it is not competitive. In the W part of Eastern Europe, the climate

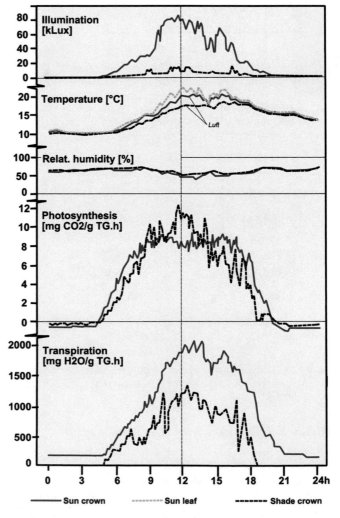

Fig. 10.23 Diurnal course of important microclimatic and ecophysiological parameters of sun and shade leaves of beech in the Solling on a fair-weather day (modified after Ellenberg et al. 1986)

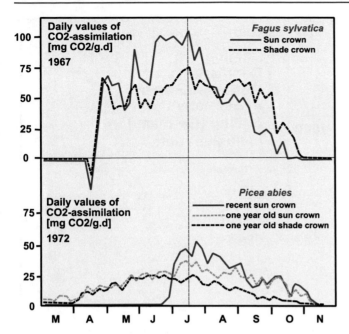

Fig. 10.24 Annual patterns of diurnal balances of CO_2 gas exchange of beech and spruce (modified after Ellenberg et al. 1986)

is too continental for the beech, which there is replaced by the hornbeam *(Carpinus betulus)* as a shade wood species, and even further east by the oak *(Quercus)*.

The mean daily temperature is 2 °C higher at the canopy in summer than at ground level, the mean daily maximum even up to 11 °C higher, while the mean daily minimum is about 2 to 3 °C lower. The mean humidity is 98% at ground level and decreases with increased height to 77%. The wind speed

is low in the forest. Since the forest floor is protected from direct radiation, it remains noticeably cooler in the forest during the day than in open stands.

The leaf area index (**LAI**), i.e. the ratio of the total leaf area of the tree stand to the ground area it covers, is very important for forest productivity. It can only reach a certain maximum, because otherwise the lower shaded leaves would not show a positive mass balance. But this maximum does not only depend on the daylight intensity, but becomes smaller with insufficient water supply and with nutrient deficiency. In pure oak stands the LAI = 5 to 6 (higher in wet years), in fresh mixed stands it can exceed 8, including all wood species and shrubs.

Dry matter production $(t \cdot ha^{-1} \cdot a^{-1})$ of a 40-year-old beech stand in Denmark:

- Gross production of assimilating leaves = 23.5
- Respiratory losses (leaves 4.6; stems 4.5 and roots 0.9) = 10.0
- Annual production (leaves 2.7; stems 1.0; litter and roots 0.2) = 3.9
- Wood production (aboveground 8.0; belowground 1.6) = 9.6

Of the maximum 8 $t \cdot ha^{-1}$ of stemwood, an average of 6 $t \cdot ha^{-1}$ can be exploited for forestry, which corresponds to 11 m^3. In the case of spruce, the timber yield is equally large in terms of weight, but volume-wise approx. 17 m^3.

How the production figures change with the age of the beech stand is shown Fig. 10.25.

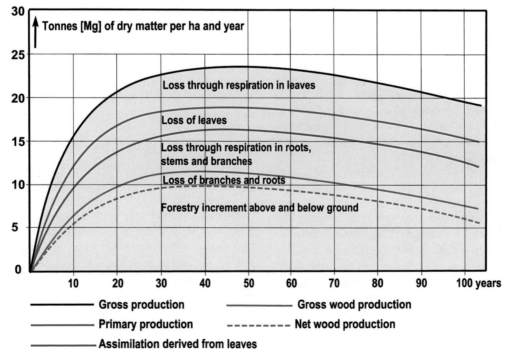

Fig. 10.25 Production curves of the beech forest (modified after Walter 1990)

The mass of dead wood going off is hardly less than the wood increment in the same period, which means that the net phytomass increment here is practically zero, as it must be in a virgin forest in the optimum phase.

The beech forest areas in the Solling also reach about 10 t·ha^{-1}·a^{-1} annual production, of which about 3 t are leaves, the blossoms and fruits account for very varying shares from year to year (mast years). The production of twigs and branches comprises about 10% of the annual production, the production of roots about 10% of the aboveground production.

Primary production per year is reported as 8.9 t·ha^{-1}, including herb layer 9.6 t·ha^{-1}. The underground production was not determined.

In accordance with the semi-arid climate, primary production is somewhat lower than in western deciduous forests.

The leaf mass and leaf area formed each year increases rapidly in the first 20 years. However, once a dense canopy is reached, leaf mass and LAI remain almost constant. The canopy is only raised more and more above the ground by the height increase of the stems. The leaves with the falling branches form the litter and together with the dying roots the total litter.

Only the wood mass produced is stored, so that the standing phytomass of the forest increases constantly but ever more slowly until old age, and can exceed 200 t·ha^{-1} for 50-year-old stands and 400 t·ha^{-1} for 200-year-old stands.

For the oak forest the following mean wood increment was found depending on the age of the stand (in brackets): 3.8 t·ha^{-1} (13), 3.6 t·ha^{-1} (22), 4.3 t·ha^{-1} (42), 4.7 t·ha^{-1} (56), 0.4 t·ha^{-1} (135), 0.0 t·ha^{-1} (220). The increasing log diameter (DBH: diameter at breast height) means a corresponding increasing wood supply in the log. This relationship is shown in Fig. 10.26.

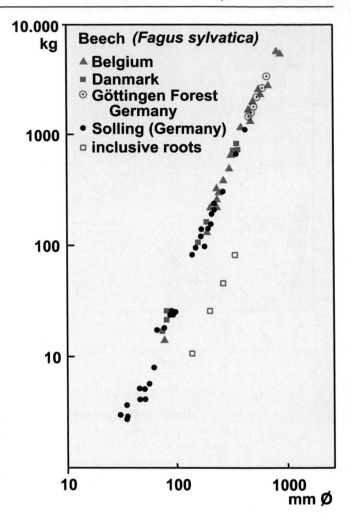

Fig. 10.26 The dependence of aboveground biomass (marked belowground with open squares) on stem diameter (DBH) in beech from different sites (modified after Ellenberg et al. 1986)

Box 10.4 Phytomass in a Mixed *Quercus* Forest

For phytomass of mixed oak forest in Russia, Goryschina (1974) gives the following values: Aboveground 306.7 t·ha^{-1} (leaves 3.7, twigs and branches 71.2 and stems 230.8); belowground 124.9 t·ha^{-1}; total 431.6 t·ha^{-1}; plus herb layer 0.7 t·ha^{-1}.

The amount of litter also accumulates in the forest until an equilibrium is reached, i.e. as much of the litter is mineralised each year as is newly added. A part of the most important nutrient elements (N, P, K, Ca) is fixed in the litter. Thick layers of raw humus are therefore unfavourable. However, litter use is particularly harmful; it removes nutrients altogether, especially lime, which rapidly depletes and acidifies forest soils, reducing timber yields. Nitrogen compounds are mineralised during litter decomposition. Most of the nutrients are available to tree roots in the lower decomposing humus layer; this is therefore always very densely rooted. Soil life is of particular importance for the forest stand, in addition to the water supply. In contrast, the proportion of animal organisms above the soil is very small; even insect feeding normally accounts for only a few percent.

The litterfall itself occurs very periodically in the deciduous trees and shows only slight differences from year to year. More than 90% of beech leaves fall in October (Fig. 10.27, top). In spruce, on the other hand, dry or yellowed needles fall almost all year round, although there is also a small maximum of needle fall here, a little later than in beech (Fig. 10.27, bottom).

Fig. 10.27 Course (in %) of annual leaf fall of beech (top, 1967–1975) and needle fall of spruce (bottom) (fine litter from three spruce plots 1968–1971) over the course of the year in the Solling shown as a plot of curves (after Ellenberg et al. 1986)

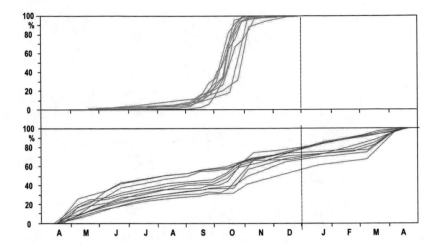

10.5.4 Ecophysiology of the Herb Layer (Synusiae)

The microclimate on the forest floor is very different from that on open sites: After the foliage of the forest, the illuminance on the forest floor is lower, the temperature conditions are more balanced, the humidity of the air as well as of the upper soil layers is higher than outside the forest. Therefore, the herbs in the forest are shade plants and hygrophytes with very low cell sap concentration, i.e. favourable hydrature of the plasma.

The light conditions at the bottom of a deciduous forest can be very heterogeneous on clear days, because individual rays of sunlight falling through the treetops create light spots on the ground. As the sun moves across the sky and the tree branches are bent back and forth by the wind, the light spots change their position and intensity within seconds.

If a leaf is hit by a spot of light, the illuminance can increase more than 30 times, which is of great importance for the photosynthesis of the herbs. Therefore, to determine the light enjoyment of herbaceous plants as a percentage of full daylight, comparative measurements are most appropriately made on bright, uniformly cloudy days. But they can only give a certain indication. It is better to have the daylight totals automatically recorded for specific locations on the forest floor. The amount of light absorbed by the herbaceous layer before the trees are in leaf, is very great, and then falls rapidly as the tree foliage develops completely.

The **spring geophytes** (*Galanthus, Leucojum, Scilla, Ficaria, Corydalis, Anemone*, etc.) take advantage of the favourable light conditions before leafing of the trees (Fig. 10.28a–d). They benefit from the fact that the little weakened solar radiation warms the litter and humus layer in which the geophytes are rooted to 25 to 30 °C already in April. The aerial litter layer has a low heat capacity and consequently a very good temperature conductivity. The

tree root in deeper layers that hardly warm up at all, which delays foliage growth.

During the short early spring period, geophytes flower and fruit, replenish their reserves in underground storage organs for the next year. When tree foliage sets in, the leaves of the geophytes turn yellow and a period of dormancy starts. This yellowing is not only due to increasing shade, but corresponds to an endogenous developmental rhythm. In the light, the geophyte leaves die even sooner. It is thus a plant group that was able to fill an existing ecological gap (niche) in the developmental process of deciduous forests.

The spring geophytes are also called **ephemeroids**; for they are characterised by a vegetation period as short as that of the annual ephemerals, but are perennial species with underground storage organs. They behave ecologically similarly and have almost the same developmental rhythm, thus forming a "working group", a functional unit which is called a **synusia**.

In the Russian mixed oak forest (Goryschina 1974; Walter 1976), the following representatives of five deciduous forest synusia were distinguished:

1. *Ephemeroids: Scilla sibirica, Ficaria verna, Corydalis solida, Anemone ranunculoides.*
2. Hemi-ephemeroids: *Dentaria bulbifera.*
3. *Early summer species: Aegopodium podagraria, Pulmonaria obscura, Asperula (Galium) odorata, Stellaria holostea.*
4. *Late summer species: Scrophularia nodosa, Stachys sylvatica, Campanula trachelium, Dactylis glomerata, Festuca gigantea.*
5. Evergreen species: *Asarum europaeum* (Fig. 10.28d), *Carex pilosa.*

The individual synusia take advantage of the different light phases on the forest floor through morphological/physiological adaptations. Thus, *Aegopodium* first forms small light

Fig. 10.28 (a) The spring geophytes as here, the liver flower *(Anemone hepatica)* take advantage of the light phase before the forest greening and complete their entire generative cycle until seed maturity (photo: Breckle). (b) The spring geophytes in an adult beech forests have a light period of a few weeks between snow melt and leaf sprouting (photo: Breckle). (c) The spring geophytes e.g. wood anemone *(Anemone nemorosa)* and the yellow flowering *Anemone ranunculoides* have long and thin rhizomes and thus can form large patches (photo: Breckle). (d) On some favoured sites the monocotylous *Paris quadrifolia* (Trilliaceae) and *Asarum europaeum* (Aristolochiaceae) with its evergreen kidney-shaped leaves may cover the herb layer (photo: Breckle)

leaves, then large shade leaves in summer and finally very small xeromorphic, cold-resistant leaves in autumn at lower temperatures, which overwinter *(Aegopodium* has no winter dormancy phase). The same takes place in *Stellaria* and *Asperula,* only the different leaves form successively on the same vertical shoot axis.

Very different in the individual synusia is the assimilate budget, that is, the way they use the assimilates:

Scilla uses up all the assimilates stored in the bulb to build up the flowering shoot and leaves; it is only towards the end of the short growing season that the newly formed assimilates are channelled into the young bulb for use the following year.

Dentaria, on the other hand, starts to fill up the reserves in the rhizome early and therefore needs more time for flowering and fruiting.

Aegopodium uses up the sparse reserves to form the light leaves, which assimilate CO_2 intensively and replenish the reserves as early as the beginning of May, while at the same time providing the assimilates for building up the shade leaves.

The late summer plant *Scrophularia* has only a few reserves in the large tuber for the formation of the first leaves.

Their assimilate yield is low in the shade, so that it takes until autumn until the shoot is fully grown and the fruits ripen.

In the assimilate household of *Asarum* it is characteristic that the previous year's (evergreen) leaves assimilate again immediately after winter, they die only in the course of early summer, long after photosynthesis of the young leaves has fully started.

The total phytomass of the herbaceous layer is not large, but its significance for the ecosystem is that it is rapidly decomposed and, in this way, promotes the turnover of matter throughout the ecosystem, while tree leaf litter decomposes slowly; the nutrients contained in the latter are not available until the following year.

Box 10.5 Synusia in Certain Ecosystems
Synusia are only subsystems within certain ecosystems. They do not have an own material cycle.

Most species of the herbaceous layer are hemicryptophytes, i.e. their renewal buds are laid at the base of the shoots and overwinter just below the soil surface, protected by the leaves that fall in autumn and any snow cover. However, very many representatives are clonally constructed, that is, they reproduce in a variety of ways by division, that is, vegetatively. The division of the mother plant, the formation of stolons, shoot or root tubers always serve to preserve the species and its distribution. Van Groenendael et al. (1996) have compiled a comparative list of the large number of clonal types; examples of species from the Central European deciduous forest are also given in Fig. 10.29.

The total biomass of the herb layer is usually very low (Fig. 10.30), but it converts rapidly. Shoots and roots do not react in the same way every year, depending on the water and nutrient supply, as Fig. 10.31 shows. The number of shoots also varies greatly from year to year (Fig. 10.31). The herb layer can adapt very quickly to changing conditions.

In mast years, such as 1971, however, the number of beech seedlings is also very high, but then declines very sharply over the next few years. The few (from one million beech nuts perhaps 0.1 to 1 seedlings and young plants grow) that remain, however, are sufficient to ensure a new generation of trees.

For many species of the herbaceous layer in the forest, illumination values have been determined. They have a illumination **maximum** (L_{max}) because they are not found in full daylight, and a illumination **minimum** (L_{min}) because they avoid the deepest forest shade. Example of the two limits in % of daylight are: *Lamium maculatum* 67 to 12%, *Lathyrus vernus* 33 to 20%, *Geranium robertianum* 74 to 4%, *Prenanthes purpurea* 10 to 5% (sterile to 3%).

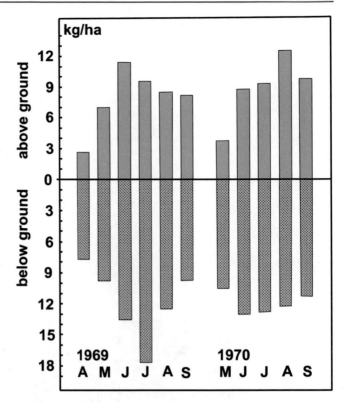

Fig. 10.30 Changes in above- and belowground biomass of the herb layer in the beech forest in Solling, in 1969 and 1970. In 1970, the growing season started later and the year was rainier (modified after Ellenberg et al. 1986)

L_{max} is dependent upon the water balance; for the hygrophilous species require a moist soil and do not tolerate high saturation deficits of the air, such as occurring at full illumination.

L_{min} is a starvation threshold for the plants. The light intensity is just sufficient to enable the production of the substances necessary for development. In general, the dead forest shade begins in the forest at 1% of daylight, in which not only the fruiting bodies of heterotrophic fungi can be found but also holosaprophytes among the flowering plants, for example the bird's nest orchid (*Neottia nidus-avis*) (Fig. 10.32a) or full-parasites, parasitic species which breach and tap the roots of woody host plants in forests, as does the "common tooth wort" (*Lathraea squamaria*) (Fig. 10.32b).

Another very important factor for the herbaceous layer is the competition of tree roots. In the dry forest areas bordering the forest steppes, the water factor is of great importance. The trees, whose cell sap concentration is higher than that of the herbs, are able to develop high suction tensions in suction roots and thus to extract water from the soil better than the herbs. As a result, the soil in such beech forests (Fagetum nudum) is "bare" (Fig. 10.15). If, on the other hand, the tree roots are cut through, thus eliminating their competition, herbs establish themselves on the forest floor, a sign that the limiting factor was not the light conditions but the water.

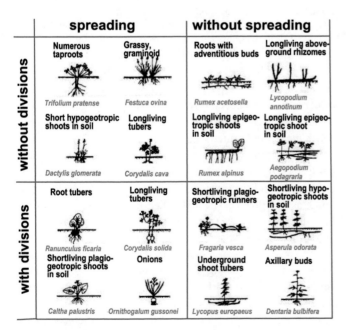

Fig. 10.29 Various clonal structures with information on the spread and life span of crown connections (modified after Van Groenendael et al. 1996)

Fig. 10.31 Different numbers of shoots of some species of the beech forest in four consecutive summer half-years from 1968 onwards. *Avenella* was not yet counted in 1968 (modified after Ellenberg et al. 1986)

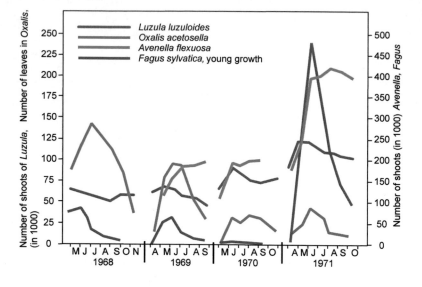

Fig. 10.32 (**a**) *Neottia nidus-avis* (Orchidaceae) is a holosaprophyte that lives in dead forest shade, e.g. in the beech forests of Saguramo-Sedaseni in Georgia, but also in many beech forests in Germany, and manages without daylight, has no chlorophyll and does not assimilate (photo: E. Fischer). (**b**) *Lathraea squamaria* (Orobanchaceae) is holoparasite, that lives in beech forests in Europe (at Bovenden near Göttingen). With its long haustorial roots it is tapping host plant roots and their vascular bundles (photo: Breckle)

In very shallow soils, the tree roots also take the nutrients for themselves, especially the nitrogen. The herbs have to make do with what is left by the tree roots. As a result, only herbs with low nutrient requirements such as *Luzula luzuloides, Avenella flexuosa, Potentilla sterilis, Vaccinium myrtillus,* etc. are found in such forests.

10.5.5 Water Balance

In the Solling, the sum of precipitation is the input variable for water turnover. The output consists of evaporation and runoff, each of which consists of partial variables (Fig. 10.33a). In the water balance equation, internal water fluxes can also be taken into account in the forest, such as crown runoff and stem runoff, which plays a role in beech but is negligible in spruce (Fig. 10.34).

Interception is the proportion of water that is retained in the canopy. In the beech forest, walking people get really wet if rain exceeds 3 mm, in the spruce forest it is only from more than 5 mm of precipitation.

Of the precipitation falling on the forest, the crowns of the beech trees retain an average of 17% (summer half-year, as deciduous), and the spruce trees retain about 27% (all year round, as evergreen). The rest either drips through or runs off the trunks. In dry years, interception ratio is significantly higher than in wet years (Fig. 10.33b).

The snow that accumulates on the forest floor in winter thaws slowly in spring and the meltwater seeps almost completely into the litter layer. It is part of the varying term dS. The transpiration of the tree layer is so strong that in summer, under forest no water is supplied to the groundwater. The water release of the herb layer is five to six times less. A well-developed deciduous forest in the forest-steppe region

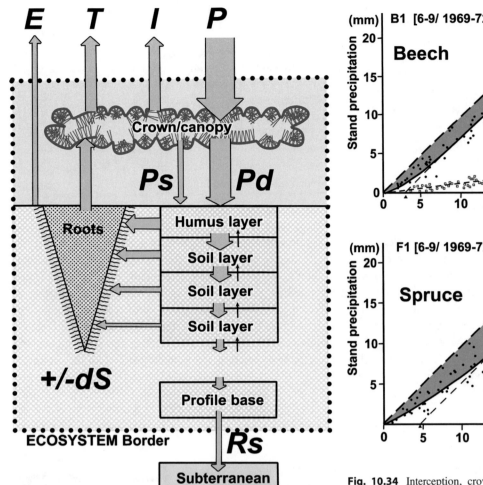

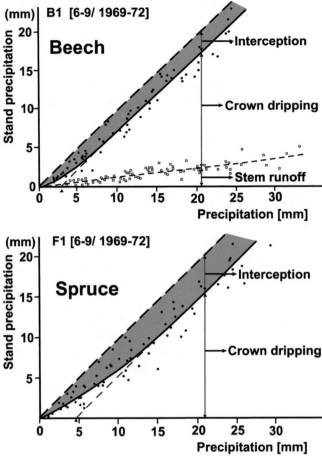

Fig. 10.34 Interception, crown emergence and stem run-off in the months of full foliage in beech and spruce as a function of outdoor precipitation. In spruce, stem run-off is negligible (modified after Ellenberg et al. 1986)

Fig. 10.33 (**a**) Compartment model of the water balance in the ecosystem with a level location in the Solling (Rs = subsurface runoff; E = evaporation; I = interception; P = precipitation; Pd = crown runoff; Ps = stem runoff; dS = storage size; T = transpiration). The width of the arrows indicates the magnitude (modified after Ellenberg et al. 1986). (**b**) Dependence of the annual values of the stand precipitation (Nb), the interception (I) and the deep infiltration (Au) on the amount of the total precipitation (N) for the beech (B) and spruce (F) old wood stocks in the Solling for the Years 1969 to 1975 (after Ellenberg et al. 1986; Lozan & Breckle 2021)

uses practically all the water supplied by precipitation, whereas a beech forest in Central Europe uses only 50 to 60% of the precipitation, with no surplus in the summer months either.

In the Solling, in accordance with the water turnover model shown in Fig. 10.33, the flows through individual soil layers were also measured using tensiometers (soil suction tension), lysimeters (with suction devices) and with tritium water. The infiltration water in the soil, as far as it is not taken up by the plant roots, penetrates 1.65 m into the depth under beech trees and 1.20 m under spruce trees in the course of one year.

If one sets the chemical energy determined during production in relation to the energy irradiated onto a hectare of forest, one obtains about 2% for gross production and 1% for primary production. One third of the irradiated energy is used for transpiration, a total of about 80% for the evaporation of water (evaporation, interception), the rest is converted into heat.

The coupling of CO_2 assimilation with transpiration, both processes being regulated by one and the same factors controlling stomatal aperture, can be expressed quantitatively with the **transpiration coefficient**. Herbaceous plants (for example wheat) consume 540 kg of water to produce 1 kg of plant matter. The beech in the Solling has an average transpiration coefficient of only 180, the spruce of 220. The way in which the beech achieves its so particularly rational performance can be seen in part from the gas exchange measurements (Fig. 10.24). Depending on illuminance, humidity, and CO_2 concentration, the leaves regulate stomatal width so rapidly that they adapt immediately to any change. Moreover, photosynthesis of the thin, shade leaves of beech is as effective relative to their dry weight as that of the sun leaves. As a result, the beech is able to make much better use of sunlight than the oak, which forms only about three layers of sun leaves, while in the beech there are an additional three to four layers of shade leaves. Below this, due to the lack of light, the substance production of the herb and moss layer is insignificant on acid soils, while it can be considerable in mixed oak forests.

The overall result is that in the Solling there has been hardly ever a lack of water for the beech.

10.5.6 The Long Cycle (Consumers)

The role of animals in an ecosystem is primarily determined by their feeding relationships. The diversity of food chains with their interconnections is so great that they have not yet been fully recorded for any ecosystem.

Plants are attacked by various parasites, mainly fungi, and a large number of insect pests. Their individual organs serve as food for various phytophages, and these in turn form the food of first order predators, namely large (birds, mammals) and small invertebrate predators. These are eaten by second

order predators, for example birds or shrews that catch predatory insects.

> **Box 10.6 Relation Between Precipitation and Infiltration in a Spruce and a Beech Forest**
> The spruce forest in Solling lets through up to 880 mm into the groundwater in wet years (1970), in dry years (1971) only 232 mm. The corresponding values for the beech forest are: 1970: 973 mm; 1971: 304 mm. These are the values that ultimately contribute to groundwater recharge in each case.

Some quantitatively significant feeding relationships in the Solling beech forest are shown in Fig. 10.35. However, it should be kept in mind that many of the relationships indicated are not unambiguous, but often only facultative or even episodic. Overall, two main pathways can be identified: a phytophage food chain that depends on living plant matter, mainly beech foliage, and a saprophage food chain that starts from the dead organic matter mainly stored on the ground.

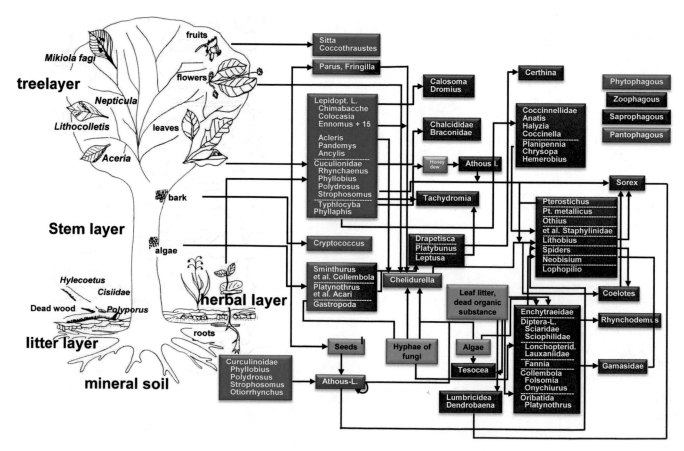

Fig. 10.35 Essential foraging relationships in the Solling beech forest (modified after Ellenberg et al. 1986)

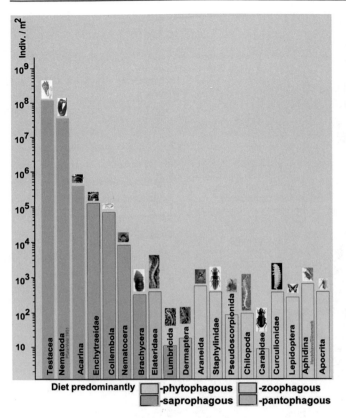

Fig. 10.36 The individual densities of individual animal groups in the Solling beech forest, indicating the predominant diet (logarithmic ordinate in individuals per m²; nematodes and aphidines with supplementary data) (modified after Ellenberg et al. 1986)

Only a very small part of the chemical energy in the food of animal organisms is converted into secondary production, i.e. animal body substance. For the most part, it is excreted with the excrement or breathed.

If you look closely at the leaves or other organs of plants, you will see how frequently they are damaged. In the case of the oak alone, one will easily find more than 20 species of insects that live on the leaves or the buds, the bark or the wood; already the number of gall-producing insects is very large in the case of the oak or the beech.

In the Solling region, many years of detailed work have gone into compiling an overview of the animal groups that occur and their food sources. In Fig. 10.36 the most important groups are shown according to their numerical occurrence of individuals. It is understandable that the microscopic groups can occur in huge numbers of individuals in the process. In Fig. 10.37, the biomass of the individual groups of organisms is contrasted in area terms as a frame of reference. Many of these species are soil or old-growth wood dwellers.

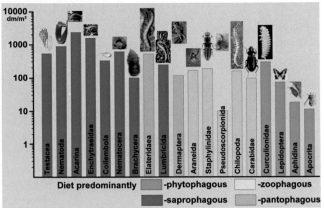

Fig. 10.37 Zoomass of individual animal groups in the Solling beech forest (logarithmic ordinate in per m²) (modified after Ellenberg et al. 1986)

An investigation of **old trees** in Bavaria revealed the following for the beetle fauna: of about 8000 native beetle species, about 2000 are inhabitants of old wood. The beetle communities change considerably in the course of an oak's life, as shown in Fig. 10.38. Many of these old wood inhabitants are relict forest species that were once widespread (six or eight tree generations ago, reckoned in evolutionary timescales a few moments ago), i.e. about 2000 years ago, and are now relegated to a few sites. At that time, dead wood and old trees were probably the most common organic substrates, which is why so many small animal species have taken over this habitat (Leicht 1996). The diversity of habitat

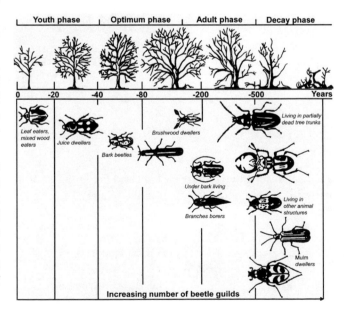

Fig. 10.38 The beetle communities occurring during the long natural life cycle of an oak (modified after Leicht 1996)

Fig. 10.39 Overview of habitat structures in old-growth and dead wood that are significant on an old tree (modified after Leicht 1996)

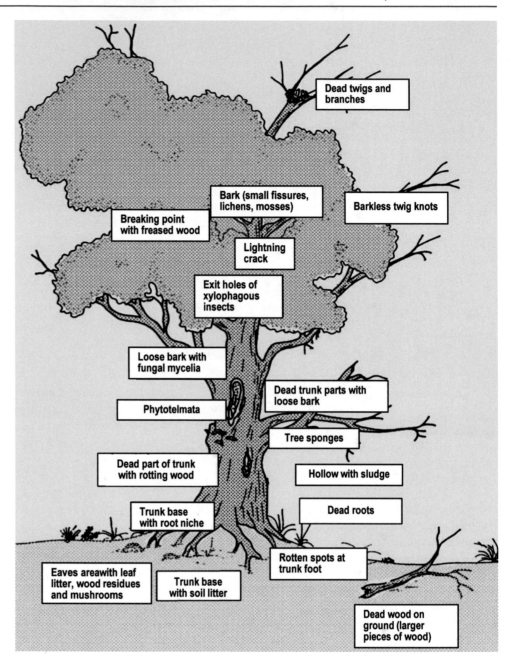

structures in an old-growth tree is indicated in Fig. 10.39. But there are hardly any old trees left in the managed forests of today.

Box 10.7 Important Factors in Forest Management
The preservation of old trees needs a lot of time and requires turning away from unnecessary "maintenance measures" and "exaggerated tree renovation". Understanding for natural processes, but also sensible handling of nature, and no insistence on the "duty of care" (Leicht 1996) is required.

10.5.7 Decomposers in Litter and Soil

Most of the annual litter in a deciduous forest consists of dead, yellowed leaves from the litter layer above the ground. It is immediately crushed by soil organisms and then attacked and broken down by microorganisms, fungi and bacteria. The small fauna of saprophages feed on the litter, and by crushing it they make it easier for microorganisms to enter. In addition to insect larvae and countless other arthropods, however, it is primarily earthworms, as explained above, in whose faecal clots the bacteria develop a lively activity.

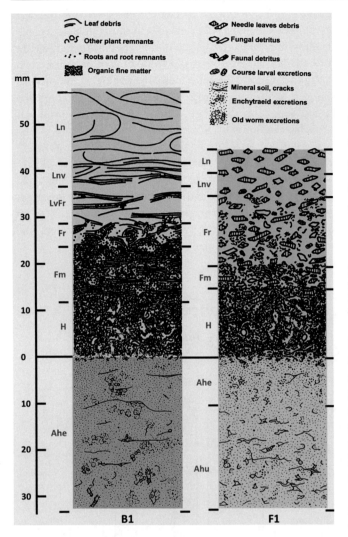

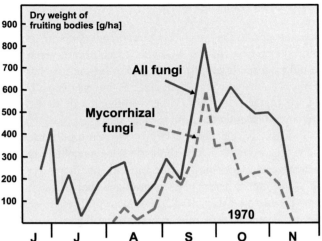

Fig. 10.41 Variations in DW (dry weight in g/ha) of fruiting bodies of mycorrhizal fungi and all higher fungi between June and November in the wet year 1970 in the Solling beech forest (modified after Ellenberg et al. 1986)

Fig. 10.40 Schematic drawings of the humus profiles in the beech forest and spruce forest of the Solling (modified after Ellenberg et al. 1986)

The activity of these animals in deciduous forest stands has been studied more precisely, including quantitative relations (Edwards et al. 1970).

Leaf fall in autumn is completed by the end of October, often within a few days. Sugars, organic acids and tannins are first leached from the litter by rain.

The smaller the C/N ratio of the litter, the faster the further decomposition takes place. By the following June, birch litter loses about 4/5 of its dry weight, linden litter half and oak litter, which is difficult to decompose, only about a quarter.

The mineralisation of the litter is not complete, but humus substances are also formed, which, when saturated with Ca, result in the mull horizon, which is rich in lumbricids (earthworms), and in the case of an acid reaction, the mould horizon with oribatids (horn mites) and collembolae (springtails). In extreme cases, a highly acidic reaction results

in a difficult-to-decompose layer of raw humus almost devoid of animal organisms but rich in fungal hyphae. The fine structure of the uppermost organic humus layer and litter reveals certain differences between beech forest and spruce forest in the Solling (Fig. 10.40). In the spruce humus, the structure is looser, with fewer fine roots, fungal mycelia and animals present. Litter decomposition to the F_m (the decay horizon of medium decomposition) takes about 4 1/2 years, in the spruce forest only 2 1/2 years.

Satchell (from Duvigneaud 1974) gives the activity, i.e. respiration in kilocalories per square metre per year, for the individual soil organism groups of an English oak forest on calcareous soil:

- Invertebrates (dipterans, collembolae, oribatids, molluscs, enchytraeids, lumbricids, nematodes, protozoa) together 361 kcal/m²/year.
- Bacteria and actinomycetes: 77 kcal/m²/year.

The most significant activity is that of fungi: in the litter layer 543, in humus 220 and in the A and B horizons 380, i.e. a total of 1143 kcal/m²/year. The mass of microorganisms is very low compared to that of invertebrates.

The dead wood lying on the ground is 90% destroyed by microorganisms, especially fungi. The fungi also have a great importance as mycorrhiza partners for the trees. In the Solling, almost half of the higher fungi can be classified as mycorrhizal fungi. The maximum of fruiting body formation is in September (Fig. 10.41); however, the other fungi also form fruiting bodies mainly in summer. This is explained by a different balance of growth substances between the fungal root and the tree during mycorrhizal symbiosis.

10.5.8　Solling Ecosystem

In conclusion to the somewhat more detailed discussion of some phenomena of the nemoral beech forest, reference should be made briefly to a comparison of the productivity of the various areas studied and to the energy flow. In Fig. 10.42, the annual net primary production of the sample plots investigated in the Solling is shown.

Surprisingly, the productivity of the fertilised meadows and that of an arable field is about the same magnitude as that of the forests. Under the climatic and soil conditions given in the Solling, the different plant stands have approximately the same productivity (Fig. 10.42). Productivity is expressed in Fig. 10.43 as a measure of energy retention. This energy flux through the major compartments (Fig. 10.43) suggests that among the heterotrophs, decomposers have the greatest energy turnover. This characterises the short cycle of organic matter, while vanishingly little is converted through the long cycle (0.5%, via herbivores and carnivores) as is also known from other terrestrial ecosystems. Interestingly, the proportion of zoophages is even greater than that of herbivores; the food pyramid is "distorted" here by a high proportion of saprophagous animals.

The Solling as a low mountain range shows in a certain way already montane features, the precipitation is clearly increased compared to the lower lying surrounding area, there are more clouds and the temperature is somewhat lower. In the other low mountain ranges and even more so in the northern Alps, however, the third dimension, the orozonal sequence, becomes clearer.

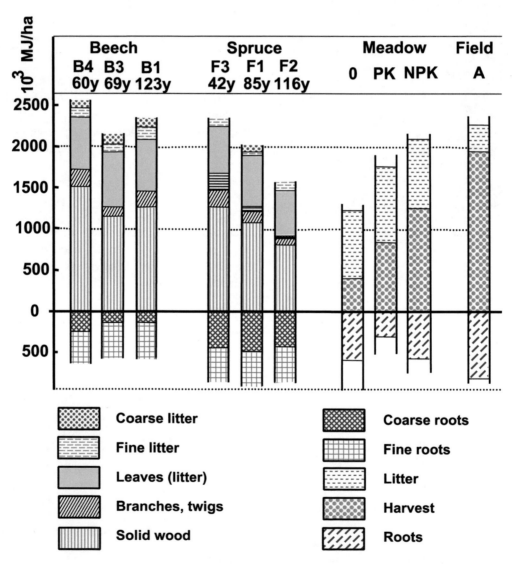

Fig. 10.42 Annual net primary production on the sample plots investigated in the Solling (beech forest, spruce forest, golden oat meadow, ryegrass meadow) given in 1000 MJ per ha (open columns: estimated values) (modified after Ellenberg et al. 1986)

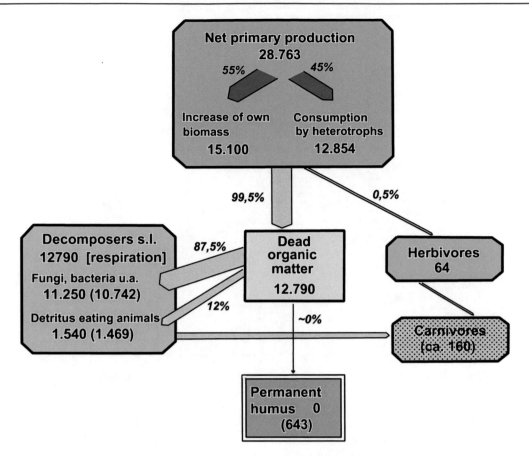

Fig. 10.43 Energy flux in the Solling beech forest (partly calculated according to respiration values (figures in kJ per m² and a; in parentheses assuming that 5% of the organic matter is converted into permanent humus; carnivores are assumed to eat about 10% of the saprophagous and herbivorous animals) (modified after Ellenberg et al. 1986)

10.6 Orobiome VI: The Northern Alps and the Alpine Forest and Tree Line

The Alps separate Central Europe (ZB VI) from Southern Europe (ZB IV) like a crossbar. Geologically, the Alps are characterised by the "crystalline" Central Alps and by limestone in the marginal Alpine chains of both the northern and southern Alps (Fig. 10.44). This has implications for flora and vegetation.

10.6.1 The Elevational Belts

The elevational belts of orobiome VI are well developed on the northern regions of the Alps. With increasing elevation, the mean annual temperature in the mountains decreases and the vegetation period shortens. Direct solar radiation increases with elevation, but diffuse radiation decreases; as a result, thermal differences between S and N slopes become sharper. Due to the wind congestion at the northern edge of the Alps, precipitation increases rapidly with elevation, for example Munich (569 m asl) 866 mm, Wendelstein (1727 m asl) 2869 mm. The vegetation of the individual elevational belts on the northern parts of the Alps changes accordingly:

Elevational belt	Vegetation
Nival	Cushion plants, mosses and lichens
Climatic snow line at about 2600 m asl	
Alpine	Alpine mats and meadows
Subalpine	Krummholz and dwarf shrubs
Timberline at about 1800 m asl	
High Montane	Spruce forest
Montane	Beech and fir forest
Submontane	Beech forest
Colline	Mixed oak forest

As the Alps are an interzonal mountain range, we have to do with an elevational sequence of Orobiom IV at the S edge of the Alps and the tree line is formed by the beech. The succession of elevational belts in the continental inner Alpine valleys is also different; the deciduous forest belts are absent, below the spruce belt there is a pine belt, above the spruce belt a larch (*Larix*)—pine (*Pinus cembra*) belt follows up to the timberline. Here, the timberline and the snowline are

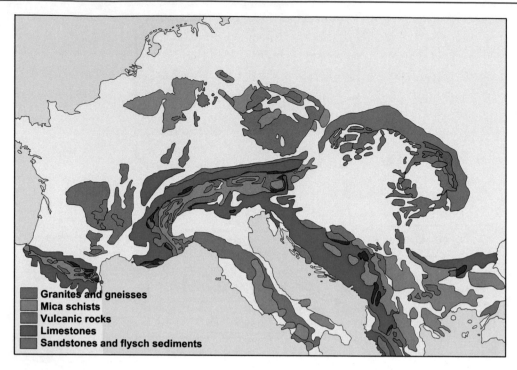

Fig. 10.44 Schematic of the geological situation of mountains in Central Europe (modified after Ozenda 1994)

400 to 600 m higher due to the stronger insolation with less cloud cover. We distinguish a Helvetic (northern edge), Penninic (central Alps) and Insubrian (southern edge) sequence of elevational belts:

Helvetic	Penninic	Insubric
(Central European)	(Continental)	(Submediterranean)
Alpine belt	Alpine belt	Alpine belt
Spruce forest	Larch-Arpine Forest	Beech forest
Beech forest	Spruce forest	Downy Oak Forest
Oak Forest	Pine forest	Sclerophyllic scrub (partly)

10.6.2 The Forest Belts

The uppermost forest belt in the Central Alps is formed by the European larch *(Larix decidua)* and the Swiss stone pine *(Pinus cembra),* which is closely related to the Siberian subspecies. *Larix decidua* is the light-loving pioneer species, which is gradually displaced by the more shade-tolerant five-needled stone pine. On avalanche tracks the larch descends to low elevations.

Above the timberline, the two-needled Scots pine *(Pinus mugo)* or mountain pine *(Pinus montana)* can be found as krummholz, which is replaced by the shrubby green alder *(Alnus viridis)* in damp locations.

On the ecology of **spruce biogeocoenes** (see below).

The question was presented whether the short summer or the long winter was responsible for the cessation of tree growth. It was found that both are of importance. When the growing season is less than three months, the young needles cannot mature properly; their cuticle does not reach its final thickness. As a consequence, during the long winter and especially in spring, when the sun is already strong but the soil is still frozen, high water losses occur, indicated by the increase of cell sap concentration up to more than 6.5 MPa. Damage due to frost-drought becomes noticeable and the needles fall off. This does not occur under a blanket of snow; therefore, the stunted forms protected by snow in winter still extend somewhat above the forest line. It is through the combined effect of both factors, the shortening vegetation period and the simultaneous aggravation of the danger of frost-drought, that the sharp boundary at a certain altitude comes about. The mountain pines growing above the spruce forest line endure a somewhat shorter growing season, but a few hundred metres higher up the same phenomenon repeats itself for them, the needles no longer ripen and suffer damage from frost-drought, the upper mountain pine tree line therefore stands out just as sharply as the forest line.

At the N part of the Alps, the causes determining the **spruce forest line** (Fig. 10.45) have been studied. The vegetation period shortens with elevation, summers become cooler, winters colder and longer. These climatic changes

Fig. 10.45 The forest and tree line in Vorarlberg near Bludenz (Austria). View from the Saladina-peak (Freiburg hut) southwards to the opposite northern slopes (photo: Breckle)

proceed steadily. In contrast, the timberline in the high mountains forms a rather sharp line. The vigour of the trees suddenly diminishes and a very narrow zone of low stunted forms (Fig. 10.46) forms the transition to the forest free alpine belt.

The causes of the polar **spruce forest line** could be of a similar nature, with the difference that solar radiation plays no role in frost-drought damage there during the polar night in winter. It is replaced by the strong and cold drying winds. Accordingly, the timberline advances in the wind-sheltered

Fig. 10.46 The spruce forest boundary with wind-swept stunted spruce, on the Tegelberg near Füssen (1600 m asl). The lower branches under winter snow cover are green, above them wind and snow blowing causes the branches to die (photo: Breckle)

valleys farther north than on the watersheds. In the Alps it is the other way round, because in the valleys, due to lakes of cold air, the temperatures reach lower degrees than on the mountain ridge from which the cold air flows.

The forest line is highest in the central Alps at 2000 to 2150 metres. Here, as we have mentioned, it is not formed by the spruce, but by the needle-throwing larch and the evergreen, relatively tender-needled stone pine *(Pinus cembra)*. Here, continuous measurements of climatic factors and photosynthesis were carried out over the course of an entire year, including the entire winter. This makes it possible to accurately compare the substance production of the larch with that of the stone pine.

In the cold winter photosynthesis is also dormant in the evergreen Swiss stone pine, but in spring the needles quickly become active, whereas the larch at this elevation does not green up until mid-June and already turns yellow at the end of September. While the Swiss stone pine has 181 days available for substance production, the larch has only 107 days. However, the needle mass of young larches is three to six times greater than that of young stone pines; moreover, despite the shorter growing season, they assimilate 47% more CO_2 per gram of needle mass per year. Therefore, the total production of a 4-year-old larch is 4.5 times and that of a 12-year-old 8.5 times greater than that of pines of the same age. It is only from the 25th year onwards that the amount of needles of larches is lower compared to that of stone pines, so that they lag behind in growth, especially on raw humus soils. With time, therefore, the Swiss stone pine also establishes itself as a shade wood species. The ratio of larch to stone pine is thus reminiscent of that of pine to spruce.

All geo-ecological boundaries were up to 400 m higher during the postglacial warm period in the Alps, as evidenced by wood finds in subfossil peat deposits in the subalpine belt. The dwarf shrubs that overwinter under snow are therefore partly remnants of the former forestation. The forest and snow line in the course of the Postglacial is given in Fig. 10.47.

As a result of the high precipitation, which falls as snow in winter, the snow cover in the alpine belt is very thick, so that for the low alpine vegetation not the air temperature plays the main role, but the snow-free period.

This is very much determined by the relief, the wind direction and the exposure: The snow is deposited in hollows and as snow cornices on the leeward side of a ridge, whereas it is blown away on its windward side. If the windward side is also on the sunny side, the snow also thaws, so that the site is open all year round. There the plants (Loiseleurietum association) are exposed to extreme frost-drought as in the mountain tundra and are also accompanied by the same lichens. On a shady windward slope there is no warming by insolation. In the case of heavy snow deposits at the foot of a slope exposed to the north, the snow-free period is reduced to a minimum

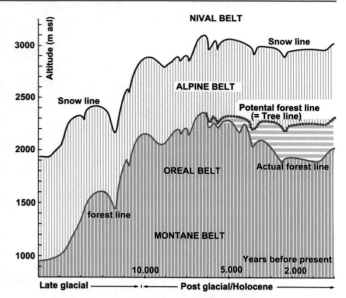

Fig. 10.47 The fluctuations of the tree and snow line in the Late and Postglacial in the Swiss Central Alps (modified after Ozenda 1994). Due to human activities, the current timberline is lower today than the potential timberline, which corresponds to the tree line

(**snow valleys**) or is absent altogether where the snow remains throughout the summer. Depending on the snowfall in the individual years at the same location, the snow-free period (Aper-time) may be longer or shorter. The average snow-free period decreases with elevation and is theoretically zero when the climatic snow line is reached (mountain glaciation). In individual cases, however, it can still be very long high above the snow line on steep faces. This is why flowering plants in the Alps on favoured sites may occur in the nival belt, i.e. above the climatic snow line.

In any case, the microclimate on sunny days is characterised by favourable temperature conditions, even at high elevation. The temperature of the leaves in the sunshine can be up to 22 K above the air temperature. There are warm niches everywhere, known to the climber, and these are exploited especially by the low-growing plants near the ground. In cloudy weather the differences balance out.

From what has been said it is clear that in the steep alpine belt there is no standard climate with regard to vegetation, but that there is a subdivision into the smallest climatic spaces; these can differ sharply at the shortest distance, for example on the sunny and shady side of a rock. Of paramount importance is the snow deposition in winter, which one must know in order to be able to judge the snow-free period; otherwise, the vegetation division remains incomprehensible to one.

A major role is played by temperature inversions and cold air lakes, which lead to a reversal of the sequence of the altitudinal belts (beech over spruce). Even in midsummer, night frosts occur in sinkholes (dolines) when there is outgoing radiation, so that no tree growth is possible at the bottom of the sinkhole.

In addition, the elevational belts are disturbed by avalanche tracks; on these, the alpine vegetation descends deep into the forest belt, because there it is not exposed to the competition of the forest vegetation. Alpine exclaves can also be found in the middle of the forest on heavily weathered dolomite with very shallow, nutrient-poor soils. Also known are the relict sites of alpine species on moors in the foothills of the Alps. At such sites the undemanding but slow-growing alpine species are less exposed to competition from others.

10.6.3   Alpine and Nival Belts

The vegetation of the Alps has been well studied ecologically. In the evergreen species, the same annual cycle of frost hardiness with hardening in late autumn and dehardening in spring can be observed as in the deciduous forests. While spruce needles are killed by frosts as low as $-7\,°C$ in summer, they can still withstand $-40\,°C$ in winter. Although the alpine species grow much higher than the spruce, their maximum frost hardiness is usually lower (below $-30\,°C$), because they overwinter under snow and are therefore not exposed to the low winter temperatures. Only *Loiseleuria*, which grows on wind-exposed sites that are impermeable in winter, has greater frost hardiness. In strong winds, the lowest temperatures are usually not reached in such locations, but the risk of frost-drought is increased. *Loiseleuria*, notwithstanding its xeromorphic leaf structure, dries out within 15 days in winter if left to hang freely. However, as it grows close to the ground in its natural snow-free habitat, even in winter sunlight thaws the snow held between its shoots, allowing water uptake in between. The dwarf shrubs hibernating under snow are not exposed to frost drought.

In summer, with frequent precipitation, the water balance is fairly equalised. The plants are only exposed to high evapotranspiration for hours when there is strong radiation or strong wind. The latter is slowed down near the ground. The water supply of the soil is always good, even at superficially dry looking scree slopes or rocky sites. At such sites, plants have extensive root systems or taproots that penetrate deep into moist rock crevices, whereas normally the root system is spread very shallowly in the upper soil layers. The favourable water balance is reflected in the low cell sap concentrations of 0.8 to 1.2 MPa. Even in xeromorphic species such as *Dryas octopetala, Carex firma* and *Androsace helvetica*, it never reaches 2.0 MPa. It is perhaps more correct to speak of peinomorphs caused by nitrogen (N) deficiency, since N uptake is more difficult at low soil temperatures. Lush hygromorphic herbs grow on nitrogen-rich sites, such as cattle paddocks. If one calculates the total water release of the plant cover of alpine grass communities, one arrives at 200 mm per year. The evaporation size depends

mainly on the wind, it is therefore conditioned by the relief, but in the opposite sense as the snow deposition.

In the alpine belt, the short vegetation period raises the question of sufficient substance production just as it does in the Arctic. The day length is shorter than in the Arctic, but radiation is stronger and the night temperature lower. Under favourable lighting conditions, 100 to 300 mg/dm^2 of CO_2 is assimilated per day. One month of good weather would be enough to accumulate sufficient reserves for the next year and for the seeds to mature. Since the growing season lasts three months, sufficient production is assured in any case. The primary production of the plant communities depends very much on the vegetation density.

> **Box 10.8 What does the snow-free period (Apertime) mean?**
> Apertime (lat. Apertus = open) is the time without snow cover.

The following annual production was found:

- closed mats: 50–276 g/m^2
- Dryadeto-firmetum (cushion sedge lawn): 91 g/m^2
- Salicetum herbaceae (herbaceous willow lawn): 85 g/m^2
- Oxyrietum (acid grassland): 15 g/m^2
- on limestone scree: 1 g/m^2

Photosynthesis in dwarf shrubs is less intense than in herbaceous species; however, because their total leaf area is larger and the growing season is longer at lower alpine belts, higher primary production occurs.

The most unfavourable conditions are found in the so called "snow hollows (Schneetälchen)" with chionophytes, i.e. where in the area of silicate rocks the snow on the northern slope thaws very slowly and gradually releases the surface from the edge. Therefore, a zonation with decreasing snow-free period can be distinguished in a very small area. The soil at such sites is rich in humus and slightly acidic, always well moistened with meltwater, but therefore also relatively cool. With a snow-free period of three months a normal *Carex curvula* lawn develops. If the growing season is shortened to two months, *Salix herbacea* becomes predominant, a species of willow that only sticks the shoot tips out of the ground, so that its leaves form a dense lawn or flat espalier. This willow, however, only fruits when the snow-free period is three months after winters with little snow. A number of very small species, such as *Gnaphalium supinum, Alchemilla pentaphylla, Arenaria biflora, Soldanella pusilla, Sibbaldia procumbens,* and others, join it. With an even shorter vegetation period, mosses can grow that do not need to form flowers and fruits, namely *Polytrichum sexangulare*

(P. norvegicum). If the snow-free period is also too short for these green mosses, only *Anthelia juratzkana,* a liverwort that looks like a mouldy coating, will grow because the moss grows in symbiosis with a fungus and feeds partly saprophytically. After snowy winters, this zone does not aperture at all.

On firn surfaces in the nival belt, the last living organism to be found is the alga *Chlamydomonas nivalis,* which gives the snow surface a pink sheen or rosy hue (Fig. 10.48).

Since bare rock soils predominate in the Alpine belt, the chemical composition of the rock plays a major role for the vegetation, it determines the soil reaction. The floristic differences between the "Limestone Alps" and the Central Alps (Silicate Alps) with siliceous rocks are very striking. Accordingly, a distinction is made between calcareous or basophilic species and limestone-avoiding or acidophilic species. Often, there are **vicariant species**, as in the well-known example of alpine roses: *Rhododendron hirsutum* on limestone, *Rh. ferrugineum* on silicate rock or acid humus soil. Other examples are: *Achillea atrata, Carex firma, Gentiana clusii, Sesleria coerulea* on limestone and *Achillea moschata, Carex curvula, G. kochiana, Sesleria disticha* on silicate soils.

From 1969 to 1976, within the framework of the International Biological Programme (IBP) on the Patscherkofel near Innsbruck, the ecosystems of the alpine dwarf shrub heaths were investigated on the following three sample areas above the present forest line (Larcher 1980):

a. *Vaccinium heath* (1980 m asl), in a wind-protected hollow with winter snow protection: *Vaccinium myrtillus 3, V. uliginosum 2, V. vitis-idaea 1, Loiseleuria procumbens 1, Calluna vulgaris 1, Melampyrum alpestre* 1, mosses 1, lichens 1.

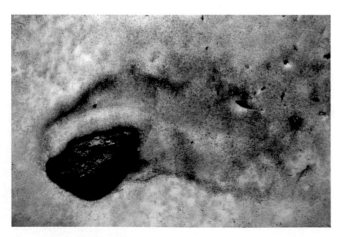

Fig. 10.48 *Chlamydomonas nivalis* as the highest-occurring organisms, as e.g. in some snowfields at 5000 m altitude on the northern slope of Shah Fuladi, Kohe Baba, Afghanistan. They are found in old snowfields in all mountains (photo: Breckle)

b. *Loiseleuria* heath (2000 m asl), dense stand often free of snow in wind-exposed position: *Loiseleuria 5, Vaccinium uliginosum 1, V. vitis-idaea 1,* others only +, lichens *(Cetraria islandica 1, Alectoria ochroleuca 1,* others only +).

c. Open, trellis-like and lichen-rich *Loiseleuria* stand (2175 m asl) in extremely wind-exposed location: *Loiseleuria 3,* stunted *Vaccinium uliginosum 2, V. vitis-idaea 1, Calluna 1,* others +, mosses +, lichens *(Cetraria islandica 2, C. cuculata 1, Alectoria ochroleuca 1, Cladonia rangiferina 1, C. pyxidata 1, Thamnolia vermicularis* and others +).

The climate is cold with an annual temperature slightly above 0 °C, frosts can occur in any month (abs. Minimum around −20 °C, but daily maxima reach +20 °C in the summer months). Snow lasts for about six months in sample plot A, about four to six months in sample plot B, whereas in sample plot C it is only sporadic and temporary. The microclimate in stands A and B is slightly warmer, while in C there are very sharp temperature differences. The CO_2 assimilation period is about 100 days for deciduous species and about 140 days for evergreens. Fig. 10.49 shows the structure of the stands and the photosynthetically active radiation (PhAR) in them, as well as the cumulative leaf area index (LAI). For further data on production ecology, see Table 10.1. Wind is strongly slowed down in dwarf shrub heath stands, even during storms, so that humidity remains high in the same. Precipitation in the area, is about 900 mm per year, with each summer month receiving an average of over 100 mm.

The soils above schistous biotite gneisses are sandy, acidic iron podsols with a thick layer of raw humus, which is only weakly developed in stand C. The soils have developed from former stone pine forest soils. The humus is mineralised very slowly (supply of nitrogen about 3 to 4 kg/ha, at C only 1/3 of it).

The phytomass should remain constant except for certain fluctuations, i.e. the stands are in ecological equilibrium with their environment, with an increase in phytomass also being prevented by browsing (game, ptarmigan, arthropods) and by certain losses of substance in winter (freezing and drying of parts protruding above the snow).

The photosynthetic capacity per leaf area is the same in deciduous and evergreen dwarf shrub species; in terms of dry leaf weight, it is similar to that of soft-leaved deciduous woody species in deciduous dwarf shrubs and comparable to that of coniferous species in evergreens. The shallow temperature optimum for photosynthesis in the Ericaceae is between 10 °C and 30 °C and corresponds to the usual temperatures in stands on cloudy and clear days; the temperature minimum for CO_2 assimilation is −5 °C to −6 °C with super-cooled leaves. Overheating of leaves rarely occurs, as

Fig. 10.49 Middle: Phytomass stratification of *Vaccinium* and *Loiseleuria* heath (assimilating parts on the left, non-assimilating and dead parts on the right). Left half: Cumulative leaf area index (LAI, red curve) and attenuation of light (PhAR) in the stand under clear (**blue curve**) and overcast (**grey curve**) skies (modified after Cernusca 1976; from Larcher 1977)

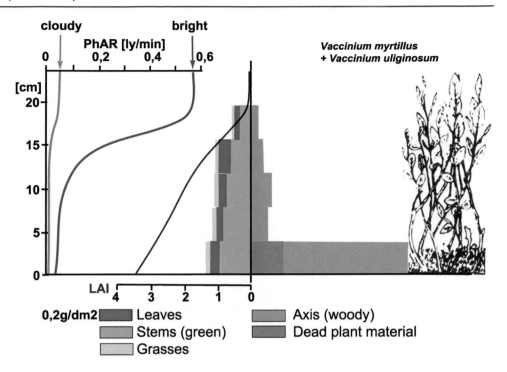

does restriction of photosynthesis due to lack of water. Although water supply is adequate during the growing season and the total transpired water volume is equivalent to 100 to 200 mm, transpiration limitation has been observed during foehn periods. In winter, cuticular transpiration is very low.

Heat damage during the summer is observed at most in individual shoots over loosely lying stones or over vegetation-free raw humus blankets. Cold damage in winter can only occur in the apical state. Hardening protects plants from frost damage; late frosts after dehardening, on the other hand, can be dangerous. Damage caused by frost-drought is difficult to prove; in most cases, the damage is caused by a combination of several factors. Completely frost-hardy are the arctic-alpine species *Loiseleuria procumbens* and *Vaccinium uliginosum.*

Respiration is markedly excessive at the time of main growth. Around this time, in the fat-storing *Loiseleuria*, the respiration coefficient drops to 0.8 to 0.9 and rises again to 1 after completion of intensive growth.

The efficiency of net primary production during the growing season is 0.9% for the dwarf shrub heath, 0.7% for the dense *Loiseleuria* heath, and 0.3% of photosynthetically active irradiance for the open stand.

The Ericaceae store abundant fat in addition to starch, but the latter is only partially mobilised; the greater part remains in the dead parts. The dwarf shrubs react immediately to the first days of frost by converting a large part of the starch into sugar, *Loiseleuria* turning a reddish colour due to anthocyanin accumulation.

Further investigations were carried out in the nival belt, i.e. above the climatic snow line, on the high Nebelkogel in

Table 10.1 Production ecological parameters of alpine vegetation units; living standing phytomass, dead parts, and litter in grams of dry matter per m² from the dwarf shrub heath (A), the dense *Loiseleuria* heath (B), and the open *Loiseleuria* stand (C)

Sample area	A	B	C
Living above-ground phytomass (max.)	983	1105	748
Adhering dead parts	263	123	72
Living underground phytomass	2443	2200	803
Dead underground parts	1549	608	56
Total living phytomass	3426	3305	1551
Together with dead parts	5238	4036	4036
Litter on the ground	819	1080	931
Shoot/root ratio	1:2,5	1:2,0	1:1,1
Percentage of assimilating parts of living phytomass	55%	68%	-?-
Above-ground net primary production (t·ha^{-1}·a^{-1})	4.8	3.2	1.,1

Fig. 10.50 Phenology of nival plants. Flower development and phenological phases at 2600–3200 m asl (modified after Moser et al. 1977)

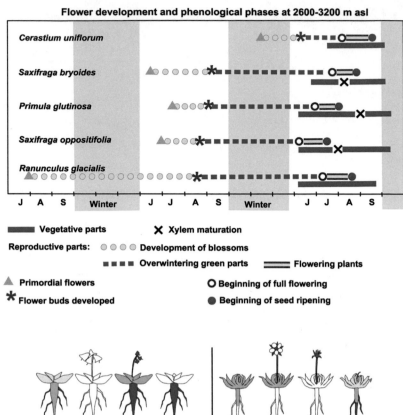

Fig. 10.51 Energy content of 2 nival species and storage of reserve materials (modified after Moser et al. 1977)

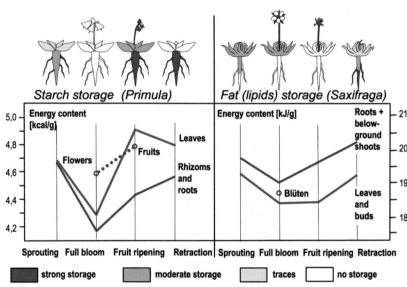

the Stubai Alps under particularly difficult conditions (Moser et al. 1977). An experimental hut had to be lowered by helicopter and carefully insulated and grounded, as it was often located in the middle of thunderclouds.

At this elevational belt there is no longer a closed plant cover. On the 0.5 ha experimental plot, at 3184 m asl, a flat ridge section with seven flowering plants and several crypto-gam species, a northern slope with very sparse vegetation and a southern slope with eleven phanerogam species on flat steps were selected.

The climatic conditions are by no means those of the high Arctic, but in summer more like those of the Páramos in the tropics. On clear days, the leaf temperature is often over 15 °C, to drop below zero at night, without photosynthetic activity suffering. In contrast, the arctic 24-hour summer day with low sun is characterised by a fairly uniform temperature.

Of the three sites selected, the southern slope has the most favourable light and temperature conditions. The phenology of the most important species is shown in Fig. 10.50. While the flowering period is advanced in *Saxifraga oppositifolia* (flowering organs frost-resistant), it is postponed furthest in *Cerastium uniflorum.*

The assimilates are stored as starch in *Primula* spp. and *Ranunculus glacialis,* which is converted into sugar in win-ter, but as fat in the *Saxifraga* species. The shift in store location is evident from Fig. 10.51. It is striking that in *Ranunculus* a precautionary translocation from the leaves to the underground storage organs takes place even in the case of a temporary deterioration of the weather, which is reversed when the weather improves. After all, any snow cover in the summer could last until the next spring. In general, the growing season on the south slope is about three months, but as a result of the often poor weather, only 60 to 70 (15 to 100) days are considered for production. At the other locations, the plants may not become snow-free at all during some years.

Half of the production in *Ranunculus glacialis* is generated during the few bright and warm days, the other half during the many cool days with low illumination due to light snow cover or fog. Photosynthesis in this species is possible in the range of $-7°$ to $38\,°C$. Assimilation efficiency is greatest at the time of full flowering and fruiting. Under optimal conditions, it reaches up to 0.056 g of dry matter per dm^2 of leaf area per day in *Ranunculus glacialis*, and 0.063 g in *Primula glutinosa*; under unfavourable weather conditions, the values are 0.015 to 0.020 g. During one growing season, the areal extent of *Androsace alpina* cushions increased by 13.5%; their average net assimilation rate during the growing season was 0.058 g of dry matter per dm^2 of cushion surface per day. As a result of low plant cover, primary production in the nival belt is extremely low. Under optimal conditions, at 10% cover, production can be estimated at 0.66 g per m^2 per day of dry matter.

> **Box 10.9 The Scientific Investigation of All the Earth's High Mountains Is Very Inhomogeneous**
> Of all the mountain ranges on Earth, none has been studied in as much ecological detail as the complex mountain system of the Alps, located in the centre of western Europe.

10.7 Zonoecotone VI/VII: Forest-Steppe

While the deciduous forests of the temperate zone are confined to the oceanic-tinted climatic regions with not too sharp temperature extremes and evenly distributed precipitation, usually with a summer maximum, in the northern hemisphere much more extensive continental parts are occupied by grass steppes and deserts. In the continental climate in Europe from west to east the temperature amplitude increases, the summers become hotter, but the winters colder to a much greater extent, so that the mean annual temperature decreases. At the same time, the annual amount of precipitation becomes less, the summers become increasingly arid.

The ZE VI/VII between the deciduous forests and the grass steppes is the **forest steppe** in Eastern Europe. It is not a homogeneous vegetation formation like the climatic, tropical savannah, but a macromosaic of deciduous forest stands and meadow steppes. At first, the former predominate and the steppes occur in island form. The more arid the climate becomes, however, the more the ratio is reversed, until finally only small islands of forest remain in a sea of steppe. In this transition zone with a climate that favours neither the forest nor the grass steppe unilaterally, the relief or soil type (Fig. 10.52) is the deciding factor. The forests are found on well-drained sites, on the slight elevations, on the slopes of river valleys, on permeable soils, while the grass steppes occupy the poorly drained level sites on relatively heavy soils. This is similar in the savannah. Competition between the sward and tree seedlings also plays a role here.

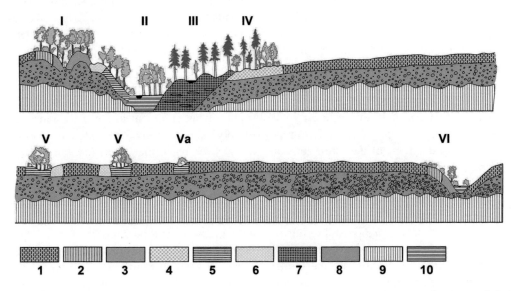

Fig. 10.52 Relationships between vegetation, soil, and relief in the forest steppe (modified after Walter 1990): **1** deep poorly drained chernosem (black earth) with meadow steppe. **2** degraded chernosem and **3** dark grey forest soils (both well drained); **4** permeable sandy-loamy forest soils; **5** light grey forest soils; **6** solonez on level terraces or around drainless depressions with sodic enrichment; **7** fluvio-glacial sands; **8** moraine deposits or loess-like loams; **9** preglacial strata; **10** alluvium in river valleys. **I** oak woodland on well-drained elevations or slopes; **II** floodplain woodland (oaks et al.); **III** pine woods on poor sands with *Sphagnum* marsh in wet depressions; **IV** pine-oak woods on loamy soils; **V** aspen groves in small depressions (pods) where water stands in spring and slowly percolates (soils depleted in central part); **Va** the same but willow scrub; **VI** ravine-oak woods, with steppe scrub at upper margin

Fig. 10.53 Steppe and scrub/forest plot mosaic in Dobruja (Romania). Some grazing by goats and sheep keeps larger parts open as steppes and species-rich dry grasslands. The bushes continue to grow outwards only slowly, regeneration is hardly given even without grazing (photo: Breckle)

If tree seedlings are protected from grass competition in the early years of afforestation experiments, they may grow in the steppe but not rejuvenate naturally. Steppes used to be favoured by grass fires started by lightning and by grazing by big game. One can only speculate about the real role of big game under natural conditions. However, the grazing density by man's grazing animals (sheep, goats, cows) is certainly much higher than that of the original big game. Nevertheless, in some regions the vegetation composition and mosaic structure are probably very similar to the original one (Fig. 10.53). Today, in many places, the steppe has been almost completely converted into arable land.

Climatically, one can easily distinguish the forest zone, the forest-steppe zone and the steppe zone in E Europe (Breckle 2021). The climate diagrams of the forest zone do not show a drought period, whereas those of the steppe zone always show a drought period. The diagrams of the forest steppe zone lack a drought period, but in contrast to the forest zone, a dry period can be recognised (Fig. 12.1).

The boundary between forest and steppe shifted in the postglacial period. In the soil under today's forest stands one can see **Krotovines** (Fig. 12.4), which are the former burrows of steppe rodents (ground squirrel = *Spermophilus*) that never inhabit forests. Therefore, it must be assumed that the forest was in the process of advancing in the period before human colonisation of the forest-steppe, because the climate became somewhat wetter after a warm optimum in the postglacial. However, due to the strong interventions of man, boundary shifts in the subsequent period can no longer be determined.

The reason for the replacement of the forest zone in the continental area by the steppe zone is the water factor. In the forest steppe the entire water turnover takes place almost only in the upper 2 m of the soil; a sinking of water to the deep groundwater does not take place. The oak forest consumes all water; the soil always remains dry at greater depths. This is the case on flat surfaces. On S slopes with runoff and high evaporation, the water content of the soil is no longer sufficient for forest, and steppe sets in. In August and September, the grass steppe dries out, because even for it the water supply is not sufficient to cover its transpiration. However, this does not cause any damage to the grass plants, but it does to the trees when the leaves dry out prematurely or entire branches die.

In a SE direction, precipitation decreases and temperatures increase in the forest steppe. Accordingly, the forest plots become more and more sparse and retreat to the N slopes, until finally, at the S border of the forest-steppe, only oak and sloe scrub *(Prunus spinosa)* remains in ravines.

Competition in the forest steppe is between grasses and tree seedlings. Clements et al. were able to show in the 1929 partly pristine long grass prairie of Nebraska (Fig. 12.8), which corresponds to the forest steppe, that planted tree seedlings only persist if all grass roots around them are removed and this tree disk is kept free.

The water consumption of forest stands increases with the age of the stand. Accordingly, reforestation experiments have shown that young, artificially created forest cultures grow relatively well, but in older ones the trees become stunted and then sprout again from below, i.e. they do not develop normally as a result of the lack of water. Good stands, on the other hand, are obtained when additional groundwater is available to the trees. Savannah-like communities are lacking in forest-steppes, because the hardwood species cannot assert themselves individually against the competition of the grasses. Only low shrubs *(Spiraea, Caragana, Amygdalus)* occur more frequently, but even these more on stony soils, which are less suitable for the steppe grasses with the intensive root system. The steppe component of the forest steppe - the meadow steppe proper - is dealt with in the next chapter (ZB VII).

Photo 38 Southern beech forest (mainly deciduous *Nothofagus* species) (ZE V/VI) on Terra del Fuego (photo: J.Tamm)

Photo 39   Mixed deciduous forest (ZB VI) during fall at Lyman Run State Park, Potter Cty, Pennsylvania, USA (photo: N.Λ. Tonellis 2009)

References

Breckle, S.-W. 2005: Möglicher Einfluss des Klimawandels auf die Vegetation Nordwestdeutschlands? LÖFB-Mitteilungen **2** (05): 12-17

Breckle, S.-W. (Hg.) 2021: Ökologie der Erde. Band 3. Spezielle Ökologie der Gemäßigten und Arktischen Zonen Euro-Nordasiens, Zonobiom VI - IX. 3. Aufl. Schweizerbart/Stuttgart 804 S.

Cernusca, A. 1976: Bestandesstruktur, Bioklima und Energiehaushalt von alpinen Zwergstrauchbeständen. Oecologia Plantarum **11**: 71-102

Clements, F.E., Weaver, J.E. & Hanson, H.C. 1929: Plant Competition. Carnegie Inst. Wash., Publ. **398**

Duvigneaud, P. 1974: La synthèse écologique. Population, communautés, écosystèmes, biosphère, noosphère, Paris, 296 p.

Edwards, C.A., Reichle, D.E. & Crossley, D.A.jr. 1970: The role of soil invertebrates in turnover of organic matter and nutrients. Ecol. Stud. **1**: 147-172

Ellenberg, L & Leuschner, C. 2010: Vegetation Mitteleuropas mit den Alpen in ökologischer, dynamischer und historischer Sicht. UTB, Ulmer-Stuttgart, 6. Aufl. 1357 S.

Ellenberg, H., Mayer, R. & Schauermann, J. 1986: Ökosystemforschung, Ergebnisse des Sollingprojektes 1966-1986. Ulmer, Stuttgart 507 S.

Goryschina, T.K. (Hrsg.) 1974: Biologische Produktion und ihre Faktoren im Eichenwald der Waldsteppe. Arb. Forstl. Versuchsst., Univ. Leningrad. „Wald an der Worskla" **6**: 1-213 (Russ.)

van Groenendael, J.M., Klimes, L., Klimesova, J. Hendriks, R.J.J. 1996: Comparative ecology of clonal plants. Phil. Trans. R. Soc. London, **B 351**: 1331-1339

Hüppe, J. 1993: Development of NE European heathlands - palaeoecological and historical aspects. Scripta Geobot. **21**: 141-146

Kaagman, M. & Fanta, J. 1993: Cyclic succession in heathland under enhanced nitrogen deposition: a case study from the Netherlands. Scripta Geobot. **21**: 29-38

Larcher, W. 1977: Ergebnisse des IPB-Projekts, Zwergstrauchheide Patscherkofel". Produktivität und Überlebensstrategien von Pflanzen und Pflanzenbeständen im Hochgebirge. Sitz.-Ber. Österr. Akad. d. Wiss., Math.-nat. Kl., Abt., Wien **I, 186**: 301-386

Larcher, W. 1980: Klimastress im Gebirge – Adaptationstraining und Selektionsfilter für Pflanzen. Rheinisch-Westfälische Akad. Wiss., N **291**: 49-88

Leicht, H. 1996: Altbäume - Tierökologische Bedeutung für die Praxis. Ber. Bayer. Landesamt Umweltschutz **132**: 86-93

Leuschner, C. 1993: Forest dynamics on sandy soils in the Lüneburger Heide area, NW Germany. Scripta Geobot. **21**: 53-60

Lozan, J.L., Breckle, S.-W. 2021: Bodenversiegelung und Änderung des natürlichen Wasserkreislaufs. In: Lozán J. L. S.-W. Breckle, H. Grassl, W. Kuttler & A. Matzarakis (Hrsg.). Warnsignal Klima: Böden und Landnutzung. pp. Xxx-yyy. Online: www.klima-warnsignale.uni-hamburg.de.

Moser, W., Brzoska, W., Zachhuber, K. & Larcher, W. 1977: Ergebnisse des IBP-Projekts "Hoher Nebelkogel 3184 m". Sitzungsber. Österr. Akad. d. Wiss., Wien, Math.-nat. Kl., Abt. I, **186**: 387-419

Ozenda, P. 1994: Végétation du continent européen. Delachaux et Nestlé/Lausanne 271 p.

Peters, R. 1997: Beech Forests. Geobotany **24**, Kluwer Acad. Publ. **169** S.

Schulze, E.-D. 1970: Der CO_2-Gaswechsel der Buche (*Fagus silvatica* L.) in Abhängigkeit von den Klimafaktoren im Freiland. Flora **159**: 177-232

Walter, H. 1976: Die ökologischen Systeme der Kontinente (Biogeosphäre). Prinzipien ihrer Gliederung mit Beispielen, Fischer, Stuttgart 131 S.

Walter, H. 1990: Vegetationszonen und Klima. 6. Aufl., Ulmer/Stuttgart 382 S.

Contents

Photo 40 Feather grass steppe (ZB VII) in the surroundings of the monastery Dawit Garedscha, Georgia (photo: E. Fischer)

Photo 41 *Iris iberica*, an endemic to the Caucasus region and one of the many geophytes of the feather grass steppes (zonobiom VII) of south Georgia near the Azerbeijan border (photo: Rafiqpoor)

Photo 42 The Amu Darya River near Koreshm (Uzbekistan) with the almost dry riverbed due to intensive water extraction for irrigation purposes in summer, which led to the eventual drying up of the Aral Sea and thus to the formation of the new Aralkum desert (zonobiome VIIa; photo: Breckle)

11.1 Climate

This continental **zonobiome VII** extends in Eurasia from the mouth of the Danube through E Europe and Asia almost to the Yellow Sea. In N America, it occupies the entire Midwest from S Canada to the gulf of Mexico. The degree of aridity varies in the different parts.

In ZB VII, the steppe, semi-desert and desert region with cold winters, six sub-zonobiomes can be distinguished (Walter 1990):

1. the semi-arid sZB with a short drought period (**steppes** or prairies, sZB VII)
2. the arid sZB with prolonged drought and winter rains [**semi-deserts**, sZB VIIa(w)]
3. the strongly arid sZB [with climate type VII (rIII)w], i.e. with as little rain as in the climate of the subtropical deserts, but with winter rain (**deserts**)
4. the arid sZB with prolonged drought and summer rains [**semi-desert**, sZB VIIa(s)].
5. the highly arid sZB, with low summer rainfall [(**deserts**) (VII (rIII)s]
6. the cold **high plateau deserts** (Tibet and Pamir) [sZB VII (tIX)].

The two semi-desert subzonobiomes are ecotones interposed between the steppes (Fig. 11.1, Chkalov) and actual deserts. These arid ecotones of semi-deserts are characterised by climate type VIIa (Fig. 11.1, Astrakhan). The semi-deserts (in N America it is the sagebrush area) are more arid than the steppes, but less arid than the deserts, and their vegetation bears transitional character; in this case the pronounced drought lasts for about 4–6 months (Fig. 8.19).

11.2 Soils of the Steppe Zone of Eastern Europe

The E European steppe is the cradle of the soil type theory, which was founded by Dokutschajev (1898) and Glinka (1914). There is no area of the same character on earth where the parallel zoning of climate, soil types and vegetation is so clearly visible, although it must be said that only very small remnants of the natural vegetation remain today. The favourable conditions for the zoning are the very flat relief and a largely uniform parent rock (loess) over large distances. The climate changes steadily from NW to SE: summer temperatures and potential evaporation increase, while precipitation decreases, i.e. aridity becomes more and more pronounced. Thus, a uniform gradient occurs over a very long distance, which makes ideal transects possible.

The boundary between forest and forest steppe corresponds to the boundary between humid and arid areas, i.e. north of this boundary annual precipitation exceeds the annual amount of potential evaporation, south of it the latter is higher (Fig. 11.2), so that runoffless depressions (endorrheic flats) can form, which then exhibit brackish soils.

The distribution of soil types is shown in a highly simplified form in Fig. 11.3. In the humid zone we find typical podsol soils and slightly podsolic grey forest soils, in the arid zone black earths (Chernozem) up to the arid chestnut and brown earths (burozem). The soil types can be recognised by their soil profiles, which are shown schematically on Fig. 11.4.

The black earths (chernozems) are A-C soils or pedocals, that is, they lack a clayey enrichment horizon (B). A distinction is made between the subzones: Northern, thick, common and southern chernozem. The humus horizon A is divided

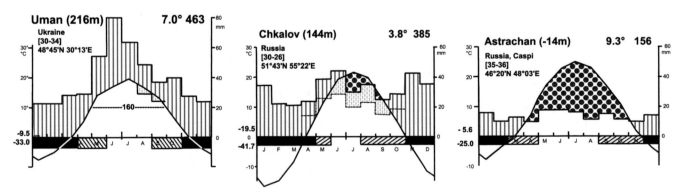

Fig. 11.1 Climate diagrams from the forest-steppe zone (with dry season), from the steppe zone (with drought and long dry season) and from the semi-desert (with long summer drought)

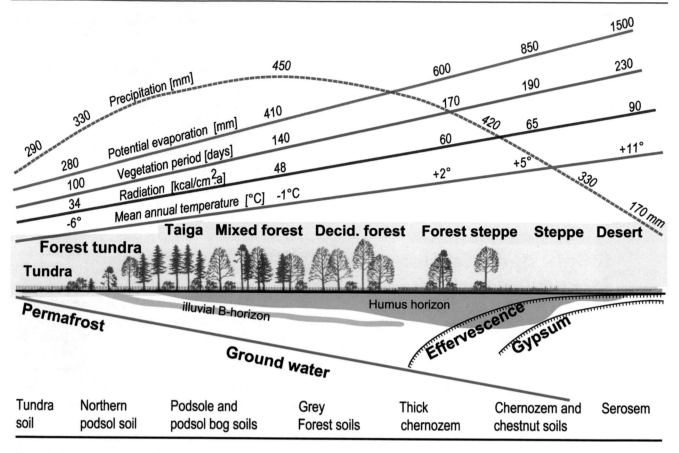

Fig. 11.2 Schematic climate, vegetation and soil transect through the E European plain from NW to SE (modified after Schennikov, from Walter and Breckle 1999). Light brown = humus horizon; yellow = illuvial B horizon; growing season in the tundra = daily mean above 0 °C, otherwise above 10 °C

into the black coloured A_1, the slightly lighter A_2 and the loess A_3, which is slightly coloured by humus. Below follows C, the unaltered loess with prismatic structure. The humus horizon of the **thick** chernozem reaches a depth of 170 cm, its thickness decreases to the north and south, the humus content of the **common** chernozem is highest with 7–8% (even higher in the eastern steppe area). There is no clay displacement in the chernozem, but the meltwater in spring washes out the lime ($CaCO_3$) from the upper horizons, so that no effervescence occurs when HCl is dripped on; this only begins deeper, and in fact the more arid the climate, the higher is the effervescence horizon. Somewhat below the effervescence horizon, the washed-out lime is precipitated, first in the form of very fine lime threads, which remind of mould (pseudomycelia), further south also as white globules (lime eyes = Bjeloglaski) and finally only as such. In addition, the cross-sections of passages of the abandoned underground squirrel burrows (Krotovines), which are filled with washed-in chernozem soil, can be seen in the profile.

All these changes in the transect take place on the soil profile in accordance with the gradual climate change quite smoothly from north to south. They reflect the increasing aridity.

Under the forest in the forest-steppe zone, the upper soil layers remain wetter; the litter forms the A_0 horizon, the mixing of the same with the mineral soil is less, the humus horizon is therefore only light grey in colour under the moist hornbeam (*Carpinus*) forest, and dark grey under the dry oak forest; its good crumbly structure is lost, it becomes platy; under the humus layer one finds mealy, bleached sand grains and underneath a compacted B horizon, all signs of incipient podsolation. This is barely indicated under the oak scrub (as the last extensions of the forest against the steppe) in the **degraded** chernozem. If one then enters the wettest form of the meadow steppes, one finds under them already the **northern** chernozem in typical formation, but with a very deep-lying effervescent horizon and without calcareous precipitates (Figs. 11.3 and 11.4).

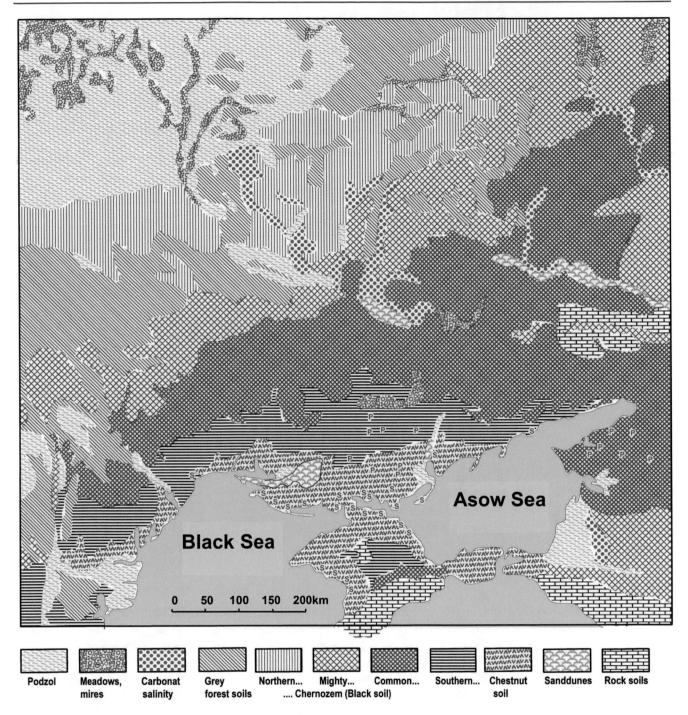

Fig. 11.3 Soil type map of the E European steppe area and adjacent forest areas. P = pod (depressions without drainage in the steppe), S = saline soils (solonchak)

Box 11.1 Relationship of Soil and Vegetation
Each soil type is associated with a particular plant community.

On the basis of the remaining remnants of natural vegetation, it was possible to demonstrate that each soil type is associated with a specific plant community, as shown in the following overview (Table 11.1).

Fig. 11.4 Schematic representation of soil profiles in the forest-steppe and steppe zone (W of the Dnieper) from N to S. Percentages = humus content of A_1, **br** with green arrow = effervescent horizon, wavy blue lines = pseudomycelia (lime), small blue dots = lime eyes, large yellow spots = Krotovines (old ground squirrel burrows)

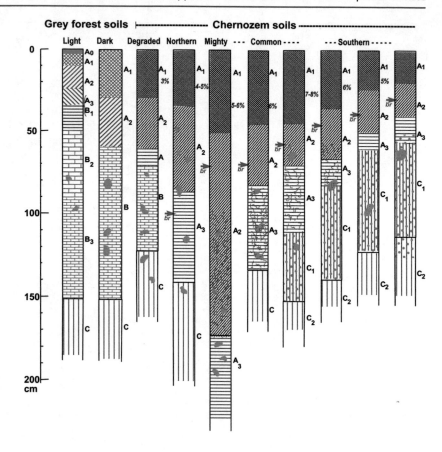

Table 11.1 Allocation of soil types to vegetation types

Soil type	Vegetation type
Grey forest soils	Oak-hornbeam and oak forest
Degraded chernozem	Oak blackthorn bush
Northern chernozem	Moist, herb-rich meadow steppe
Mighty chernozem	Typical meadow steppe
Common chernozem	Herb-rich feather grass (Stipa) steppe
Southern chernozem	Dry, herb-poor Stipa steppe

This assignment makes it possible to reconstruct the former vegetation structure on the basis of the soil map.

11.3 Meadow Steppes on Mighty Chernozem and the Feather Grass Steppes

The word **steppe** comes from the Russian term "stepj". One should therefore use it only for grass steppes of the temperate zone, which resemble the Eastern European steppes, such as the **Prairies** and the **Pampas**. In the tropics there are no steppes in this sense, so it is better to speak of "tropical grasslands". The term steppe is often associated with the idea of barren, poor vegetation, for example when one speaks of a "desertification" of the landscape. However, the opposite is true for the northern variants of the E European steppes. They are now the most fertile parts of the country with the best chernozem; in their natural state they surpass the most luxuriant meadows of the temperate zone in flowering splendour; only in autumn do they make a dry impression.

The forest steppe is a macromosaic of deciduous forests and meadow steppes. The seasonal development of the meadow steppes is explained in Figs. 11.5 and 11.6 and is described below.

After the snow melts, the soil in the steppe is well moistened and temperatures rise, so that a rich spring flora develops. At the end of April, the purple flowers of *Pulsatilla patens* appear, *Carex humilis* also begins to dust, joined in early May by the large golden stars of *Adonis vernalis* and the light blue inflorescences of *Hyacinthus leucophaeus*. In mid-May the steppe turns green; among the sprouting grasses are *Lathyrus pannonicus*, *Iris aphylla* and *Anemone sylvestris*. *In* early June the most colourful stage is reached with quantities of flowering *Myosotis sylvatica*, *Senecio*

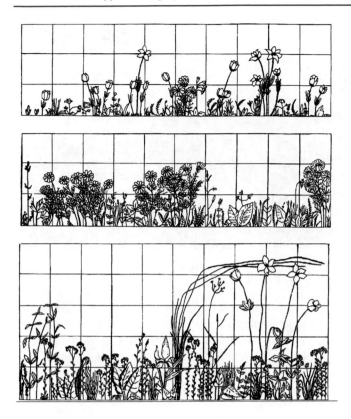

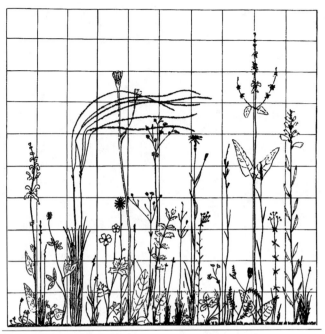

Fig. 11.6 Early summer aspect (from Walter and Alechin 1936). Besides the feather grass *Stipa ioannis,* many herbs (over 40 cm high are: *Salvia pratensis, Hypochoeris maculata, Filipendula hexapetala, Scorzonera purpurea, Phlomis tuberosa* and *Echium rubrum*)

Fig. 11.5 Spring aspects of the meadow steppe (after Pokrovskaja, from Walter 1968). Vertical projection, squares in dm. Top, early April: brown aspect with purple *Pulsatilla patens patches, Carex humilis* dusting. Mid - late April: yellow *Adonis vernalis aspect,* pale blue *Hyacinthus leucophaeus.* Below, late May: Blue *Myosotis sylvatica aspect,* white *Anemone sylvestris,* yellow *Senecio campestris,* some flowering *Stipa*

campestris and *Ranunculus polyanthemus*; at the same time the first feather tails of *Stipa joannis* appear (Fig. 11.7). In early summer, the feathery long awns of *Stipa* species move undulating in the wind, and panicles of *Bromus riparius* (close to *B. erectus*) stretch out; in between, *Salvia pratensis* and *Tragopogon pratensis* flower. Towards the end of June, the steppe turns white with the flowers of *Trifolium montanum, Chrysanthemum leucanthemum, Filipendula hexapetala,* to which *Campanula sibirica* and *C. persicifolia, Knautia arvensis* and *Echium rubrum* form a colour contrast. At the beginning of July, the blaze of colour comes to an end with the pink-flowering *Onobrychis arenaria* and the yellow *Galium verum.*

From the mid of July, the plants begin to dry up; the dark blue panicles of *Delphinium litwinowi* still appear, and later the brown-red candles of *Veratrum nigrum.* From August the steppe looks dry and remains so until the snow covers it.

This description shows that the dry meadows and steppe heaths in Central Europe are something like the poor extrazonal outposts of the meadow steppe in the humid climate on dry, shallow sites. The floristic composition is very similar, except that in central Europe sub-Mediterranean elements are added, such as the orchids, which the steppe lacks.

Further south, the feather grass steppes begin on **common** and **southern** chernozem. In these, various *Stipa* species predominate. The less drought-resistant herbs are not competitive in the increasing drought and are receding more and more. The density of the plant cover decreases, so that the soil is partly covered by the moss *Tortula (Syntrichia) ruralis* and the alga *Nostoc.* In spring, geophytes (*Iris, Gagea, Tulipa*) and some winter annuals (*Draba verna, Holosteum umbellatum*) are more abundant. *Paeonia tenuifolia* is particularly conspicuous. In summer other herbs (*Salvia nutans, S. nemorosa, Serratula, Jurinea, Phlomis* and others) appear, and in late summer Apiaceae (*Peucedanum, Ferula, Seseli, Falcaria*) and Compositae (*Linosyris) are* added. Still further south, the vegetation density continues to decrease; in addition to the long-grass feather grasses, the long-grass *Stipa capillata* and *Festuca sulcata* play a greater role, and among the herbs, those with deep taproots (*Eryngium campestre, Phlomis pungens, Centaurea, Limonium, Onosma*).

On the chestnut-coloured soils, wormwood (*Artemisia*) species become more prominent, initiating the transition to the *Artemisia* semi-desert. This is again interposed as sZB VIIa between the steppe and the even more arid desert [sZB VII (rIII)].

Fig. 11.7 May aspect of feather grass steppes in the vicinity of Dawit Garedcha in Georgia (photos: E. Fischer)

11.4 North American Prairie

The conditions in the prairie correspond to those of the steppe; they are only more complicated. While the steppe extends around the 50th parallel from the foothills of the Carpathians eastwards far beyond Europe, the prairie in Canada also begins south of the 55th parallel, but the zones run more in a N-S direction as far south as over the 30th parallel and merge into *Prosopis* savannahs. In addition, the vast plains in N America rise slowly from E to W to 1500 m asl. Precipitation decreases from E to W (Fig. 11.8), but

Fig. 11.8 Transect through the prairies of North America showing vegetation and soil gradients (partly modified after Burrows 1990)

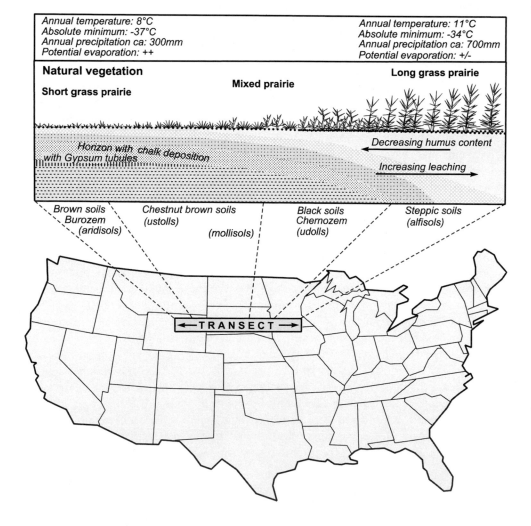

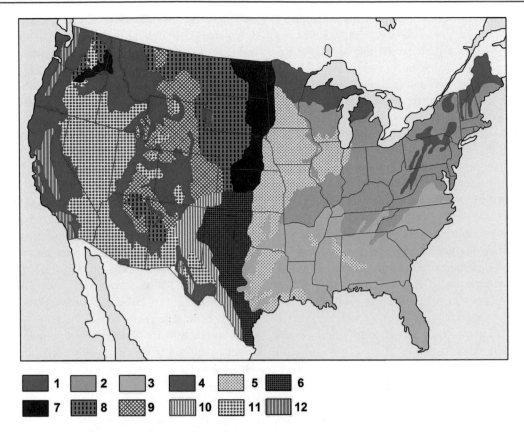

Fig. 11.9 Soil type map of the USA (modified from US Dept. of Agric). **1** podsol soils, **2** gray-brown forest soils, **3** yellow and red forest soils, **4** mountain soils (general), **5** prairie soils, **6** southern black and dark brown soils, **7** northern black soils, **8** chestnut brown soils, **9** northern brown soils, **10** southern brown soils, **11** gray soils (serosem), **12** Pacific valley soils. Brown earth = Burosem. Corresponding to: **1** the coniferous forest zone, **2** and **3** the mixed forest and deciduous forest zone, **5** the long grass prairie, **6–10** the mixed and short grass prairie, **11** in the northern part the sagebrush semi-desert, in the southern part other types

temperature increases from N to S. This does not result in such a clear soil zonation, but more in a checkerboard arrangement of soil types (Fig. 11.9).

Box 11.2 Arrangement of the Steppe Regions of N America

The individual vegetation zones, such as long grass prairie, mixed prairie, and short grass prairie, follow one another as one goes west from the east in the direction of increasing aridity, but in each zone, there is a floristic gradient from N to S.

A more important role than *Stipa* is played by *Andropogon* species, i.e. grass species of southern origin.

The transitional zone of forest-steppe is also present in N America with forests on valley slopes or on light soils and grasslands on level watersheds with heavy soils. The long grass prairie corresponds to the northern meadow steppe on the **mighty** chernozem, but the prairie soils are wetter, the limestone is all washed out, and an effervescent horizon is absent. The question why the prairie is nevertheless treeless was answered experimentally by planting tree seedlings with and without competition of grass roots. The result was that tree growth is quite possible when grass competition is eliminated.

If prairie fires are prevented, the forest advances slowly, about 1 m in three to five years, against the prairie with a shrub zone as a vanguard. But accurate statistics for 1965 showed that on average there is one lightning fire per year for every 5000 ha of prairie area; fire is thus a natural environmental factor in the prairie area in favour of the grasses. It must also be borne in mind that prairie vegetation was formerly favoured by grazing large herds of bison. Add to this, as a natural experiment, the catastrophic drought of 1934–1941, the effect of which on prairie vegetation was still evident in 1953. Such periodic droughts, recurring every century, are certainly partly responsible for the treelessness of the prairies.

The **long grass prairie** is just as herb-rich as the meadow steppe, floristically even more species-rich. During the main flowering period in June, 70 species flower simultaneously. However, the main grasses *(Andropogon scoparius* and

A. gerardi) have their flowering time as southern elements with a C4 photosynthesis only in late summer; with the deep moisture penetration of the prairie soils they do not suffer from water shortage in normal years.

These grasses grow 40–100 cm tall, with inflorescences 1–2 m tall. In the **mixed prairie zone,** besides the long grasses *(Andropogon scoparius, Stipa comata)*, also abundant short grasses *(Bouteloua gracilis, Buchloe dactyloides)* occur, which then dominate in the short grass prairie alone; the herbs recede completely, whereas *Opuntia polyacantha* becomes frequent, especially on overgrazed areas (Küchler 1974). Grazing generally changes the character of the prairie slightly in the direction of an apparent greater aridity, i.e. the **long grass prairie** becomes the **mixed prairie**, the latter the **short grass prairie**. The increasing aridity to the west is expressed by the calcareous precipitates in the soil profile. In the short grass prairie, the calcareous eyes already appear at a depth of 25 cm, the humus horizon is only slightly thick, the root depth decreases, because the roots hardly go into the horizon with the calcareous precipitates, as this indicates the average depth of the soil moisture.

As part of the US IBP, French (1979) published ten papers with ecological studies on production, consumers and grazing problems.

Net production is approximately 2 t/ha per year in short grass prairie, 3 t/ha in mixed and over 5 t/ha and year in long grass prairie.

Like the steppes of E Europe, the N American prairies have also been taken under the plough. There are hardly any original prairies left. Cultivation has led to a severe loss of humus, and the proportion of soluble organic carbon (oC) in

the soil has halved since cultivation (Fig. 11.10). Not only the strong changes in humus content caused by the conversion of prairie to cropland, but also the possible further development to ensure the preservation of the carrying capacity are shown in Fig. 11.10. Extensive investigations serve the purpose of stopping the humus losses or even increasing the humus contents of the soils again by appropriate techniques. This is to be achieved, for example, by reduced cultivation intensity and by recycling litter.

Towards the N in the Canadian area, as well as in the mountains, increasingly larger forest islands indicate the transition of a forest steppe to the forests of the taiga or the mountain forests. In N America *Populus tremuloides* plays an important role. This species is able to form large clones by extensive root runners. Under certain circumstances, entire groves can grow from one plant (Fig. 11.11). This species is also frequently found in mixed forests with pines (*Pinus banksiana*) (Fig. 11.12), but is then also particularly favoured by beavers for dam construction.

As an example of a N American orobiome in the semi-arid climatic region VII we bring the succession of elevational belts of the Front Range of the Rocky Mts. near Colorado Springs: The short grass prairie at the foot of the mountains at 1.500 m asl is first followed by a belt of long grass prairie, then by a belt of deciduous scrub only 50 m wide and *Pinus edulis* and *Juniperus* (pinyon stage), followed by a broad forest belt in which *Pinus ponderosa, Pseudotsuga menziesii* and *Picea engelmannii* successively become predominant. At 3700 m asl the timberline is reached; after a narrow subalpine belt with *Picea briars* and *Dasiphora (Potentilla) fruticosa* scrub the alpine belt begins.

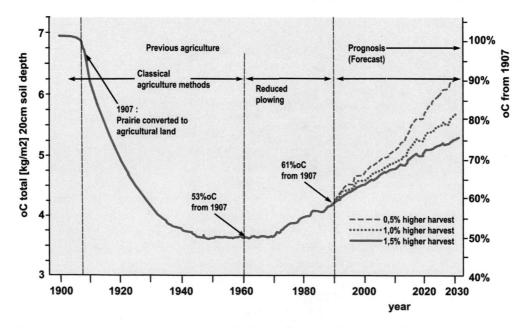

Fig. 11.10 The effects of prairie land use on the humus (organic C) content of the Great Plains in the USA (modified from Donigian et al. 1995)

Fig. 11.11 Insular small groves of *Populus tremuloides* and other woody plants in the Wasatch Mountains (Utah) (photo: Breckle)

Fig. 11.12 Open steppe forest with *Pinus banksiana* and *Populus tremuloides in* autumn in Prince Albert Parc (Saskatchewan, Canada) (photo: Breckle)

11.5   Ecophysiology of Steppe and Prairie Species

The growing season of the steppe plants is limited by the cold winter on the one hand and the drought in late summer and autumn on the other. The plants have about four months with very favourable growing conditions in spring and early summer. The species are for the most part hemicryptophytes and must establish a large producing leaf area in this short time with as little material as possible. Precise determinations of leaf area indices are not available, but they are likely to be similar to those of deciduous forest in grassland steppe. However, total leaf area varies greatly from year to year depending on precipitation levels. For the low herbaceous feather grass steppe, an aboveground phytomass of

4530–6250 kg/ha is reported in wet years compared to 710–2700 kg/ha in dry years. Thus, there is a reduction in transpiring leaf area under unfavourable water supply and this results in lower production. In contrast, below ground phytomass remains unchanged. It is much larger compared to the above-ground.

The annually dying above-ground mass forms the litter layer (steppe felt) on the soil surface, which reaches 8–10 t/ha in the meadow steppe, and only 3 t/ha in the dry steppe. The dying underground mass is transformed into humus by soil organisms. The litter layer is subject to strong decomposition in spring and summer; it shows a minimum at the beginning of drought and a maximum at the beginning of winter. The seasonal change of environmental factors and biotic variables for the largest steppe reserve is shown in Fig. 11.13.

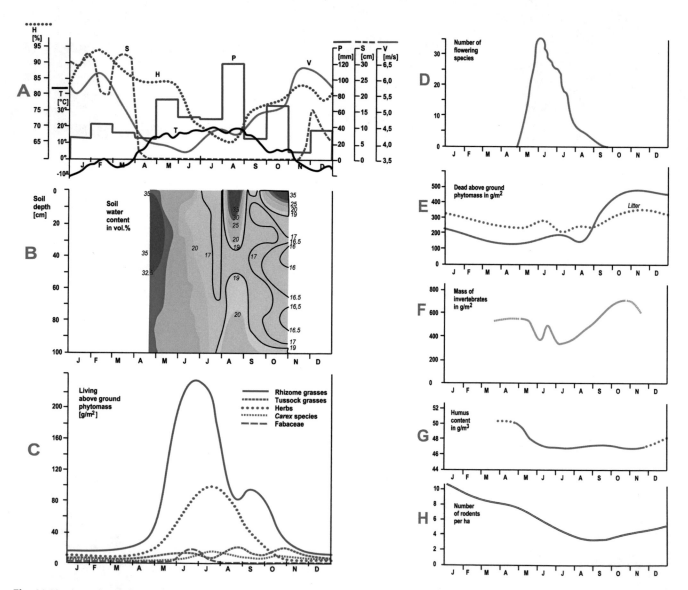

Fig. 11.13   Annual variation of abiotic (**a**, **b**) and biotic variables (**c–h**) of a meadow-steppe ecosystem in the chernozem Earth Reserve in 1957. (**a**) meteorological factors, (**b**) water content of soil, (**c**) aboveground phytomass, (**d**) phenology of plants, (**e**) dead aboveground plant parts. (**f**) number of invertebrates, (**g**) humus mass, (**h**) number of rodents (predominant vertebrates) (modified after Walter 1990)

Excessive enrichment of the litter, for example in protected areas, has an unfavourable effect. The regeneration of grasses becomes more difficult, gaps appear in the plant cover, and "weed" plants such as *Artemisia, Centaurea* and thistles establish themselves. A certain amount of grazing is therefore necessary for the typical development of steppe vegetation, as was normal in the primeval steppe by gazelles and saiga antelopes, the wild horse and the wild ass, and above all by the countless steppe rodents (gopher, ground squirels etc.) and grasshoppers. The burrows of the steppe rodents, along with the earthworms, contribute to the good mixing of the humus with the mineral soil. Occasional natural steppe fires also resulted in the destruction of the enriched litter. In the steppe reserves, people help themselves by mowing the areas every three years.

There is a similar ecological balance between the steppe grasses and herbs as there is between the woody plants and the grasses in the savannah. All grasses have a very intensive, finely branched root system, while the herbs have an extensive root system, often with a deep taproot, so they root in different horizons in the soil.

According to their water balance, the steppe herbs belonged to the group of malacophyllous xerophytes. In spring, the cell sap concentration is very low. Temporary dry periods cause wilting with a steep increase in cell sap concentration. In the late flowering species, at the time of flowering, when drought begins, transpiration is restricted by withering of the leaves; the flowers and the ripe fruits consume little water and obtain the restorative substances from the yellowing parts of the plant.

Very typical of the wide open steppes are the **tumbleweeds** (*Eryngium, Falcaria, Seseli, Phlomis, Centaurea*, etc.). In these, the stiffened stem with the dry fruits remains as a spherical structure; at the neck of the root there is a weak point where the stem breaks off and is rolled across the steppe by the wind, scattering the seeds; often the tumbleweed becomes interlocked, forming together metre-sized balls that chase along in high leaps with the wind at great speed across the steppe (Fig. 11.14).

In *Stipa* species, transpiration is regulated not only by stomatal closure, but also by leaf curling, which influences photosynthesis. The individual species are adapted to certain site conditions, which determines their distribution.

Many works dealt with the water balance of the so-called **steppe heath** in C Europe (Gradmann 1950). This is an extrazonal relict vegetation from a xerothermic period of

Fig. 11.14 Wind has accumulated *Salsola kali* tumbleweeds as aggressive neophytes along a pasture fence in eastern Ceduna, Australia (photo: Breckle)

the postglacial time. The steppe heath is bound to warm and dry sites on loess and lime slopes or sandy soils and consists of malacophyllous steppe species, which are very hydrolabile. The aridity in central Europe is not caused by the climate, but by the low field capacity of the soils and the high potential evapotranspiration on the south-exposed slopes. The long drought of late autumn is absent, but short dry periods occur more frequently when precipitation is absent for a while.

11.6 Asian Steppes

The eastern European steppe zone continues in the more continental climate of Asia, bypassing the southern Urals; however, it is interrupted east of Lake Baikal by many mountains and is more restricted to the basin landscapes and broad valleys. Only in Outer Mongolia and Manchuria is it again formed as a zone. The West Siberian steppes bear the same character as the European ones with certain floristic differences. *Lilium martagon* ssp. occurs frequently in the steppes, as does *Hemerocallis* spp.

The Transbaikalian steppes, on the other hand, have an extremely continental climate with very snowless winters and dry springs. Spring flora is almost absent, strongly represented by *Filifolium (Tanacetum) sibiricum,* which turns bright red in autumn. The strong admixture of alpine elements (species of the genera *Leontopodium, Androsace, Arenaria, Kobresia,* etc.) is also striking, which is related to the particularly continental climate (Walter 1975).

In very flat terrain of the northern Siberian-Kazakh steppe countless small lakes appear as a result of the semi-arid climate, as paradoxical as it seems. In the humid climate every depression overflows and a river system develops. In the semi-arid flats this is not the case. Each small depression has its small drainage basin. The depressions form where pools stand after rain, and the water slowly penetrates deep into the soil. As a result, the soil particles are rearranged and move closer together, the soil volume decreases, the surface becomes denser, and the depression deepens as a result. Such lake plates or pattern can be seen from the air in other semi-arid regions: in N Dakota (USA), in the Pampas of Argentina and in W Australia. If these lakes have a weak subterranean drainage, the water remains sweet; if all the water evaporates, the lakes become brackish (soda brackish = soda = $Na_2CO_3 \cdot 10H_2O$) in the weakly arid region, otherwise chloride-sulfate brackish).

In the eastern part of the European steppes, but also in N Dakota, aspens (*Populus tremula* resp. *P. tremuloides)* colonise the edges of small mostly circular lakes. The steppe appears dotted with small groves. Such aspen groves (in Siberia with much birch) form the forest component in the forest-steppe where in the continental climate the nemoral

zone wedges out and the steppe borders on the boreal coniferous forest zone (i.e. in zonoecotone VII/VIII), as is the case in W Siberia and in the Canadian steppe area.

11.7 Wildlife of the Steppes

The primeval steppe was the realm of the **big game**, as was the American prairie. Until the eighteenth century, the tarpan (*Equus gmelini)* was still represented in the E European-Asian steppe, the last specimen being delivered to the zoological garden in Moscow in 1866. More strongly represented were the cloven-hoofed animals. It is assumed that the aurochs (*Bos primigenius)* also originally inhabited the steppe and then retreated into the forests before humans. The antelope (*Saiga tatarica)* lasted longer and is even today represented in some places (protected area near Astrakhan, in Kazakhstan etc.). Deer and roe deer used to live in the forest-steppe, and the wild boar (*Sus scrofa)* stayed around waterholes and in the reeds. Light grazing is part of the maintenance of the steppe vegetation. But the big game and the predators have been completely destroyed by man. What remains are the numerous steppe rodents. Today the animal world can only be studied in the few steppe reserves and partly still in the Siberian and Mongolian steppes, as far as they are still grazed by nomads.

Most important are the soil organisms (Edaphon), which are of great importance in the formation of the chernozem. **Earthworms** should be emphasised: the large ones (*Dendrobaena mariupolensis)* traverse the soil with their tunnels in all directions to great depths; 525 tunnels/m^2 were counted in the upper metre, and 110 tunnels/m^2 were still counted at 8 m depth. The small earthworms (*Allophora* spp.) are more restricted to the upper soil layers. The earthworms mix the soil and enrich the lower layers with organic material. The tunnels make it easier for the roots to penetrate the soil.

Second are the **ants**, which also promote soil mixing, and third are the **rodents** with underground burrows. Their activity can be seen on any chernozem profile by the **Krotovines**, the cross-sections of abandoned tunnels that have been filled in from above with humus soil and appear in the loess soil on the profile as dark circles. The soil ejected by the rodents came from the deeper layers. 175 heaps/ha were counted, occupying 0.5–2% of the area. Over time, they create a micro relief with small successions of vegetation. Steppe shrubs (*Caragana frutex, Amygdalus nana* and others) often settle on these piles of earth, which are protected here from the competition of grass roots. The rodents also contribute to the loosening and mixing of the soil.

As long as the steppes were only lightly grazed by the herds of nomads, who constantly changed their location, as was the case little more than two centuries ago, the steppe vegetation remained almost unchanged. However, the conversion of the steppe into arable land or intensive cattle grazing in the last two centuries has destroyed the entire steppe ecosystem and the fauna has suffered severe losses. Only some animal organisms have adapted to the new conditions. Rodents have become a plague for arable farming; pests that were harmless in the steppe have been transferred to cereals and sugar beet; they now appear in masses and have to be controlled with chemical agents.

Livestock grazing degrades the plant cover of the steppe. Regeneration of steppe vegetation on protected areas is slow. However, we cannot go into these secondary successions here.

11.8 Steppes of the Southern Hemisphere

Compared to the grasslands of the northern hemisphere, those of the southern occupy only a relatively small area. The largest contiguous area is the eastern Argentine Pampa in the province of Buenos Aires with parts of the neighbouring provinces. It could also be considered a semi-arid variant of zonobiome V with relatively high precipitation during the hot summer season: the Pampa lies between 32 and 38° S latitude, extends over about 500,000 km^2 and borders directly on the coast of the Atlantic Ocean. It is thus located in the warm-temperate region and corresponds to the southernmost part of the prairies in Oklahoma and Texas. Precipitation reaches 1000 mm in the NE of the Pampas region and drops to less than 500 mm in the SW at the drought limit. These values seem very high, but it should not be forgotten that temperatures, and therefore potential evaporation, are also high (Buenos Aires: mean annual temperature 16.1 °C). Yet, the climate of the Pampa was considered humid, and the question of why the Pampa is a treeless grassland was always raised. The simplest, but uncritical, assumption is that it is an anthropogenic vegetation, evolved from an earlier forest vegetation by man-made fires.

It has been shown, however, that the assessment of the climate is not correct. Even in the wettest parts of the very flat Pampas one can observe many shallow lakes without drainage (here called Lagoons) and also countless small pans which contain water in spring but dry up in summer. The water in the Lagoons is strongly alkaline, containing soda; sodic soils (solonez) are common around the pans, with the typical grass of the brackish soils *(Distichlis)*. This suggests that the climate is semi-arid, similar to the forest-steppe of eastern Europe. In fact, measurements of potential evaporation show that evaporation immediately on the banks of the La Plata is equal to precipitation, while in the Pampa it exceeds the latter. The negative water balance is about 100 mm in the humid Pampa and up to 700 mm in the driest

parts of the Pampa. Potential evaporation is particularly high in the months of January to February, which also have a rainfall minimum. During these months, heavy thunderstorms occasionally occur in the evenings and at night, while during the day the radiation is very intense. The vegetation, which is abundantly supplied with water in spring, burns out strongly in January.

In a forest-steppe climate, woody vegetation can be expected on well-drained soils. Such woody islets with *Celtis spinosa* occur near the coast on small mounds with permeable calcareous or sandy soils, while the poorly drained soils support grassy vegetation. Scrub grows on stony elevations (Frangi 1975). Almost nothing remains of the original Pampas vegetation. European grasses, softer than the Pampas grasses and preferred by European livestock breeds, are found in the grazed areas. Many planted exotic trees, whose roots are protected from the competition of the intensive grass root system, grow well everywhere.

On the basis of some small remains in ungrazed places, it can be said that a *Stipa-Bothriochloa laguroides* steppe predominated in the wet NE part of the Pampa, composed of about 23 graminids and 46 herbs (Lewis and Collantes 1975). The soil profile beneath this Pampa has a humus horizon up to 1.5 m thick, reminiscent of the mighty chernozem or prairie soils, but suggesting a strong alternating moisture and leading to the subtropical grassland soils of S Brazil. There are no signs of previous forestation. At higher groundwater levels, stands with dense horsts of *Paspalum quadrifarium* are found, which transition to the sodic soils (pH = 8 to 9) with *Distichlis* at very high groundwater levels.

The dry SW Pampa used to be a tussock grassland with *Stipa brachychaeta* and *St. trichotoma,* almost without herbs. Tussock refers to a growth habit of grasses that is absent in the northern hemisphere but very common in the southern hemisphere with its mild winters. They are tufted clumps, which can grow over 1 m tall, of old hard leaves with the young green ones between them; the tussock grassland is therefore always yellowish in colour (Fig. 11.15). These grasses have little grazing value and attempts are therefore made to plough them up and replace them with European ones.

When precipitation in the W falls below 500 mm per year and falls mainly in summer, the pampa is replaced by sparse xerophytic *Prosopis caldenia* woody plants. Light sandy soils also occur instead of loess soils. With even lower rainfall, one enters the *Prosopis* savannah (Fig. 11.16), which is very reminiscent of the *Acacia* savannah in SW Africa (ZE II/VII). At the same time, extensive saline soils with halophytes begin. With less than 200 mm of rain per year, the *Larrea* semi-desert is found on stony soils with many broom-like shrubs belonging to different families (Caesalpiniaceae, Scrophulariaceae, Capparaceae, Asteraceae). The small transpiring area of these stands and

Fig. 11.15 Tussock grassland of *Festuca dolichophylla* *in* the upper Charazani Valley in NE Bolivia as an example of a tussock grass formation in S America (photo: Rafiqpoor); see ▶Fig. 11.16, where individual tussock grasses also occur

Fig. 11.16 Group of trees with *Prosopis denudans* and the tussock grasses *Stipa tenuissima* and S. *gynerioides* in La Pampa National Park near Luro (Argentina) (photo: J. Hager)

the severe throttling of transpiration during the six-month drought period allow the semi-desert vegetation to survive with the meagre amounts of water available in the soil at its site. These amount to about 50–80 mm on the surface, and only 25–55 mm on the slopes, whereas in the valleys with inflow they exceed 140 mm.

Fig. 11.17 Wind erosion area in the Patagonian desert near Mamuel Choique, Depártamento Ñorquincó, Argentina. A fully blown *Chuquiraga aureum* cushion, in front of it half dead hard cushion of *Azorella caespitosa* (photo: J. Hager)

The *Larrea* semi-desert stretches along the E foot of the Andes chain as far as Patagonia, where S of the 40th parallel the constant stormy westerly winds begin, blowing over the here lower Andes chain (pass height about 1000 m). But they are fall winds (foehn) that are dry.

While the eastern edge of the mountains still receives 4000 mm of rain and supports *Nothofagus* forests, these change eastwards in the lee into dry *Austrocedrus* forests and then into a scrub with the splendid red flowering proteaceous *Embotrium coccineum*, whereupon the woody plants disappear and the Patagonian steppe begins. Only 100 km from the Andes, precipitation is 300 mm a year and drops further to 160 mm. Only the western edge of Patagonia is a steppe, where low tussock grasses *(Stipa* and *Festuca)* predominate; otherwise it is more correct to speak of the **Patagonian semi-desert,** for which xerophytic cushion plants, again belonging to quite different families (Asteraceae, Apiaceae, Verbenaceae, Rubiaceae, and others), are characteristic (Fig. 11.17). Often the ground is 60–70% bare. The cushion form is probably an adaptation to the constant strong wind (mean wind speed 4–5 m/sec); within the cushion a favourable microclimate is established in the windbreak (Hager 1987).

The Patagonian tussock grassland bears much resemblance to that in Otago in the S Island of New Zealand. This too is 40° S and in the lee of the New Zealand Alps with rainfall around 300 mm. Low tussock grasses *(Festuca novae-zelandiae, Poa caespitosa)* predominate. They are replaced by 1.5 (to 2) m high tussock grasses *(Chionochloa = Danthonia)* at altitudes of 750–2000 m, where snow remains for two to three months. In part, as a result of fire and grazing, tussock grasslands have expanded greatly at the expense of former *Nothofagus* forests. Ecophysiological studies have not yet been carried out in these grasslands (or are not known to us).

11.9 Sub-Zonobiome of the Semi-Deserts

11.9.1 Distribution

The semi-desert differs from the desert in its diffuse vegetation, but with a low cover of about 25%. In the desert, the vegetation density is even lower, and at the same time the transition to contracted vegetation is taking place there (Fig. 11.18).

The plant cover of the semi-deserts is very diverse (Djamali et al. 2012). In the frost-free subtropics or tropics, woody plants and succulents are mostly found. In contrast, in the temperate zone with cold winters, it is mostly semi-shrubs, especially of the genus *Artemisia*. This is the case in Eurasia as well as in N America. A whole range of other species can dominate such semi-deserts, for example the genus *Atriplex* (*A. confertifolia* in Utah), but also *Ceratoides* (= *Eurotia*) in the Midwest of the USA, *C. lanata* (Fig. 11.19), in C Asia *C. latens* and *C. papposa* (= *Krascheninnikovia ceratoides*) as well as other species. In windy Patagonia we had become acquainted with the cushion plants as a characteristic element.

Fig. 11.18 View from an inselberg near Delaram (S Afghanistan). Contracted desert vegetation caused by variations in water availability and soil texture. Only in the small furrows where water accumulation in the soil is possible, sparse vegetation can grow. Outside these furrows, the desert is almost devoid of vegetation (photo: Breckle)

Fig. 11.19 Dwarf shrub semi-desert covered with the woolly grey *Ceratoides lanata* in Rush Valley near Tooele (Utah, USA) (photo: Breckle)

The semi-desert occupies large areas in Kazakhstan between the Siberian steppes in the N and the deserts in the S. In **N America,** the sagebrush zone with *Artemisia* *tridentata* corresponds to them (Fig. 11.20). *Artemisia* *tridentata* is a 1.5–2 m tall semi-shrub that lives 25–50 years. The taproot penetrates up to 3 m deep probably

Fig. 11.20 Sagebrush semi-desert from *Artemisia tridentata* in E Utah, USA (photo: Breckle)

deeper into the soil. Shallow and far-reaching lateral roots extend from it. The water supply is good in spring after the snow melts. The cell sap concentration is then very low at 1.0–1.5 MPa. It soon increases to 2.0–3.5 MPa. During acute water shortage in summer, it may reach 7.0 MPa. At this stage, as with all malacophylls, the older leaves are shed.

Sagebrush semi-desert is tied to brown semi-desert soils that are free of salts. *Artemisia tridentata* is the dominant species. The dwarf shrub *Chrysothamnus* (Asteraceae) is often found as a companion. The area belongs to the arid ZB VIIa. However, in the arid climate, the drainless depressions are always brackish. These are salt pans and salt lakes as remnants of much larger Pleistocene lakes, for example Lake Bonneville, whose level was 310 m above that of the present Great Salt Lake and the adjacent very extensive vegetated salt desert in Utah. The area of Lake Bonneville was 32,000 km^2 with a maximum length of 586 km and width of 233 km. Today's salt desert extends 161 km in length and 80 km in width. At the 1906 high water mark, the corresponding figures for the salt lake were 120 and 56 km. Its contours vary widely, the mean depth is little over 5 m, and the salinity ranges from 13.7% to 27.7%. About 80% of the salts are NaCl, the remaining 20% are MgCl$_2$, Na$_2$SO$_4$, K$_2$SO$_4$, MgSO$_4$, etc. Many species of halophytes occur widely around the salt flats.

In keeping with the greater aridity of the semi-deserts, brackish soils are widespread. This is particularly striking in E Europe in the Ukraine, where the vast expanses of the Lazy Sea (Lake Syvash) north of the Crimea dry out in summer and cover themselves with a crust of salt, the salt dust being blown northwards by the wind and deposited in the zone of southern chernozem and **chestnut soils**. This supply of Na salts leads to the **solonisation** of the soils. The salt is washed out of the upper soil layers by the melt water in spring, whereby the humus brines formed (soda formation) also take the sesquioxides (Fe$_2$O$_3$, Al$_2$O$_3$) with them into the deeper soil layers, where precipitation takes place and a compacted B horizon with a strongly alkaline reaction is formed (Fig. 11.21).

Towards the south the salt supply increases constantly; from the A horizon the humus substances are completely leached out, while the strongly alkaline B horizon becomes more and more compacted and takes on a columnar structure due to the deswelling in summer and swelling during the humid season. This **columnar solonez** is in some respects reminiscent of the podsol soils, which, however, exhibit a strongly acidic reaction because in them the peptizing action is triggered by H ions. Under the B horizon of the solonez soil (black alkali soil), the poorly soluble CaCO$_3$ precipitates first (lime leaches), then the gypsum (CaSO$_4$) as tubes or druses, while the easily soluble salts are washed into the groundwater.

If the groundwater rises, which is the case, for example, on the northern coast of the Black Sea, which is slowly sinking,

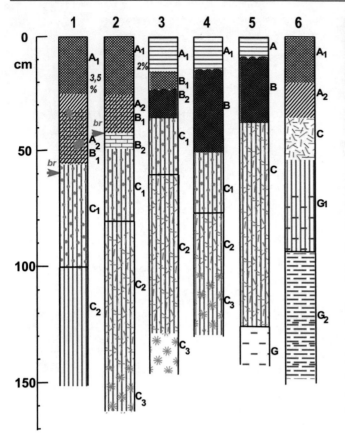

Fig. 11.21 Soil profiles in Eastern Europe of weakly to strongly brackish soils. **1** Weakly solonised southern chernozem with slight compaction (**A₂B**), **2** Dark chestnut brown soil with B horizon, **3** Light chestnut brown soil strongly solonised (**A** low in humus and platy, **B** columnar and very dense), **4** Typical columnar solonez soil, **5** Solonez altered by rising groundwater, **6** Typical solonchak with high groundwater and humus-rich **A₁**. Calcareous eyes **C₁**, gypsum tubes **C₂** at **2–4** and **C** at **5–6**, gypsum druses **C₃**, gley horizon (groundwater) **G**, **G₁** and **G₂**

Fig. 11.22 *Collema*, a gallert lichen with symbiontic *Nostoc* (Cyanobacteriae) on clay soil in the Pamir semi-desert (photo: M. Keusgen)

have a humus content of only 2–3% and are brown in colour; the flare horizon is 25 cm deep, and the plant cover is less than 50%. The vegetation consists of *Festuca sulcata* and the low semi-shrubs *Pyrethrum achilleifolium, Kochia prostrata* and *Artemisia maritima-incana,* which avoids saline soils. *Stipa* species are found only sporadically, but in spring many ephemerals develop. In the Caspian lowland both societies often form a micromosaic on burosem and on solonez soils, caused by the microrelief. On quite wet solonchak *Salicornia* and *Halocnemum* predominate, on less wet *Suaeda, Obione, Petrosimonia, Limonium caspica, Atriplex verrucifera* and many others (cf. Levina 1964; Walter and Box 1983).

The S part of the Caspian lowland was covered with alluvial sands after the retreat of the sea in the area of delta formation of the Volga-Ural river system. These were originally covered with *Artemisia maritima-incana, Agropyron cristatum, Festuca sulcata, Koeleria glauca* and others. Grazing destroyed the vegetation cover and the sand began to move, creating large vegetationless shifting sand dunes (barchans). If the sand movement subsides, *Elymus giganteus* and the Chenopodiacee *Agriophyllum arenarium* appear as pioneers, followed by *Salsola* and *Corispermum* species. In the dune valleys appear *Aristida pennata, Artemisia scoparia,* etc. Slowly the zonal vegetation re-establishes itself.

The sand dunes, especially those without vegetation, are water reservoirs. There is always groundwater in their lower parts and beneath them, which forms small freshwater lakes in the dune valleys, around which the olive willow (*Elaeagnus angustifolia*), willows (*Salix*) and poplars (*Populus*) grow. Attempts have been made to reforest the sandy areas with poplars and willows (*Salix acuminata*). They thrive well at first at the expense of the water supplies in the soil; but these are used up in the course of four years, and the crops scant or die.

then a wet saline soil is formed, which is called **solonchak.** The groundwater is capillarily drawn up to the surface of the soil and evaporates. Accordingly, above the gley horizons we find one with gypsum tubes and above it the humus horizons with a white salt crust on the surface in the dry season. With the high salt concentration, the formation of humus sols does not occur, because the humus substances are flocculated.

On the solonez soils the steppe grasses recede. In addition to *Artemisia maritima-salina* and *A. pauciflora,* species of the genera *Camphorosma, Limonium, Kochia, Petrosimonia* and others are found (in North America *Ceratoides lanata, Atriplex confertifolia, Kochia* spp. etc.); in addition, there are soil lichens *(Aspicilia),* and the liverworts *Riccia* and *Nostoc* (cyanobacteria that form spherical or gelatinous colonies) (Fig. 11.22).

On the small elevations, which are not salinated, the semi-desert brown earth (Burosem) is formed. The upper horizons

The halophytes are all Chenopodiaceae with the exception of the descending grass *(Distichlis).* The total area is a vast halobiome with biogeocoene complexes. This succession also largely corresponds to the zonation of halophyte types as it occurs in Central Asia.

The climate of Utah is reminiscent of that of Ankara. The strong predominance of *Artemisia* in Anatolia is the result of overgrazing; previously grasses *(Agropyron, Stipa and Festuca* species) were common.

11.9.2 Vegetation in Afghanistan

Areas of semi-desert and desert vegetation are also quite extensive in Afghanistan (Freitag 1971; Breckle & Rafiqpoor 2010). It surrounds the central mountain ranges in a wide circle. Accordingly, very different expressions are formed in the N, W and S. As can be seen from the vegetation map (▶ Fig. 8.57), they are chenopodiaceous, ephemeral-rich or

Fig. 11.24 The spring aspect of the ephemeral deserts and semi-deserts of N Afghanistan with numerous geophytes (photo: M. Flader)

scrubby semi-deserts with *Amygdalus* and other micro-shrubs. Almost regularly various species of wormwood *(Artemisia)* or also Lamiaceae are admixed; the semi-desert therefore always has an aromatic smell.

Along the mountains, partly also in the dry mountain valleys, these semi-deserts are usually grazed very intensively. There are hardly any slopes on which one cattle track is not visible densely next to the other (Fig. 11.23). Accordingly, many thorny or poisonous plants also occur. In spring after the rains, ephemerals appear for a short time, there are numerous species of geophytic bulbous and tuberous plants among fast-growing annuals (Fig. 11.24).

The major semi-desert and desert formations (Freitag 1971; Freitag et al. 2010) are discussed below as a remarkable example of vegetation mosaic.

Calligonum-Stipagrostis Communities of Sandy Deserts (Fig. 8.57: 1a)

Sandy deserts with dunes are found only in the low and driest parts of S and N Afghanistan, where annual rainfall is below 150 mm. The dunes are partly fixed by scattered shrubs (up to 20% cover) of *Haloxylon persicum, Xylosalsola richteri* and some *Calligonum* species, and by the tall perennial grasses *Stipagrostis karelinii* (only in the N and NW) and *S. pennata,* associated with quite a few semi-shrubs and deep-rooted annuals. The uprooting of the perennial plants, which have a very effective sand-binding capacity due to their deep roots, almost always leads to a renewed secondary activation of dune formation, usually by the formation of sickle dunes (barchans), which are completely devoid of vegetation and move slowly with the wind, covering irrigation fields, blocking roads and threatening villages. These are the very typical signs of desertification.

Fig. 11.23 The overuse of grazing areas leads to the degradation of the landscape. Cattle grazing in Dare-Ajdahar (Valley of the Dragon) in Bamyan, Afghanistan is a good example of this (photo: Breckle)

Haloxylon Salicornicum Communities in Gravel Deserts (Fig. 8.57: 1b)

The gravelly plains of the distinctly dry and hot regions of SW and S Afghanistan, which have high levels of gypsum and soluble salts in the topsoil, generally support *Haloxylon salicornicum* communities. They also have some other semi-shrubs mostly belonging to the Chenopodiaceae and a whole range of xerohalophytic annuals. These include attractively variegated species such as the Chenopodiaceae *Halarchon vesiculosus* (endemic, with pink blister-like appendages on the anthers that can be blown away, giving the landscape a purple hue), *Halocharis* species and the remarkable endemic Cruciferae *Veselskya griffithiana* with its bright purple flowers. After good winter rains these species appear in great masses. This community is closest to a true desert because it is concentrated in large areas on the shallow gullies or depressions (contracted vegetation), forming a typical habitat mosaic. This open desert vegetation seems little influenced by man, except for occasional collection of the semi-shrubs and occasional grazing.

Other Shrubby and Semi-Shrubby Chenopodiaceous Deserts and Semi-Deserts (Fig. 8.57: 1c)

Quite a few very different plant communities with dominance or high abundance of chenopodiaceous shrubs and semi-shrubs occur somewhat more inland, including at low elevations in S Afghanistan, as well as on the W and N periphery of the mountains where rainfall rarely exceeds 150–200 mm. They also partially dominate the dry interior valleys of C and E Afghanistan, where gypsum is locally enriched from surface rocks, as in the Bamyan and Ajar valleys, parts of the Ghorband valley, and the lower Gomali valley. Important species include, shrubby and semi-shrubby Chenopodiaceae, such as *Halothamnus subaphyllus*, *Xylosalsola [Salsola] arbuscula*, *Salsola montana*, *Caroxylon gemmascens*, *Salsola [Seidlitzia] rosmarinus*, several species of *Artemisia*, especially *A. sieberi* and *A. oliveriana*, and the shrubs *Zygophyllum atriplicoides*, *Z. eurypterum*, *Ephedra strobilacea*, *E. sarcocarpa* and *Cousinia deserti*. Depending on the soil structure, ephemerals and hemicryptophytes can considerably increase the diversity of these communities in spring.

On sand-covered soils, these communities can look very spectacular when the large umbellifers *Ferula assa-foetida* and *Dorema aitchisonii* sprout their hapaxanth inflorescences and flower. Most of these vegetation units occur near the mountains, on the foothills, which are also often quite densely populated. They are not infrequently violently degraded, often to barren stages without any woody plants, or to stages with few semi-shrubs such as *Haloxylon griffithii*, *Kaviria [Salsola] tomentosa,* and *Artemisia* species, which have a high regenerative capacity. But for most of the year these areas appeared almost completely bare and devoid of vegetation.

Ephemeral Semi-Desert on Loess Soils (Fig. 8.57: 1d)

The low-lying parts of the loess belt in N Afghanistan, at altitudes up to 600 m and with rainfall of 150–300 mm, are characterised by communities of typical C Asian ephemeral semi-deserts. In March and April, they can almost look like an English golf lawn, towards May they almost resemble a lush, colourful meadow or steppe, but during June they dry out rapidly and only a few species remain active for a while, such as the annuals *Salsola leptoclada* and *Diarthron vesiculosus*. The plant cover is mainly composed of shallow-rooted perennials such as *Poa bulbosa*, *Carex pachystylis* and *C. stenophyllus*, often associated with thistle-down *Cousinia microcarpa*, *C. olgae* and *Gundelia* aff. *tournefortii* (the latter is a curious composite with variegated globose heads of numerous basket flowers, occurring only in the W and NW), with numerous annual grasses of the genera *Aegilops*, *Bromus*, *Eremopyrum* and *Vulpia*, with crucifers of the genera *Malcolmia* and *Torularia*, with annual legumes such as *Trigonella grandiflora* and with an abundance of geophytes of the genera *Tulipa*, *Gagea*, *Iris*, *Ixiolirion* and *Allium*.

These ecosystems are the main grazing areas in winter and spring. In good years, additional hay is made and fuel is collected and stacked in large piles of hay around the nomads' winter camps or brought into villages and piled in thick layers on flat roofs. While these ecosystems remain intact under a moderate grazing regime, severe overgrazing can lead to the proliferation of inedible pasture "weeds" such as the perennial *Cullen (Psoralea) drupacea* and *Ammothamnus lehmannii*.

Shrubby *Amygdalus* Semi-Desert (Fig. 8.57: 1e)

In and around the foothills of the mountains in S and W Afghanistan as well as in the dry valleys of the Hari Rud, the Kokcha and Surkhab etc., where the annual rainfall is between 150 and 250 mm, various closely related thorny *Amygdalus* species occur, usually growing to 0.5–1.5 m high. They are the typical plants of this open shrubby semi-desert. They are accompanied by a number of other low shrubs such as *Ephedra intermedia*, dwarf shrubs such as *Acanthophyllum*, *Acantholimon*, *Cousinia* and *Artemisia*, perennial grasses such as *Stipa hohenackeriana*, and numerous annuals and geophytes.

Overgrazing effects are usually not very noticeable in these communities; because of the thorny nature, the woody plants are well protected against grazing livestock as well as against fuel-gathering inhabitants – the result of century long selection. Sometimes the density of thorny shrubs is even increased, especially in the case of the widespread *Cousinia stocksii* community in the S and W of the country.

11.10   Subzonobiome of the Central Asian Deserts

This area is located N of the border of date cultivation. It is characterised by regular frosts.

The Tsaidam Basin (Fig. 11.25) leads to the high mountain deserts of Tibet with Pamir in the far west.

Central Asia still receives cyclonic rains from the Atlantic Ocean, which fall as winter rains in the S part, and more in spring and summer in the N; in any case, in this part the ground is always moist in spring after the melting of the snow. The rainfall decreases from W to E. Floristically, the Irano-Turanian element is strongly represented. In contrast, in C Asia the moisture comes from the foothills of the E Asian summer monsoon. Winter and spring are extremely dry (Fig. 11.26 Murghab). The different rainfall distribution means that E Chinese-Mongolian elements predominate in the flora.

> **Box 11.5 Difference Between the Deserts of Central and Middle Asia**
>
> A distinction is made between the central Asian deserts and the middle Asian deserts (Fig. **11.27**). The first include the Irano-Turanian desert area, which occupies

> **Box 11.5** (continued)
>
> the S part of the Aralo-Caspian lowlands, and the S part of Kazakhstan with the Dsungarei. The middle Asian deserts include in part the Dsungarei, the Gobi desert, the W part of Ordos in the great knee of the Hwang-ho, ala-Shan, Pei-Shan, the Tarim Basin (Kashgaria) with the Takla-Makan desert, and the already higher Tsaidam Basin.

Of these areas, the vegetation of the Middle Asian deserts in the Aralo-Caspian lowlands (former Turkestan) has been studied ecologically in the great detail. The whole area receives less than 250 mm of precipitation. As a result of cold winters, evaporation is very low in this season. Therefore, the annual evaporation of the Bay of Bugaz reaches only 1100 mm. The different vegetation types are determined by the soils (biogeocene complexes):

Ephemeral desert is found on loess-like, salt-free soils that are very wet in spring but dry from May onward. Annual species and geophytes develop during the short growing season from early March to mid-May. The most important species are *Carex hostii (C. stenophylla)* and *Poa bulbosa. In* places, the 2 m tall *Ferula foetida* occurs, and the 40–50 annual species reach seed maturity in 30–45 days. In good rainy years, the desert resembles a meadow and produces 0.5–2.5 t/ha of dry matter; it can be grazed for three months but is completely lifeless for nine months (Fig. 11.24).

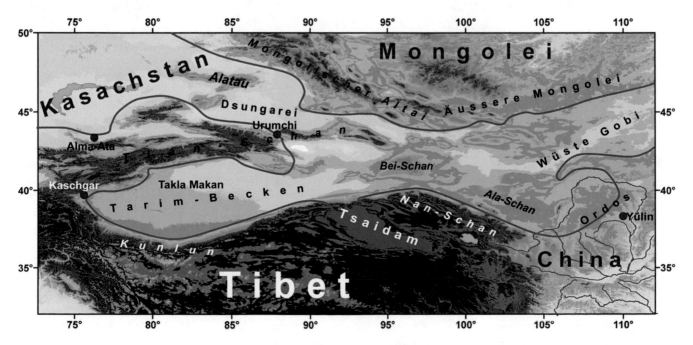

Fig. 11.25   The division of the Central Asian desert regions (data from Walter 1990). The Dzungaria is a transitional area to Middle Asia

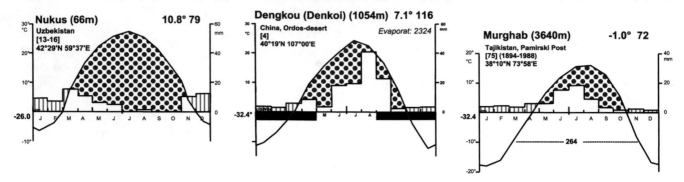

Fig. 11.26 Climate diagrams of Nukuss in Middle Asia with winter rains, Denkou in C Asia with summer rains, and Murghab (= Pamirski Post) in the cold desert (here only 264 daily means above −10 °C)

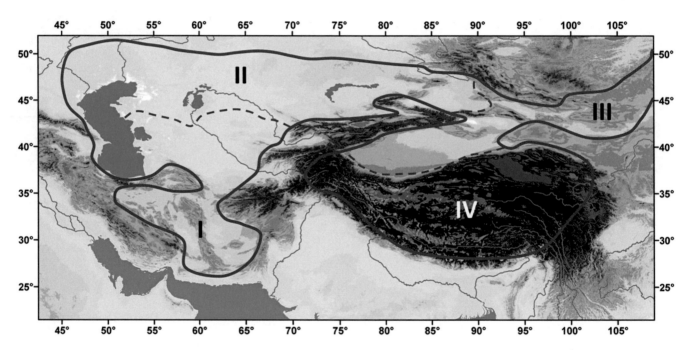

Fig. 11.27 Asian deserts of the temperate climate zone (data from Walter 1990). Middle Asian deserts: **I** Irano-Turanian (partly almost subtropical) and **II** Kazakh-Dzungarian. Central Asian deserts: **III** in the narrow sense (hot summers) and **IV** Tibetan cold high mountain desert

The **Gypsum desert** is a rocky desert (Hamada) on the plateaus of the Table Mountains. The soils contain up to 50% gypsum, which retains moisture. The conditions are reminiscent to that in the Sahara. Therophytes develop in spring; otherwise, gypsum plants cover 0.1% and stand more densely only in erosional gullies (Fig. 11.28). Some halophytes also occur.

On groundwater near the lower reaches of rivers, in depressions (Schory) or around salt lakes, a **halophyte desert** is more widespread. Most species are hygrohalophytes (*Salicornia, Halocnemum, Haloxylon, Seidlitzia,* etc.). (Fig. 11.29).

Takyre are apparently vegetationless clayey flat areas that are flooded in spring by surface water draining from the mountains but soon dry out again (Fig. 11.30). In the shallow pools, which warm rapidly, 92 Cyanobacteriae (blue-green algae), 38 Chlorophyceae (green algae), and other algae are found; they produce 0.5 t/ha of dry matter with an N content of 4.5% (binding atmospheric nitrogen). Lichens (*Diploschistes* et al.) settle on slightly higher areas. Flowering plants are rare.

Sand deserts play a particularly important role: Karakum (Black Sand) between Kaspi and Amu Darya, Kysylkum (Red Sand) between Amu- and Syr Darya. The sandy soil allows for denser vegetation. Detailed ecological studies have been carried out at the Repetek desert station (Fig. 11.31) in the Karakum desert since 1912. This desert is therefore discussed subsequently.

Fig. 11.28 Nearly vegetationless desert with cobble soils over gypsum layers in the Dashte Margo in S Afghanistan, with small *Stipagrostis* horsts and *Suaeda* dwarf shrubs, in whose lee some fine material accumulates (photo: Breckle)

Fig. 11.29 Clayey alluvial plain in the semi-desert near Bukhara (Uzbekistan), with dense growth of vigorous saxaul shrubs and small trees *(Haloxylon aphyllum)* up to about 6 m high (photo: Breckle)

Fig. 11.30 Dry Takyr surface north of Lake Balkhash, Kazakhstan. The low-salinity, cracked clay surface is split into polygons. There is little plant growth on the dry clay surface; here some *Anabasis* species (photo: Breckle)

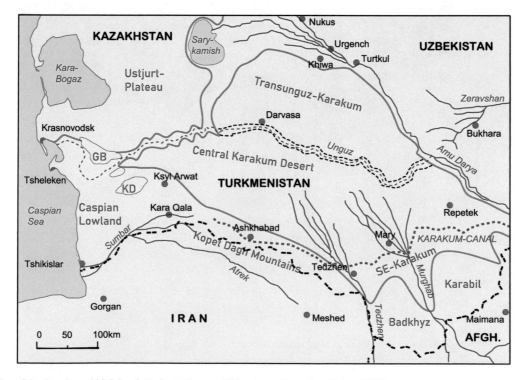

Fig. 11.31 Map of the Karakum (thick border). Scale 1 cm = 75 km (modified after Walter 1976)

11.11 The Karakum Sand Desert

Covering an area of 350,000 km², the Karakum occupies the S part of the Turanian lowlands between the Caspian Sea to the W and Amu Darya to the E, the 2600-km-long stream that rises at nearly 5000 m asl in the Pamirs (Fig. 11.31).

The sand desert is a geographically well-defined biome of the subzonobiome of the temperate deserts of ZB VII. It is a large basin that was filled with alluvial unconsolidated rock materials by the Amu Darya since the Tertiary; the latter subsequently underwent rearrangement by wind. In the process, the dust particles were deposited in the S on the Kopetdag slope as loess, while the sands formed a dune landscape (Fig. 11.32).

The Amu Darya River originally emptied into the Caspian Sea, but was displaced eastward by the delta deposits of the Murghab and Tedshen Rivers coming from the S, so that today it flows to the Aral Sea. It is probable that the lower course of the Amu Darya has changed its bed considerably several times in the past millennia. But it is still determinant

Fig. 11.32 Karakum desert with slightly undulating dune relief and sparse shrub cover *(Haloxylon persicum* and others). Soil covered with ephemerals and ephemeroids only in spring (photo: https://t1p.de/opsv)

of the Karakum today, because its waters feed by infiltration a groundwater lake, the level of which lies under the whole of the central Karakum with a slight inclination towards the Caspian Sea. Only in places does groundwater come to the surface, resulting in the formation of salt pans.

One can establish the water balance for the entire biome (Tables 11.2 and 11.3).

In the protected area near the Repetek desert station (SE Karakum, Fig. 11.31), ecological surveys were carried out over many years, which we summarise very briefly. The area is 34,000 ha in size (14,000 ha vegetationless barchan fields, 18,000 ha vegetated dunes, and 2000 ha dune valleys). Characteristic of the sandy deserts are the tree shrubs *Haloxylon persicum* (White Saksaul) and *H. ammodendron = aphyllum* (Black Saksaul). Although the groundwater moves slowly from E to W, observations show that no water drains into the Caspian Sea. The groundwater is slightly brackish, but freshwater lenses float on it under vegetationless dunes. They are formed by seeping rainwater, even with annual precipitation around only 100 mm. In such places, wells provide good drinking water.

The climate of the Karakum can be seen from the climate diagram of Nukuss (Fig. 11.26). 50–70% of the precipitation falls in spring; it is the most favourable season. Winter is cold, but no permanent snow cover remains. In the hot summer there is an extreme drought. The potential evaporation is 1500–2500 mm, that is, 10–20 times the precipitation.

In the psammophyte complex, a distinction is made between:

1. The biogeocoene of the Ammodendretum conollyi aristidosum on slightly mobile sand with a pioneer synusia of *Aristida karelinii* on dune crests and the shrubs (*Ammodendron conollyi, Calligonum arborescens, Eremosparton* and others) on the upper slope
2. The biogeocoene of Haloxyletum persici caricosum on immobile sand on the lower slope and on the sandy areas with the synusia of spring and summer ephemerals and ephemeroids (143 species, including 24 common ones). In addition
3. The biogeocoene of the deep dune valleys with groundwater at a depth of 5–8 m reached by the roots of the salt-tolerant *Haloxylon ammodendron*, a 5 to 9 m tall woody tree. Here, too, several synusia can be distinguished in the understory (ephemerals, halophytes).

The most widespread is Haloxyletum persici caricosum (Fig. 11.32) with an open 3–5 m high shrub layer (100–300 specimens/ha, in which the aphyllous *Calligonum* spp. (Polygonaceae) also occur. The age structure of the shrubs is listed in Table 11.4.

Table 11.2 Water inflow to groundwater in Karakum

Inflow to groundwater through	Quantity
Groundwater infiltration from the Amu Darya in the mean	150 m³/sec
Rainwater seepage in the barchan area	30 m³/sec
Infiltration from the Murghab and Tedshen	21 m³/sec
Underground inflow from Kopetdag (from the south)	20 m³/sec
Seepage from the hills and Takyrs	1 m³/sec
Total inflow to the groundwater lake	**222 m³/sec**

Table 11.3 Water consumption in Karakum

Loss due to	Quantity
Evaporation from wet salt pans with high groundwater levels	165 m³/sec
Groundwater losses due to phreatophytes (plants bound to groundwater)	57 m³/sec
Total losses	**222 m³/sec**

Table 11.4 Age structure of shrub species in the Saxaul semi-desert

Age in years	1	2	3–5	6–10	**11–15**	16–20	>20	Dead
Specimens in%	8	3	1	14	20	41	11	2

Box 11.6 Biogeocoene complexes of the Karakum desert
Based on the vegetation and soils, the following biogeocoenic complexes can be distinguished in the Karakum:

- Psammophyte complex occupying more than 80% of the area
- Takyr complex
- Halophyte complex

In-depth ecological studies showed that the shrubs are active throughout the year because the upper 2 m of the sand always contains absorbable water. The osmotic potential decreases somewhat during drought, water deficits are never high, the very intense transpiration is reduced to half or a third during drought, photosynthesis is not interrupted in summer; it is particularly intense in ephemerals during the short growing season.

In the *Ammodendron (Ammodendron conollyi)* plots on mobile sand, 83 mm of stored water was found in the upper 2 m of the sand, with 34 mm remaining at the end of the growing season; of the 49 mm lost, 37 mm was used for plant transpiration, leaving only 12 mm evaporating from the soil.

The conditions are somewhat less favourable in the more densely vegetated immobile sand: the shrubs transpired 30 mm *(Haloxylon persicum* alone 16 mm), the dense *Carex physodes* undergrowth 17 mm, together 47 mm. Soil stored 62 mm of water in the spring, and only 8 mm remained in the fall. Water loss was thus 54 mm, of which 47 mm was through plant transpiration and thus 7 mm through evaporation from the soil.

In the deep dune valleys above groundwater with woody plants, transpiration totaled 149 mm (*Haloxylon aphyllum* 108 mm, other shrubs 30 mm, the sparse *Carex physodes* carpet 11 mm). Stored in the soil was only 76 mm. If we assume that *H. aphyllum* takes the 108 mm it transpires from groundwater, then the 41 mm of all other species is amply covered by water supplies in the upper 2 m. In *H. aphyllum*, 14% of the precipitation runs off the stem; in rainy years, this makes it easier for the taproot to penetrate to greater depths.

We thus see that, notwithstanding the very high transpiration intensity of the Karakum shrubs per gram of fresh weight, the total transpiration per area is low as a result of the small total leaf area, and can be met by rainfall. The main adaptation of the plants consists in aphylly and dropping the small leaves during the drought period, as far as such are formed in spring.

The aboveground phytomass was determined: on still mobile sand 80 kg/ha (of which 25% *Calligonum* and 12% *Aristida),* on vegetated sand 2.4 t/ha (of which *Haloxylon persicum* 85%, *Carex physodes* 10%). Much larger is the subterranean phytomass, which was determined only in dune valleys with *Haloxylon aphyllum:* above ground 6.4 t/ha (of which *Haloxylon* 82%), below ground 19.4 t/ha (of which *Haloxylon* 49%).

The annual primary production for *Haloxylon aphyllum* was 1.17 t/ha above ground and 2.11 t/ha below ground. This high production in deserts is in this case only made possible by water uptake from groundwater.

The Takyr biogeocoene complex is found where rainwater runoff runs out on a wide plain and covers a clay layer over which it stands until it evaporates. Masses of algae, mainly cyanobacteriae (also N-binding), develop in the water, leaving an algal skinny membrane after drying. The soil develops drying cracks, which quickly close by swelling the next time they are wetted. In places that are only moist but not flooded, lichens develop, upslope also ephemerals. The phytomass (partly primary production) in these three biogeocoenes is: algae = 0.1 t/ha, lichens = 0.3 t/ha and ephemerals = 1.2 to 1.6 t/ha.

The halophyte-biogeocoene complex occurs where salt pans are formed as a result of a high groundwater level. The biogeocoenes, often consisting of only one species, arrange themselves in circles around the central part with a salt crust. The pioneer species *Halocnemum strobilaceum* was found to have a phytomass of 1.76 $t \cdot ha^{-1} \cdot a^{-1}$ (of which roots 1.04 $t \cdot ha^{-1} \cdot a^{-1}$) and an annual production of 0.5–0.7 $t \cdot ha^{-1} \cdot a^{-1}$.

Lush liana-rich *Populus-Halimodendron* floodplain forests (referred to as Tugai, with *Lonicera, Vitis, Clematis* and many other species) and extensive reed *(Phragmites)* stands grew formerly in the Amu Darya delta area, from which aboveground phytomass and primary production were determined, respectively: Floodplain 77.8 $t \cdot ha^{-1} \cdot a^{-1}$ and 11.4 $t \cdot ha^{-1} \cdot a^{-1}$ respectively, reed 35 $t \cdot ha^{-1} \cdot a^{-1}$ and an exceptional production of 18 $t \cdot ha^{-1} \cdot a^{-1}$. Today, these stands are deprived of groundwater due to the drying up of the Aral Sea (Breckle et al. 1998; Klötzli 1997).

The animal world plays a very important role. The former herds of antelopes, wild horses and donkeys have now been replaced by three million Karakul sheep, which graze the sandy desert all year round and provide Persian furs. The cattle tread prevents the sandy areas from becoming overgrown with *Carex physodes* and the moss *Tortula desertorum;* at the same time, the seeds of the shrub seedlings are trampled into the sand, which facilitates germination. The many rodents also rummage through the soil, as do the large tortoises (100 per hectare), which feed on the ephemerals for ten weeks to four months in spring and otherwise sleep in the soil. In the Tugai forests, tigers, leopards, Bactrian deer and other large animal species were still common in past centuries.

However, as usual, the zoomass is not large in the three biogeocoenes: mammals 0.3 to 1.4 kg/ha, birds 0.02 to 0.07 kg/ha, reptiles 0.21 to 0.7 kg/ha (excluding turtles), invertebrates maximum 15 kg/ha (above and below ground).

The cycles of mineral elements were also studied, but the decomposers were only summarily considered.

A more detailed account of the Karakum desert can be found in Walter (1976) and Walter & Box (1983).

11.12 The Aralkum Desert

The Aral Sea was once the fourth largest lake on Earth with a surface area of about 68,300 km^2 until about 1960. Essentially, there are 3 factors that can be held responsible for the water level fluctuations of the endorrheic (drainless) Aral Sea in historical times:

1. Development, expansion and intensification of irrigated agriculture with ever increasing water consumption in the river oases, mainly since 1960
2. Climate fluctuations with fluctuations of the water inflow into the Aral Basin hydrotope
3. Variation in sediment transport and sedimentation, which has led to multiple changes in river flows (Breckle & Geldyeva 2012).

However, in the past decades since 1960, the increasing irrigation areas and the enormous water consumption in many regions of Central Asia are primarily responsible for the catastrophic drying up of the Aral Sea. Beginning around 1960, the lake level of the Aral Sea has steadily declined (Fig. 11.33). The separation of the northern Aral Sea (Small Aral Sea, North Aral Sea) from the residual water area began in the 1990s. Currently, only 4 residual basins remain (Fig. 11.33): the North Aral Sea, now a separate water body, separated by an artificial dam 20 km long between the former island of Kokaral and the Syr Darya estuary, the South Aral Sea with a western, deep basin, which will continue to exist as a water body in the future, plus a western, very shallow basin, which is currently an extensive huge salt marsh, but in rainy years is also partially refilled with water, plus the small Chebas basin in the NW, N of Kulandi. We refer to the entire drained areas of the former Aral Sea as the Aralkum desert (Agachanjanz & Breckle 1994; Breckle et al. 2001; Breckle & Geldyeva 2012), which today covers an area of about 60,000 km^2.

The Aralkum is a new surface where terrestrial plants, including their seed bank and terrestrial animals, were not previously present; it is a surface that is being actively recolonised by organisms. The formation of plant communities, soils, the aquifers and water table, all the components and processes of the terrestrial ecosystem are newly formed and must settle, this is all happening in parallel and simultaneously. These are typical processes of primary succession (Wucherer et al. 2012). The dried lake bottom of the Aral Sea is the largest area in the world where primary succession is currently occurring. It is (more or less) unintentionally the largest succession experiment of mankind at

Fig. 11.33 The drying stages of the Aral Sea and the formation stages of the new Aralkum desert

present. It has been playing out on the dry lake bed for several decades. The older areas, which became dry between 1960 and 1980, have extensive sandy deserts with dune formations (Fig. 11.34a); the younger areas are predominantly salt deserts with salt crusts (Fig. 11.34b) composed of various salts, including carbonates, which react alkaline to form a powdery loose crust that is easily blown away by the wind. This regularly leads to devastating salt dust storms in Uzbekistan and Kazakhstan, although some of the salt that is blown up may also originate from abandoned salinised fields.

Spontaneous immigration of 410 species of vascular plants has occurred in the Aralkum so far; they form a new open desert vegetation with a highly variable composition from year to year. However, the succession is ultimately towards a *Haloxylon aphyllum* community. The decline of the lake water level can be recognised very well on the somewhat steeper north coast of the North Aral Sea due to the distinct respective coastlines marked there by *Tamarix* belts (Fig. 11.35).

The North Aral Lake has partially refilled with water since an artificial dam was completed in the Kokaral area in 2006. Thus, the Syr Darya water flows completely initially only into this lake, whose level has increased from about 33–42 m. The eventual overflow is channeled through a low-pressure power plant and then flows towards the Chebas basin.

Fig. 11.34 Sand desert (**a**) and salt crust and deflation traces (**b**) in the Aralkum (photos: Breckle)

Fig. 11.35 (**a** and **b**) Belt-shaped arrangement of vegetation zones on today's northern shore of the Aral Sea after its drying up (photos: Breckle)

11.13 Orobiome VII (rIII) in Middle Asia

11.13.1 Tien Shan

Particularly interesting conditions show the altitudes of this orobiome in Middle Asia, where it is located in the climatic zone ZB VII (rIII). The mountains belong to the Pamiro-Alai and Tien Shan systems and rise to over 7000 m asl. Almost every step sequence here has its peculiarity depending on the nature of the local gradient winds; however, two main types can be distinguished: (1) arid step sequences without forest belt and (2) more humid with one or two forest belts (Stanjukovitsch 1973).

In the extreme case of the Central Tien Shan, the semi-desert belt from 2000 m asl is followed by a mountain steppe belt up to 2900 m, with alpine species (*Leontopodium*

alpinum, Polygonum viviparum, Thalictrum alpinum, etc.) mixed in at the subalpine belt (from 2600 m). The steppe elements extend into the lower alpine belt *(Kobresia* grassland*)* up to 3500 m asl and disappear only above 3500 m. The high montane, subalpine and alpine belts merge completely smoothly. The explanation for this strange mixture of steppe and alpine elements is to be sought in the fact that the steppe plants requires a favourable vegetation period of four months, which is given in the arid mountain climate by the strong radiation still up to 3500 m asl in summer. The remaining eight months can be arid or even cold. The alpine elements manage with a shorter vegetation period, but grow even with a longer one, if it is humid enough to compete with the steppe plants (Walter 1975).

In less extreme cases, *Juniperus* tree forests occur in the subalpine zone, especially on N slopes, which take on a trellis form in the alpine zone.

Fig. 11.36 Steppe forest in the Transili-Alatau, south of Almaty (Kazakhstan) near Medeo at ca. 2000 m asl, exclusively with *Picea schrenkiana* (photo: Breckle)

Fig. 11.37 Steppe forest in the deep valleys of the eastern Hindu Kush (upper Nuristan Valley) with *Pinus gerardiana* on the slopes and flood-plain scrub on the valley floor (photo: Breckle)

Particularly interesting are the humid belts, where above a semi-desert with xerophilous woodlands *(Pistacia, Crataegus)* and mountain steppes there is a hardwood belt with wild fruit species *(Amygdalus communis, Juglans regia, Malus sieversii* with large fruits, *Pyrus* and *Prunus* species). The capital of Kazakhstan Almaty, (formerly Alma-Ata), is called "Apple-father", because above it in the Transili-Alatau the *Malus* belt is very clearly formed. The population takes advantage of this in the fruiting season and cooks applesauce and apple slices on bonfires. Above the hardwood belt follows a coniferous belt (Fig. 11.36) of *Picea schrenkiana*, possibly with *Abies semenovii*. Above this the alpine belt follows.

11.13.2 The High Mountains of Afghanistan

A mountainous country with similar vegetation conditions is Afghanistan with the Hindu Kush Mountains, where, however, monsoon influences already occur in the east (Freitag 1971; Breckle 1971, 1973, 1974, 1983, 2021; Breckle et al. 2013). On the other hand, winter rains predominate, so a general overview was already given in Chap. 8 (Fig. 11.37).

The species and endemic richness are much greater in the Hindu Kush than in the Alatau. Nuristan in eastern Afghanistan can be cited as an example of a particularly species-rich area (Fig. 11.38), where Himalayan floral elements are already radiating in. On the other hand, the peaks of the Hindu Kush reach over 7000 m elevation, and above 5000 m about 40 species of higher plants still occur (Breckle 1971, 1973, 1974; Breckle et al. 2013). Despite the summer drought, the alpine-nival belt benefits from the large snow and ice reserves of the glaciers.

The dependence of species numbers on elevation is shown in Fig. 11.39. This shows that the maximum species numbers of the montane altitudes of the Hindu Kush decrease sharply

from 2800 m, and more slowly from 3500 m onwards. After that, the number of species decreases only gradually.

In the eastern slope of the Hindu Kush, Sino-Japanese floral elements radiate from the Himalayas into Afghanistan, where they have different ecological niches. A large number of semi-xerophytic and mesophytic trees and shrubs belong to this region: *Cedrus deodara, Pinus gerardiana, P. wallichiana, Quercus balloot, Qu. semicarpifolia, Indigofera gerardiana, Plectranthus rugosus* and *Syringa emodi*, as well as the grasses *Stipa brandisii* and *Piptatherum munroi*. Some of the elements from this floristic region are endemic to eastern Afghanistan, such as *Gymnospermium sylvaticum, Pertya aichisonii* and *Saussurea afghana*, which occur as understory in the mesophytic oak woodlands, as well as the sub-endemic *Rhododendron afghanicum* in the high montane conifer woodlands, and *Rh. collettianum* in the subalpine *Juniperus* belt of the SE slope. From the S slope of

Fig. 11.38 High mountain peak (Kohe Baba Tangi, 6800 m asl) in the north-eastern Hindu Kush (Wakhan, Afghanistan) with glaciated flanks and high reaching scree slopes with open alpine-nival vegetation up to over 5000 m asl (photo: Breckle)

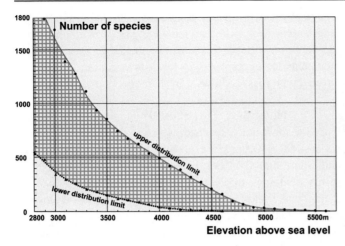

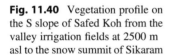

Fig. 11.39 Floristic drainage of the mountains: decrease in species numbers from 2800 m asl upwards to 5600 m asl in Afghanistan. The lower limit of species distribution is also marked. The difference between these two curves gives the real species numbers for the corresponding elevation

Fig. 11.41 Subalpine tragacanthic hard-cushion plants from the subalpine elevational zone of Kohe Baba, C Afghanistan, at 3200 m asl with *Acantholimon auganum, A. cabulicum, A. leucochlorum* and *Artemisia glanduligera* (photo: Stefan Michel)

Safed Koh, the vegetation transect is shown in Fig. 11.40 from the Sikaram summit to the valley sites. The species-rich forest belts are joined upward by Krummholz stands and thorn cushions.

These thorn cushion floras are particularly prominent in the subalpine region (Fig. 11.41, Fig. 8.59, legend **7a**), even more outside the monsoon-influenced mountainous parts.

Plant communities dominated by thorn cushion plants represent a broad vegetation belt in most of the higher mountain systems of Afghanistan, where summer rains are absent and the water supply of the plants depends to a large extent on the water reserves in the soil derived from winter snowmelt. These very conspicuous thorn cushion floras extend from the tree line, which is to be fixed at about 2800–2900 m, in the W, and in the NE and E at 3300–3500 m to about 3800–4000 m, and play an important role as summer

pastures. The most common species are thorny representatives of the genera *Cousinia, Astragalus, Onobrychis, Acantholimon, Acanthophyllum* and *Cicer,* but species composition can vary greatly between different mountain systems, and the proportion of endemics is comparatively very high. *Astragalus* is the genus in Afghanistan with the largest number of species (320). Other common dwarf shrubs include *Artemisia species, Ephedra gerardiana, Rhamnus prostrata* and *Krascheninnikovia ceratoides.* The herbaceous layer includes much valued by livestock grasses such as, *Piptatherum laterale, Poa araratica, Koeleria* spp. and *Festuca* spp. among others, together with legumes of the genera *Trigonella, Astragalus* and *Oxytropis,* they are an important grazing resource. One of the most conspicuous plants is the large steppe-lily *Eremurus kauffmannii.* It must

Fig. 11.40 Vegetation profile on the S slope of Safed Koh from the valley irrigation fields at 2500 m asl to the snow summit of Sikaram

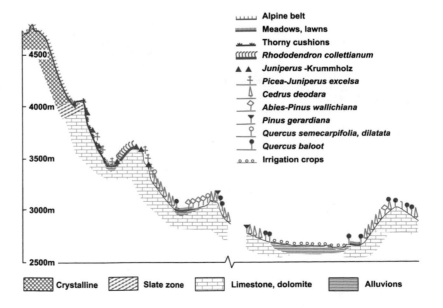

be assumed that the often high grazing pressure has led to the fact that the thorn cushions as well as the tussock-grass *Leucopoa karatavica* have a significantly larger area and density compared to the original vegetation.

Alpine Semi-Deserts, Steppes and Meadows (Fig. 8.59: 7b)

The boundary between the subalpine and alpine vegetation is very blurred and difficult to define, except in the wetter mountainous regions of eastern Afghanistan, where the upper limit of *Juniperus* scrub can serve as a marker. In most regions the thorny cushions are very gradually replaced by smaller species of the same genera, and the composition of the herbaceous species also changes only very gradually. True alpine grass-dominated meadows with a large number of herbaceous species are confined to the mountainous parts of the central and eastern Hindu Kush and the Pamirs, where certain proportions of summer rains already occur. Accordingly, they are heavily used for summer grazing during the short summer (2 months), mostly by the nomads (Kuchis). In the other parts of the mountains, even on deeper soils, the vegetation is fairly open, except in damp places, along gullies and streams, in swamps where the winter snowmelt water collects, where lawns and meadows then occur. Because of the steep topography in many places, delayed soil formation, and locally longstanding snowfields, a wide variety of plant communities can form on a small scale. However, large areas of high mountains look completely devoid of vegetation as they are covered with rocks, boulders, loose moraines and screes.

Nival Belt (Fig. 8.59:8)

Once the snow line is reached, which in the Hindu Kush is at about 4800–5000 m (N exposure) to 5400 m (S exposure) and where the Nival belt begins, the cover and number of plants continues to decline sharply. However, on south-facing rocky sites in the Hindu Kush, even woody plants such as *Juniperus semiglobosa* and *Lonicera microphylla* reach heights of over 5000 m, as does the fern *Cystopteris dickieana*.

The elevational record of vascular plants in Afghanistan is held by the showy *Primula macrophylla* in the C Hindu Kush at 5600 m. *Sibbaldia cuneata* has been recorded from 5500 m. However, mosses and lichens still occur on the highest peaks of the mountains. A contiguous nival range occurs only in the highest mountains, in the Pamirs, the Hindu Kush and the Kohe Baba, where extensive snowfields with penitential snow are often found (Fig. 11.42), in old snow not infrequently with snow algae (Fig. 10.48).

Fig. 11.42 Nival belt with forms of the Penitentes snow at Tölzer Köpfl, S of Mir Samir, C Hindu Kush at 5100 m asl, the tips of the snow columns are oriented towards the sun; they were formed by sublimation (photos: Breckle)

Ecophysiological Data from Afghan Mountains

In some alpine and subalpine areas, ecophysiological studies have been carried out with measurements of diurnal variations of essential microclimatic parameters. As a first example, a diurnal variation of these parameters from the Dashte Nawor (3120 m) is explained (Fig. 11.43a–i). In June, radiation (Fig. 11.43e) is very strong, and air temperature reaches a maximum of about 21 °C shortly after noon (Fig. 11.43f); in the morning, it is close to 0 °C. The leaf temperature of *Nepeta* leaves, however, can rise over 10 K above the air temperature to over 25 °C. The soil temperature at the surface can reach 60 °C at noon.

Accordingly, the evaporation rate is very high around noon (it can reach 1.5 ml/h for 3 cm Piche disks, Fig. 11.43h), The air saturation deficit is larger than 20 mbar at noon (Fig. 11.43i), while the wind speed remains quite low on that day (Fig. 11.43g).

The ecophysiological parameters of *Ceratoides* (*Krascheninnikovia ceratoides*) and *Nepeta* (Fig. 11.44a–d), Cl content, electrical conductivity, refractometer reading, and cell sap osmotic potential show a clear diurnal cycle that is more pronounced in *Ceratoides*.

The second example explains a diurnal course of micrometeorological data from a high-alpine site in the Wakhan area on the Wazit Pass (4620 m asl) (Fig. 11.44a–h).

Relative humidity at Wazit Pass (Fig. 2.44a, RF) is very low in the afternoon, reaching values below 15%. In the vegetation layer, e.g. between *Nepeta pamirensis* leaves, the air is slightly less dry. At night values close to 100% are reached, but dewfall was not detectable. Leaf temperatures of *Nepeta* sun leaves and shade leaves and flowers are higher than air temperature during the whole measuring day, the difference can be up to 10 K (Fig. 11.44b), at night there are mostly sub-temperatures at leaves of 2–3 K. Air temperature (Fig. 11.44c, T_1) together with soil temperature at different soil depths (1 cm, 10 cm) and at the surface reveals the effect of high radiation. At the soil surface, temperature extremes of 0 °C at 6 h and 40 °C at 15 h occur within one day. Rosette plants must be adapted to these high fluctuations. The measurement day was relatively cloudy with up to 75% cloud cover of the sky (Fig. 11.44d) approaching from the south, probably monsoon influenced. Accordingly, the radiation was not always very high (Fig. 11.44e) (measured in lux). The wind (Fig. 11.44f) was mostly from NE during the day, shifting to SW in the evening. The air saturation deficit (Fig. 11.44g) was highest around 16 h with 10 Torr; Piche evaporation (Fig. 11.44h) reached about 1 ml/h shortly after noon, directly above vegetation (10 cm) as well as higher (100 cm).

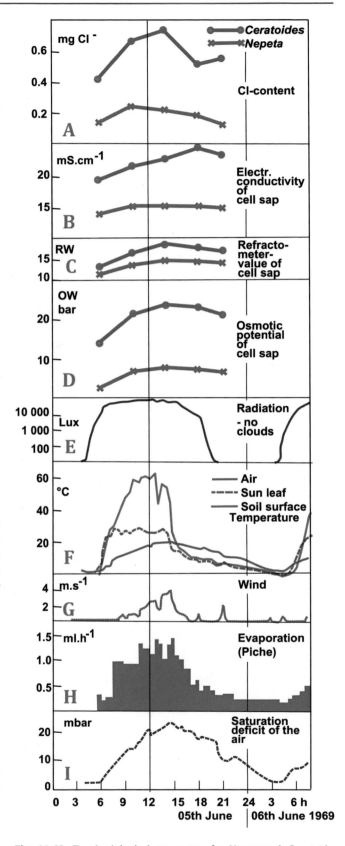

Fig. 11.43 Ecophysiological parameters for *Nepeta* and *Ceratoides* measured at 3120 m asl at Dashte Nawor, Ghazni Province, Afghanistan, on 5 June 1969

Fig. 11.44 Diurnal course of some ecophysiological parameters for a site in the high alpine stage of the Wakhan (Wazit Pass, 4620 m asl, on 8 August 1968

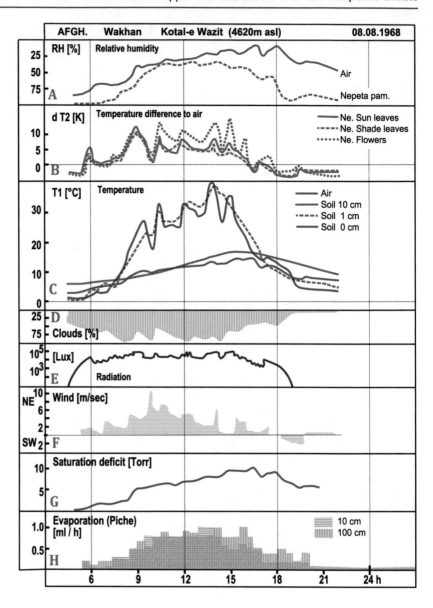

11.14 Subzonobiome of the Central Asian Deserts

As mentioned, the last remnants (cyclones) of the East Asian low pressure disturbances reach into this area. Precipitation therefore falls in summer and decreases from E (Ordos 250 mm) to W (Lop-Nor depression 11 mm). Winter and spring are dry, and the spring ephemerals so characteristic of Middle Asia are entirely absent in Central Asia. The flora is poor; shrubby psammophytes *(Caragana, Hedysarum, Artemisia,* and others) predominate among the East China-Mongolia elements. *Stipa* is also represented by C Asian species. Common sea buckthorn *(Hippophäe rhamnoides)* and tall grass Tschij *(Lasiagrostis splendens)* are widespread. In floodplain forests, *Ulmus pumila in* particular is found alongside *Populus diversifolia* and *Elaeagnus.* Examples of

halophytes are *Nitraria schoberi, Zygophyllum, Reaumuria, Kalidium and Lycium* species.

The geological structure and rock types are important for the character of the deserts. The geographical position of the deserts can be seen from Fig. 11.25.

1. Ordos. The area lies in the knee of the Hwang-Ho north of the Great Wall of China, which runs along the edge of the shifting sand dune area. It adjoins the steppe area of the loosening plain of the upper Hwang-Ho, which is now cultivated and cut by erosion gorges. It was a *Stipa steppe,* but the dry spring makes it very different from the Eastern European one. In the Ordos proper, soft sandstones are present, which led to the formation of wide sand and dune areas. On these, the *Artemisia ordosica* semi-desert with *Pycnostelma* (Asclepiadaceae) is widespread (cover 30 to 40%). In the drainless central part saline lakes with Na_2CO_3 and NaCl are found.

2. Ala-Shan. It is a desert consisting mainly of sandy plains with barchans; it adjoins the Hwang-Ho to the W and is bounded to the S by the Njan-Shan Mountains (▶ Fig. 11.25). To the N, it borders the Gobi at Gushun-Nor. Precipitation in this area decreases from 219 mm in the E to 68 mm in the W, with potential evaporation increasing from 2400 mm to 3700 mm. The rainfall maximum is in August; the continental climate has a mean temperature of 8 °C, the minima −25 to −32 °C. Groundwater is present in the dune area. The marginal mountains receive higher precipitation. Above a desert and steppe elevational belt, mesophilous shrubs with *Lonicera, Rosa, Rhamnus, Dasiphora (Potentilla) fruticosa, and* others begin at 1900–2500 m; above this, a coniferous forest with *Picea asperata, Pinus tabulaeformis,* and *Juniperus rigida* grows to 3000 m, followed by subalpine shrubs and alpine mats.

3. Bei-Shan. This adjoining area represents an ancient uplifted block and rises from 1000 m to 2791 m. It is bounded on the W by the Turpan and Lop-Nor depressions and Hami (Fig. 11.45). Precipitation ranges from 39 to 85 mm, and potential evaporation is 3000 mm. The vegetation is a low shrub desert of Central Asian species with some halophytes. *Picea asperata* already occurs on the highest elevations.

4. Tarim Basin with Takla-Makan. The basin has a length of 1300 km and a width of 500 km and is framed on three sides by high, snow-capped mountains. It is the most arid part of Central Asia with hot summers and cold winters (min −27.6 °C). Despite this, it is rich in groundwater fed by mountain rivers. The 2000 km Tarim River carries an average of 1200 m³/sec of water and forms extensive floodplains. In its lower reaches it is a nomadic river,

constantly shifting its course and percolating into the central sandy desert, as do tributaries to the Turpan Depression (Fig. 11.45). The Lop Nor was once not infrequently a salt lake 100 km in diameter, sometimes almost dry, now almost completely dry because the water is almost entirely diverted for irrigation. The Takla-Makan sand desert is devoid of vegetation, but groundwater can be reached by ditching in the dune valleys.

5. **Tsaidam**. It is a basin situated at an altitude of 2700–3000 m, surrounded on all sides by much higher mountains. The Altyntag River separates it from the Lop Nor depression. The mean annual temperature is 0 °C, the minimum below −30 °C. This basin is also irrigated by the marginal mountains. The central part was a large lake in the Pleistocene and is now a salt desert without vegetation. At the foot of the mountains an *Artemisia* semi-desert is found on sandy soils. Tsaidam forms the transition to the still higher Tibet.

6. **Gobi** (Mongolian = desert). This area extends N from the deserts mentioned so far and occupies the whole S part of outer Mongolia. It is separated from the forests and steppes in the E by the Chingan Mountains; in the W it abuts the Dsungaria, which already receives precipitation from Atlantic cyclones and therefore has Middle Asian features. In the N the Gobi gradually merges into the Mongolian *Stipa* steppes with *Aneurolepidium (Agropyron)* and *Artemisia* species. In the desert, saline and gypsum soils are common; the C part is devoid of vegetation and covered by a stone pavement, even otherwise the plant cover is sparse; the production of dry matter hardly reaches 100–200 kg/ha, in the northern parts of the steppe 400–500 kg/ha. *Nitraria sibirica, Lasiagrostis, Kalidium,* etc. grow on the brackish lowlands, and the Saksaul *Haloxylon ammodendron* grows on sand-covered areas. In the whole W part, the groundwater does not come to the surface. In the E there are some oases. The Mongolian Altai mountain range extends into the Gobi Desert from the NW, continuing into the Gobi Altai. In the latter, only a steppe elevational level is reached; in the former, a coniferous forest belt is already present, namely in N exposure, which, however, with *Larix* already bears a completely Siberian character.

Fig. 11.45 The Turpan Depression (tectonic depression down to -100 m asl) with irrigation crops between loess plains (photo: Breckle)

11.15 Subzonobiome of the Cold High Plateau Deserts of Tibet and Pamir (sZB VII, tIX)

Between the mountain wall of the Himalayas in the S and the Kuen-Lun and Altyntag in the N lies Tibet, the largest mass elevation on earth with an average altitude of 4200 to 4800 m asl (Miehe 2021). The plateau has a length of 2000 km from E to W and a width of 1200 km from N to S. It consists of debris-filled basins bounded by numerous mountain ranges another 1000 m higher (Figs. 11.46 and 11.47)

The S and E parts are still under the influence of the monsoon (Fig. 11.47, Nepal), and SE Chinese and Himalayan forest elements occur in the deeply incised valleys that form the headwaters of the major S and E Asian river systems (Chang 1981; Mosbrugger et al. 2018).

The larger W and C part, the Changtang Desert (Chang Tan), is characterised by an extreme climate. The average annual temperature is −5 °C, only July has a positive average of +8 °C. Daily temperature fluctuations of 37 K occur, precipitation rarely exceeds 100 mm. The pronounced precipitation gradient from S to N is evident from Fig. 11.47. The poor flora migrated only after the ice age, i.e. very young; they are Central Asian elements (*Krascheninnikovia ceratoides, Kochia, Reaumuria, Rheum, Ephedra, Tanacetum, Myricaria,* etc.).

The western end of the plateau, from which the high mountain ranges emanate, is the Pamir with the Pamir Biological Station (Chechekti) located at an altitude of 3864 m, where many Russian researchers conducted ecophysiological studies. The average precipitation here is 66 mm, mainly in May to August. The air is dry, the solar radiation is 90% of the solar constant, so the ground surface heats up to over 52 °C in the summer months. Only 10–13 nights a year are frost-free (Fig. 11.26 Pamirski Post, now Murghab). A closed snow cover is not present. The soils are so dry that they do not freeze.

Dwarf shrubs 10–15 cm high grow on the desert-like sites: *Krascheninnikovia ceratoides, Artemisia rhodantha, Tanacetum pamiricum* or *Stipa glareosa* as well as the

cushion plant *Acantholimon diapensioides.* The boundaries between high mountain deserts, desert steppes, and the temperate steppes and also other vegetation units shown in Fig. 11.47 are often very blurred. Very clearly delineated are the alpine meadows occurring along streams in the valleys (Fig. 11.48) and the open *Kobresia* meadows at high elevations (Fig. 11.47). The melt waters flowing off form such wetted areas, frost swamps with the sedge *Kobresia tibetica,* partly also "salt lakes". There are also sand dune areas.

Growth in the dry locations is extremely slow. *Krascheninnikovia ceratoides* only comes to flower after 25 years, but reaches an age of 100–300 years. The root systems are strongly developed and their mass is ten to twelve times greater than that of the shoot systems. Most roots are found at a depth of 0–40 cm, i.e. in the soil layers that warm up to over 10 °C in summer. Laterally, the roots reach more than 2 m. The supply of usable water in the upper 100 cm of the skeleton-rich soil is a maximum of 26 mm and a minimum of 5 mm. This is very little, but is sufficient for quite intensive transpiration given the sparse vegetation. Photosynthesis is lively only in the morning hours. The daily yield is reported to be 25 mg/dm^2 leaf area. The low night temperatures prevent major respiration losses.

Three biogeocoenes were studied (Table 11.5): one with predominant *Krascheninnikovia ceratoides* on rubble soils, a second more steppe-like on clayey soils with *Artemisia* and *Stipa glareosa,* and a third with the low herbaceous cushions on stony soils with relatively favourable water conditions at gullies.

It can be seen that, notwithstanding the relatively high transpiration intensity, the water consumption of the biogeocoenes in millimetres is so small that it is covered by

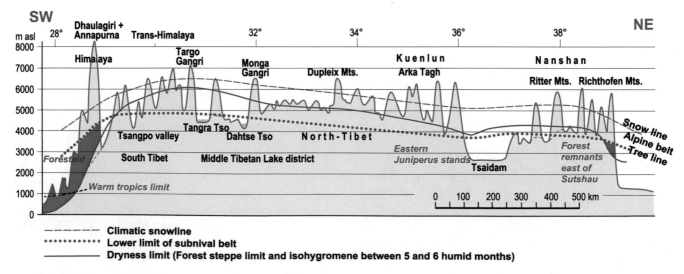

Fig. 11.46   Profile through Tibet from SW to NE and the slightly lower Tsaidam Desert (7½ times superelevation). The temperature-related tree line is only a theoretical one in Tibet and in Tsaidam, as the forest line is conditioned by drought. Forest is found only on the southern slope of the Himalayas and as a narrow elevation step in the Richthofen Mountains (Nanshan), (modified after von Wissmann 1961)

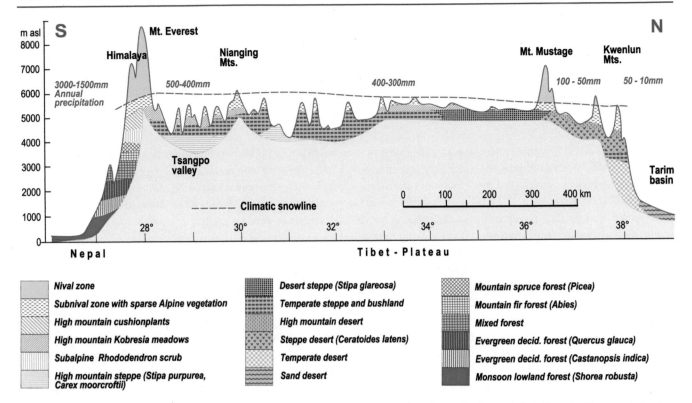

Fig. 11.47 Profile through Tibet from S to N between the Himalayan foothills of Nepal and the Tarim basin with vegetation formations. The decrease in precipitation from the monsoon-influenced south to the desert basin landscapes in the north is indicated with numerical values (modified from Chang 1981)

Fig. 11.48 High mountain desert in the Eastern Pamirs (4000 m asl) with sparse growth of *Krascheninnikovia ceratoides* (photo: E. Klein)

Table 11.5 Phytomass and water consumption in Eastern Pamirs

Factors	Biogeocoenes		
	Desert	Steppe	Cushion plants
Coverage of the plants in %.	5–18	15–20	15–30
Phytomass in t/ha	0.14–0.54	0.09-0.48	0.4-0.89
Transpiration (g·gFG^{-1}·h^{-1})	0.3-0.9	0.1-0.7	0.1-0.19
Water consumption in mm during the growing season	8–40	6–87	25–446

precipitation. Only the herbaceous cushion plants near the stream channels receive additional water through inflow. Generally, one-half to one-third of precipitation is consumed to meet plant transpiration.

The conditions for the orobiomes are complicated because the elevational grading depends very much on precipitation. In areas with less than 100 mm, there is no actual snow line because the small amount of snow evaporates even above 5500 m asl in the strong radiation (Fig. 11.42). Up to the upper distribution limit of plant growth, the desert persists, whereas otherwise alpine steppes or even alpine meadows occur at higher altitudes above 500 mm of rain. In the upper alpine belt, mostly cushion plants play a significant role.

11.16 Man in the Steppe and Cold Desert

The steppes covered vast areas of the northern hemisphere. They fed countless wild grazing animals, the buffalo on the prairies in N America, the wild cattle and wild horses in Eurasia, and probably many other herbivores (Lozán et al. 2016a, b). In C Asia, the true camel (two-humped) still plays a major role today. In earlier millennia and into the Middle Ages, the nomadic way of life and the possibilities of long-distance migrations led to Asian equestrian peoples being able to advance as far as Europe.

Only recently, the steppe and prairie was almost completely ploughed for agriculture. Dust erosion of the prairies in the thirties, in Kazakhstan in the sixties have caused great damage and desertification.

Crucial for living in these areas is the availability of water, either from natural springs (and thus oases) or through deep wells, which are drilled in many places today and feed cisterns. In addition, the sparse vegetation is exploited by herds of cattle (mostly many sheep with a few goats, but also cattle), which need large areas for action. Grazing on the steppes and semi-deserts has become an acute problem due to overgrazing, degradation and soil erosion. One solution most likely to remedy the situation would be to replace domestic herds with wild animals (Campbell 1985). Steppe herbivores are excellently adapted to their environment; they

could restore balance to ecosystems that are prone to disturbance. On the same area of land, if properly managed, they could provide more meat than domestic livestock. Cattle should be reserved for temperate latitudes, only on land that can tolerate intensive grazing.

11.17 Zonoecotone VI/VIII: Boreo-Nemoral Zone

In contrast to the zones discussed so far, the boreal coniferous forest zone with a **cold temperate climate** in the northern hemisphere in northern Eurasia and North America (interrupted by seas) extends almost around the entire globe. In the S, this zone borders on the nemoral deciduous forest zone in the oceanic climate zone, but on the arid steppes or semi-deserts in the continental climate zone (Fig. 11.49).

The boundary between the deciduous forest zone and the coniferous forest zone is not sharp; rather, a boreo-nemoral ecotone VI/VIII is involved here, in which either mixed stands of some coniferous species (mostly pines) with deciduous species occur, or else a macromosaic-like intermixing takes place with pure deciduous forests on favourable sites and on better soils and pure coniferous forests on unfavourable sites with poor soils. In E North America, it is various *species of Pinus* that occur in mixed stands; in the Great Lakes region, *Pinus strobus* above all, but also *Tsuga canadensis;* in the SE, *Juniperus virginiana*. Often pines are pioneer species after forest fires or on abandoned cultivated land. They grow upward more rapidly than hardwoods on poor soils and therefore form an upper tree layer. However, regeneration of pines in such mixed stands is difficult when the hardwood understory is dense. Therefore, pine persists only where fire is a recurrent factor. It has been shown that fires in these forests, especially on sandy soils with a litter layer that is dry in summer, are frequently caused by lightning, even without human intervention. In Europe the conditions are much simpler: on the poor fluvioglacial sands, which extend as a broad strip in front of the terminal moraines in C and E Europe, pure pine forests are found in

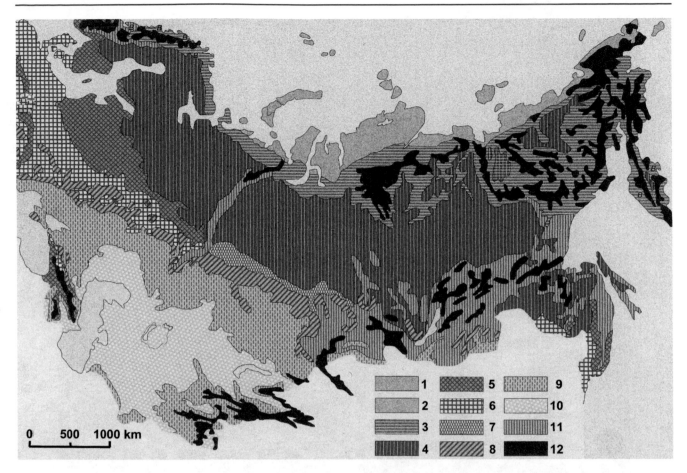

Fig. 11.49 Vegetation zones of Euro-Siberia. 1 Arctic desert; 2 tundra; 3 dwarf shrub and forest tundra; 4 boreal coniferous forest zone; 5 mixed forest zone; 6 deciduous forest zone; 7 small-leaved deciduous forests; 8 forest steppe; 9 grass steppe; 10 semi-deserts and deserts; 11 mountain coniferous forests; 12 alpine zone

the area which climatically still belongs to the hardwood zone, which in E Europe are called "bor". On better loamy and sandy soils oaks occur as the lower tree layer, a Querceto-Pinetum (mixed oak-pine forest) or 'subor'.

On loamy soils, the hornbeam (*Carpinus betulus*) is added, so that the forests called 'sugrudki' are three-layered (Carpineto-Querceto-Pinetum, oak-hornbeam-pine mixed forest). Finally, on loess we have the zonal deciduous forests with oak in the upper and *Carpinus* in the lower tree layer, called 'grud'. Human interventions change these forests very much: forest fires and firewood production by felling the hardwoods promote pine; exploitation of pine, which is valuable as timber, leads to pure deciduous forests. In addition, there is forest grazing. In C Europe, forestry has created extensive pine forests even in former pure deciduous forest areas, for example in the Upper Rhine Plain. Further N (in S Scandinavia, C and E Europe), spruce (*Picea abies*) and oak (*Quercus robur*) play a greater role, usually intermixing macromosaically (Klötzli 1975). As the better soils (the oak forest sites) are now mostly cultivated, the proportion of spruce in the remaining forests has increased; it is also favoured by forestry. In C Europe proper, the spruce forests at low altitudes have all been artificially created. The "fencing" of the landscape had increased more and more, because the forestry industry wanted to achieve higher yields. But in the last decades, as a result of the "acid rain" and other pollutant inputs as well as additional soil impoverishment, not only firs but also spruces and the deciduous trees became more and more diseased. Most recently drought years and bark beetle have caused enormous damages in the managed forests in C Europe.

The boundary between the boreo-nemoral and the boreal zone proper corresponds in Europe to the N limit of distribution of the oak. It runs in S Sweden at 60 degrees latitude, then along the S coast of Finland and from there to the middle Kama, where the steppe borders the boreal zone.

Photo 43 *Fritillaria imperialis* (Liliaceae), the "Imperial crown" is not rare in the mountains of Iran and Afghanistan (Orobiom VII), here from the Panshir valley from 2800 m elevation (photo: M Keusgen)

Photo 44 High mountain desert in Eastern Pamir (Orobiom VII [rIII]) with puny hemicryptophyte growth (photo: C. Opp)

Photo 45   Patagonian steppe (ZB VII) near Torres del Paine   (photo: J. Tamm)

References

Agachanjanz, O.E. & Breckle S.-W. 1994: Umweltsituation in der ehemaligen Sowjetunion. Naturwiss. Rundschau 47: 99–106.

Breckle, S.-W. 1971: Vegetation in alpine regions of Afghanistan. In: Davis, P.H. et al. (eds.): Plant Life of South-West Asia. Proceed. of the Symposium 1970. Edinburgh: 107–116.

Breckle, S.-W. 1973: Mikroklimatische Messungen und ökologische Beobachtungen in der alpinen Stufe des afghanischen Hindukusch. Bot. Jahrb. System. 93: 25–55.

Breckle, S.-W. 1974: Notes on alpine and nival flora of the Hindu Kush, East Afghanistan. Bot. Notiser (Lund) 127: 278–284.

Breckle, S.-W. 1983: Temperate Deserts and Semideserts of Afghanistan and Iran. In West, N.E. (ed.): Temperate Deserts and Semideserts. Ecosystems of the World (ed.: Goodall, D. W.), Elsevier, Amsterdam 5: 271–319.

Breckle, S.-W. (ed.) 2021: Ökologie der Erde. Band 3. Spezielle Ökologie der Gemäßigten und Arktischen Zonen Euro-Nordasiens. 3. Aufl. Schweizerbart/Stuttgart 804 pp.

Breckle, S.-W., Agachanjanz, O.E. & Wucherer, W. 1998: Der Aralsee: Geoökologische Probleme. Naturwissen. Rundschau 51: 347–355.

Breckle S-W, Rafiqpoor MD 2010: Field Guide Afghanistan - Flora and Vegetation. Scientia Bonnensis, Bonn, Manama, New York, Floríanópolis: 864 pp.

Breckle, S.-W. & Geldyeva, G.V. 2012: Dynamics of the Aral Sea in geological and historical times. In: Breckle, S.-W., Dimeyeva, L., Wucherer, W. & Ogar, N.P. (eds.): Aralkum – a man-made desert. The desiccated floor of the Aral Sea (Central Asia). Ecol. Studies 218: 13–35.

Breckle, S.-W., Hedge, I.C. & Rafiqpoor, M.D. 2013: Vascular Plants of Afghanistan – an augmented Checklist. Scientia Bonnensis, Bonn, Manama, New York, Florianópolis, 598 S.

Breckle, S.-W., Wucherer, W., Agachanjanz, O.E. & Geldyev, B.V. 2001: The Aral See crisis region. In: Breckle, S.-W., Veste, M. & Wucherer, W. (eds.): Sustainable land-use in desert. Springer, Berlin: 27–37.

Burrows, C. J. 1990: Processes of vegetation change. U. Hyman/London 551 S.

Campbell, B. 1985: Ökologie des Menschen. Harnack, München 232 S.

Chang, D.H.S. (1981) The vegetation zonation of the Tibetan Plateau. Mountain Research and Development 1 (1): 29–48. DOI: https://doi.org/10.2307/3672945.

Djamali, M., Brewer, S., Breckle, S.-W., Jackson, S.T. 2012: Climatic determinism in phyto-geographic regionalization: a test from the lrano-Turanian region. SW and Central Asia. Flon-Morphology, Distibution, Functional Ecology of plants 207: 237–249.

Dokutschajev, U.V. 1898: Zur Lehre über die Naturzonen. St. Petersburg (Russ.)

Donigian, A.S.j. et al. 1995: Modeling the impacts of agricultural management practices on soil carbon in the central U.S. Soil management and greenhouse effect. CRC Press, Boca Raton: 121–135.

Frangi J 1975: Sinopsis de la communidades vegetales y el medio de las sierras de Tandil (Provincia de Buenos Aires). Bol Soc Argentina de Botan. 16: 293-319

Freitag, H. 1971: Die natürliche Vegetation des südostspanischen Trockengebiets. Bot. Jahrb. 91: 147–208.

Freitag H, Hedge IC, Rafiqpoor MD, Breckle S-W 2010: Flora and Vegetation Geography of Afghanistan. In: Breckle S-W, Rafiqpoor MD: Field Guide Afghanistan - Flora and Vegetation. Scientia Bonnensis, Bonn, Manama, New York, Floríanópolis, 79-115.

French, N. R. (ed.) 1979: Perspectives in grassland ecology. Ecol. Stud. 32: 204 S.

Glinka, K.D. 1914: Die Typen der Bodenbildung. Berlin 215 S.

Gradmann, R. 1950: Das Pflanzenleben der Schwäbischen Alb. Albverein, Stuttgart, 449 P.

Hager, J. 1987: Polsterförmige Zwergsträucher auf Kreta und in Patagonien. Natur und Museum 117: 105–132.

Klötzli, F. 1975: Edellaubwälder im Bereich der südlichen Nadelwälder Schwedens. Ber. Geobot. Inst. Rübel 43: 23–53.

Klötzli, S. 1997: Umweltzerstörung und Politik in Zentralasien - eine ökoregionale Systemuntersuchung. Europäische Hochschulschriften 4 (17); Peter Lang, Bern 292 S.

Küchler, A. W. 1974: A new vegetation map of Kansas. Ecology 55: 586–604.

Levina, F.J. 1964: Die Halbwüstenvegetation der nördlichen Kaspischen Ebene. 344 S., Moskau-Leningrad (Russ.).

Lewis, J.P. & Collantes, M.B. 1975: La vegetacion de la Provincia de Santa Fé. Bol. Soc. Argentina de Bot. 16: 151–179.

Lozán, J.L., Breckle, S.-W, Müller, R. & Rachor, E 2016a: Warnsignal Klima: Die Biodiversität. Wissenschaftliche Auswertungen. www.warnsignal-klima.de

Lozán, J.L., Breckle, S.-W, & Rachor, E 2016b: Vom Mensch bedingte Biodiversitätsänderungen seit Ende der letzten Eiszeit. In: Lozán, J. L., Breckle, S.-W., Müller, R. & Rachor, E 2016 (see above): 68–74.

Miehe, G. 2021: Extrem kalt arides Subzonobiom VII (tIX) der Kälte- und Hochlandwüsten Zentralasiens: Tibet. In: Breckle, S.-W. (ed.) 2021: Ökologie der Erde. Band 3. Spezielle Ökologie der Gemäßigten und Arktischen Zonen Euro-Nordasiens. 3. Aufl. Schweizerbart/Stuttgart 405–454.

Mosbrugger, V., Favre, A., Mueller-Riehl, A.N, Päckert, M. et al. 2018: Cenozoic Evolution of Geobiodiversity in the Tibeto-Himalayan Region. In: Hoorn, C., Perrigo, A. & Antonelli, A. (eds.): Mountains, Climate and Biodiversity. J. Wiley & Sons Ltd.: 429–448.

Stanjukovitsch, K.V. 1973: Die Gebirge der USSR. 412 S. Duschanbe (Russ.)

von Wissmann, H. 1961: Stufen und Gürtel der Vegetation und des Klimas in Hochasien und seinen Randgebieten (Teil B). Erdkunde 15: 19–44.

Walter, H. 1968: Die Vegetation der Erde, Bd. II: Gemäßigte und arktische Zonen. 1001 S., Fischer, Jena-Stuttgart.

Walter, H. 1975: Über ökologische Beziehungen zwischen Steppenpflanzen und alpinen Elementen. Flora 164: 339–346.

Walter, H. 1976: Die ökologischen Systeme der Kontinente (Biogeosphäre). Prinzipien ihrer Gliederung mit Beispielen, Fischer, Stuttgart 131 S.

Walter, H. 1990: Vegetationszonen und Klima. 6. Aufl., Ulmer/Stuttgart 382 S.

Walter, H. & Alechin, W.W. 1936: Grundlagen der Pflanzengeographie. Moskau-Leningrad (Russ.)

Walter, H. & Box, E.O. 1983: Overview of Eurasian continental deserts and semideserts. In: Ecosystems of the World vol. 5: 3–269, Amsterdam.

Walter, H. & Breckle, S.-W. 1999: Vegetation und Klimazonen. 7. Auflage, Ulmer-Stuttgart, 544 S.

Wucherer, W., Breckle, S.-W. & Buras, A. 2012: Primary Succession in the Aralkum. In: Breckle, S.-W., Dimeyeva, L., Wucherer, W. & Ogar, N.P. (eds.): Aralkum – a man-made desert. The desiccated floor of the Aral Sea (Central Asia). Ecol. Studies 218: 161–198.

Part K: ZB VIII—Zonobiome of the Taiga or of the Cold Temperate Boreal Climate

12

Contents

Photo 46 Secondary pine forest (taiga, ZB VIII) with fire traces near Baikal Lake area, Siberia (photo: U. Kull)

Photo 47 Pine-spruce taiga (ZB VIII) in S Finland at Nuuksio (photo: Breckle)

Photo 48 *Linnaea borealis* (Caprifoliacedae), in the herb layer of the N taiga (ZB VIII) in Finland (photo: Breckle)

Photo 49 Lichen pine forest (ZB VIII), a taiga with a dense carpet of fruticose (mini-shrub like) lichens *(Cetraria, Cladonia, Alectoria* etc.) in the herb layer near Begna-Stormyrhaugen, E Norway (photo: E. Fischer)

12.1 Climate and Soils

The actual boreal zone (Fig. 11.49), **Zonobiome VIII**, begins where the climate becomes too unfavourable for the hardwood-deciduous species, i.e. where the summers become too short and the winters too long. It can be recognised in the climate diagram by the fact that the duration of time with daily means above 10 °C falls below 120 days and the cold season lasts more than six months (Fig. 12.1). This zone is identified on the map of ecoclimates (Fig. 2.50) as a cold-temperate region with 3–4 thermal months of vegetation and is divided into four categories according to humidity levels despite the uniform vegetation cover. The N boundary of the boreal zone against the Arctic is where about only 30 days with daily means above 10 °C and a cold season of eight months are typical for the climate.

However, with the wide extension of this zone, one cannot speak of a uniform climate, but one must distinguish a more cold-oceanic climate with a relatively low amplitude of temperatures and a cold-continental one, where in extreme cases the range between the temperature maximum (+30 °C) and minimum (−70 °C) can reach 100 K. Likewise, the temperature ratios change from N to S.

12.2 The Coniferous Species of the Boreal Zone

Due to the climatic conditions, the following subzonobiomes can be distinguished: a northern, a middle, a southern (each with evergreen conifers) and an extreme continental (with the deciduous *Larix*). Separated from these are the oceanic forms with birch trees in NW Europe and NE Asia.

Similarly structured, but not as extensive, is the taiga in Canada and Alaska.

In the oceanic area, birch species *(Betula)* play an important role (Ahti and Jalas 1968). The floristic composition of the tree layer is naturally different across the vast distances. For coniferous forests, the number of coniferous species is very large in N America and E Asia, but very small in the Euro-Siberian region. As a cause, the same vegetation history is reflected here as already in ZB VI.

In N America we have plenty of species of the genera *Pinus, Picea, Abies, Larix,* but also *Tsuga, Thuja, Chamaecyparis* and *Juniperus,* which, however, belong more to the transition zone. The species of these genera on the Pacific coast are different from those in the E part. Only the Norway spruce *(Picea glauca)* goes from Newfoundland to the Bering Strait. Black spruce *(Picea mariana),* which otherwise occurs mostly on

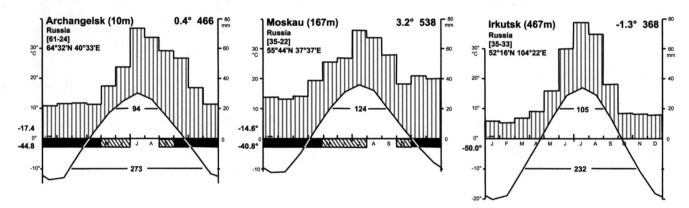

Fig. 12.1 Climate diagrams from the boreal zone of N Europe (Archangalsk), the mixed forest zone (Moscow) and the boreal zone of Siberia (Irkutsk). With horizontal lines number of days with mean above +10 °C (top) and above −10 °C (bottom)

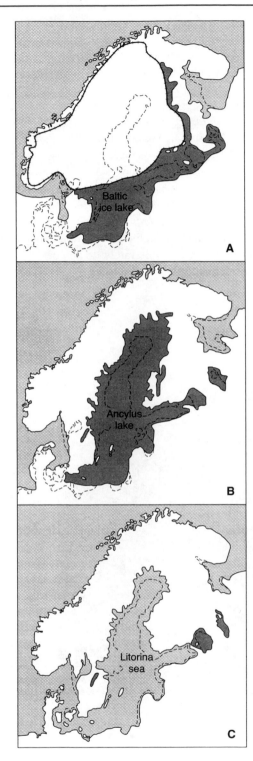

Fig. 12.2 Late glacial and Holocene development of the Baltic Sea. (**a**) Baltic Ice Lake with fresh water (10,200 years ago); (**b**) Ancylus Lake with fresh water (8000 years ago); (**c**) Litorina Sea (about 5000 years ago) (from Dierßen 1996)

poor soils, also grows at the tree line toward the Arctic, as does *Larix laricina* in the continental areas. In addition, *Abies balsamea* and *Thuja occidentalis* grow, as does *Pinus banksiana,* the latter especially on burnt areas. Very diverse species are found in the coniferous forest belt of the mountains.

In contrast, only the spruce *(Picea abies)* and the pine *(Pinus sylvestris)* play a role in the boreal zone of Europe. Only in the E part is the spruce replaced by the closely related Siberian species, *Picea obovata,* and *Abies sibirica, Larix sibirica* and *Pinus sibirica,* a subspecies of the Swiss stone pine *(Pinus cembra)* are added. The proportion of spruce decreases, and in continental E Siberia it is completely absent. At the same time *Larix sibirica* is replaced there by *L. dahurica.* Larch forests alone cover 2.5 million km^2 in Siberia. In the N part of Japan, the number of coniferous species is increasing again.

In N Europe, the traces of the Ice Age are omnipresent. Especially in the area of the European taiga (Scandinavia), the transformation of the landscape by the large inland ice masses from last glacial and their melting can be seen above all in the formation of the Baltic Sea (Fig. 12.2). It was only a few thousand years ago that the Baltic Sea was formed in its present shape. In the surrounding landscapes, the glacial traces can be read in the formation of countless dead-ice waters (Fig. 12.3) and in the various deposits (moraines, glacial soils, bedloads, etc.).

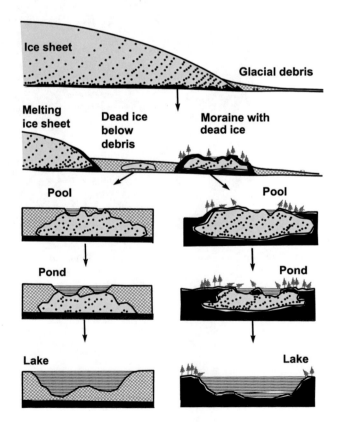

Fig. 12.3 Formation of water bodies after melting of the dead ice in the bedload (left), the deposits from the advancing ice, and in the moraine material (right)

12.3   The Oceanic Birch Forests in ZB VIII

The strong differences in continentality across the taiga zone of Eurasia are clearly noticeable floristically, as already discussed. In the oceanic climate zone on the Atlantic in Norway on the one hand, and on the Pacific in Kamchatka on the other, sparse forests of birch and pine occur. In Norway, spruce is almost entirely absent, so a taiga zone is not actually developed. The tree line in N Scandinavia is formed by *Betula tortuosa* (closely related to *B. alba*) (Fig. 12.4). *Betula tortuosa* is low-growing with an irregularly curved trunk ('drunken trees') (Fig. 12.5a).

Similar looking sparse forests occur in Kamchatka (Fig. 12.5b). There *Betula ermanii*, partly with the crooked pine *Pinus pumila,* again forms the polar forest boundary. The forests of Kamchatka are dominated by *Betula ermanii*, which forms very heavy wood (specific gravity >1; hence the name stone birch) and can live for several hundred years. Its distribution is very wide, it also reaches Japan and Korea. Other birches *(B. japonica, B. middendorfii)* and *Larix* species (L. *gmelinii, L. kamtschatica, L. cajanderii)* are sporadically associated. Kamchatka forests also do not belong to the taiga in the strict sense. Only rarely do spruce forests occur (with *Picea ajanensis).*

Fig. 12.4 Forest tundra in a mosaic of forest plots of *Betula tortuosa* and dwarf shrub tundra near Abisko (N Sweden) (photo: Breckle)

Fig. 12.5 (**a**) Low-growing trees of *Betula tortuosa* with an irregularly curved trunk ('drunken trees') in the tree tundra of the Kola Peninsula (photo: O. Agachanjanz). (**b**) Curved trunks of *Betula ermanii* in the subarctic region in Kamchatka, N of Petropavlovsk (photo: Breckle). (**c**) The species rich herb layer in the stone birch forest of Kamchatka has attractive species like *Cypripedium jatabeanum* (photo: Breckle). (**d**) *Trillium camchatkense* in the herb layer of *Betula ermanii* forests in Kamchatka (photo: Breckle). (**e**) *Trollius riederana* in the herb layer of open stone birch forests in Kamchatka is common on less shady sites (photo: Breckle). (**f**) There are many genera which are the same in Europe as well as in Kamchatka but represented by other species like *Geranium erianthum* in the *Betula ermanii* forest (photos: Breckle)

12.4   The European Boreal Forest Zone

Typical for ZB VIII in N Europe is the dark spruce forest, called taiga, on podsol soils with a layer of raw humus, a bleached horizon and a compacted B horizon (Fig. 12.6). Such soils form in the humid boreal zone on any parent rock, but become more pronounced the more base-poor it is. The litter of spruce is difficult to decompose and lies above the A_0 horizon, which consists of organic matter interwoven by the rhizomes and roots of dwarf shrubs as well as fungal mycelia and is called the raw humus layer. It can be easily lifted off from the underlying A_1 horizon (mineral soil containing humus, hence also called overlying humus or dry peat) as a soft layer. The humic acids formed in the raw humus migrate with the rainwater into the depth and on their way cause a complete leaching of the bases and sesquioxides (Fe_2O_3, Al_2O_3), so that only bleached fine quartz sand remains in the A_2 horizon (bleached horizon). At the boundary of the unleached subsoil, the humus brine with the sesquioxides are precipitated as a result of the decrease in acidity or by the water withdrawal of the tree roots. The B horizon is formed, which is dark brown (humus podsole) or rusty red (iron podsole) in colour.

In the spruce forest (Piceetum typicum), in addition to the tree layer, a herb layer and a closed moss layer can often be distinguished. The herb layer is dominated by blueberry (*Vaccinium myrtillus*), in dry forest types also by cranberry (*Vaccinium vitis-idaea*) or in the S zone often by wood sorrel (*Oxalis acetosella*). *Lycopodium annotinum, Maianthemum bifolium, Linnaea borealis, Listera cordata, Pyrola (Moneses) uniflora*, etc. are also very characteristic. At high groundwater levels and high precipitation, the accumulation of raw humus increases, leads to peat formation, and results in the formation of raised bogs (Fig. 12.18). The moss layer is first dominated by *Polytrichum* and in the later stage by

Sphagnum. If waterlogging occurs due to flowing, oxygen-rich groundwater, the spruce forests change into floodplain forests.

In addition to spruce forests, the proportion of pine forests is always very large in the boreal zone. The pine (*Pinus sylvestris*) displaces the spruce on drier sites. The herb layer of these sparse forests is formed by heather (*Calluna vulgaris*) together with lingonberry (*Vaccinium vitis-idaea*), many lichens (*Cladonia, Cetraria*) are found in the moss layer; characteristic species of the herb layer are *Pyrola species, Goodyera repens, Lycopodium complanatum*, etc. However, pine is often also widespread on sites favourable for spruce, but only after forest fires, which can also be caused by lightning. On burnt areas there is often a mass development of *Molinia coerulea, Calamagrostis epigeios* or *Pteridium aquilinum*, according to an increasing dryness of the habitat.

On such burnt sites of the tree species, the birch and aspen grow up most rapidly; they are then displaced by the pine. Under the pine the spruce grows slowly. In N Sweden the birch stage lasts about 150 years, the pine stage 500 years. New fire often occurs before the spruce stage corresponding to the zonal vegetation is reached. The large proportion of pine is therefore understandable. Pine is absent only on moist sites with low fire danger. Corresponding forests are found in North America, only they are somewhat richer floristically. Several species are distributed throughout the taiga (circumboreal). Others differ regionally, typically with each genus having several distinct species, each occupying different regions of the taiga.

The forests of the taiga in N America are as in Siberia largely coniferous, dominated by larch, spruce, fir and pine. The woodland mix varies according to geography and climate so for example the Eastern Canadian forests ecoregion of the higher elevations of the Laurentian Mountains and the northern Appalachian Mountains in Canada is dominated by balsam fir *Abies balsamea*, while further north the Eastern Canadian Shield taiga of northern Quebec and Labrador is notably black spruce *Picea mariana* and tamarack larch *Larix laricina*.

Fig. 12.6 Podsol soils form in the humid boreal zone on any parent rock. Their special feature is the formation of the A_2 bleaching horizon (photo: Breckle)

12.5   On the Ecology of the Coniferous Forest

The denser the stand, the less the sun's rays penetrate to the ground. Under a spruce forest, the ground is therefore 2 K colder than in open places. The snow cover is also less thick, so that the ground freezes deeper. The frost depth was 85 cm in the soil of a dense stand compared to 50 cm in the thinned stand, where the ground frost disappeared at the beginning of June, while in the dense stand it persisted until the beginning of August.

The spruce roots very shallowly in the upper 20 cm and even shallower at high groundwater levels (Fig. 12.7). A constantly good water supply at a medium groundwater level is necessary for a high production of spruce forests. The deeper-rooted pine is not so sensitive to soil dryness. The total annual water output of typical spruce forests is about 250 mm in the northern taiga, 350 mm in the middle, and 450 mm in the southern. The average production of organic matter is 5.5 t/ha per year, wood increment is 3 t/ha, whereas in the southern taiga it is up to 5 t/ha. The greatest annual increment is reached by forest stands in the north only at the age of 60 years, in the south already at the age of 30–40 years. The phytomass of the tree layer in pine forests is a maximum of 270 t/ha, that of the understory in old stands up to 20 t/ha. Similar data from a pine forest in central Sweden are compiled in Fig. 12.8.

The quantity of litter produced during the growth of old stands may exceed 1000 t/ha; however, it is not accumulated, but continuously decomposed until a state of equilibrium between inflow and outflow is reached at a litter mass of 50 t/ha. Only at very wet sites by peat formation is organic mass stored. Under such unfavourable conditions, the annual increase in dry mass of the tree layer is often lower than that of the other layers, for example, in the herbaceous spruce swamp forest for the tree layer 850 kg/ha (total 1906 kg/ha), and in the pine raised bog for the tree layer 104 kg/ha (total 1780 kg/ha). The leaf area index is relatively high because there are at least two years of needles (for pine forests of the boreo-nemoral zone LAI = 9 to 10, for spruce forests of the taiga above 11).

The conifers always have an **ectotrophic mycorrhiza**, whereby the area of the root system is greatly expanded by the fungal hyphae. In this way, the nutrients contained in the raw humus layer are more easily accessible to the trees. Tree root competition is very high for herbaceous layer species. On shallow granitic soils, pines can consume all water, so the herb layer is completely absent and the soil is covered only with lichens (Fig. 12.9). Under these circumstances, young pine growth cannot occur, although the light conditions are favourable. It only occurs where an old tree dies (Fig. 12.7), forming a gap and root competition is lacking. With greater soil moisture, root competition makes itself felt by competing for the nitrogen that tree roots take up, so that only dwarf shrubs with extremely low nutrient requirements (*Vaccinium myrtillus*) can persist. However, if the tree roots are cut to eliminate their competition, more demanding species such as *Oxalis acetosella* or even the nitrophilous raspberry (*Rubus idaeus),* which otherwise only occurs in clearings where tree-root competition is also absent, are established under unchanged light conditions. Thus, it is often not the light factor that determines the composition of the herb layer, but the amount of nutrients available to the herbs under competitive root conditions.

The following information is given on the water balance of a spruce forest in Sweden:

Fig. 12.7 Pine-spruce taiga, south of Saint Petersburg, with large windthrows and thus upright root plates (for size comparison: Prof. Okmir Agachanjanz). The herb layer is dominated by *Vaccinium* species, deeper parts are boggy, there are small *Sphagnum* depressions (photo: Breckle)

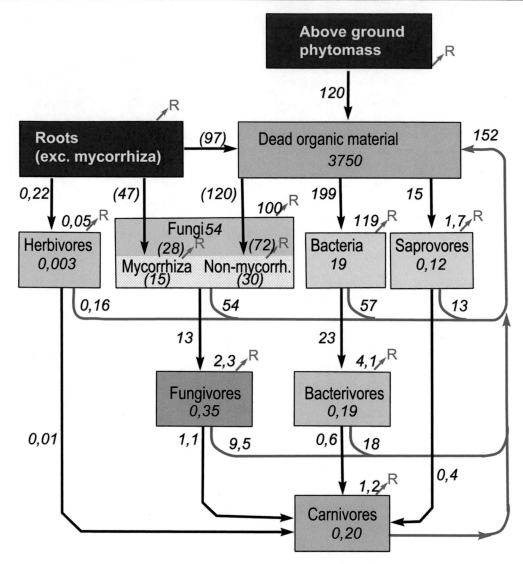

Fig. 12.8 Carbon stocks (in boxes in g C/m²) and carbon fluxes (arrows in g C·m⁻²·J⁻¹) in a pine forest in central Sweden. Mineral soil is included to a depth of 30 cm. Arrows with R: respiration losses. Blue lines indicate transport via dead organisms or via excreta. Values in brackets are uncertain (modified after Dierßen 1996)

A large proportion of the precipitation is lost through wetting of the crowns (interception), amounting to just over 50% (in the less dense E European stands it is only 30%). The moss and litter layer also retains further water, so that only about the 1/3 precipitation is available to the roots. There was 90 mm in the summer months, and 202 mm in the others, making a total of 292 mm. These are almost completely consumed by transpiration of the 40-year-old stand. In humid locations, as much as 378 mm are released to the atmosphere by transpiration; for this reason, part of the water losses must be covered by drawn from the groundwater.

Most ecophysiological studies have been carried out in the spruce stage of the Alps, but conditions in the boreal zone are likely to be at least partly analogous.

The active transpiration goes in parallel with a correspondingly intensive photosynthesis. In spruce one can distinguish between sun and shade needles. The conditions are reminiscent of those of the beech. In contrast to the beech, the active period of the evergreen spruce starts very early in spring and continues in autumn until the occurrence of occasional frosts. The seasons with low night temperatures and thus low respiration losses are particularly favourable for the net gain of dry matter. After a night of frost, however, photosynthesis is temporarily inhibited, but it is only after the onset of the actual cold period that the spruce falls into permanent dormancy and no longer assimilates even on sunny days.

At the same time, however, respiration drops to such low levels that it is hardly measurable and does not cause any

Fig. 12.9 Dry lichen pine forest in central Norway near Grimsdalen (350 mm annual precipitation). The herb layer consists almost exclusively of a dense carpet of shrub lichens *(Cetraria, Cladonia, Alectoria,* etc.) up to 25 cm high (photo: Breckle)

significant loss of substances. During this time, the needles lose their fresh green coloration, and the chloroplasts are difficult to see under the microscope.

After a long cold period, photosynthesis needs a certain start-up time in spring until it returns to normal. The photosynthetic apparatus must first be reactivated. In the case of mountain pines, it was found that young saplings overwinter under snow with green needles and then immediately start CO_2 assimilation in spring at higher temperatures.

The transition to winter dormancy is associated with hardening, that is, with a strong increase in frost resistance. The same processes as in defoliated deciduous trees can be observed in the evergreen conifers of the boreal zone. The hardening is still much more definite. While spruce needles in the unhardened state are killed by frosts as low as $-7\,^{\circ}$C in autumn, they can withstand temperatures of nearly $-40\,^{\circ}$C in winter without damage. Very sensitive to even light frost are the very young spruce shoots in spring. They can therefore be damaged by late frosts.

The frost hardiness of the needles can also be changed artificially, namely hardening by exposure to low temperature especially in late autumn and spring, dehardening by normal room temperature especially in December and late winter. Hardening causes that frost damage of conifers is not observed in their natural habitat, even in Siberia at temperatures below $-60\,^{\circ}$C. As a result of hibernation, conifers can withstand polar winters in complete darkness. Adaptability is species-specific, which is reflected in the

distribution of individual species. Only a few species can withstand the extremely continental Siberian winters (Fig. 12.13), the needle-throwing larch better than the evergreen species. Certainly, there are also differences within a species according to provenance. Spruces from the Alps will behave differently than those from the northern boreal zone, spruces from the upper tree line differently than those from lower altitudes. Even the shape of the tree is different; the more extreme the conditions, the more conical, pointed-crowned the trees become, i.e. the growth of the lateral branches is more inhibited than that of the main shoot. The same can be observed with the pine in the polar region.

It is difficult to say whether this tree form is due to the selection of mutants which suffer less from the danger of snow-breakage, for the same phenomenon is observed in fir in Albania at the lower, i.e. the dry limit, where there is no danger of snow-breakage; it seems rather that under generally unfavourable conditions the inhibition of the lateral branches occurs earlier than in the main shoot (under unfavourable light conditions it is the reverse). In Utah (N America), for example, *Picea, Abies,* and *Pseudotsuga* had extremely pointed crowns on dry slopes, but blunt crowns on the valley floor, with the same risk of snow breakage.

Another phenomenon must be mentioned here: the **waves of regeneration**. In coniferous forests in Japan, it has long been observed that broad strips of dead trees gradually move on and are replaced by a new wave of young growth of almost the same age. This so-called **Shimagare**

Fig. 12.10 Schematic transect through an *Abies balsamea* forest (Maine, NE USA) showing a regeneration wave, as is also known as the Shimagare phenomenon (▶ Fig. 12.11) from Japan (modified after Sprugel 1976, from Burrows 1990)

Fig. 12.11 *Abies veitchii forest* on the Shimagare slope (Japanese Alps) with a dieback wave. A regeneration wave starts from the left (photo: Breckle)

phenomenon (Figs. 12.10 and 12.11) only occurs in monospecific forests with a nearly equal-age structure of trees, which usually grow up very densely as a natural monoculture.

This natural forest dieback not only presupposes certain, albeit rare, events (storms, fire streaks) that contribute to age cohort synchronisation, but it is also an example of spatial self-organisation, whereby cohorts of trees of the same age die synchronously and young growth follows synchronously; from other regions this phenomenon also was described (Mueller-Dombois 1987; Iwasa et al. 1991; Jeltsch 1992).

12.6 The Siberian Taiga

Throughout the Eurasian taiga region, the pine *(Pinus sylvestris)* occurs, of which many forms are distinguished. However, it does not form a zonal vegetation, but only fills gaps, for example on burnt areas, on poor sandy soils and on boggy soils. Instead, *Picea obovata*, which is closely related to the European *P. abies,* often occurs dominantly. Together with the Swiss stone pine *(Pinus sibirica,* closely related to the alpine *P. cembra)* and with *Abies sibirica* it forms the

Fig. 12.12 Original pine-birch taiga between Irkutsk and Kultuk, with *Rhododendron dahuricum* and *Ledum palustre* in the low shrub layer (photo: U. Kull)

Dark taiga. However, pure stands of *Abies sibirica* also occur and are called **Black** or **Gloomy Taiga.** Conversely, in extremely continental E Siberia *Larix gmelinii* (= *L. dahurica*) occurs in pure stands, forming the **Light Taiga**. All these taiga types also contain *Larix sibirica* in the more N zone, which forms the polar tree line in W Siberia, and *L. gmelinii* takes its place in E Siberia. But also in the Siberian taiga, birch repeatedly appears in open places (Fig. 12.12). It not only forms pioneer transition stages on fire or storm areas, but also persists for a long time in undisturbed stands.

Fire is of the most important factors shaping the composition and development of boreal forest stands, it is the dominant stand-renewing disturbance through much of the Canadian taiga (Amiro et al. 2001). The average time within a fire regime to burn an area equivalent to the total area of an ecosystem is its fire rotation (Heinselman 1973) or fire cycle

(Van Wagner 1978). However, as Heinselman (1981) noted, each physiographic site tends to have its own return interval, so that some areas are skipped for long periods, while others might burn two-times or more often during a nominal fire rotation.

The dominant fire regime in the taiga is high-intensity crown fires or severe surface fires of very large size, often more than 100 km^2, up to 4000 km^2. Such fires kill entire stands. Fire rotations in the drier regions of W Canada and Alaska average 50–100 years, shorter than in the moister climates of E Canada, where they may average 200 years or more. Fire cycles also tend to be long near the tree line in the subarctic spruce-lichen woodlands. The longest cycles, possibly 300 years, probably occur in the W boreal taiga in floodplain white spruce.

Amiro et al. (2001) calculated the mean fire cycle for the period 1980 to 1999 in the Canadian taiga) at 126 years. Increased fire activity has been predicted for W Canada, but parts of E Canada may experience less fire in future because of greater precipitation in a warmer climate (Flannigan et al. 1998).

12.7   Extreme Continental Larch Forests of Eastern Siberia with Thermokarst Phenomena

The shady coniferous forests of W Siberia with *Picea obovata, Abies sibirica* and *Pinus sibirica* (Dark Taiga) differ substantially from the very light forests of needle-throwing *Larix dahurica* (Light Taiga) in E Siberia. This is a vast subzonobiome with an extremely continental boreal climate (absolute annual temperature variation up to 100 K), which can be seen from the climate diagrams on Fig. 12.13. In N America, a similar limited but somewhat less extreme climate area is present around Fort Yukon (Alaska).

Precipitation in this area is very low (less than 250 mm); however, this is compensated for by the slowly thawing upper layer of permafrost soil. The roots absorb the meltwater so that a forest can grow. The larch forests usually have a dwarf shrub understory of *Vaccinium uliginosum, Arctous alpina,* on dry soils *Vaccinium vitis-idaea, Dryas crenulata,* on moist soils *Ledum palustre* and on very dry only a soil layer of lichens. Further north, the sparse forests change into open tree meadows (Redkolesje) and then into a dwarf shrub tundra (ZE VIII/IX) with *Betula exilis (*knee-high) and *Rhododendron parviflorum.*

The extent of thermokarst phenomena in the coldest part of the northern hemisphere is particularly impressive here.

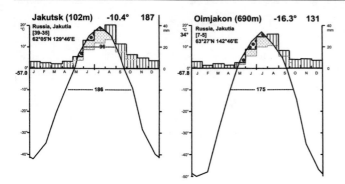

Fig. 12.13 Climate diagrams from the extremely cold continental region of E Siberia. Oimjakon is the cold pole of the N hemisphere

Burkhard Frenzel reported on thermokarst (oral communication):

> The permafrost (Fig. 12.14) of Siberia, probably also of Alaska, developed since the early glacial periods. Each ice age contributed to its expansion, whereas in the warm periods its area and thickness were reduced. However, warm-period climates are also favourable to the formation of new permafrost in these landscapes, even if the thickness of the seasonal thaw is greater than during cold or glacial periods: decay and formation of permafrost go hand in hand. These appearance and dissolution processes become ecologically and geomorphologically effective especially on fine-grained sedimentary rocks.

During the ice ages, loesses (an aeolian sediment) and their derivatives were formed over large areas of what was then the extremely cold winter climate zone. In today's highly continental climates of the boreal coniferous zone, they are filled with permafrost ice up to 80 percent by volume. Local disturbances of the radiation balance and of the heat flow between atmosphere and soil, for instance by the natural forest fires occurring per area approximately every 180 to 240 years, by river erosion, etc., initially lead to an increase in thickness of the summer thaw soil. Since the rock had previously been far oversaturated with ice, this layer now slides away on sloping land surfaces. The soil is literally eaten up, which is why in Siberia **we** speak of the '**Jedom**' series ('**Yedoma**'), i.e. the (Russian) 'eating away' loose rocks. Thus, due to increased summer thawing of the topsoil, a decrease in volume has occurred. It is commonly referred to

Fig. 12.14 The permafrost cover at Herschel Island, in the western Canadian Arctic in Yucon Territory. In summer, only a few cm of the upper layer of this thick permafrost cover thaws. The vegetation uses the meltwater for its growth (photo: Boris Radosavljevic, Alfred Wegener Institute for Polar Research: https://t1p.de/o03l)

as **Thermokarst** (Fig. 12.15). On horizontal surfaces, when thermokarst develops, the soil collapses: Drainless depressions up to several kilometres in size are formed, the so-called 'Alases', in which the rise of groundwater drowns the forest that may have existed previously.

These phenomena have already been described by the travellers who roamed Siberia in the mid-seventeenth century, a clear indication that Alas formation is a natural process, the beginning of which can now be traced back to the end of the Late Glacial (about 12,000–10,000 years before present). At present, Alas formation is promoted by clearing and construction activity.

Alases occur particularly frequently in the Viljuij depression, which is subject to a highly continental climate, and in its peripheral areas of C and E Yakutia. If the Alas are first as a rule in their centre water-filled, then the steep, up to 50 m high edges offer due to improved natural drainage and intensified irradiation very multi-coloured steppe societies suitable starting points. About 1/3 of the approximately 900 kinds of higher plants of Jakutia belong to such plant communities, whose total area amounts to only few per cent of the country (Troeva et al. 2010).

If the lowering of the soils of the Alas proceeds slowly and does not accumulate too much water, then natural meadow communities, which are important today for the local cattle breeding, take the place of the destroyed larch or pine forests. Their species composition often refers to salinated sites.

Alas lakes can silt up. Hereby again the heat flow is changed and the permafrost spreads. However, as there is now a great deal of water in the Alases, large, ice-core-filled hills form in them, the 'Bulgunnjachi' or 'Pingos', which can also occur in the tundra (ZB IX) (Fig. 12.16). They grow in height until summer. Insolation prevents further growth of their ice cores, or until they rupture as a result of their high growth and summer heat can act deep into the Bulgunnjachi, leading to their disintegration. Such mounds have a life span of a few decades to a few millennia at most. However, it is always true that they also contribute to the increase of biotope diversity, since their steep slopes often provide abundant starting points for colourful steppe communities due to good drainage and increased insolation.

The occurrence of steppe communities in Yakutia is characteristic for all dry and in summer very warm, steep southern slopes, for example on the flanks sloping down to the large rivers.

In general, the climate of Yakutia can be described as at least semi-arid. This is clearly evident from the climate diagram on Fig. 12.12, i.e. the potential evaporation is higher than the low annual precipitation even on euclimatopes.

In accordance with this, in the larch forests one finds treeless places—the "**Tscharany**" (Charani)—on which some salt accumulation takes place due to the strong evaporation. On such brackish, solonised soils grow salt plants, which also occur on sea coasts, such as *Atriplex litoralis,*

Fig. 12.15 Thermokarst-induced destruction of coniferous forest in the Canadian taiga. The top 5–8 m of the soil supporting the forest thaws, and the waterlogged mud transports the fallen trees downslope (photo: http://bit.do/bFRC6)

Fig. 12.16 Bulgunnjachs or Pingos in a water-filled Alas in the tundra near Tuktoyaktut, NW Territories, Canada. The slopes of the Pingo, already split again and thus decaying, are occupied by grassy steppe vegetation (photo: Emma Pike: http://bit.do/bFT5R)

Spergularia marina and *Salicornia europaea,* on wet salt soils also the grasses *Puccinellia tenuiflora* and *Hordeum brevisubulatum* (Walter 1974).

Since in the far north the steep S slopes are struck perpendicularly by the rays of the low sun at noon, individual steppe species can even still grow on Wrangel Island (71° N) (Yurtsev 1981). The following typical species are cited: *Ephedra monostachya, Stipa krylovii, Koeleria cristata, Festuca* spp. and other grasses, *Pulsatilla* spp., *Potentilla* spp., *Astragalus* and *Oxytropis* spp., *Linum perenne, Veronica incana, Galium verum, Artemisia frigida, Leontopodium campestre, Aster alpinus* (typical for Siberian steppes) and others.

These steppe islands occur today extrazonally on warm S slopes. They are relics of zonal steppes of glacial times, when the climate was more continental. At that time, foehn-like, strongly warming downdrafts came off from the huge icecap in summer, deflected to the east and blew over the ice-free periglacial surfaces, depositing the thick loess layers. The summers were obviously so hot that large dry cracks formed in the loess, in which the permafrost caused ripening and filling with ice (Jedome series). On the basis of recent Russian investigations, it must be assumed that during the glacial period such periglacial steppes extended zonally over the whole of Eurasia and North America and made possible a rich steppe fauna with steppe rodents, antelopes, wild horses up to the woolly rhinoceros and mammoth. And it was only

with the appearance of man that the large mammals disappeared.

The tundra vegetation was probably only bound to boggy and swampy places around lakes, i.e. as pedobiomes to the lower parts of the relief. The frequent occurrence of *Ephedra* and *Artemisia* pollen in the pollen spectra of peat samples from the glacial period proves that steppes with these species grew all around, cold steppes, as they are found today rather restricted in a few places, especially in Yakutia. It could be called a very arid sZB VIIIa, which was widespread during glacial period but very restricted today.

Box 12.3 The Dynamics of Permafrost Regions
The permafrost area, where all life seems to be so reduced because of the winter cold, is full of dynamics.

Only when the ice melted in the postglacial period, the sea level rose, the Arctic Ocean was formed, the land connection between East Asia and Alaska was interrupted, and the Gulf Stream in the North Atlantic brought warm water to the Arctic Ocean, did the entire atmospheric circulation change. The air fronts moving westward from the Aleutian Islands on the one hand and from Iceland on the other reshaped the climate of the northern latitudes; it became humid and strongly oceanic in tint on the western flanks of the land masses. In the N part of the former periglacial steppes, tundra vegetation now spread and conquered the areas that had become ice-free, followed by forest vegetation, starting from the refugia, until the tundra and forest zones assumed their present position.

The periglacial steppe vegetation retreated into the arid area of today's continental steppes, and with it the corresponding fauna. But the animal species that could not keep up with the changes, parallel to the appearance of man, became extinct. These were just the largest forms such as mammoth, woolly *Rhinoceros*, giant deer *(Megaloceros)* and others (cf. Walter & Breckle 1991).

These remarks are intended to suggest that today's zonal tundra vegetation, but also the boreal coniferous forest zone with the many bogs in their present form, are in geological terms rather recent new formations. Today's raised bogs probably did not exist in the past either. Certain relics of the periglacial steppes can be found on chalk cliffs in central Russia. *Carex humilis* with its scattered occurrence in the steppe heaths is also considered a periglacial relict. In the alpine mats grow many species that genetically belong to typical steppe genera, such as *Astragalus, Oxytropis, Potentilla, Pulsatilla, Festuca, Avena* s.l., especially also *Artemisia,* the edelweiss *(Leontopodium)* and *Aster alpinus.*

12.8   Orobiome VIII: Mountain Tundra

The elevational sequence is very short in these N latitudes of ZB VIII. Already at low altitude the forest line is reached, formed by *Picea, Pinus sibirica* or *Larix,* depending on the geographical position. Above this in the alpine belt, however, one does not find a typical tundra, but a mountain tundra (Stanjukovitsch 1973).

In the Alps, the first snow falls on unfrozen ground and the temperature on the ground remains at about 0 °C throughout the winter under a thick blanket of snow. The perennial herbs are therefore not exposed to deep frosts or frost-drought and the vegetation consists of dense alpine mats.

It is different in the **mountain tundra**: the snow falls on ground that is already frozen, the snow cover is thin and blown off from the summits. There is permafrost, which is almost non-existent in the Alps. Winter storms are very strong, frost weathering is very intense; the debris moves slowly downwards (solifluction), and the fine soil is blown out. All this results in the mountain peaks in the mountain tundra being bare and called 'Golzy' (Russian golyj = bare). They are covered only by lichens and a few mosses, as well as isolated dwarf shrubs between the rocks. The conditions are reminiscent of the wind-swept ridges of the Alps with *Loiseleuria* and the same lichens.

Conditions are somewhat more favourable in the subalpine or 'Podgolez' belt, where the drifted snow can accumulate. The mountain tundra is found in the continental climatic area as far S as 50° N, even still in the Altai.

In the oceanic area of the boreal zone (Scandinavia, Kamchatka) mountain tundra is absent, and the alpine belt is somewhat more reminiscent of conditions in the Alps. Winters here are very snowy. The timberline is formed by birches *(Betula ermanii, B. tortuosa).*

12.9   Mire Types of the Boreal Zone (Peinohelobiome)

The climate of the boreal zone is largely humid, i.e. precipitation exceeds potential evaporation due to lack of energy for evaporation, so the water balance is positive despite low precipitation. When the runoff of excess water to rivers is impeded, the water table rises and bogging occurs. Since the soils in the boreal zone are poor and acidic (podsols), the groundwater also has an acidic reaction and contains few mineral components. Mostly the groundwater is brown coloured by humus brine. Only in the case of limestone the conditions are different. Since large areas of the boreal zone in both Eurosiberia and N America are very flat, the groundwater table is high. As long as it remains more than 50 cm below the ground surface for most of the year, tree growth is possible, otherwise it is inhibited and the forests turn into bogs.

> **Box 12.4 Boreal Peatlands**
> Extensive areas in the boreal zone are not occupied by the zonal vegetation of coniferous forests, but by peatlands.

In sub-regions of Finland, peatlands account for more than 40%, and in some cases even more than 60%, of the total area. The same applies to the boreal zone of E Europe and especially W Siberia, which is entirely covered by peatlands except for the parts near rivers. Similar conditions are found to some extent in Kamchatka (Fig. 12.17), Alaska, and Labrador and in the areas south of Hudson Bay. It is therefore necessary to treat the pedobiome of bogs following the coniferous forests. Often the boundary between coniferous forest and bog is difficult to draw. The already mentioned spruce forests with *Polytrichum* and *Sphagnum* already show strong peat formation.

In the geological sense, peat bogs (mire, "Moor" in German) are deposits of peat with a layer thickness of at least 20 to 30 cm. If the peat layer is thinner or the content of combustible substance is only 15 to 30%, we speak of anmoor bogs. In the ecological sense, bogs are communities of life that are bound to high groundwater, regardless of the thickness of the peat layer on which they grow. Because of

Fig. 12.17 Bog lakes in W Kamchatka with species-rich bog banks, partly with *Myrica tomentosa, Rubus chamaemorus* between low *Salix fuscescens* trellis, but mainly Cyperaceae such as *Carex rotundata,* *C. middendorfii,* in addition *Comarum palustre, Drosera rotundifolia, Pedicularis labradorica, Hammarbya paludosa, Platanthera tipuloides* and *Sphagnum,* in the water floating *Sparganium* (photo: Breckle)

the poor aeration of the soil, the bog plants root very shallowly, so that only the condition of the uppermost peat layers is important for them. The following types of peatland can be distinguished:

1. **Topogenous bogs**, which are bound to a very high water table and therefore occupy the lowest parts of the relief or occur where spring water escapes. This subheading includes fens of various kinds.
2. **Ombrogenous bogs**, which are wetted exclusively by rainwater falling on the surface and which rise above the surrounding area. These are **raised bogs** (Fig. 12.18).
3. **Soligenous bogs**, which are also wetted by precipitation, but do not rise above the surrounding area and are additionally overflowed by water that runs off the slopes when the snow melts.

The groundwater of topogenous peatlands may contain many mineral substances and can be rich in nutrients. Such bogs are therefore eutrophic or minerotrophic. Rainwater, on the other hand, is very pure and nutrient-poor; therefore, ombrogenic peatlands are oligotrophic or ombrotrophic. The trickle water that the soligenic peatlands receive is, if it is not only meltwater, again more nutrient-rich; these peatlands are therefore mostly minerotrophic, otherwise moderate oligotrophic.

In the boreal zone the groundwater is low in mineral salts, so that it is difficult to distinguish between fens and raised bogs; one often speaks of mesotrophic transition bogs. If the water contains less than 1 mg Ca/litre, then one already finds the less demanding species of oligotrophic bogs.

> **Box 12.5 The Bog Types**
> Topogenous, ombrogenous and soligenous bogs can be distinguished according to the origin and composition of the water in the bog soil.

> **Box 12.6 Climatic Types of Oligotrophic Peatlands**
> The nutrient-poor, i.e. oligotrophic bogs, which are found only in cool to cold humid climates, are peinohelobiomes. According to their structure and topography, several types linked to specific climatic conditions can be distinguished (Fig. 12.19): Blanket bogs—raised bogs—Aapa mires—palsa bogs.

The eutrophic bogs, in which sedges (*Carex* species) play the main role, occur in the temperate zone, regardless of climate, wherever the soil is wetted by calcareous but not

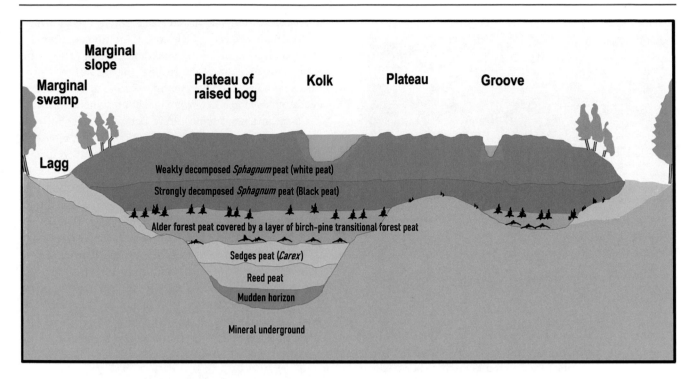

Fig. 12.18 Schematic representation of a raised bog and the formation of the stratification

Fig. 12.19 Bog lakes and taiga alternate, C Sweden near Sunnersta (photo: Breckle)

brackish groundwater. They all belong to the pedobiomes, namely the helobiomes.

1. **Blanket bogs**. We had already mentioned these in the extreme oceanic climate of the Atlantic heath region of the British Isles and all along the west coast of Scandinavia. They cover the whole terrain.

2. **Raised bogs**. They are characteristic of the somewhat less oceanic NW corner of C Europe with heath areas, the whole boreo-nemoral zone and the S part of the boreal

Fig. 12.20 Forest raised bog in the Saamaa region in Estonia (photo: Breckle)

zone (Figs. 12.18 and 12.19). When typically formed, they are treeless. However, if the climate becomes more continental and drier, pines are also found on these bogs, which are then referred to as forest raised bogs (Fig. 12.20). They run along the entire S border of the boreal upland bog area (Fig. 12.21).

3. **Aapa mires or string bogs**. They are found N of the high bog zone in Fennoscandia and in W Siberia. They are soligenous bogs with shallow slopes. They consist of somewhat elevated ridges perpendicular to the slope and ombrotrophic; between them are elongated deepened areas filled with minerotrophic water (Finnish "rimpis", Swedish "flarke"). The whole bog slopes down in steps,

reminiscent of terraces in rice cultivation. The thrusting effect of the ice sheet, which covers the rimpis in winter and expands them in a horizontal direction, plays a role in the upward curvature of the strings (Fig. 12.22).

4. **Palsa** peat **bogs of** the peat hummock tundra. These occur already outside the boreal zone in the forest tundra in areas with a mean annual temperature below -1 °C. Soil ice plays a significant role in the formation of peat mounds, which can be 20 to 35 m long and 10 to 15 m wide and reach a height of 2 to 3 m (up to 7 m). If less snow is deposited on slightly elevated areas, frost penetrates the peat soil more rapidly. Ice layers form, and these attract water from the unfrozen surrounding peat soil. The ice lenses become thicker and lift the peat up. Since not all the ice melts in summer, some of the elevation remains. As a result, the next year the snow cover is even less, the ground freezes even more rapidly; the ice masses grow larger year by year, and the peat mound with the ice core grows higher. In summer the whole thing sinks in, creating a ditch-like depression around the peat mound, filled with water, in which the dwarf birch (*Betula nana*) and cotton grasses (*Eriophorum*) grow (Fig. 12.23). The top of the peat mounds (Palsen) may dry out in summer, in which case it develops cracks. It is then eroded by the wind and may be worn away completely. Most of the Palsen are thought to be subfossil formations from a colder climatic period and are in the process of disintegration. They may be considered as thermokarst phenomena of smaller extent (Chap. 11).

5. **Polygonal bogs:** These are typical of the Arctic (ZB IX). They are discussed further below.

Fig. 12.21 Distribution area of mire types in N Europe (modified after Walter 1990). **1** Palsen bogs; **2** Aapa bogs; **3** typical raised bogs; **4** blanket bogs; **5** forest raised bogs; **6** mountain bogs. Light areas of the S regions with predominantly topogenous bogs

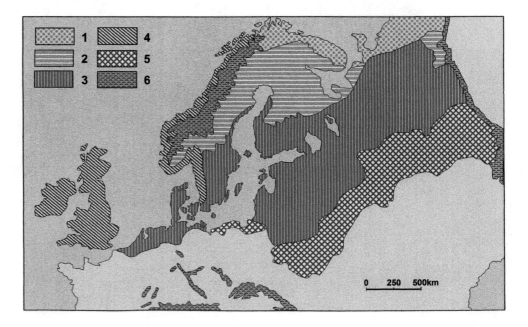

Fig. 12.22 Endless expanse of the West Siberian moors: Noyabrsk Moor with a very large proportion of open water areas (photo: M. Succow)

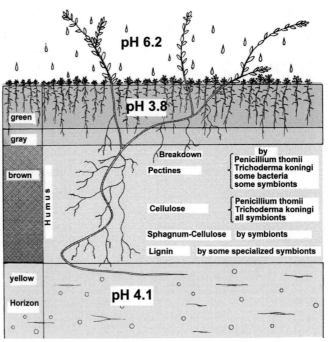

Fig. 12.24 Peat profile of a typical raised bog with indication of biological processes at different depths (modified after Burgeff 1961). The circles in the yellow layer indicate gas bubbles (mainly methane)

mosses settle only on poor acidic soils; podsol soils are very suitable for this purpose. Raised bogs therefore often emerge from waterlogged coniferous forests.

> **Box 12.7 Dependence of Peatlands on Climate**
> The true raised bogs are bound to an oceanic climate in both Eurasia and N America, whereas Aapa and Palsen bogs have a circumpolar distribution.

Fig. 12.23 Palsen or peat hummock bogs near Abisko in Sweden (photo: Breckle)

12.9.1 Ecology of Raised Bogs

The most important plants that cause the development of a raised bog are the *Sphagnum* mosses (*Sphagnum* species). Since they consist for the most part of large dead cells which easily fill with water by capillary action, they act like sponges in the cushion-like growth and hold many times their dry weight in water. At the upper end, they grow upward; at the lower end, they die and perish and are converted into peat (Fig. 12.24). The cushions grow larger and larger, merging together, and eventually a watch-glass-shaped raised bog arches over the surface (Fig. 12.18). Because *Sphagnum* mosses do not tolerate drying out, evenly moist and cool summers are a prerequisite for raised bog formation. Peat

Figure 12.25 illustrates the structure of a typical raised bog. In a large growing raised bog, a distinction is made between the very wet and little arched upland surface, the better drained and relatively steeply sloping marginal slope, and a minerotrophic bog surrounding the raised bog, called a **Lagg.** The plateau is not completely flat, but consists of small elevations, the **Bults,** which protrude above the moss surface, and of **depressions ("schlenken")** sunk into the moss carpet, in which the water stands close to the surface; in them grow hygrophilous *Sphagnum* mosses as well as *Carex limosa* or *Scheuchzeria.* When several depressions join, bog pools, called **"blänken"** or "kolks", **are** formed (Fig. 12.19). Their depth is usually only 1.5 to 2 m; however, they have no solid bottom but are filled with soft detritus. Excess water drains from the upland surface in small channels called gullies ("rüllen"). The types of raised bogs are shown in Fig. 12.26.

Fig. 12.25 A typical raised bog
(Augstumalmoor in the
Memeldelta, after Weber,
modified from Walter 1968) with
Blänken (raised bog oaks),
gullies, and a Lagg (mesotrophic
intermediate bog merging into a
eutrophic low bog, fen)

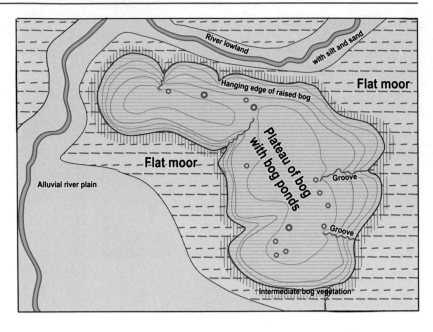

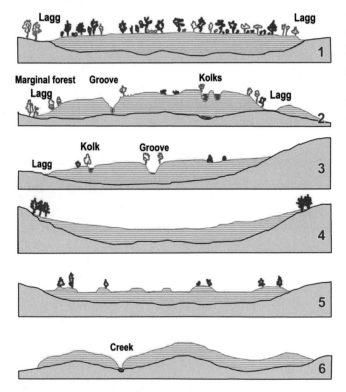

Fig. 12.26 Schematic representation of the different oligotrophic bogs
(after Osvald, modified from Overbeck 1975): **1** = forest raised bog,
2 = typical raised bog, **3** = plan raised bog, **4** = soligenous bog,
5 = Aapamoor (string bog, a northern type), **6** blanket bog. From 1 to
6 increasing humidity of climate or waterlogging

The number of flowering plants growing on raised bogs is
not large, they are extremely undemanding species in terms
of nutrients: *Eriophorum vaginatum, Trichophorum
caespitosum* and the dwarf shrubs *Andromeda polifolia,*

*Vaccinium oxycoccus, V. vitis-idaea, V. uliginosum, Calluna
vulgaris* and *Empetrum;* in the Atlantic area *Narthecium,* in
the E *Ledum palustre* and *Chamaedaphne calyculata, in* the
N *Rubus chamaemorus, Betula nana* and *Scheuchzeria
palustris* are added.

In addition to nutrient poverty, the second ecological
factor that determines the distribution of the species is the
overgrowth by *Sphagnum* mosses. The substrate on which
the flowering plants germinate is the growing living tips of
the *Sphagnum* mosses. Depending on the water supply, the
height growth of the *Sphagnum* mosses is 3.5–10 cm per
year. By this amount, the flowering plants must raise their
shoot base each year by stretching the rhizomes or forming
adventitious roots, otherwise they will be overgrown by the
Sphagnum mosses (Fig. 12.27). They can escape overgrowth
all the more easily the slower the mosses grow, which is the
case on the relatively dry hummocks or on the well-drained
marginal slopes. It is in these places that most dwarf shrubs
are found. Each hummock shows a certain zoning:
Eriophorum vaginatum and *Andromeda* grow at the base,
Vaccinium oxycoccus higher up, other dwarf shrubs at the
very top. Often the top of the hummock is so dry that other
mosses *(Polytrichum strictum, Entodon schreberi)* or even
lichens *(Cladonia species, Cetraria)* thrive instead of
Sphagnum.

The trees (pine, spruce) that have the most difficulty are
those that are unable to relocate their trunk base and show
only slight height growth on the poor substrate (Fig. 12.28).
Often, only the uppermost branch tips protrude from the bog
hummocks. Forest raised bogs are therefore only found
where, as a result of the dry climate, the growth of peat
mosses is low. As soon as the bogs are drained and the
growth of the *Sphagnum* mosses comes to a standstill, a

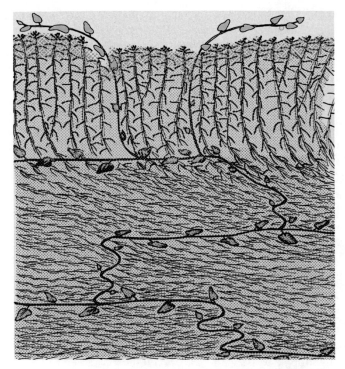

Fig. 12.27 Growth of *Vaccinium oxycoccus* in a *Sphagnum* bog. In spring, the young shoots first grow vertically out of the *Sphagnum* layer that overgrew them the previous year and then lay down on the moss surface. The older shoots are pressed together kinkily by the compaction of the peat (after Grosse-Brauckmann, modified from Walter 1968)

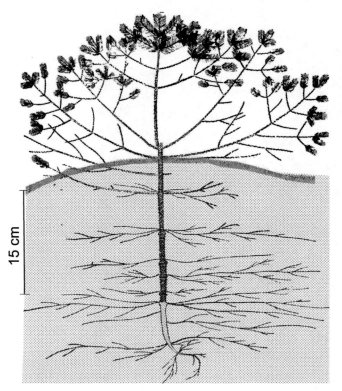

Fig. 12.28 An approximately 9-year-old spruce whose trunk is encased 20 cm deep by the peaty moss layer (after Bertsch, modified from Walter 1927)

rapid heathland formation occurs, i.e. the dwarf shrubs come to dominate. Soon, tree species such as birch, pine or spruce are added. Most peatlands in Central Europe are in this state, and this effect is intensified by the additional nitrogen input from the atmosphere.

It is striking that along the gullies or at the edge of the funnels often species of the minerotrophic soil grow, although the water is just as nutrient-poor as in the rest of the bog. It can be seen that flowing water or water moved by waves provides a more favourable supply of nutrients than still water, in which only diffusion of nutrients takes place. With the high water content of bog soils, they warm up very slowly. Bogs are therefore cold sites, and it is understandable that Nordic-Arctic floral elements, including many relics of the glacial period, can persist on them; in addition, on raised bogs they are protected from competition from fast-growing demanding species.

With the exception of the *Drosera species* (or *Utricularia* in the bog ponds), which supplement their nitrogen supply by digesting insects caught on the leaves, all other species are xeromorphic, although water is available to them in abundance. This is attributed to a lack of nitrogen supply. It has generally been shown that **xeromorphy** is observed when the growth of plants is inhibited, for example by a lack of water, but also when there is an excess of water, i.e. a lack of oxygen in the soil, by low soil temperatures which make the uptake of nitrogen difficult, or by a direct lack of nitrogen. The seemingly xeromorphoses here are therefore deficiency symptoms; it is therefore more appropriate to speak of **peinomorphoses.**

A summary account of the peatlands in NW Europe was given by Overbeck (1975), a new overview by Succow & Joosten (2001).

> **Box 12.8 Western Siberia: The Largest Bog on Earth**
> 40% of the peat deposits of the whole earth are located in the west Siberian lowlands. The peatlands, with over 100,000 peat lakes, store an amount of water said to be equivalent to the two-year runoff of the vast Ob-Irtysh River system.

12.9.2 The West Siberian Lowlands, the Largest Marshland on Earth

This area with the Ob-Irtysh basin is a peinohelobiome of a scale difficult to imagine (Tanneberger et al. 2003). It extends from the forest tundra in the N to the steppe in the S over 800 km and from the Urals in the W to the Yenisey in the E over about 1800 km.

In this moorland area the settlements are located only along the rivers that serve the traffic. The actual marshland was explored in detail only by Popov (1971–1975).

The causes of the peatlands can be called the topography, climate and hydrological conditions. The bogs have replaced the dark taiga.

The great basin is underlain by Meso-Neozoic strata. The Pleistocene ice ages had little effect. Deposition of alluvial sediments, some of which were impermeable to water, occurred, promoting waterlogging. Peat mosses (*Sphagnum* species) easily settle on the wetted nutrient-poor podsol soils, which initiate peat formation. The climate is humid with annual precipitation of 500 mm, as evaporation is only 240–300 mm and runoff 127–270 mm. In terms of temperature, the climate is very continental; the frost-free period is 174 days, yet the daily averages are above 10 °C on 100 days. Summers are therefore relatively warm and, as a result, plant production and peat growth are considerable. Even short hot droughts occur, so that forest fires are not necessarily excluded. The burnt areas easily become boggy.

However, the hydrological conditions are of particular importance. The little incised rivers meander strongly, which inhibits the discharge. The spring flood on the upper reaches of the Ob and Irtysh begins 1.5 months earlier than the snowmelt on the lower reaches, i.e. when the rivers are still covered by ice in the N. When the ice melts, high ice dams are formed; upstream of these, the water is additionally dammed. Since the sources of the Ob are fed by the glaciers

of the Altai Mountains, the summer flood follows immediately, i.e. the high water level of the rivers (12 m above low water) lasts practically the whole Siberian summer. The low watersheds are also flooded and a single large body of water is formed together with the bog lakes.

The peatland formation already began in the subarctic period of the postglacial period. The starting point was formed by wide shallow depressions with water low in mineral salts. In them, *Scheuchzeria* bogs with *Eriophorum vaginatum* and various *Sphagnum* species developed. Corresponding mesotrophic *Scheuchzeria* peats are found at the base of the oldest 4–7 m deep peat profiles. The oligotrophic phase is indicated by the occurrence of the main peat moss species, *Sphagnum fuscum*. It begins in the middle postglacial. The peatlands bulged upwards and the water table was raised. This led to the wetting of the adjacent forests; *Sphagnum* species colonised under the dying trees, and the bogs spread rapidly in a horizontal direction. All the younger bog profiles, and they are the majority, have a peat thickness of 3–4 m and always have peats with a lot of pine and bark remains in the lowest horizon; just after that the oligotrophic phase with *Sphagnum fuscum* peat begins.

The bogs of western Siberia mostly are string bogs (Fig. 12.29, Fig. 12.20) with an average inclination of only 0.0008 to 0.004°. On the more or less wide strings grow

Fig. 12.29 Partially forested parts of the Wasjugan Marshes in W Siberia (photo: Succow)

Pinus sylvestris (in the stunted form *P. willkommii*) and *Ledum palustre* as well as the dwarf shrubs *Chamaedaphne calyculata, Andromeda polifolia, Oxycoccus microcarpus,* scattered *Rubus chamaemorus* as well as *Drosera rotundifolia.* The moss layer consists of *Sphagnum fuscum;* patches with lichens *(Cladonia* spp., *Cetraria)* are rare.

In the hollows (Schlenken) one finds *Eriophorum vaginatum* with *Sphagnum balticum* or *Scheuchzeria,* respectively *Carex limosa* with *Sphagnum majus,* but also *Rhynchospora alba* with *Sphagnum cuspidatum* occur.

The string bogs usually undergo regression on the wetted watersheds, leading to the formation of bog lakes (Fig. 12.22). This occurs particularly wherever recent tectonic movements are associated with subsidence. Recent aero-geological surveys showed subsidence of 0.07 to 0.25 mm per year in several areas. This is sufficient to disturb the very unstable equilibrium between strings and sinks and lead to increasing waterlogging. This excess water initiates the regression phenomena. It leads to oxygen depletion even in the upper peat layers and to the formation of methane gas.

> **Box 12.9 The Peatlands of Western Siberia**
> Thus, the rivers of Western Siberia do not drain the lowlands, but on the contrary, their waters become dammed, overflow them with water and promote bog formation

When drilling in such places, the escaping methane gas causes fountains of liquid peat. When the gases escape naturally, the plant cover dies. Dead bog areas form, which become bog lakes. The lakes, which are small at first, merge into larger ones, where wave action causes the peat banks to collapse, resulting in the formation of ever larger bodies of water. The bog lakes of various sizes, together with the bog depressions, all form a single hydrological system—an ecological unit that, because it is low in nutrients (ash content of peat only 2–4%) and wet, we call the peinohydrobiome (Fig. 12.20, Walter 1977).

In places, the string bogs can also dry out if such a wetted area forms its own drainage system and the bog streams cut into the peat so that the banks are better drained. On such banks a narrow strip of forest can develop with pine, birch and the stone pine *(Pinus cembra* ssp. *sibirica).*

The description of peatlands given here applies to the taiga zone. The peat thickness decreases towards the N because of the shortening of the vegetation period and the lower plant production. South of the taiga the types of peatlands change.

In the area of forest steppes, i.e. in Siberia in the ZE VII-VIII with birch-aspen forests, the Ca content of the groundwater is already high, and slightly domed eutrophic hypnaceous peatlands with *Carex* species (sedges) prevail. However, oligotrophic peatlands can also form on these forest peat islands ("Ryami"). Peat growth is almost inhibited by the greater aridity of the climate. In the S part there are only lowland peatlands in the lowest parts of the relief, especially in the wide river valleys. The ash content of the peat there can be very high (19%). Typical lowland bogs are often found, the hummocks consisting of old clumps of *Carex caespitosa* and *C. omskiana.* Such fens are helobiomes.

Even further S in the N steppe zone, the climate is semi-arid. In the Baraba lowlands no complete river system is formed, but like in the Pampas there are countless small drainless lakes, some of which are brackish. Around these one finds eutrophic mires or even halophytic swamps with salt plants, thus already transitions to a halo-helobiome (ZE VIII/VII).

12.10 Man in the Taiga

The vast expanse of the taiga has not prevented it from being cut up in many places during this century, and large areas have even been destroyed, these days ever increasing. This is caused by large-scale oil and gas prospecting, pipelines and ore extraction and the land-intensive infrastructure required for this (Fig. 12.30). Exploitation has reached huge proportions; biodiversity has not been irreversibly destroyed, as in some tropical regions.

In past centuries, the Siberian taiga was a difficult area to access. Fur hunters and individual settlers roamed the area, which in many places, however, was also populated by original tribes on a small scale at wide intervals.

This original use of the taiga as a habitat for wandering nomads and widely scattered individual settlements has been sustainable in every respect. On the other hand, it must be noted that many large mammals survived the entire ice ages and only became extinct in the Late Glacial. There is much evidence to support that this was already an effect of the destructive influence of man. Today it is oil, gas, mining, iron, gold, diamonds, hunting, logging, timber!

Fig. 12.30 The Mirny diamond mine in Sakha (Yakutia) with a diameter of 1200 m and a depth of 500 m as well as the associated infrastructure buildings with extensive overburden. Large-scale destruction of the taiga can be seen here (photo: Breckle)

12.11 Zonoecotone VIII/IX (Forest Tundra) and the Polar Forest and Tree Line

Similarly as the forest steppe inserts itself between forest and steppe as ZE VI/VII, we have also between the boreal forest zone and the treeless tundra as ZE VIII/IX the forest tundra, in which forest and tundra are macromosaic-like interlocked. At first, individual treeless patches occur in the forest area, mostly on elevations. They increase to the N until only individual islands of the forest remain which finally consist only of bushy cripples. In the mountains this stunted zone is quite narrow, but here in the flat terrain it can extend for hundreds of kilometres. The tree species in the oceanic area are birches (Fig. 12.31), in the extremely continental area larches (Fig. 12.32), otherwise spruces.

As causes for the occurrence of the polar timberline we can assume the same as for the alpine timberline. The frost-drought are increased by the winter storms. The forest advances furthest on the slopes of river valleys, where it is protected from wind and snow, where the well-drained soils thaw more deeply in summer and the rivers coming from the S carry warmer water. However, lack of regeneration is also cited as a cause. At the N limit of distribution, trees rarely produce germinable seeds. In addition, most are eaten by animals. Storms can transport them (sliding on the snow surface) far N, where development is no longer possible.

> **Box 12.10 The Destruction of the Taiga**
> The encroachment and extent of destruction in the Siberian taiga sometimes exceeds that in the tropical rainforests.

Also, dense lichen and moss cover is present in the forest tundra, which is an unfavourable germination bed. The importance of humans and their reindeer herds is very great

Fig. 12.31 The arctic-polar tree line with *Betula tortuosa* above Torneträsk (oceanic expression) in northern Sweden (photo: Breckle)

Fig. 12.32 The Arctic-Polar tree line with *Larix gmelinii* (extreme continental expression) in northeastern Siberia, wedging light forest in the Moma valley of the Chersky Range (photo: O. Agachanjanz)

Fig. 12.33 Cross-section through the trunk (just under 10 cm ⌀) of a (c.104-year-old) *Larix gmelinii* (= *L. dahurica*) from Arymas (72° 30′ N) with very narrow annual growth rings (photo: Agachanjanz/Breckle)

(Fig. 12.4). In addition to damage by animals, timber logging is namely important, because the natural increment of woody plants is extremely low. In most cases, a tree seedling only succeeds in gaining a foothold if particularly favourable temperature conditions prevail for two years in succession. Even then, further growth is extremely slow. 20- to 25-year-old saplings hardly protrude from the herb layer; the annual height increase is 1 to 2 cm. Tree thickness growth shows a very close correlation to July temperatures. The northernmost true forests, a taiga of trees 2 to 5 m high, are now found on the Taimyr Peninsula, at Arymas, beside the Chatanga estuary at 72° 30′ N, with *Larix gmelinii*. Growth is very slow, for example, a 104-year-old trunk has a diameter of 9.5 cm (Fig. 12.33) (Walter & Breckle 1990).

The open areas in the forest tundra are mostly occupied by the dwarf tundra. This also forms the S subzone of the true tundra (Fig. 11.49).

The timberline was significantly further N during the postglacial warm period. The tree stumps trapped in the peat in today's tundra serve as evidence. The consequences of the Climate change that has been constantly taking place over the last millennia are particularly evident in the forest tundra.

Photo 50   Beavers are constructing huge dams predominantly from *Populus tremuloides* stems and branches at water creeks, thus changing landscape creating ponds between the Canadian Taiga at Prince Albert Park (photo: Breckle)

Photo 51 Canadian taiga, dominated by *Picea mariana* on moist sites, Quebec (Wikipedia)

Photo 52 The five-needle pine *Pinus pumila* is forming dense scrub at the subalpine belt above the timberline, with *Ledum decumbens*, *Rhododendron aureum, Vaccinium* species and some other dwarf shrubs, Kamchatka (photo: Breckle)

References

Ahti, T. L., & Jalas, J. 1968: Vegetation zones and their sections in Northwestern Europe. Ann. Bot. Fenn. **5**: 169-211

Amiro, B. D., Stocks, B. J., Alexander, M. E., et al. 2001: Fire, climate change, carbon and fuel management in the Canadian boreal forest. Internat. J. Wildland Fire **10**: 405–13

Burgeff, H. 1961: Mikrobiologie des Hochmoores. Fischer/Stuttgart 207 S.

Burrows, C. J. 1990: Processes of vegetation change. U. Hyman/London 551 S.

Dierßen, K. 1996: Vegetation Nordeuropas. Ulmer, Stuttgart 838 S.

Flannigan, M. D., Bergeron, Y., Engelmark, O., Wotton, B. M. 1998: Future wildfire in circumboreal forests in relation to global warming. J. Veg. Sci. **9**: 469–76

Heinselman, M. L. 1973: Fire in the virgin forests of the Boundary Waters Canoe Area, Minnesota. Quat. Res. **3**: 329–82

Heinselman, M. L. 1981: Fire intensity and frequency as factors in the distribution and structure of northern ecosystems". Proceedings of the Conference: Fire Regimes in Ecosystem Properties, Dec. 1978, Honolulu, Hawaii. USDA. For. Serv., Washington DC, Gen. Tech. Rep. WO-26. pp. 7–57

Iwasa, Y., Sato, K. & Nakashima, S. 1991: Dynamic modeling of wave regeneration (Shimagare) in subalpine *Abies* forests. J. of Theor. Biol. **152** (2): 143-158

Jeltsch, F. 1992: Modelle zu natürlichen Waldsterbephänomenen. Dissertation Univ. Marburg.

Mueller-Dombois, D. 1987: Natural dieback in forests. BioScience **37**: 575-583

Overbeck, F. 1975: Botanisch-geologische Moorkunde. 719 S., Neumünster

Popov, A. I. (Hrsg.) 1971-75: Die natürlichen Verhältnisse Westsibiriens. Lief. I-V. Verlag d. Moskauer Univ. (Russ.)

Sprugel, D.G. 1976: Dynamic structure of wave-generated *Abies balsamea* forest in the northeastern United States. J. Ecology **64**: 889-912

Stanjukovitsch, K.V. 1973: Die Gebirge der USSR. 412 S. Duschanbe (Russ.)

Succow, M. & Joosten, H. 2001 Landschaftsökologische Moorkunde. Verlag Nägele & Obermiller, Stuttgart. 622 S. ISBN 978-3-510-65198-6

Tanneberger, F., Hahne, W., Joosten, H. 2003: Wohin auch das Auge blicket: Moore, Moorforschung und Moorschutz in Westsibirien. Telma **33**: 209-229

Troeva, E.I., Isaev, A.P., Cherosov, M.M., Karpov, N.S. 2010: The Far North: Plant biodiversity and ecology of Yakutia. Springer Science 385 pp.

Van Wagner, C. E. 1978: Age-class distribution and the forest cycle. Can. J. For. Res. **8**: 220–27

Walter, H. 1927: Einführung in die allgemeine Pflanzengeographie Deutschlands. Fischer, Jena 458 p.

Walter, H. 1968: Die Vegetation der Erde, Bd. II: Gemäßigte und arktische Zonen. 1001 S., Fischer, Jena-Stuttgart

Walter, H. 1974: Die Vegetation Osteuropas, Nord- und Zentralasiens., Vegetationsmonographien, Fischer, Stuttgart. 452 S.

Walter, H. 1977: The oligotrophic peatlands of Western Siberia - The largest peino-helobiom in the world. Vegetatio **34**: 167-178

Walter, H. 1990: Vegetationszonen und Klima. 6. Aufl., Ulmer/Stuttgart 382 S.

Walter, H. & Breckle, S.-W. 1990: Ökologie der Erde, Bd. **1**: Ökologische Grundlagen in globaler Sicht. UTB Große Reihe, 2. Aufl. Fischer, Stuttgart. 238 S.

Walter, H. & Breckle, S.-W. 1991: Ökologie der Erde, Bd. 4: Spezielle Ökologie der Gemäßigten und Arktischen Zonen außerhalb Euro-Nordasiens. UTB Große Reihe, Fischer, Stuttgart. 586 S.

Yurtsev, B. A. 1981: Relikte von Steppenkomplexen in Nordostasien. 168S. „Nauka", Novosibirsk (Russ.)

Part L: ZB IX—Zonobiom of the Tundra, the Arctic Climate

13

Contents

© Springer-Verlag GmbH Germany, part of Springer Nature 2022
S.-W. Breckle, M. D. Rafiqpoor, *Vegetation and Climate*, https://doi.org/10.1007/978-3-662-64036-4_13

Photo 53 Striped tundra (ZB IX) with espalier *Salix* species and frost heap strips on the Finland–Norway border (photo: Breckle)

Photo 54 *Cassiope hypnoides* (Ericaceae) between *Salix polaris* in the tundra of Finland (ZB IX, photo: Breckle)

Photo 55 Tundra (ZB IX) in the vicinity of Soresby Sund, East Greenland (photo: Hannes Grobe, https://commons.wikimedia.org/wiki/File: Greenland-sydkap-hg.jpg)

13.1 Climate and Soils

Zonobiome IX comprises two very widely separated sub-areas, which differ greatly from each other due to the very different distribution of land and ocean areas in the N and S and in the corresponding latitudinal positions. In each case, the continentality decreases sharply from S to N.

The largest forestless tundra area occupies an area of three million square kilometres in N Siberia. The number of days with a temperature average above 0 °C there is 188 to only 55. This is related to the always low position of the sun. However, the low summer warmth is partly due to the heat consumed in thawing the snow and the permafrost ground. Winters are quite mild in the oceanic region and extremely cold in the continental one (Fig. 13.1). However, the cold pole at Oimjakon (near Werchojansk) is still in the forest area, although the mean annual temperature there is −16.3 °C and permafrost reaches deep into the ground (Fig. 12.14). Soil freezing does not have a major influence on vegetation conditions. It depends only on the thickness of the upper soil layers thawing in summer.

The growing season in the S tundra begins in June and ends in September. Of great importance is the wind, due to which the irregular deposition of snow is caused, which is the precondition for the mosaic of vegetation. The storms in winter reach 15–30 m per sec. Precipitation is low, often even less than 20 mm per month.

Nevertheless, the climate is humid with the very low potential evaporation. The excess water cannot seep into the soil due to the permafrost. The consequence is a strong siltation; but there is no significant peat formation, because the production of the plants is very low or too low. The snow depth is 20–50 cm, exposing the elevations, so that snow and ice abrasion play a major role as mechanical factors for vegetation.

With the low position of the sun in summer, steep, stony S-facing slopes are heated relatively strongly. Therefore, they often form real small scale "flower gardens". Such S slopes and the banks of streams and rivers are the most favourable locations. Flat elevations with stony patterned soils (polygon soils, stripes soils) are only weakly colonized, as are gentle slopes subject to solifluction. The soils are often mini-podsoles and rankers (Fig. 13.2).

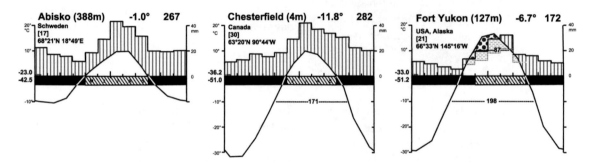

Fig. 13.1 Climate diagrams from the forest tundra of Sweden (oceanic), from the tundra of N America, and from the extreme continental boreal region of Alaska (Fig. 2.5, Werchojansk and Fig. 12.13)

Fig. 13.2 A mini- or dwarf podsol of low profile thickness in the subpolar regions of the Northern Hemisphere. It forms on different bedrocks and is common in association with permafrost (photo: Breckle)

13.2   The Vegetation of the Tundra

In the tundra, endless areas are covered with dwarf birch and willow, as well as *Eriophorum* and *Carex* species. On dry soils one finds a pure lichen tundra, on moist ones mosses play a major role, but not the *Sphagnum* species. Measurements of air temperature taken at a height of 2 m are not decisive for the low plant carpet. When the air temperature reaches 0 °C, the soil has already thawed half a meter and vegetation development is in full swing. The temperature of the plants during the day at the ground is often 10 K above the air temperature. Nevertheless, the short summer period is often not sufficient for the seeds to mature. This is why, for example in Greenland, half of all species have their flowers produced the year before, so that flowering can take place very early. The buds and also the green leaves usually overwinter under the snow, but the open flowers die.

> **Box 13.1 Distribution of Tundra in both Hemispheres**
> ZB IX comprises the areas around the poles of the earth. In the north, these are the tundra areas and the cold deserts N of the Arctic timberline; in the south, S of the Antarctic timberline, tundra occurs only in small areas and on individual islands. Antarctica itself is covered by a huge ice desert.

Particularly interesting are the aperiodic species, such as the small Brassicaceae *Braya humilis*. Their development is extended over several years and temporarily interrupted during the winter at any stage. These species are thus independent of the short summer and flower either at the beginning of the growing season or later, although the buds may have been laid down two years earlier.

Fruit or seed dispersal is by wind (gliding on snow) in 84% of species, and by water in 10%. Berry fruits occur only in the forest tundra. With low productivity in the tundra, seeds are small; in 75% of species they weigh less than 1 mg. Most plants are frost germinators, that is, they acquire the ability to germinate only after the action of the low winter temperature, then germinate immediately in the spring and have time to accumulate some reserves until the autumn. Viviparous are 1.5% of species, various grasses, but also *Polygonum, Stellaria, Cerastium* species, etc. With abundant seed production, open places, for example, on the lower Lena, are quickly colonized. Most species are hemicryptophytes and chamaephytes. Annual species (therophytes) are only *Koenigia islandica*, three *Gentiana* species, *Montia lamprosperma*, two *Pedicularis* species and few others. The short vegetation period here with low temperatures is not favourable for the annuals (compare, on the other hand, the desert). Most species have thick roots as reserve storage. The

age of a single plant can exceed 100 years even in herbaceous species. For dwarf shrubs it is between 40 and 200 years.

The nitrogen balance plays a major role. Mineralization and nitrogen uptake are very inhibited due to the low temperatures. The legumes *(Oxytropis, Hedysarum, Astragalus)* have root nodules that lie directly under the warming soil surface. Where there is almost no nitrogen in the soil, only mosses and lichens are found. Fertilization by animal excrement is of importance. *Dryas drummondii,* which is a pioneer species growing in Alaska, is stated to have root nodules similar to *Alnus*. During the pioneer stage of colonization, soil nitrogen content increases from 33 kg/ha to 400 kg/ha.

Climatic conditions that differ from the rest of the Arctic are found in some trough valleys in the interior of Peary Land (N Greenland) at latitude 80°. In summer there is a lack of precipitation here due to the descendings winds blowing from the inland, desert-like conditions prevail with salt efflorescence on the soil surface with alkaline soils, where even some halophytes occur. Even otherwise, vegetation is not entirely absent, as drifting snow accumulates from the mountains in winter and melts in spring. The water seeps away as the soils thaw to a depth of 1 m. Accordingly, *Braya purpurascens* also has a taproot over 1 m long. The number of frost-free days reaches 59, the July temperature is 6 °C.

13.3   Ecophysiological Investigations

The temperature of the low plants and the soil is fairly uniform during the polar day when the sun is low for 24 h, but the direction of irradiation still has an effect. The differences to the air temperature can then become very clear on fair weather days (Fig. 13.3). Sufficient temperatures are a prerequisite for active metabolic processes in plants.

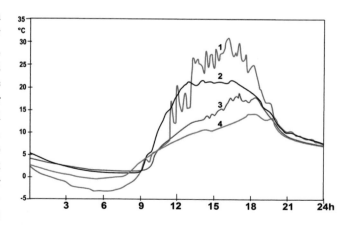

Fig. 13.3 Diurnal changes in soil surface temperature in a catena from Carici rupestris-Dryadetum (CD) to Salicetum polaris on 29.8.1990, a fair weather day at Liefdefjord in NW Spitsbergen (90 m asl) (modified after Dierßen 1996). **1** CD/*Carex nardina* facies; **2** CD/*Dryas* facies; **3** CD/*Carex misandra* facies; **4** Salicetum polaris

The water balance of the arctic plants is well-balanced, their cell sap concentration is 0.7 to 2.0 MPa. If the species nevertheless often show xeromorphic traits in their structure, they are probably inherited peinomorphs evolutionary caused by nitrogen deficiency, just as in the case of the raised bog plants.

Particularly important is the question of photosynthesis and thus the production of substances. The maximum intensity of CO_2 assimilation is 12 mg/dm^2.h^{-1}. On cloudy days, CO_2 uptake temporarily falls below zero. However, since it can usually continue for 24 h, with a minimum parallel to the light minimum at midnight, the yield on a summer day reaches 100 mg CO_2/dm^2 = about 60 mg starch.

This yield is sufficient to build up sufficient reserves of material in the summer. The primary production of vegetation cover in one year is 2500 kg/ha in the subarctic region in Swedish Lapland near Abisko (vegetation period 111 days), 830 kg/ha in Alaska (vegetation period 70 days), and only 30 kg/ha in the high Arctic (vegetation period 60 days). The phytomass of an arctic willow scrub on Greenland reaches 5.5 t/ha.

The "tundra biome" (ZB IX) has been studied very intensively in Alaska within the framework of the I.B.P. (cf. Bliss and Wielgolaski 1973).

13.4 Wildlife of the Arctic Tundra

The vast tundra expanses of Siberia are one of the few areas on our planet where you can still find original wildlife reasonably undisturbed by humans and thus study their influence on the vegetation. In winter most of the large vertebrates leave the tundra, the birds migrate south. Only the lemmings *(Lemmus)* and ground squirrel *(Spermophilus lateralis)*

remain in the tundra. Arctic fox and snowy owl retreat from the northernmost, low prey areas.

The lemmings do not go into hibernation, nor do they accumulate food stocks, but remain active under the hard shell of the snow cover and feed mainly on the renewal buds of the Cyperaceae. A lemming, although weighing only 50 g, needs about 40–50 kg of fresh plant matter per year. It usually inhabits well-drained southern slopes and builds a nest of cyperaceous shoots near its grazing area in winter, which is about 100–200 m^2 for a family. An entire settlement covers about 1 to 1.5 ha, where 90–94% of the plants are grazed. *Eriophorum angustifolium* does not reach flowering on such areas. A maximum of lemmings occurs on average every three years. The dry plant parts are not eaten; in spring they form the "hay" (1–2 t/ha), which is washed together and piles up into peaty bulbs. After leaving their winter quarters, the lemmings make their burrows on higher ground, throwing out up to 250 kg/ha of soil.

At such disturbed sites one finds a characteristic plant community that initiates secondary succession. The same applies to squirrel burrows, comparable in Central Europe to a mole meadow. In this way, a constant dynamic is maintained within the plant cover. The flocks of waterfowl, especially geese, that come in spring also destroy 50–80% of the plant cover by biting off the young shoots of *Oxytropis* and tearing out the starchy rhizomes of *Eriophorum*. On the bare ground, solifluction sets in until a dense moss cover develops.

The nesting and gathering places of the birds are heavily fertilized, so that nitrophilous species *(Rhodiola, Stellaria, Polemonium, Myosotis, Draba, Papaver* and others) are established.

The animal kingdom of the tundra also includes the reindeer *(Rangifer tarandus)* (Fig. 13.4), which remains in the

Fig. 13.4 (**a, b**) The reindeer *(Rangifer tarandus)* is a mammal of the deer family. It lives in the circumpolar region in the tundra in summer and in the taiga in winter (photos **a**: Breckle; **b**: E. Fischer)

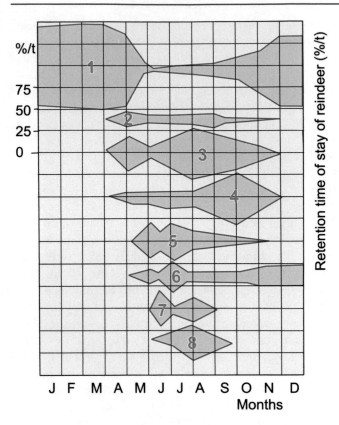

Fig. 13.5 Seasonal foraging behaviour of wild reindeer in Hardangervidda (modified after Skogland 1983) **1** Loiseleurio-diapension; **2** Cladonio-juncetum trifidi; **3** Phyllodoco-Vaccinion myrtilli and Potentillo-Polygonion vivipari; **4** Nardo-Caricion gigelowi; **5** Adenostylion alliariae; **6** Caricion nigrae; **7** Ranunculo-Salicetum herbaceae; **8** Cassiopo-Salicetum herbaceae

tundra in winter only if areas remain partly "aper", i.e. not covered by snow.

In summer, the reindeer graze dispersedly and have little impact on the vegetation. However, when they gather into large herds in autumn, their trampling becomes noticeable. In the process, the lichens and dwarf shrubs are eaten and destroyed, while the grass communities with *Deschampsia* and *Poa* spread. However, feeding behaviour is highly adaptable according to the food supply (Fig. 13.5). The number of wild reindeer is now decreasing in favour of domesticated ones. Reindeer are the main herbivorous animal in the tundra; in the North American tundra it is the caribou *(Rangifer caribou)*, whereas in the Euro-Siberian tundra it is the reindeer proper *(Rangifer tarandus)*. The effect of predators (arctic fox, sporadically bear and lynx) on the flora is small.

It is now known that numerous species of megafauna have only become extinct in the last 20,000 years (Martin 1984; Simmons 1996). It is even assumed that up to 200 genera of large mammals and birds disappeared by the end of the last ice age. In North America, about 2/3 of the large mammals that can still be traced toward the end of the last ice age (i.e.,

about 13,000 years ago) are extinct. These include 3 elephant-like species, 15 ungulate species, numerous large rodents and predators, 6 edentate species (giant sloth, armadillo, anteater, etc.). There is no evidence of similar extinction rates from earlier glacial epochs and interglacials that would suggest glacial climatic changes as the cause. Rather, it must be assumed that, especially at the end of the last ice age, the great waves of Indian immigration across the Bering Strait were the main cause of this mass exodus. The conquest of the continent from Canada to Mexico may have taken place in 350–500 years. Within a period of 500 years, most species also became extinct. Similar extinction periods are known from New Zealand, from Madagascar, from Java, each after the arrival of man.

In Eurasia, the extinction rate has not been quite so dramatic. While at least 24 genera have disappeared in North America, there were probably nine species in Eurasia, including the mammoth *(Mammuthus primigenius)*, the woolly rhinoceros *(Coelodonta antiquitatis)*, the great elk *(Megaloceros giganteus)*, the musk ox *(Ovibus moschatus)*, the steppe bison *(Bison priscus)*, a buffalo species *(Homoioceros antiquus)*, and three carnivores (Simmons 1996). Musk oxen have been reintroduced in some places today.

In addition to improved hunting techniques during the heyday of prehistoric man, rapid warming and the advance of forest vegetation may also be partly responsible for the suddenly high extinction rate of megafauna.

13.5   Man in the Tundra

The seasonal migration of reindeer herds has had a lasting effect on the hunting behaviour of the people of the tundra. Domesticated reindeer herds, for example among the Tungus (a tribe in Siberia), migrate from the taiga to the tundra in summer and back to the taiga in early autumn. In the tundra of North America live different Eskimo tribes, which have adapted themselves completely to the arctic conditions. The individual tribes also trade among themselves, some hunt mainly whales and seals, others are active inland as caribou hunters, but also capture mountain sheep, moose, beavers, bears, snow hares, ducks and geese. Whale meat, blubber and fish oil are then traded for caribou meat and berries (Campbell 1985).

The way of life of the Eskimo in N Alaska and especially their way of living in mostly circular huts is an example of the possible way of life of the people during the ice age in Europe, for example during the Magdalenian period.

The huts are built semi-underground, thickly packed with grass sods, with a roof of whale ribs covered with animal skins, providing excellent thermal insulation.

From south to north, three subzonobiomes can be distinguished in the Arctic tundra:

1. The dwarf shrub tundra in the area of postglacial forestation,
2. The actual moss and lichen tundra and
3. The cold desert, commencing where plant growth becomes very sparse.

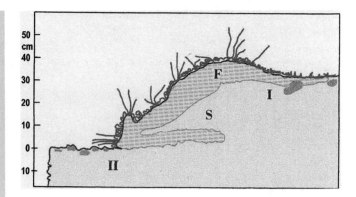

Fig. 13.6 Earth flow on a shallow slope in the Arctic (Alaska). The fibrous peat layer (F) with the living plant cover has moved about 30 cm from I to II, forming a fold in which the free silt soil (S) is partially enclosed (modified after Hanson 1950)

Similar construction methods are known from numerous excavations from the Magdalenian period, with first evidence of Cro-Magnon man as early as 30,000 before today, with a heyday in southern France (Dordogne) between 19,000 and 13,000 years before today. Even today there are such constructions in Iceland or northern Finland (Lapland).

The reindeer was also the main source of meat for Cro-Magnon man. But also remains of bison *(Bison bonasus),* mammoths, horses, wild cattle have been excavated from the dwellings. It is very likely that the systematic exploitation of the rich wild animal population was forced by the Cro-Magnon man. In any case, by the end of the Upper Pleistocene, this form of foraging, based primarily on a wild animal species, was already fully developed. It is still evident today in a similar way among tribes of the tundra, albeit today with improved technological methods and the availability of dogs, boats, and sleds. However, both the Stone Age people in the ice-age tundra of Central Europe and the Eskimos had highly developed technical aids: thermally insulated huts, clothing, traps, etc., and also already simple machines such as harpoons and spear-throwers.

Today, Western civilization has caused profound changes. Alcohol is a big problem. Snowmobiles replace dog sleds, hunters use rifles. Today, a few Eskimos can decimate an entire herd of caribou in a day. Hunting has become so easy that it is no longer important to use all the parts of a shot animal; only the best pieces are taken. Surplus game is sold.

13.6 Arctic Cold Desert and the Solifluction

The Arctic Cold Desert is the northernmost of the three subzonobiomes of ZB IX. Here, too, oceanic and continental areas can be additionally distinguished (Aleksandrova 1971).

In the cold desert, days of alternating frost, on which the temperature exceeds zero twice, are very frequent; this gives rise to the phenomenon of solifluction, the flowing of the soil. Already in the tundra itself, as well as in the alpine and nival altitudes of the mountains, the peat mounds and frost hummocks or bulbous tundra are formed by local formation of ice with a great increase of volume under the cover of

plants in the wet soil. Even on slopes with very little incline, the soil is slowly pushed downward, the slope taking on an appearance as if it were covered with cattle steps. The frost stairs are low steps that run parallel to the isohypses. Fig. 13.6 shows the cross section through such a step. Such soil movements become more conspicuous towards the north.

Where, in autumn, an unfrozen wetted layer is compressed between the permafrost soil below and a freezing layer above, it bursts the upper freezing layer in places and pours out as a liquid clay mush over the plant cover, forming a vegetationless patch several centimetres higher (Fig. 13.7). The spotted tundra is formed.

A consequence of the frost is also the pressing and sorting out of stones from the soil. Figure 13.8 explains this process: when the upper layer of soil freezes, it sucks in water from below and increases in volume; in the process, it lifts up the stones that are stuck in the freezing layer. A cavity forms under the stone into which fine sand falls; after thawing,

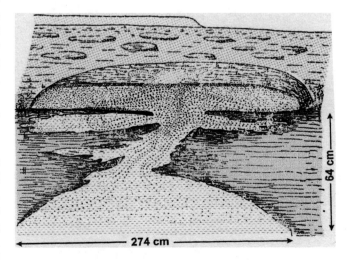

Fig. 13.7 Spotted tundra with vertical section through a spot. The spots are formed by pressing up the liquid loam trapped between the permafrost soil below and the freezing layer above, which then pours over the surface and forms a vegetationless loam spot a few cm higher (modified after Walter 1990)

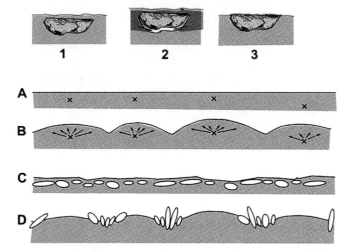

Fig. 13.8 Top: Schematic representation of the processes involved in freezing and thawing the soil. **1** before freezing; **2** soil frozen on top, stone is lifted; **3** after thawing, stone has moved up to the surface. Bottom: stone net formation. (**a**) at x freezing centre; (**b**) arrows show the direction in which the stones move; (**c**) original position of the stones in the soil; (**d**) their final position when the frost stone network or polygonal soil (in section) has formed (after Walter 1990)

therefore, the stone remains at a slightly higher level than before. If this is repeated many times on the days when the frost changes, the stone eventually lies above the surface of the ground. It may happen in winter also in gardens of temperate regions where stones are accumulating on the surface. Usually, the freezing of the soils starts from single points, which are one or more meters apart. Then the stones are not only lifted out, but at the same time pushed aside. In the final result, they form a stone network pattern between the freezing centres, that is, a polygonal soil (Fig. 13.9). The plants find sporadic refuge between the stones of the

polygonal floor, where the movement is least. If this process takes place on a slope, the stones are not only lifted but also pushed down the slope; the stone streams or slide, patterned strip soils are then formed (Fig. 5.67).

This constant soil movement in the Arctic does not allow the plant cover to come to rest and has an unfavourable effect. This can already be observed on Iceland (Lötschert 1974), much more clearly on Spitsbergen.

Solifluction is of equal importance in the mountains, too, in the upper alpine and subnival belts, but only locally and not over such wide areas as in the Arctic (Fig. 5.67).

As far as the composition of the vegetation is concerned, the floristic differences around the whole North Pole are relatively small. A relatively large percentage of the few species are distributed circumpolar.

13.7 Antarctica and Subantarctic Islands

Only two flowering plants have been found native on the edges of the ice-covered Antarctic continent: *Colobanthus crassifolius* (Caryophyllaceae) (Fig. 13.10) and the grass *Deschampsia antarctica* (Fig. 13.11). Recently, *Poa pratensis* and *P. annua* have been imported, as well as *Stellaria media, Ranunculaus repens* and *Carex aquatilis*. Otherwise, only mosses, many lichens (Øvstedal & Lewis Smith 2001), and land algae occur, a total of several hundred species. They are restricted to intermittently snow-free sites on the coast, steep cliffs, and scree slopes (Fig. 13.12). Quantitatively, their biomass is of very little importance.

Where higher plants are generally absent, due to lack of water or low temperatures or human influences, so-called "biological soil crusts" made up of prokaryotic bacteria and

Fig. 13.9 Tundra with larger stone rings (**a**) and earth polygons (**b**) in Spitsbergen (photos: Jaroslav Obu; Alfred-Wegener-Institute for Polar Research, https://t1p.de/o03l)

Fig. 13.10 *Deschampsia antarctica,* the only grass species in Antarctica, Penguin Island, Antarctic Peninsula (photo: O. Krüger)

cyanobacteria, eukaryotic algae, microfungi, lichens and mosses can instead develop extensive microecosystems. In the Antarctic, they regionally cover up to 55% of the open ground. They are joined by rock-colonizing organisms such as cyanobacteria, green algae and lichens, which live as so-called endoliths in the upper millimetres of rocks or as hypoliths on the underside of quartz pebbles. The astonishing diversity of these "primitive" communities is only revealed by looking through the microscope or looking at molecular genetic data (Kanz et al. 2020).

Bacteria and fungi were also detected in soil samples and rock surfaces. For the animal world, this low phytomass plays no role. The penguins and many other animals that occur intermittently in the coastal area of Antarctica have their food base in the sea. Whether there are invertebrates that live on the crustose lichens or the rock algae is not known to us.

Fig. 13.12 Vegetation of Antarctica mainly consists of lichens and mosses. Those cushions have grown distinctly faster the last decades and have conquered new areas on rocks due to global warming. On snowpatches red coloured green-algae flourish (photo: C. Colesi/AWI Bremerhaven- www.klimareporter.de/erdsystem/gruene-antarktis)

In the sea around Antarctica, with its constant westerly storms, are scattered many small islands, most of them south of the 50th parallel. They are all characterized by their treelessness, for the summers are cool, the winters not cold; almost isothermy prevails on these islands, for example, the values of temperature vary almost throughout the year on the Macquarie Islands (54° 3′ S) only between 2.8 °C and 7.7 °C (Fig. 13.13). Drizzle and fog are typical of the weather. One has spoken of a **wind desert**; for only in the windbreak is the vegetation more luxuriant.

The most common plant on the Kerguelen Islands is the dense rosette-forming *Azorella selago* (Apiaceae). In the past, the Kerguelen cabbage, *Pringlea antiscorbutica* (Brassicaceae) with its large leaves, was used by sailors as a fresh vegetable against scurvy (vitamin C deficiency disease) (Fig. 13.14). *Acaena species* (Rosaceae) are common on all islands. Tussock grassland *(Festuca* and *Poa* species) also occurs, as well as many mosses, ferns, and lichens. Various cushion-forming species are characteristic of the subantarctic, as always of very windy sites.

Fig. 13.11 *Colobanthus crassifolius* (Caryophyllaceae), Hannah Point, Livingston Island, Antarctic Peninsula (photo: O. Krüger)

Fig. 13.13 Thermoisopleth diagram of Macquarie Islands with the course of isotherms parallel to the Y axis as an indication of a distinct seasonal climate of the polar regions (modified after Troll 1943)

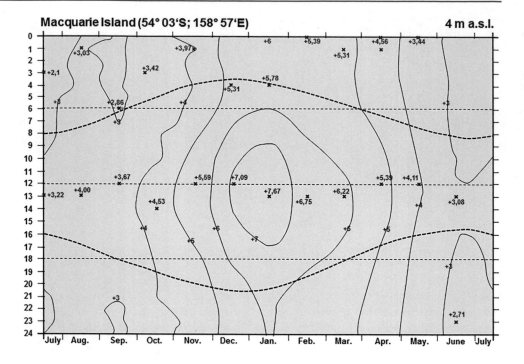

Fig. 13.14 The Kerguelen cabbage *(Pringlea antiscorbutica*, Brassicaceae), which used to be a vital vegetable for the seafarers' vitamin C supply. (**a**) whole plants at subantarctic landscape; (**b**) close-up of green rosette (photo https://t1p.de/blzc)

Photo 56 Freezing and thawing of soil at day and nights during summer in ZB IX causes a strong assortment of pebbles and stones, forming smaller or larger polygons, here seen in the highlands of Iceland (photo: Breckle)

Photo 57 The microtopographic variability of polygonal bog landscapes of NE-Siberia controls hydrological heterogeneities, the polygonal features are formed by expanding ice wedges exhibiting typical water-filled troughs and centers (photo: M. Succow)

Photo 58 Green slopes and hills in ice-free Antarctica (ZB IX) are expanding by rising temperatures during last decades by increasing thick highly diverse moss-covers, Ardley island (photo: M. Amesbury, Univ. of Exeter)

References

Aleksandrova, V.D. 1971: On the principles of zonal subdivision of arctic vegetation. Bot. Z. 56: 3–21 (Russ.)

Bliss, L.C. & Wielgolaski, F.E. (eds.) 1973: Primary production and production process, Tundra Biome. Proc. Conf. Dublin, Swedish IBP Comm., Stockholm 250 S.

Campbell, B. 1985: Ökologie des Menschen. Harnack, München 232 S.

Dierßen, K. 1996: Vegetation Nordeuropas. Ulmer, Stuttgart 838 S.

Hanson H.C. 1950: Vegetation and soil profiles in some solifluction and mound areas in Alaska. Ecol. 31: 606-630

Kanz, B., Büdel, B., Jung, P., et al. 2020: Leben zwischen Eis und Felsen. Biologische Bodenkrusten in der Antarktis. Biologie in unserer Zeit 50: 122–133.

Lötschert, W. 1974: Über die Vegetation frostgeformter Böden auf Island. Ber. Forschungsstation Neori As (Island) 16: 1–15.

Martin, P.S. 1984: Prehistoric overkill: the global model. In: Martin, P.S. & Klein, R.G. (eds.): Quaternary extinctions: a prehistoric revolution. Tucson, Univ. of Arizona Press: 354–403.

Øvstedal, D.O., Lewis Smith, R.I. 2001: Lichens of Antarctica and South Georgia. A guide to their identification and ecology, Cambridge University Press, Cambridge 411 pp.

Simmons, I.G. 1996: Changing the face of the earth. Blackwell Public., Oxford.

Skogland T 1983: Wild reindeer foraging-niche organization. Holarct Ecol 7: 345-379

Troll, C. 1943: Thermische Klimatypen der Erde. In: Petermanns Mitteilungen 89: 81–89.

Walter, H. 1990: Vegetationszonen und Klima. 6. Aufl., Ulmer/Stuttgart 382 S.

Part M: Summary, Conclusions

14

Contents

© Springer-Verlag GmbH Germany, part of Springer Nature 2022
S.-W. Breckle, M. D. Rafiqpoor, *Vegetation and Climate*, https://doi.org/10.1007/978-3-662-64036-4_14

Photo 59 The mega-city Tokyo as an example of urbanisation and globalisation. A metropolis that is constantly expanding not only outwards to the peripheries, but also into the third dimension upwards (photo: Breckle)

Photo 60 Coal-fired power station near Leipzig operated by opencast lignite mining, one of the anthropogenic contributions to the additional greenhouse effect (photo: Breckle)

Photo 61   Garden art in the Oriental city of Shiraz, Iran (photo: Breckle)

14.1 Phytomass and Primary Production of the Individual Vegetation Zones and the Entire Biosphere

The geo-biosphere covers the Earth's surface as a thin shell, as the thinnest little skin; it includes the uppermost rooted soil layer and the air layer near the ground, as far as the organisms protrude into it, as well as all waters. The entire biological material cycle thus also takes place in it.

Of the total **biomass** on land, **phytomass** accounts for over 99%, so that we can confine ourselves to the distribution of the same in our considerations. It shows clear relationships with the zonobiomes.

The exact determination of phytomass and primary production encounters difficulties. As early as 1970, Bazilevich et al. published calculations, evaluating the relevant literature, for the individual thermal zones and bioclimatic areas of the Earth. The mean phytomass and the mean annual primary production per hectare (t/ha) are calculated for the individual areas as dry mass in tons (t). After measuring the area of each region, not including the area of rivers, lakes, glaciers and firns, the total phytomass and the total annual primary production are also given for each region. Summing these figures gives the phytomass and annual production of the Earth's land surface. To this, the Table 14.1 also adds the corresponding data for water bodies. These are potential values, i.e. based on natural vegetation that has not been altered by humans.

Bazilevich et al. (1970) distinguish five thermal zones: **1.** polar (arctic), **2.** boreal, **3.** temperate, **4.** subtropical, and **5.** tropical. The first two zones have a humid climate, while three areas are distinguished for each of the other three: a humid (**h**), a semiarid (**s**), and an arid (**a**) (cf. map in Fig. 14.1, and Table 14.1).

This outline differs somewhat from the zonobiome outline, as the comparison shows (Table 14.2).

If we compare the conditions on land with those in the oceans, we see that the production of the latter at $60 \cdot 10^9$t is only about 1/3 of that on land, although its area is almost three times greater. It is also noticeable that the phytomass in the oceans is infinitesimal, especially in relation to the primary production, which is 300 times greater. This is understandable if one takes into account that the producers, the plants of the plankton, consist of unicellular organisms that are constantly dividing. In contrast, the primary production on land is only about 7% of the phytomass.

If we add up the total active carbon stocks of the Earth, we get about 800 Gt C (mainly CO_2) in the atmosphere, about 700 Gt C in the biosphere, 38,000 Gt C in the hydrosphere and about 1600 Gt C in the pedosphere. The proportion of carbon stocks in the lithosphere is enormous (Table 14.3). If one asks about the mass of consumers and decomposers, only $20 \cdot 10^9$ t of dry mass is given for all continents together, i.e. less than 1% of the phytomass, whereas in the oceans one reckons with about $3 \cdot 10^9$ t, which is more than 15 times the phytomass there. Unlike unicellular plants, consumers in

Table 14.1 Distribution of the potential productivity of the Earth (after **Bazilevich** et al. 1970)

Climate zones	Area [10^6 km^2]	Phytomass Total [10^9 t]	Phytomass Mean [10^9 t·ha^{-1}]	Primary production Total [10^9 t·ha^{-1}]	Primary production Mean [10^9 t·ha^{-1}·a^{-1}]
Polar	8.05	13.8	17.1	1.33	1.6
Boreal	2.20	439	189	15.2	6.5
Temperate					
Humid	7.39	254	342	9.34	12.8
Semiarid	8.1	16.8	20.8	6.64	8.2
Arid	7.04	8.24	11.7	1.99	2.8
Subtropical					
Humid	6.24	228	366	15.9	25.5
Semiarid	8.29	81.9	98.7	11.5	13.8
Arid	9.73	13.6	13.9	7.14	7.3
Tropical					
Humid	26.5	1166	440	77.3	29.2
Semiarid	16.0	172	107	22.6	14.1
Arid	12.8	9.01	7.0	2.62	2.0
Geo-biosphere					
Landmass	133	2400	180	172	12.8
Glaciers	13.9	0	0	0	0
Hydro-biosphere					
Lakes/Rivers	2.0	0.04	0.2	1.0	5.0
Oceans	361	0.17	0.005	60.0	1.7

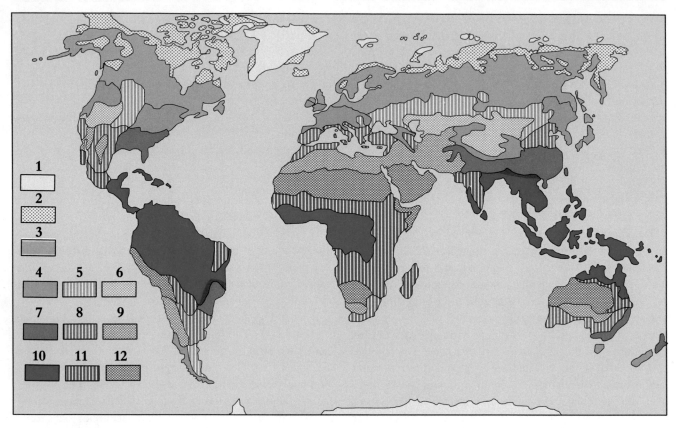

Fig. 14.1 Thermal zones and bioclimatic areas (modified from Bazilevich et al. (1970): **1** glaciers and firn areas; **2** Arctic zone; **3** boreal zone; **4–6** temperate zone: **4** humid areas, **5** semiarid areas, **6** arid areas; **7–9** subtropical zone: **7** humid areas, **8** semiarid areas, **9** arid areas; **10–12** tropical zone: **10** humid areas, **11** semiarid areas, **12** arid areas)

the oceans are also large animal organisms that are exploited for human consumption.

We have shown by various examples how small, in contrast, is the zoomass of the great terrestrial consumers. The phytomass on land consists mainly of the wood mass in the forests, which accounts for 82% of the total phytomass on all continents, although the forests occupy only 39% of the area. The bulk of forest phytomass, about 50%, is found in tropical forests, about 20% in boreal forests, and about 15% each in subtropical and temperate forests. These figures should also be kept in mind for the "global change" discussion.

> **Box 14.1 Primary Production of Different Ecosystems**
> The total annual potential primary production of the biosphere on land, in the oceans and lakes and rivers is about $233 \cdot 10^9$ t. of this, the land mass accounts for $172 \cdot 10^9$ t, the lakes and rivers for $1 \cdot 10^9$ t, and the oceans for $60 \cdot 10^9$ t.

The phytomass of deserts is very low (0.8%) compared to the large area of more than 22% that they occupy in the total land area.

The mean phytomass in t/ha of the forests (humid areas) increases steadily from 189 t/ha in the boreal zone to 440 t/ha in the tropics with increasingly favourable temperature conditions. In contrast, the mean phytomass in the tropical arid regions is lowest at 7 t/ha; this is because drought with permanently high temperatures is particularly unfavourable for plant growth.

If we look at the average annual primary production, it is more than seven times higher on land than in the oceans, at 12.8 t/ha·a, and about two and a half times that in lakes and rivers with their aquatic and marsh plant populations.

The primary production of the humid areas per hectare also increases equatorwards on land, doubling from the boreal to the temperate zone and from the latter to the subtropical in each case, but then increasing little further to the tropical zone. The differences between the humid and semiarid zones are not as great as in the case of the values for phytomass, since the wood masses in the forests do not produce and it is the leaf area that is more important (comparison of meadow and forest in the Solling). The relatively high production in the subtropical semi-arid and arid areas with 13.8 and 7.3 $t \cdot ha^{-1} \cdot a^{-1}$, respectively, is striking; it is due to the often very lush and productive ephemeral vegetation that can develop during the favourable cooler season.

Table 14.2 Comparison of Bazilevich's (1970) thermal climate zones with zonobiomes

Thermal zones and climatic regions	Zonobiome
Zone 1	ZB IX
Zone 2	ZB VIII
Zone 3 h, s, a	ZB VI and VII
Zone 4 h, s, a	ZB V, IV and III (outside the tropics)
Zone 5 h, s, a	ZB I, II and III (within the tropics)

Table 14.3 The carbon quantities [Gt C] in the individual global reservoirs (after **Ittekkot** et al. 2002)

Storage	Form	Amount of CO_2 [Gt C]
Atmosphere	Carbon dioxide (CO_2), carbon monoxide (CO), methane (CH_4)	790
Biosphere	Organic compounds in terrestrial And in marine organisms respectively	700 3 3
Hydrosphere	Carbon dioxide (CO_2), hydrogen carbonate (HCO_3^-), carbonate (CO_3^{2-})	35.000
Lithosphere	Calcium carbonate ($CaCO_3$, calcite), calcium-magnesium carbonate ($CaMg(CO_3)_2$, dolomite	At least 60,000,000
Sediment	Kerogen[a] Gas hydrates	At least 15,000,000 10.000
Fossil fuels	Coal, oil, natural gas	4.100
Pedosphere Soil	Dead biomass (humus, peat)	1.500

[a]**Kerogen**: The polymeric organic material from which hydrocarbons are formed as geological subsidence and heating increases

Lieth & Whittaker (1975) arrived at somewhat different values. They start from the vegetation formations and do not calculate the potential but rather the real production taking into account the cultivated areas. Therefore, the values for terrestrial production are lower. Lieth gives a primary production of $121.7 \cdot 10^9$ t of dry matter on a land area of $149 \cdot 10^6$ km^2 as the most accurate figure.

Finally, if we ask ourselves how high the consumption of mankind was with a population of three billion with a biomass of $0.2 \cdot 10^9$ t, we can put it approximately equal to the total agricultural production at that time, which accounted for 0.7% of the primary production of the biosphere. Energy consumption is given as $2.8 \cdot 10^{18}$ cal, since only part of the energy taken in with food is utilised. These figures do not seem high, but consumption has now risen sharply with the rapid increase in population to about eight billion.

14.2 Conclusions from an Ecological Point of View

The preceding chapters provide a concise overview of the major natural ecological interrelationships of the geo-biosphere. Their knowledge is the prerequisite for a correct assessment of the dangers arising from the increasing human intervention in natural processes and ecosystem structures (Breckle 2019).

These are so manifold and profound that they cannot be dealt with within the framework of this overview. Thanks to his intellectual abilities, man has built up his own, seemingly independent world alongside the natural one, that of a technically oriented world economy.

He has become more and more alienated from nature through progressive urbanisation. In the process, he loses the ground under his feet, considers everything technically feasible and believes in unlimited economic growth.

The Club of Rome (http://www.clubofrome.org) pointed out the utopian nature of this attitude as early as 1972 on the basis of many studies and predicted an economic crisis if countermeasures were not taken immediately (cf. also Gruhl 1975). But nothing substantial happened. The crisis has occurred in the meantime, locally for a long time, regionally in many places. People are still on the lookout for the first pink streaks on the horizon of economic growth. While "ecology" is on everyone's lips, there is no fundamental shift in mindset. The so-called economic constraints still have priority. Almost all economic theories assume necessary growth. How is constant growth to be sustained? It cannot be sustainable. Only an economic system in dynamic equilibrium (steady state) without exploitation of nature can last in the long run. The destruction of the environment, on which man's existence depends, continues almost unabated throughout the world. Attempts are made only to cover up local damage by cosmetic means. But these are global problems. The two greatest dangers must be briefly pointed out here:

- the population explosion
- over-technification

The question must be asked and clarified how sustainable land use is possible, i.e. land use that preserves the basis of human life for many generations of mankind (i.e. for centuries to millennia). This can only be done through education, rational insight, modesty and humility.

14.3   The Population Explosion

In 1981, Aurelio Peccio, President of the Club of Rome, in the German edition of his paper "Die Zukunft in unserer Hand" (The Future in Our Hands), again pointed out that the world population is increasing at such a rate that something must be done about it immediately. According to Peccio, 223 children are born in the world every minute, which is 321,000 in a day or 120 million in a year. However, this number is much higher in 2021! However, the exponential increase has leveled off somewhat in recent years. Between 2010 and 2014, the population increased by 82 million per year, so that at the end of 2014, there were exactly 7 billion people living on our planet. This number has increased by about 160 million at the end of 2016 (PRB 2016).

At the moment, the population of the Earth is increasing by one billion in less than 15 years. If it were possible to keep all the children born alive, and that is what we are trying to do, then in barely 10 years there would be 1.2 billion children under the age of 10 in the world. These would have to be fed and educated. After another 10 years, it would be necessary to obtain jobs for them, and soon after that they in turn would bring more children into the world.

The population explosion occurs with exponential growth (Fig. 14.2). 2000 years ago, there were an estimated 200 to 300 million people across the globe. Until the year 1800, the population increased only slowly.

Today, the situation in very many countries is catastrophic. Diseases and epidemics were also successfully fought there. The mortality rate fell, but hardly the birth rate. As a result, the population increased and continues to increase rapidly, and this within a few decades. In the context of human development, this is only a moment. If millions are malnourished or starving as a result of today's catastrophic situation in these countries, the population explosion is the direct cause that must be addressed first and foremost. Starvation is only a symptom—the natural law consequence that applies to all living beings in ecological systems, including humans, that no species may reproduce indefinitely at the expense of other living beings. No development aid can eliminate this law. Man with his earthly body, which must

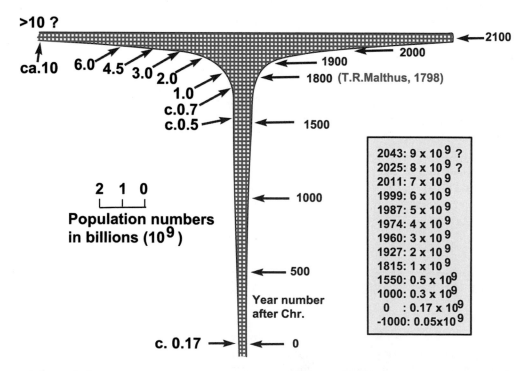

Fig. 14.2 The population explosion on Earth symbolized as an atomic bomb mushroom cloud from the year zero of our era: It took millions of years since the appearance of man until there were a billion people on Earth in the middle of the nineteenth century. After another 100 years there were already 2 billion, then after another 37 years 3 billion, but already 13 years later 4 billion. Around the year 2000 there were 6 billion and around the year 2025 one must reckon with 8 billion; new calculations give already the year 2023. The number around 2100 could not be shown, because if the increase remained the same, one would have to double the width above to over 20 billion. According to recent calculations, however, the number should settle at 11–14 billion as early as 2050 (BIRG, oral comm.), while in Europe the population is already shrinking as of 2005 (without immigration). Both is very hypothetic

be nourished, is and remains a part of nature. That is why well-intentioned food aid is particularly harmful, because it fuels the population increase even more, as if one wanted to extinguish a fire with oil (Walter 1990).

A graduate farmer at the university of Hohenheim, who later became a university professor of agricultural sciences in his country, exclaimed in a radio lecture: "Hands off the developing countries, they must carry out their own recovery; all development aid prevents that".

Remarkably honest is also in this respect the statement of the development adviser of the World Council of Churches Jonathan Freyers on the basis of his experiences. It caused outrage in wide unsuspecting circles. For in a newspaper article, he expressed the view that food shipments cause devastating damage. They would be distributed to the poorest, the buyers in the markets would stay away, causing the arable farmers to stop struggling and become aid recipients as well. Domestic production would collapse and the number of aid recipients would increase more and more— a vicious circle!

If one issues the slogan "help for self-help" or "stimulation for self-initiative", then one again misjudges the attitude as well as the way of thinking of many natives. Through thousands of years their way of life was regulated by strict moral laws, and these were optimally adapted to the environment, according to the cultural level, thus sustainable. Otherwise, survival would not have been possible for thousands of years.

Even colonial rule did little to change this, the power struggles among the tribes were stopped, farms or plantations with certain earning opportunities for the workers were established on still unsettled areas. A slow incorporation into the European economic system was starting and in the offing. The hasty release into independence with the requirement to create unified states according to democratic rules within the former colonial borders led to chaos and tribal struggles everywhere. The unschooled masses could not cope with the leapfrogging of a development that had taken over a millennium in the West. Our ancestors at the beginning of our era probably wouldn't have been able to accomplish this. The very strict moral and cultural rules regulating sexual intercourse and population were abolished, the unbridled urge to multiply set in, and with it the enormous increase in the number of births. Private property in our sense was unknown, everything belonged to the great clan and was regulated by it; thus there was no incentive for individual initiative. Most development workers return deeply disappointed. As long as one gives the necessary guidance in a project, people work very willingly and eagerly. However, if the guidance stops, then in most cases nothing happens anymore, only the outer facade is maintained, but this is not enough. "But you have to do something, you have to help the developing countries", is the argument of those who have never worked practically in the developing countries themselves. However, we must not forget that these are sovereign states that are very suspicious and immediately interpret incisive advice as neo-colonialism. This is particularly true with regard to advice on curbing the population explosion. However, until this problem is solved, any help will be in vain or harmful. Of course, the establishment of commercial enterprises, SOS Children's Villages, care for the blind, etc., is a help and laudable for the people covered in the process. But it does not change the catastrophic overall situation, which is getting worse and worse and is triggering an avalanche of refugees into the industrialised countries. Even if it is pointed out (more in passing) that the purpose of development aid is to open up new markets for our industrial products, for which there is an unlimited demand in the developing countries, this calculation is unlikely to work out.

Box 14.2 Exponential Economic Growth as an "Interglacial Fallacy"

Colonialism and communism, as well as capitalism with its dogma of exponential economic growth, are perhaps nothing other than an "interglacial error" (to paraphrase **Succow**).

Delivery can only be made on credit, with little expectation of repayment or interest. The examples of countries rich in raw materials such as Brazil and Mexico make this clear. In addition, an economic form and monopolisation is being imposed on the developing countries, of which it is impossible to say today whether it will guarantee man a lasting existence at all, or whether it will not itself burst like a shimmering soap bubble. All civilisations of the past that were alienated from nature collapsed and were replaced by "barbarians" who were close to nature. Today, however, alienation, and the problem of boundless egoism, is no longer merely regional, but global.

14.4 The Overtechnification

Technical development makes it possible to raise the so-called standard of living in the industrialised countries more and more, which is seen as great progress. This progress is measured by the level of the gross national product (including, for example, all accident car repairs) or the mean per capita income (which per month ranges from barely 500€ pension to outrageous and impudent millionaire's bonus fees in Europe and USA).

The aim is also to reduce working hours as much as possible and thus to extend leisure time in order to give everyone the opportunity for "self-realisation".

This ideal had already been achieved, but not in an industrialised country, but by the small island nation on the coral island of Nauru in the Pacific Ocean (about 2° S and 164° E). The happiest people should live there. Their median per capita income exceeded that of the richest industrialised countries. Weekly working hours were zero, free time all year (IWZ report, January 8–14, 1983). The children are born there as pensioners.

The island is 21.4 km² in size and rises up to 60 m above sea level. It is inhabited by 4000 Nauruan people (2015: 10,000). On it were many metres thick fossil guano deposits. These are the purest phosphate deposits known.

These were discovered in 1900 by the German colonial administration, which also began mining. After the First World War, Great Britain, Australia and New Zealand took turns to continue mining on an increased scale. In 1968, Nauruan chief Hammer de Roburt succeeded in asserting the island's independence within the British Commonwealth, and in 1979 the phosphate deposits became the property of the Nauruans. Since then, no Nauruan has needed to work. This was done by guest workers from Australia, New Zealand, Hong Kong, Taiwan and others, but they were not allowed to obtain citizenship. About two million tons of phosphate were mined annually and sold at the world market price.

The main occupation of Nauruans was sleeping, eating (corpulence is the ideal of beauty) and sitting in front of the TV (Mickey Mouse, Wild West and Australian commercials were the most popular). Sports are too strenuous with the fullness of the body. Nauru is the country with the highest percentage of diabetes sufferers in the world. People drove around the island in the most modern car models on the 18 km long car road. Empty beer cans decorate the landscape. A hobby was fishing with high-powered motorboats. One allowed oneself the luxury of a loss-making "Air-Nauru" with six jets flown by Australian pilots to Melbourne, Hong Kong, Manila and Samoa, and a luxurious shipping line. Currently (2015) it now looks like this (according to Wikipedia), *"The people of Nauru were long able to live off the mining of the rich phosphate deposits. When this ran out, it becomes apparent that the state and most of the citizens had not invested the profits in a future-proof way. Nauru, which at the time of phosphate mining still boasted the highest per capita income in the world, became increasingly impoverished after the complete depletion of its only resource. As a result, the state's finances regularly hover on the brink of bankruptcy, but have been stabilised in recent years by support measures coordinated by the Pacific Islands Forum"*.

Nauru was only accessible by sea from December 2005 to September 2006, as Air Nauru, the only airline serving the island, had to cease operations. In September 2006, however, the airline, which was simultaneously renamed Our Airlines (now Nauru Airlines), was able to resume operations with the help of Taiwanese funding.

To secure the future, two thirds of the income was transferred to the Nauru Royalties Trust and supposedly safely invested abroad in land, hotels and commercial buildings. The "Nauru House" in Melbourne with 50 floors is the highest commercial building in Australia. But there is a "but": According to estimates, the phosphate deposits will only last a few more years, some new deposits have been discovered, but what remains is a sterile coral landscape with 10 to 20 m high tooth-shaped rocks. When asked why mining is not done more sparingly, the answer is that the Nauruans are no different from the rest of the world, they love money like the Europeans and Americans and live selfishly into the day as long as they have it.

In fact, it is not that much different in the industrialised countries. All the warnings that resources are running out have changed almost nothing; people are only thinking until the next election date and putting off unpleasant decisions, including the increasingly pressing environmental and climate problems.

It is hard to deny that most people in industrialised countries do not know how to use leisure time properly themselves. Leisure time is organized and commercialised. Leisure activities have become a lucrative business ("tourism industry"). Just think of the many travel agencies and the mass accommodations in the rapidly growing domestic and foreign "resorts" with their entertainment venues. The vacationer does not have to take care of anything, he can passively let everything happen to him and only has to pay the price. In foreign countries he lives in a ghetto, as much as possible as he is used to, although the misery in the developing countries cannot be overlooked.

What is the profit of this mass tourism? The cost of memory cards for digital cameras. Otherwise only a passive taking in like the stream of manipulated information through the mass media. It rushes by and cannot be processed at all. The same applies to teaching, both in schools and universities. The amount of information is constantly growing, and there is not enough time to critically process the problems.

Independent thinking is not only not stimulated, but deliberately prevented by new study reforms. Thinking, many believe, can be left to the computer. And electricity comes from the socket. Science, fragmented into special subjects, threatens to become a Tower of Babel. "Massification" means that fruitful discussion in smaller circles is no longer possible. A mass lecture is not much different from a television presentation. Listeners passively let everything wash over them, cramming only a few weeks before the exam. Knowledge that doesn't last long. Comprehension is lacking. Besides all these shortcomings, in some federal states (e.g. in NRW) students are even exempted by law (http://is.gd/

qwf7fL) from compulsory attendance at courses under the pretext of "freedom of study", and you can no longer fail exams.

One draws attention, however, to an increasingly hostile attitude of people towards technology. But it would be more correct to speak of an increasingly anti-human mechanisation of all areas of life. Technology, which should help people to make the course of life easier and more pleasant, has developed a momentum of its own and forces the masses of people more and more under its spell and into a relationship of dependence.

It must be remembered that the purpose of technology has always been primarily the manufacture of weapons. Acts of war always gave technology the greatest impetus for further development. New inventions were immediately used for weapons technology. Without the two world wars, technology and mass production would not have reached their present state. Although the stockpile of weapons of destruction is sufficient to wipe out humanity ten times over, rearmament still continues and there is no end in sight. Unfortunately, experience teaches that newly developed weapons have mostly been used, more recently in self-created conflict areas such as the Middle East.

Military budget (or **military expenditure**), also known as a **defence budget**, is the amount of financial resources dedicated by a state to raising and maintaining armed forces or other methods essential for defence purposes. Generally excluded from military expenditures is spending on internal law enforcement and disabled veteran rehabilitation. Despite that the expenditures are incredibly high. How could this enormous and still growing amount be used for peaceful and ecological and sustainable purposes?

In 2018, the United States spent 3.2% of its GDP on its military, while China 1.9%, Russia 3.9%, France 2.3%, United Kingdom 1.8%, India 2.4%, Israel 4.3%, South Korea 2.6% and Germany spent 1.2% of its GDP on defence (acc. to the Stockholm International Peace Research Institute). In absolute figures it is shown in Table 14.4.

The misanthropy of technology is also expressed in the destruction of the environment. While almost everywhere in the world large areas of forest are falling victim to technology every year, in Japan environmentally conscious large corporations are trying to increase the area of forest: All the steel mills of Nippon Steel Coop, all the operating complexes and research centres of Honda Motors Co. and Topay Industries, the power plants of Tokyo Electric Co. and Kansai Electric Co. among others are reforesting the areas around their operating complexes as air filters and recreational areas. Native tree species have already reached a height of 10 m (Miyawaki 1983). However, wood can be fetched in Borneo, after all.

In Germany, the concrete blocks are surrounded by asphalt parking lots and mostly bare lawns. The rest of the

remaining environment is often poisoned. Although maximum values for individual toxins should not be exceeded, no one knows whether they are still valid when many toxins are added together. Think of the increase in allergies or the accumulation of pollutants and heavy metals in cultivated soils. In the 1970s the pollutants were finally so high even in mother's milk that one did not advise against breastfeeding children only because the substitutes were not lower in pollutants or the advantages also had to be considered.

In the 1990s, however, the contamination of breast milk dropped to considerably lower levels thanks to a correspondingly more conscious diet and lower limit values. It has been done quite a lot, but achieved not really much.

The mechanisation of agriculture is particularly serious. The larger, largely self-sufficient farms, which managed without external energy, were the only farms that were in a certain harmonious balance with the environment. They are now being replaced by agricultural factories with huge masses of organic waste that are difficult to dispose of. Cleared, monotonous landscapes were created. Fauna is more and more on the „Red List".

Box 14.3 Modesty Is Needed

A new modesty is needed: Better poorer and healthy than rich and half dead. In other words derived from unknown American Indian sources:

"Only after the last tree has been cut down
Only after the last river has been poisoned
Only after the last fish has been caught
Then will you find that money cannot be eaten."

One of the serious consequences is the greatly accelerated decline in the diversity of native species since 1960, also in Germany. The "Red Lists" show not only an increasing endangerment of rare, mostly specialised species, but also a decline of formerly widespread and common species within very few decades (Ruckdeschel 1996). Agriculture, but also forestry and hunting, tourism, open-cast mining, and industry are the main causes of a sharp decline in the diversity of small habitats (Fig. 11.35 and Fig. 11.39).

Farms have been drawn into the vortex of the global economy, with the result that agriculture is losing its resistance to crisis. Technology is increasingly depriving man of the natural basis of life. Therefore, ecologists cannot be expected to be friendly to and to accept excessive technology. It is their duty to point out the impending dangers again and again.

Man can, if he must, do without much and get along with very little, but he needs pure air to breathe, clean water to drink, and a poison-free diet, as well as a natural use of his physical powers.

Table 14.4 According to the Stockholm International Peace Research Institute, in 2018, total world military expenditure amounted to 1822 billion US$, the main countries are listed. But now those data are completely obsolete by recent senseless wars of conquest and insane destruction of civilian infrastructure, false claims and fake-news and confusion of causes and effects

Ranking No.	Country	Military budget 10^9 US-$ (*italics: estimate figures*)
1	USA	649
2	China	*250*
3	Saudi-Arabia	*67.6*
4	India	66.5
5	France	63.8
6	Russia	61.4
7	Great Britain	50.0
8	Germany	49.5
9	Japan	46.6
10	S Korea	43.1
11	Iran	12.2

What technology produces are, for the most part, things that are not essential to life and serve only convenience or prestige. Often it is frippery. Needs are artificially fuelled by worldwide propaganda and intrusive advertising. Everyone should be able to have everything. Millions of tons of food are destroyed in Germany every year, and use of food from rubbish bins is forbidden. Not the interests of the people are in the foreground of technology, but the profit thinking and the purely economic interests, especially of the large corporations. More and more the human being is pushed out of the production process as a worker by rationalisation (robots, microelectronics) and degraded to a mere consumer of mass production, or even a slave of digitization. But how are the masses supposed to buy industrial products if they are not guaranteed an income and become unemployed? One speaks of economic constraints of competition—a vicious circle!

People have become neither happier nor healthier as a result of technology. Civilisation diseases of a physical or psychological nature are constantly on the increase. If average life span is rising, it is through more and more medicines and expensive treatments, the cost of which is rising immeasurably.

Walter commented as follows: "If one were to look over the eight decades of one's own life and pass judgment on the blessings of technology, it can only be a very subjective one. By what criteria should it be done? In any case, there was a lack of stress. Also, the crossing of the oceans in the research voyages by ship were a nice rest before and after work and allowed for a slow changeover, whereas in today's air travel that is not the case; even the changeover to the different climate, the different environment, the different time of day is too sudden" (Walter 1989).

Without technology, "massification" would not have been possible. It has now led to the growing number of unemployed, who are a heavy burden on the future. The solution to this problem is a simple milkmaid's calculation, which would only require a little more solidarity and less egoism in order to divide the available "cake" more equally. But even in Europe the population is too large.

14.5 Sustainable Land Use

Every living being is influenced by its environment, but conversely every living being also influences its environment. The latter becomes more obvious the greater the population density of a species.

Man has now reached a frightening population density. Consider the future plans for the city of Beijing. The impact on the environment increases exponentially with population density (Fig. 14.2 and Fig. 14.3).

Land use always changes soils and promotes erosion. However, soil formation is a lengthy process. Soil erosion destroys valuable resources for centuries or millennia. However, this varies greatly between zonobiomes. A worldwide problem is water shortage, soil erosion and salinisation (Breckle 2009, 2021a, 2021b).

Such sustainable land use can only be achieved if population density (including in cities) does not exceed a certain level and if land use methods for arable farming, livestock breeding and forestry are based on natural processes, i.e. if a circular economy is consistently introduced at all levels. This necessarily includes industrial processes.

> **Box 14.4 Sustainable Land Use Systems**
> Only in those areas where it is possible to use, settle and live in harmony with the existing vegetation and fauna over many generations can one speak of sustainable land use. This also includes the preservation of the yield capacity of the soils over long periods of time.

The global changes that have become apparent in the meantime (global change and climate warming), the main effects of which are summarised in Fig. 14.3, include in

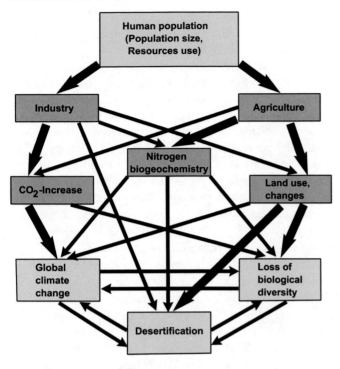

Fig. 14.3 The components of global change. The thick arrows indicate strong effects (modified after Vitousek 1994)

been known for a long time, but the steady increase in CO$_2$ content from about 280 ppm (in pre-industrial times) to currently (2021/2022) about 420 ppm CO$_2$ is detectable worldwide. The fact that, in addition to this, other global material cycles are now also undergoing a change as a result of growing anthropogenic activities has so far been taken less notice of. Figure 14.5 compares the natural N fixation with that caused by humans.

In addition to the changes in global material cycles, the effects of climate-relevant gases in the atmosphere (H$_2$O, CO$_2$, CH$_4$, N$_2$O etc.) must be taken into account not only for the effects on the climate system (which have always been effective), but also for the effects on the equilibrium in the cryosphere. The changes in ice masses on land surfaces (in the Arctic, Antarctic, high mountain glaciers) due to temperature and precipitation changes in recent decades has led to a significant sea level rise being observed since the middle of the nineteenth century. Sea level rise is mainly due to two phenomena: Warming of the oceans leads to the expansion of water (expansion coefficient 1.0002 per K, roughly three times in terms of volume), and increased air temperatures lead to the melting of glaciers and ice sheets, releasing water from the mainland into the oceans.

Between 1901 and 2010, sea level rose by 1.7 mm per year, and by an average of 3.2 mm per year between 1993 and 2010 (IPCC 2014). For 2018, the record value of 3.7 mm was measured (Nerem et al. 2018).

At the end of the last ice age (20,000 years ago), sea level was about 130 m lower than today, allowing for diverse floral and faunal exchanges via land bridges (Dogger Bank, Bering

particular the changes in the chemical composition of the atmosphere. The increase in CO$_2$ and also in other trace gases (CH$_4$, N$_2$O, CFCs, etc.) must lead to a change in the equilibrium of the Earth's radiation budget. The seasonal fluctuations through the seasons in the northern hemisphere, shown by the so-called Mauna Loa curve (Fig. 14.4), have

Fig. 14.4 The concentration of CO$_2$ in the atmosphere at Mauna Loa in Hawaii (**a**) and at the South Pole (**b**). The annual oscillations are caused by the seasonal activity of land plants in the northern hemisphere, the steady increase by the burning of fossil fuels and deforestation (added and modified after Keeling & Whorf 1994)

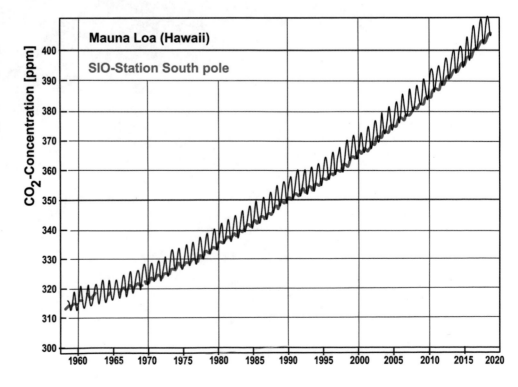

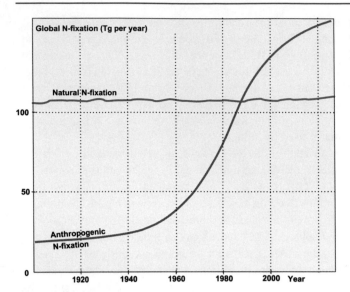

Fig. 14.5 The global N budget is characterised by the largely constant natural N fixation (biological nitrogen fixation in terrestrial ecosystems and binding of N in electrical discharges) as well as by the strongly increased anthropogenic N fixation (industrial fertilizer production e.g. Haber-Bosch process, N fixation in fossil fuel combustion and N fixation by legume cultivation) (after Vitousek 1994)

Strait, etc.). Today, with many large cities and densely populated areas directly on the coasts, a rise of only 1 or even more 2.5 metres is a huge problem; a problem that will have to be dealt with until 2100. Sea-level rise is a particular

threat to island states and countries with broad coastlines, as well as low-lying hinterlands such as Bangladesh and the Netherlands.

Logically, exponential economic growth will not be able to offer any solutions; only a rethinking of globally sustainable and more modest economic practices can help.

Sustainable use of forests is possible for centuries in temperate latitudes under climatic conditions that do not exhibit any particular extremes, but even in Central Europe there are the problems of game damage by too high game density. In regions with more variable moisture, the rate of erosion on deforested land is a major problem. Soils are heavily washed away after logging, making reforestation more difficult. In the tropical rainforests, logging with European methods is economically a nonsense.

Almost two thirds of the earth's original primary forests as a whole have been lost forever. Of the $8.08 \cdot 10^9$ ha that were still covered by forest about 8000 years ago, only $3.04 \cdot 10^9$ ha remain today (Fig. 14.6). This is the alarming conclusion of a study by the WWF (World Wild Life Fund) on the global state of forests. The preservation of what remains is by no means certain. Today, some $17 \cdot 10^6$ ha of virgin forests are destroyed each year by large-scale clearing, industrial logging, road construction and other human interventions, or are replaced by species-poor timber plantations or cattle ranches or soybean fields of little ecological value. Of particular

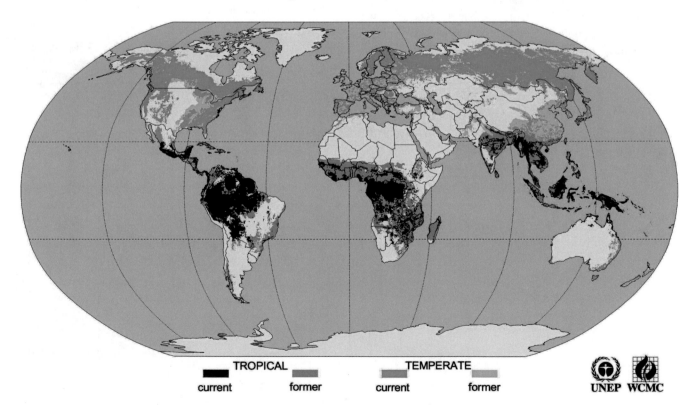

Fig. 14.6 The original and present forest cover on Earth (source: http://tinyurl.com/n4p6bm9)

concern is the fact that destruction has accelerated rather than decreased in recent years. The WWF therefore proposes the creation of a global network of protected zones, each comprising 10% of tropical and subtropical forests and 10% of temperate and boreal forests. For Europe alone, 100 forest areas are proposed for this purpose. Who can enforce this?

Box 14.5 Are Species Conservation and Climate Protection Mutually Exclusive?

Where is the insight that species protection is even more important than climate protection? Climate changes are reversible, but extinct species have irreversibly disappeared!

14.6 Confessions

How can we remedy the situation and achieve a development towards sustainable use that will enable our children's children to live a life worth living?

This is not a problem of natural science and ecology, but a question of the sociological-political system as well as of the prevailing respective religion. Nature and its preservation play a significant role in the scriptures and ideas of all world religions (Barthlott 2019). Religions consciously or unconsciously shape the value systems and actions of many people. But there is no impression of either politics or religion to treat the mass of the population as mature citizens. One shies away from the long overdue drastic measures and prefers instead to constantly take out "loans" in the broadest sense at the expense of the grandchildren and regards the state as a self-service shop. Positive role models have become very rare, as have ground-breaking parliamentary resolutions and court decisions. Remedial action is only possible if everyone pulls together and treats each other with respect and empathy, putting aside claims to sole representation. Remedial action is possible at the level of the individual; at the level of the intact family, sustainable behaviour can be handed down from generation to generation.

However, the loss of values and tradition is leading to worrying developments. In Hungary, Taiwan, Afghanistan and many other countries, around 90% of the population consider children to be a prerequisite for a fulfilled life. In the USA (46%) and in Germany (49%), only just under half of adults attribute a meaningful meaning to children. But this so-called self-realisation of the individual, the exaggerated liberalisation and boundless pleasure lead to chaos.

In the last few years, neurophysiological research in the USA has been able to show how enormous the imprintability of the human brain is in the early stages of development and unfolding.

In earlier times it was absolutely clear that the profession of the mother in an intact family is the best guarantor for the spiritual and mental prosperity of the children and thus for the future of humanity. This has been forgotten in some Western countries in recent decades.

The humanity and the close contact between people is more and more lost, it is limited to short and mostly meaningless telephone conversations or even to short e-mail notes and SMSs, even more in pandemic times. The individual human being becomes a number: a personnel number, a tax number, a health insurance number, many customer numbers, etc. The name only appears on letters, which have now rapidly lost their significance. How much longer? What used to mean the relationship to nature to most people, today's youth in landscapes saturated with technology no longer knows; they do not know what has been taken from them. For it is not the material standard of living that matters, but the **quality of life**—not the outward appearance, but the **inner being**.

The state, but even more so the individual, is challenged. Walter formulated thoughts on this at an advanced age, which will be taken up briefly in the concluding section. They are more relevant than ever.

Standard of living and quality of life need not be opposites. But experience teaches that **the more one attaches importance to outward appearances, the more one usually impoverishes the inner life**, of which one therefore does not speak.

Quality of life is also expressed outwardly, through a healthy and natural way of life, the sensible use of one's life forces and the renunciation of all addictive substances, the preference for a quiet way of life in modesty and with self-control, even more in pandemic times. For this, one does not need a God and certainly not wars waged in his name. He who is truly connected with nature and knows it in all its diversity and vastness does not feel himself to be the centre of creation. He knows that he is only a tiny blob of protein in the infinity of the universe.

To this, Walter says, mutatis mutandis:

Not only the **outer world** (which is the subject of this book) to which we belong with our body and which we explore with our thinking intelligence, but also the other side of man, his inner world, which is not subject to logic, for which various complicated terms are used by philosophers, but which is commonly called "soul", opens up to him. This cannot be put into words, nor can it be proved. To profess it is an act of free choice on the part of each individual, without which there is no true freedom for man. Only it gives him independence from the judgment of others and thus inner security, calm and serenity, and inner cheerfulness. It is not a question of this world or the next. The Absolute knows no boundaries. It is within us and also outside us. This is the most important conclusion for the youth searching for the meaning of life, the result of a long life dedicated to the exploration of the living all over the Earth, a life full of miracles, in an age that does not believe in miracles and has lost touch with the centre of all things. One must always swim against the polluted or dirty current until one reaches the pure source that comes from the deep (Walter 1989: "Confessions of an Ecologist").

Photo 62 *Adenium obesum* (Apocynaceae) in the Dhofar Mountains (Zonobiome III) in southern Oman (photo: Breckle)

Photo 63   Cultural landsape in the middle Moselle valley (Zonobiom VI) with intensive vine cultivation, Rhineland-Palatinate (photo: Rafiqpoor)

Photo 64 *Clianthus formosus* (Leguminosae) is the floral emblem of South Australia. The short-lived species has evergreen leaves, it occurs in the dry central parts of Australia (ZE III/II). (photo: Breckle)

Box 14.6 The Foundation of a Person's personality Is Laid in the Early Stages of Their Development
Whether clever or lame, whether vigorous or feeble, whether mentally resilient or impaired, whether strong-willed or prone to addictions, criminality and mental illness, whether optimistic or despondent, whether, in other words, happy or lifelong unhappy—this depends largely on what impressions the human brain stores in its early phases of development.

References

Barthlott, W. 2019: Naturschutz und Religion – Gedanken zu einer mächtigen Partnerschaft beim Erhalt der Biodiversität. Ein streitbares Essay. In: Koenigiana **13** (1): 35-42

Bazilevich, N.I., Rodin, L.E. & Rozov, N.N. 1970: Untersuchungen der biologischen Produktivität in geographischer Sicht. V. Tag. Geogr. Ges. USSR, Leningrad (Russ.)

Breckle, S.-W. 2009: Is sustainable agriculture with seawater realistic? In: Aschraf, M., Ozturk, M. & Athar, H.R, (eds.): Tasks for Vegetation Science, Vol. **44**: 187-196

Breckle S-W 2019: Vegetation, climate and soil. 50 years of global ecology. Progress Botany **80**: 1-63

Breckle, S.-W. 2021a: Bodenversalzung (Meeresspiegelanstieg, Aridität und Fehler bei der Bewässerung. In: Lozán J. L. S.-W. Breckle, H. Grassl, W. Kuttler & A. Matzarakis (Hrsg.). Warnsignal Klima: Böden und Landnutzung. pp. Xxx-yyy. Online: www.klima-warnsignale.uni-hamburg.de.

Breckle, S.-W. 2021b: Grüne Mauern („green walls") gegen die Wüstenausbreitung als Maßnahme gegen Desertifikation. In: Lozán J. L. S.-W. Breckle, H. Grassl, W. Kuttler & A. Matzarakis (Hrsg.). Warnsignal Klima: Böden und Landnutzung. pp. Xxx-yyy. Online: www.klima-warnsignale.uni-hamburg.de.

Gruhl H 1975: Ein Planet wird geplündert. Die Schreckensbilanz unserer Politik. Fischer/Frankfurt 376 pp.

IPCC, 2014: Climate Change 2014: Synthesis Report. Contribution of Working Groups I, II and III to the Fifth Assessment Report of the Intergovernmental Panel on Climate Change [Core Writing Team, R.K. Pachauri & L.A. Meyer (eds.)]. IPCC, Genf, S. 42

Ittekkot, V., Rixen, T., Suthhof, A. & Unger, D. 2002: Der globale Kohlenstoffkreislauf. In: Wefer, G. (ed.): Expedition Erde. Beiträge zum Jahr der Geowissenschaften. Universität Bremen: 202-209

Keeling, C.D. & Whorf, T.P. 1994: Atmospheric CO_2 records from sites in the SIO air sampling network. p. 16-26. In: Boden, T.A. et al. (eds.): Trends 93: A compendium of data on global change. ORNL/CDIAC-65. Carbon Dioxide Information Analysis Center, Oak Ridge Nat. Lab., Oak Ridge

Lieth, H. & Whittaker, R.H. (eds.) 1975: Primary productivity of the biosphere. Ecol. Stud. **14**

Miyawaki, A. 1983: Conservation and recreation of vegetation and its importance to human existence. Look Japan **28** (323): 10.2.83

Nerem, R.S., Beckley, B.D., Fasullo, J.T., Hamlington, B.D. et al. 2018: Climate-change driven accelerated sea-level rise detected in the altimeter era. PNAS **115**: 2022-2025; https://doi.org/10.1073/pnas.1717312115

PRB 2016: World Population Data Sheet from the Population Reference Bureau (PRB) - https://scorecard.prb.org/.

Ruckdeschel, W. 1996: Landbewirtschaftung als ökologische Schlüsselfunktion. Ber. Bayer. Landesamt für Umweltschutz **132**: 61-72

Vitousek, P.M. 1994: Beyond global warming: ecology and global change. Ecology **75**: 1861-1876

Walter, H. 1989: Bekenntnisse eines Ökologen. Erlebtes in acht Jahrzehnten und auf Forschungsreisen in allen Erdteilen. 6. Aufl., Fischer, Stuttgart. 353 S.

Walter, H. 1990: Vegetationszonen und Klima. 6. Aufl., Ulmer/Stuttgart 382S.

Taxonomic Index

Subject Index

A

Aapa, 464
Aapa mire, 462
Abandoned tunnels, 410
Abiotic factors, 302
Abisko, 465, 483
Aboriginal settlements, 194
Aborigines, 258, 294
Aboveground biomass, 370
Aboveground phytomass, 116
Absolute humidity, 42
Absorption, 12, 279
Acacia savannah, 411
Access and Benefit-Sharing (ABS), 71
Accumulation, 25
Accumulation layers, 167
Acidification, 25
Acidophilic, 386
Acid rain, 437
Aconcagua, 280
Activity of water, 19
Acts of war, 503
Adaptive radiation, 297
Adherent water, 226
Adhesive water, 225, 226
Adiabatic gradient, 88
Adiabatic heating, 87
Aeolian sediment, 458
Aeration, 200, 462
Aerial roots, 132
Aerosol, 30
Affodill vegetation, 273
Affodil site, 273
Afforestation, 390
Afghan camel drivers, 194
Afghan flora, 307
Afghanistan, 4, 28, 30, 43, 63, 73, 75, 194, 267, 268, 302–312, 414, 417, 421, 428, 431
Afghan mountains, 304
Africa, 51, 106
African savannah, 194, 195
African volcanoes, 151
Afroalpine belt, 153
Agadir, 265
Age of a single plant, 482
Age of the trees, 117
Aggregate states, 42
Aging phase, 154
Agricultural factories, 503
Agricultural landscapes, 191
Agriculture, 258, 436

Aichi Target, 71
Air Nauru, 502
Air pollution, 361
Air saturation deficit, 431
Air temperature, 114
Aishihik, 52
Ajar, 418
Alas, 459
Ala-Shan, 433
Alaska, 485
Alas lakes, 459
Alatau, 428
Albania, 455
Alexander the Great, 207
Alfisols, 115
Algeciras, 333
Alkali halophytes, 28, 29
Alkaline, 29
Alkaline B horizon, 415
Alkaline soils, 482
Alkaloids, 169
Allelopathy, 59
Allergies, 503
Allochthonous water, 44
Alluvial deposits, 228
Alluvion, 76
Almaty, 428
Almonds, 310
Alpine, 381
    altitudinal belt, 256
    belt, 63, 204, 205, 294, 301, 309, 312, 383–385
    debris, 281
    mats, 461
    meadows, 434
    nival belt, 428
    species, 427
    vegetation, 63, 430
Alps, 381–389
Alternating frost, 485
Altiplano, 204
Altitudinal belts, 149
Altitudinal gradient, 141
Altitudinal limit, 151
Altitudinal sequence, 238
Altitudinal zones, 88
Altyntag, 433
Altyntag River, 433
Aluminium oxides, 25
Amanus Mountains, 295
Amazon, 197
Amazon basin, 113, 114, 125, 136, 137, 167

© Springer-Verlag GmbH Germany, part of Springer Nature 2022
S.-W. Breckle, M. D. Rafiqpoor, *Vegetation and Climate*, https://doi.org/10.1007/978-3-662-64036-4